Progress in Soil Science

Progress in Soil Science series publishes books that contain novel approaches in soil science in its broadest sense – books in the series should focus on true progress in a particular area of the soil science discipline. The scope of the series is to publish books that enhance the understanding of the functioning and diversity of soils in all parts of the globe. The series includes multidisciplinary approaches to soil studies and welcomes contributions of all soil science subdisciplines. Key themes: soil science - soil genesis, geography and classification - soil chemistry, soil physics, soil biology, soil mineralogy - soil fertility and plant nutrition - soil and water conservation - pedometrics - digital soil mapping - proximal soil sensing - soils and land use change - global soil change - natural resources and the environment.*Submit a proposal* : Proposals for the series will be considered by the Series Editors. An initial author/editor questionnaire and instructions for authors can be obtained from the Publisher, Dr. Robert K. Doe (robert.doe@springer.com).

More information about this series at https://link.springer.com/bookseries/8746

Andrew W. Rate
Editor

Urban Soils

Principles and Practice

Springer

Editor
Andrew W. Rate
School of Agriculture and Environment
University of Western Australia
Crawley, WA, Australia

ISSN 2352-4774 ISSN 2352-4782 (electronic)
Progress in Soil Science
ISBN 978-3-030-87315-8 ISBN 978-3-030-87316-5 (eBook)
https://doi.org/10.1007/978-3-030-87316-5

This Springer imprint is published by the registered company Springer Nature Switzerland AG
The registered company address is: Gewerbestrasse 11, 6330 Cham, Switzerland

Preface

This book was rewarding, and even fun, to write, and started a learning experience for me that I hope will be similarly rewarding for students interested in urban soils everywhere, especially the ones who I teach! A large proportion of the book was written during a 6-month sabbatical leave which was generously approved by the School of Agriculture and Environment at The University of Western Australia. Much of the remainder was written during the COVID-19 pandemic, while at the same time teaching undergraduate and postgraduate courses which informed my writing (and *vice versa*).

My background and passion is in chemistry, and so as the main author, the emphasis of the book reflects my chemistry bias; my other bias is towards environmental science as a broader discipline. As readers will notice, however, my passion for both subjects took me to new and interesting places in the field of environmental science, and soil science in particular. Writing this book also made it necessary to learn some new and interesting ways to look at environmental data, and to write the necessary code in R. Interested readers should contact me by email for the R code to conduct the data analyses or prepare many of the graphics and maps in this book.

Notes

Words or phrases in *italics* are the first instances of terms in the Glossary, except terms that are traditionally written in italic font such as Latin phrases. International conventions have been used where appropriate, for example soil classification, soil texture categories, SI units, and IUPAC chemical names. This may clash sometimes with the use of standard Australian English (*e.g.* in the spelling of 'sulfur').

We have endeavoured to make referencing formal and complete; this is not always the case in textbooks, but we think it's good to set an example of what we expect from our students. As far as possible we have tried to cite literature that specifically deals with urban environments and soils, rather than the environmental science or soil science literature in general. As a result, the sources cited may not be the most seminal publications on any particular topic, but may be more useful for readers wishing to learn more about urban soils.

We have used examples from as many places in the world as possible, with the hope that this will increase our readers' connectedness to the material. Where relevant, coordinates are provided for geographic locations in decimal degrees (by convention, degrees south or west are negative), suitable for searching using software such as Google Earth or equivalent.

Perth, Western Australia
June 2021

Andrew W. Rate

Photo Credits for Opening Images of Each Chapter

Chapter 1: François Molle (CC-BY-2.0).
Chapter 2: (composite image) left, Augusto Malta (Public Domain); right, Andrew W. Rate.
Chapter 3: (composite image) left, Lav Ulv (public domain); right, Anna Paltseva and Zhongqi Cheng (CC-BY-4.0).
Chapter 4: (composite image) left, Roger Thurow (public domain); right, Chika Watanabe (CC-BY-2.0).
Chapter 5: (composite image) left, Andrew W. Rate; right, Katie Hetrick, UC Davis Arboretum and Public Garden (CC-BY-2.0).
Chapter 6: (composite image) left, Andrew W. Rate; right, Andrew W. Rate.
Chapter 7: (composite image) left, USDA-NRCS (public domain); right, Andrew W. Rate.
Chapter 8: (composite image) top, Franck Michel (CC-BY-2.0); bottom left, Steve K. (CC-BY-2.0); bottom centre, Chika Watanabe (CC-BY-2.0); bottom right, Armen Phelps (CC-BY-4.0).
Chapter 9: (composite image) left, https://pxhere.com/en/photo/986452 (Public Domain); right, Sergio Souza (CC-BY-4.0).
Chapter 10: (composite image) top, UNEP (Public Domain); left, Laura Kraft (Public Domain); right, Ja Ma (Public Domain).
Chapter 11: (composite image) both images by Marco Verch (CC-BY-2.0).
Chapter 12: (composite image) left, Romon Logov (public domain); centre, Willian-Justen-De-Vasconcellos (public domain); right, Anaya Katlego (public domain).

Acknowledgements

First, thanks to Dr Ian Francis who was our initial contact at Springer Nature and who first approached Andrew about writing this book, and handled the review of the textbook proposal.

We are grateful to the following people, who generously donated their time to read and comment on drafts: Assoc. Prof. Louise Barton, Cassandra Bond, Kai-Ying Khoo, Giles Knight, Jessica Petersen-Jones, Annelies Rate, Dr Ming Wu, and Peter Zafiropoulos.

Thanks also to the environmental science community on Twitter and other social media who provided comments, ideas, and encouragement: Dr Rivka Fidel, Prof. Anas Ghadouani, Dr Dylan Irvine, Dr Gavan McGrath, Joanne Moo, Dr Colby Moorberg, Dr Bree Morgan, Assoc. Prof. Suzie Reichman, Dr Matthew Ricker, and Jeremy Smith. Also to the many students at The University of Western Australia who looked forward to the book coming out, and said so.

Prasad Gurunadham, the project coordinator at Springer Nature, deserves special mention for his patience in dealing with Andrew's occasionally impatient emails and at the same time tolerating his frequent delays in providing material.

Finally, Andrew greatly thanks his wife and two daughters, who endured his intense focus on, and devotion of inordinate amounts of time to, this writing project.

Contents

Chapter 1
Urban Ecosystems: Soils and the Rise and Fall of Cities

Andrew W. Rate

Abstract Soils, as a crucial component of the critical zone of the Earth's surface that sustains life on land, are as essential in cities and peri-urban areas as they are in natural or agricultural environments. In this chapter we explore the idea that the properties and geographical location of soils have influenced the historical and present location of urban environments. The importance of cities to human survival and well-being is established by analysing global trends in urbanisation and urban migration and by noting that urban areas will become increasingly important for food production. The remainder of this introductory chapter presents the main topics which are covered in this textbook. Human impacts on urban soils are described in several ways, through an investigation of soil-forming factors in urban environments (Chap. 2), soil variability and data analysis (Chap. 3), ecosystem functions provided by soils (Chap. 4), changes in soil physical properties (Chap. 5), chemical pollution of soils (Chaps. 6 and 7), and soil biological phenomena in urban soil (Chap. 8). We start to consider the management of soils in urban environments in Chap. 9, which covers urban soils as a source and sink for material and some basics of risk analysis. The management theme is continued in Chap. 10 which analyses the role of soils in sustaining human health within the framework of the United Nations' Sustainable Development Goals and in Chap. 11 which presents remediation options for degraded urban soils. Finally, Chap. 12 looks to the future of soil science in an urban context.

Keywords Urban soils · Soils · Urbanisation · Population · Human impact · Ecosystem functions

---

A. W. Rate (✉)
School of Agriculture and Environment, University of Western Australia, Crawley, WA, Australia
e-mail: andrew.rate@uwa.edu.au

A. W. Rate (ed.), *Urban Soils*, Progress in Soil Science,
https://doi.org/10.1007/978-3-030-87316-5_1

What you could learn from this chapter:

- Whether you want to keep reading this textbook!
- How soils and urban ecosystems are interlinked, in both the past and the present.
- The context and relevance of urbanisation of human populations.
- Why urban soils are important, and how they are different from 'other' soils (and therefore worth studying).

1.1 Introduction

Soils are the living core of the *critical zone*, the zone encompassing the upper layers of Earth's land surface which are responsible for sustaining terrestrial life. As much in cities as in land used for food production, forests, or pristine reserves, the properties and functions of soil are crucial for human survival and, indeed, for the healthy functioning of all terrestrial ecosystems. The critical zone is comprised of components of the near-surface environment, from the lowest accessible groundwater to the upper vegetation canopy, in which complex interactions involving rock, soil, water, air, and living organisms provide essential ecosystem services (National Research Council et al. 2001; Brantley et al. 2006). Even despite issues like pollution, surface sealing, and other consequences of mistreatment of urban soils, soils in cities remain both mundane and mysterious, delicate but degraded, and ignored yet indispensable.

It may also be in cities, however, that soils are, paradoxically, least understood. For example, there is some evidence that the general population (including regulators, legislators and planners) makes very little use of soil knowledge (Teixeira da Silva et al. 2018). A proportion of the urban population, though, has close contact with and both personal and financial investment in urban soils; in urban and *peri-urban* environments, gardeners and urban farmers have a detailed knowledge of urban soils (Wakefield et al. 2007; Reséndiz-Paz et al. 2013). Many scientists and engineers also understand urban soils; in order to rehabilitate natural environments, clean up pollution, prevent erosion, and of course build on or using soil, a thorough understanding of soil properties and functions in an urban context is needed. The following passage from McDonald et al. (2019) expresses the impacts of urbanisation on biodiversity (and, therefore, humanity) very compellingly:

> By 2030, an additional 1.2 billion people are forecast in urban areas globally. We review the scientific literature (n = 922 studies) to assess direct and indirect impacts of urban growth on habitat and biodiversity. Direct impacts are cumulatively substantial, with 290,000 km^2 of natural habitat forecast to be converted to urban land uses between 2000 and 2030. Studies of direct impact are disproportionately from high-income countries. Indirect urban impacts on biodiversity, such as food consumption, affect a greater area than direct impacts, but comparatively few studies (34%) have quantified urban indirect impacts on biodiversity. (From the Abstract in McDonald et al. 2019)

At this stage in this book, we need to define what we mean by 'urban'. For this definition we are guided by the analysis presented by Liu et al. (2014), who discussed definitions of urbanisation in the context of how much of Earth's land area is occupied by cities. We will use the definition of 'urban' in the widest context described by Liu et al. (2014) that of land (and therefore soil) within the administrative boundaries of cities, which encompasses nearly 3% of global land area. In many cases, this definition extends to peri-urban or even rural land as well, which is appropriate for our purposes. Most of our examples will be drawn from built-up urban land closer to city centres, but in some cases of industrial or even degraded rural land, it makes more sense to use the wider definition. Sometimes industrial activities, which have many similar effects on soils as cities, occur in rural areas. Similarly, sometimes historical cities have ceased to exist as built areas or centres of human population but still leave an imprint on the properties and composition of the soil.

1.2 The Influence of Soils on Cities

Urbanisation (the movement of human populations into cities) and soils have been linked phenomena for a very long time. Cities, and civilisations, arose following the development of agriculture in areas where there were both fertile soils and accessible fresh water (McNeill and Winiwarter 2004; Hillel 2006), although several other factors were involved. Figure 1.1 expresses this graphically and lists several other factors including proximity to trade routes or topography and climate. From independent beginnings in places like Mesopotamia, China, and the Americas, the fertile land which was required to grow crops nourished a growing human population. The stability which agriculture – dependent on soil fertility – allowed was followed by the establishment of permanent settlements, which ultimately became towns and cities (Hillel 1991; Redman 2011). In today's world, the locations of contemporary cities, and the diverse range of urban environments, share much in common with the cities which developed millennia ago. Kummu et al. (2011), for instance, find that more than half of the world's population lives within 3 km of fresh surface water.

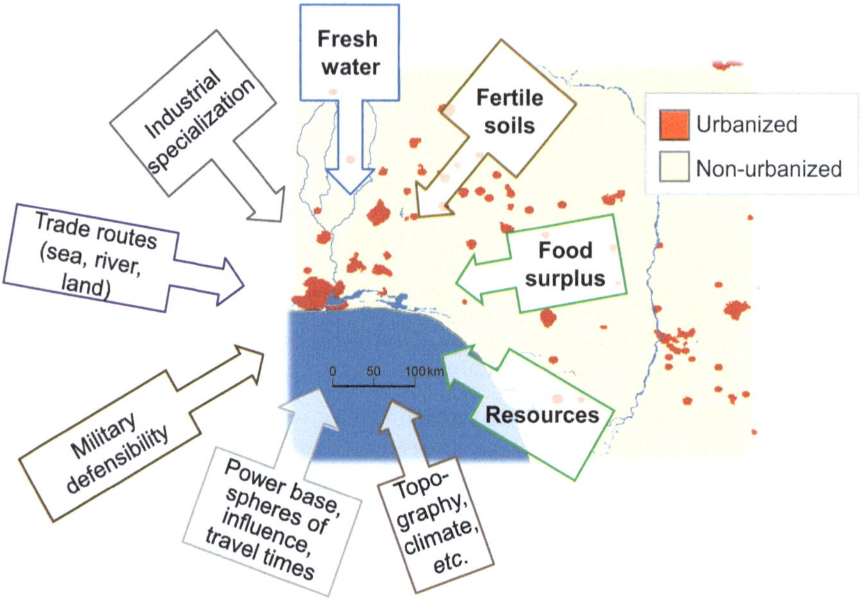

Fig. 1.1 Factors affecting the location of cities (bold text shows factors allowing the commencement of urbanisation). Graphic by Andrew W. Rate

The benefits of living close to fresh water are not restricted to water availability alone; as Fang et al. (2018) point out, rivers are associated with fertile soils and also provide transport routes and the conveniently low relief of alluvial plain environments, especially near the coast. Figure 1.2 shows the clustering of large, contemporary cities along coastlines, on major rivers, and especially in areas with fertile soils. In an ominous foreshadowing of contemporary environmental problems, even ancient cities, like the cities of the twenty-first century, contained polluted soils or soils which were degraded in other ways (Zhang et al. 2005). Urban environments are considered sufficiently different from other environments that some authors consider them as a separate Earth subsystem, the *urbosphere* (a term probably first used by Voloshyn 2004).

1.3 Global Trends in Urbanisation

Since approximately 2007, urban environments have been the most populous human habitat on Earth, a trend (Fig. 1.3) that is projected to increase the proportion of human population in urban areas until at least 2050 (Foley et al. 2005; Grimm et al. 2008; Lyons and Harmon 2012; United Nations 2018). The trend towards greatly increased urbanisation has been occurring steadily since about 1800 AD (Grimm et al. 2008) in many nations worldwide, with some countries (e.g. Brazil, much of

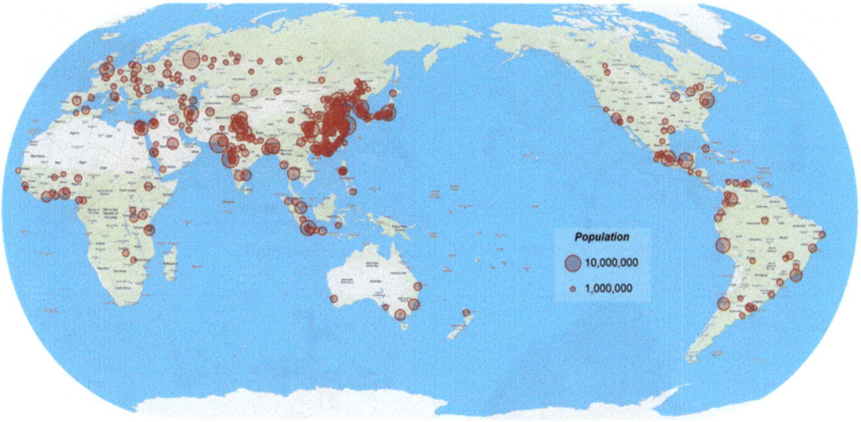

Fig. 1.2 World map showing locations of cities (red-coloured circles with area proportional to population) with population greater than one million inhabitants. (Robinson Conformal Projection using data from Mapbox (2019) and UNdata (2019); graphic by Andrew W. Rate)

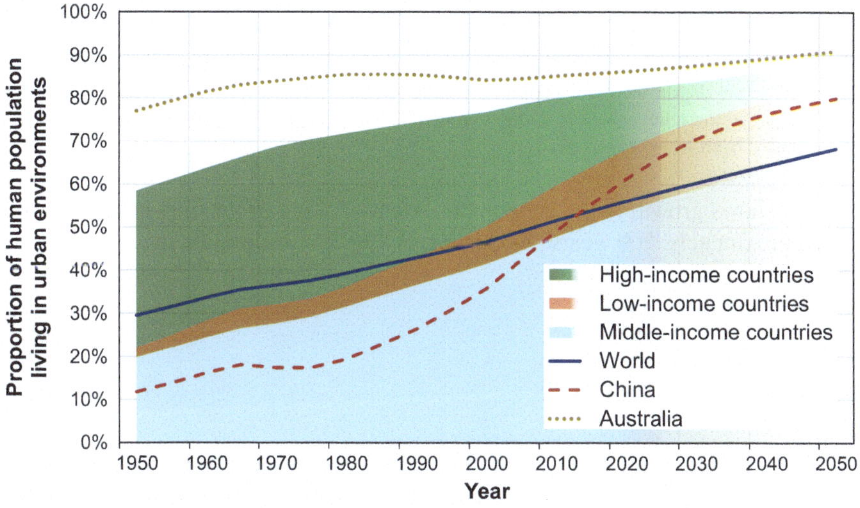

Fig. 1.3 Changes in the proportion of Earth's human population living in urban areas, 1950–2050. Data and projections from United Nations (2017, 2018); graphic by Andrew W. Rate

eastern Europe) showing their greatest urbanisation rates in the mid-twentieth century. The late twentieth century has also seen accelerations in urbanisation rates in several countries, including China, Indonesia, and South Korea (Luo et al. 2012; United Nations 2018). In contrast, countries such as Australia and the United Kingdom had already become highly urbanised by the beginning of the twentieth century, with low, stable rural populations and population growth only in urban areas. Some of the urban growth is manifested in the increasing sizes of cities;

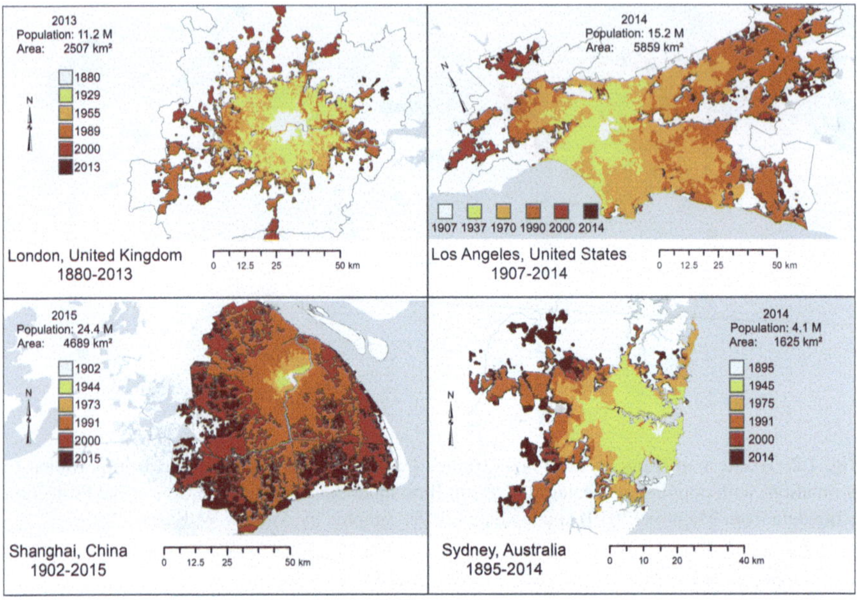

Fig. 1.4 Expansion of different urban areas showing the growth of four selected cities as a function of time (the darker the map shading, the more recently an area was converted to urban land use). The larger the relative area of darker-coloured shading, the more recent the expansion. Maps are compiled from Angel et al. (2016)

Fig. 1.4 shows growth in the geographic extent of some contrasting cities over the last approximately 200 years. The maps in Fig. 1.4 exemplify the rapid recent change in countries like China.

1.4 Human Use of Soils

The fertile soils that allowed cities to flourish also mean that urban areas are becoming increasingly important areas for food production (Satterthwaite et al. 2010). In 2000, about 6% of the world's croplands were in urban or peri-urban areas; the largest areas of land growing irrigated urban crops are in Southern and South East Asia, with mainly rain-fed urban crops grown in North America and Europe (Thebo et al. 2014) (Fig. 1.5). In many cities worldwide, the area of land dedicated to gardening in cities is increasing, with a resurgence in developed countries, with projects like urban community gardens providing not only a source of food but also health and social benefits (Wakefield et al. 2007; Laidlaw et al. 2018). As urban populations continue to increase, a greater proportion of food will need to be grown in urban or peri-urban areas (Thebo et al. 2014).

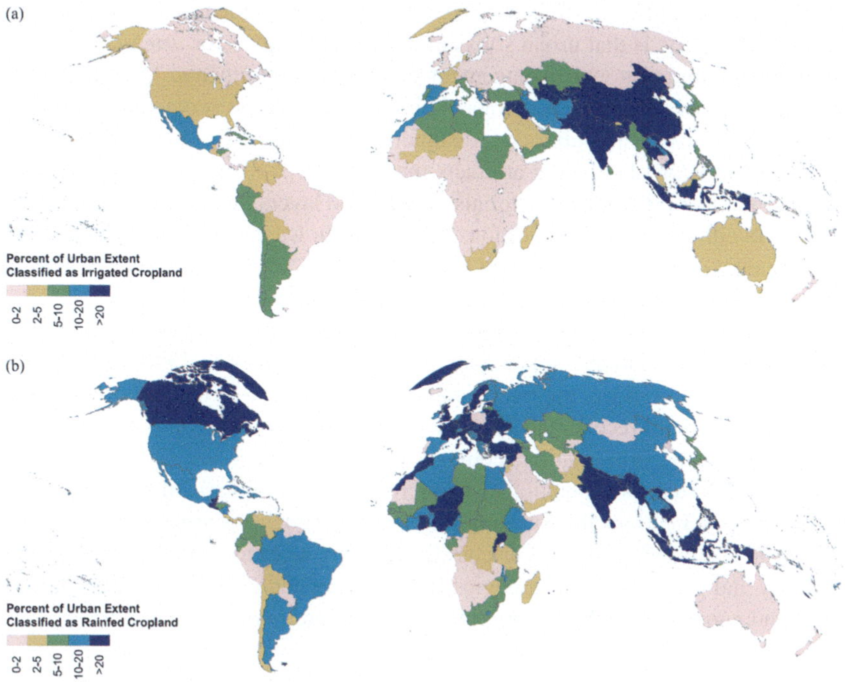

Fig. 1.5 Percentages of urban extent occupied by irrigated or non-irrigated cropland (recoloured from Thebo et al. (2014); used within the terms of a Creative Commons (CC BY 3.0) license)

Soils in cities are commonly much different from natural, unmodified soils, reflecting the importance of human factors in terms of their original *parent material* and soil-forming processes. The *geomorphology* of cities can also be highly modified, for example, by flattening of topography, redirection of streams and rivers, or drainage of wetlands (Alexandrovskaya and Alexandrovskiy 2000; Paul and Meyer 2001). Soils subject to a dominant human, or *anthropogenic* effect, are recognised explicitly in theories of soil formation (Amundson and Jenny 1991; Pouyat et al. 2008) and have been formalised into many of the widely used *soil classifications* (such as those presented in Isbell 1996; IUSS Working Group WRB 2014). Soils modified by humans may even provide a clear marker of the *Anthropocene*, the most recent geological period defined by profound human impact (Crutzen 2002). Certini and Scalenghe (2011) argue that the beginning of the Anthropocene is most appropriately and accurately marked by the presence of anthropogenic soils. These anthropogenic soils are called Anthroposols and Technosols in the United Nations' World Reference Base for Soil Resources (IUSS Working Group WRB 2014). Anthroposols are not confined to cities, nor are all urban soils Anthroposols or Technosols, but understanding anthropogenic soils is very instructive for understanding soils in urban environments. The formation and properties of urban soils, including Anthroposols and Technosols, will be covered in Chap. 2.

One of the most obvious differences between anthropogenic urban soils and their natural counterparts is that urban soils contain buried objects, or *artefacts*, derived from human habitation and management of urban environments. In some cases, these artefacts may be of archaeological significance and preserve valuable information about historical cultures (Alexandrovskaya and Alexandrovskiy 2000; Dubay 2012). Alternatively, the artefacts may simply be more recent additions of building rubble or other waste materials, engineered layers, or simply relocated soil material. In either case, soil conditions may result in *weathering* and loss of some anthropogenic signals (Howard and Olszewska 2011). Chapter 2 will also address the archaeological significance of urban soils and sediments.

Urban soils can be extremely heterogeneous, with steep gradients in soil properties observed over relatively small spatial scales (Schleuß et al. 1998; Rate 2018). This heterogeneity reflects the overprinting of natural soil-forming processes and parent materials with recent, and in some cases substantial, additions and losses of material caused by urbanisation, coupled with similarly recent changes in soil-forming factors such as vegetation and (micro)climate. Soil variability in any environment has implications for soil sampling and subsequent measurements and how soil data are analysed. Chapter 3 discusses the nature and implications of urban soil variability and presents some strategies for sampling and analysis of soils and statistical and numerical methods for exploration and interpretation of soil data.

1.5 Human Impacts on Soils

Numerous differences in *soil physical conditions* exist between urban and non-urban environments. Some of the most obvious of these are compaction, surface sealing, and changes in energy balance due to urban *heat island* effects. The physical properties of, and processes in, urban soils will be discussed in Chaps. 4 and 5, along with the implications of changes in soil physical conditions and geomorphology for urban hydrology (the behaviour of water in urban landscapes).

The human signature in urban soils may also be present as changes in the *chemical composition* of soils relative to their natural or 'background' state. These changes are *geochemical* records, separate from physical artefacts, which contain information on the history of urban land use (e.g. Zhang et al. 2005; Davidson et al. 2006; Albanese and Cicchella 2012; Appleton and Cave 2018; Rate 2018). The distinct chemical composition (i.e. a characteristic combination of concentrations of the chemical elements) or *geochemical signature* in geochemical records of urban soils can be interpreted with *spatial analysis* and *statistical analysis* of soil chemical composition, and we will address some of the issues involved with these analyses in following chapters. The changes in chemical composition of soils caused by urbanisation extend tens of kilometres to more than 100 km from urban centres, depending on the size of a city (Pouyat et al. 2008), and so represent a potentially powerful tool for understanding the effects of urbanisation on soils.

Changes in soil composition due to human activity, however, are often due to various forms of *pollution*. If substances are added to a soil (by pollution, or otherwise) in amounts or concentrations that are considered harmful to humans or other organisms, a soil is considered to be *contaminated* (National Environment Protection Council 2013). Soil contamination degrades the very resource(s) that originally made the cities possible and is an inherent consequence of the wide range of human activities that occur in urbanised areas. Chemical pollution in soils will be discussed in Chap. 6 (inorganic contaminants) and Chap. 7 (organic contaminants). Chapter 4 will also present essential concepts for understanding the chemical composition and reactions in soil systems, with a focus on processes encountered in urban soil environments.

Urban soils can provide the same *ecosystem services* and perform the same functions as non-urban soils, regardless of the multiple differences in the properties, starting materials, and formation processes (Sauerwein 2011; Lal and Stewart 2017; Teixeira da Silva et al. 2018). In reality, urban soils exist on a continuum between highly modified anthroposols and unaltered natural soils, and there is no clear boundary between them (De Kimpe and Morel 2000; Byrne 2007). In many cases their origins and *pedogenesis* (i.e. formation processes) are similar to 'natural' soils (Giusti 2013). The range of soil functions and ecosystem services provided by soils will be discussed, with emphasis on urban systems, in Chap. 4. In addition, since many important functions of soils are mediated by microorganisms and because microbial ecology is sensitive to changes in soil conditions, Chap. 8 expands on below-ground biological processes in soils and reviews the effects of urban soil changes on the abundance and diversity of soil organisms. Although we do not have a separate chapter relating to the influence of urban soils on above-ground vegetation (and nor is there a separate section or non-soil fauna), this discussion is integrated into several of the other chapters.

The pollution and subsequent soil contamination which are widespread in cities also mean that the urbosphere itself has become a source of pollution. Contaminated urban soils can be sources of contaminants for surface water and groundwater (Wong et al. 2012). Transfer of material from soil to air by wind is a recognised factor in decreasing urban air quality (Almeida et al. 2005; Sillanpää et al. 2006), and soil to atmosphere fluxes may also introduce contaminants to other ecosystem compartments (Laidlaw and Filippelli 2008; Reis et al. 2013). Possibly the most critical effect on atmospheric air composition is from the emission of greenhouse gases; urban areas are the single major contributor of global anthropogenic greenhouse gas emissions, with soils adding to overall urban emissions (Bellucci et al. 2012). The ability of urban soils to act as both a sink and source for pollutants will be covered in Chap. 9, along with frameworks for estimating risks from polluted urban soils.

The health of soils relates positively to human health, in terms of both physical and psychological well-being (Swartjes 2015; Brevik and Pereg 2017). In urban environments, people's quality of life is improved by soil-related activities such as gardening, which provide benefits including improved mental health and community connectedness (Wakefield et al. 2007; Soga et al. 2017; Laidlaw et al. 2018). Conversely, degraded soils, especially if contaminated, pose multiple ongoing threats to human health (Filippelli et al. 2012; Pepper 2013). The health effects of soils are unevenly distributed in cities, and this is one focus of the *environmental justice* movement. For example, lower socioeconomic groups in a city are known to be exposed to, or affected, more severely by soil pollution and can also have less access to beneficial soil-related activities such as use of public open space (Zhuo et al. 2012; McClintock 2015). This book will explore the relationships between soils and human health in urban systems, in the context of the United Nations' Sustainable Development Goals, in Chap. 10.

There is, however, still cause for optimism about the health of urban soils. Along with the multitude of forms that soil degradation can take in cities, there are an increasing number of options for *soil remediation* (Bolan et al. 2014), from simple revegetation to highly technical methods such as electrokinetic migration (Ho et al. 1995). Many soil remediation techniques have been developed for urban environments, but others were developed for environments such as waste facilities and mine sites. As a result, some urban soils are becoming, or being made, cleaner as pollution sources are removed or better controlled and as land is rehabilitated (Andersson et al. 2010). The prevalence of degraded soils in urban environments makes cities ideal environments to study the options for land rehabilitation, and this topic is covered in Chap. 11.

Although urban environments and soils are very distinctive, it will become apparent that we can use many of the concepts in traditional soil science to understand and analyse urban soils (Fig. 1.6). It will also become clear, however, that urban soils show multiple differences from their non-urban counterparts and this, together with existing in distinctive and increasingly crucial urban ecosystems, makes urban soils an essential subject to study within the wider discipline of soil science.

1.6 Additional Reading

Grimm NB, Faeth SH, Golubiewski NE, Redman CL, Wu J, Bai X, Briggs JM (2008) Global change and the ecology of cities. Science 319:756–760

Hillel D (1991) Out of the earth: civilization and the life of the soil. Free Press, New York

Satterthwaite D, McGranahan G, Tacoli C (2010) Urbanization and its implications for food and farming. Philos Trans R Soc B Biol Sci 365:2809–2820

1.7 Summary

- Globally, soils are a keystone of the critical zone, the upper layers of Earth's land surface which sustain terrestrial life.

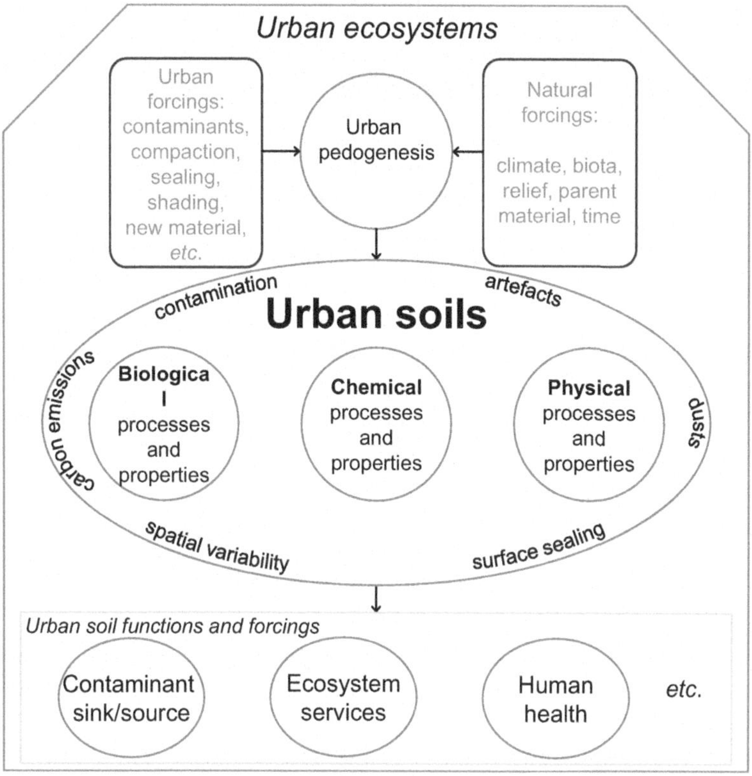

Fig. 1.6 Conceptual diagram for urban soils within urban ecosystems, showing selected examples of their key drivers (forcings), properties, and ecosystem functions (graphic by Andrew W. Rate)

- The beneficial properties of soils have in many cases been a factor determining the location of cities; conversely, urbanisation has affected, and often degraded, the properties of soils.
- Human populations in urban areas are increasing rapidly, and urban soils will increasingly be needed to support these populations, while simultaneously being affected by human activities.
- Urban soils may show multiple differences from natural soils but may also perform many of the same ecosystem functions. Conventional soil science concepts, therefore, still have a part to play in understanding urban soil environments.

1.8 Questions

1.8.1 Checking your Understanding

1. What is the critical zone, and how does is relate to urban soils?
2. Give some examples of how humans use soil in cities or are affected in some way by urban soils.
3. Historically, how did the fertility of soils affect the development and location of cities?
4. What are some of the distinctive characteristics of urban soils?
5. List some of the ways in which urban soils have their chemical and physical composition changed.
6. What biological changes might you expect in urban soils as modification and disturbance caused by humans increases?

1.8.2 Thinking about the Issues

7. If increasing areas of urban and peri-urban land will need to be used for food production, what are some of the soil-related issues that might limit this kind of use?
8. Is the proposed existence of the Anthropocene a useful concept for understanding urban soils? Discuss your reasoning.
9. Is it possible that urbanisation could improve the properties of some soils? How?

1.8.3 Contemplating Urban Soils Creatively

10. Would you expect the environmental effects of urbanisation to differ for cities which have expanded in geographical extent at different times (e.g. expansion in the first versus the second half of the twentieth century)? Explain your reasoning and what the difference might be.

References

Albanese S, Cicchella D (2012) Legacy problems in urban geochemistry. Elements 8:423–428. https://doi.org/10.2113/gselements.8.6.423

Alexandrovskaya EI, Alexandrovskiy AL (2000) History of the cultural layer in Moscow and accumulation of anthropogenic substances in it. Catena 41:249–259. https://doi.org/10.1016/S0341-8162(00)00107-7

Almeida SM, Pio CA, Freitas MC, Reis MA, Trancoso MA (2005) Source apportionment of fine and coarse particulate matter in a sub-urban area at the Western European coast. Atmos Environ 39:3127–3138. https://doi.org/10.1016/j.atmosenv.2005.01.048
Amundson R, Jenny H (1991) The place of humans in the state factor theory of ecosystems and their soils. Soil Sci 151:99–109. https://doi.org/10.1097/00010694-199101000-00012
Andersson M, Ottesen RT, Langedal M (2010) Geochemistry of urban surface soils – monitoring in Trondheim, Norway. Geoderma 156:112–118
Angel S, Blei AM, Parent J, Lamson-Hall P, Sánchez NG (2016) Atlas of urban expansion – 2016 edition, volume 1: areas and densities. New York University, UN-Habitat, and Lincoln Institute of Land Policy, New York/Nairobi/Cambridge. http://atlasofurbanexpansion.org/file-manager/userfiles/data_page/Full%20Volumes/Atlas%20of%20Urban%20Expansion%202016%20Volume%201%20-%20Areas%20and%20Densities.pdf
Appleton JD, Cave MR (2018) Variation in soil chemistry related to different classes and eras of urbanisation in the London area. Appl Geochem 90:13–24. https://doi.org/10.1016/j.apgeochem.2017.12.024
Bellucci F, Bogner JE, Sturchio NC (2012) Greenhouse gas emissions at the urban scale. Elements 8:445–450. https://doi.org/10.2113/gselements.8.6.445
Bolan N, Kunhikrishnan A, Thangarajan R, Kumpiene J, Park J, Makino T et al (2014) Remediation of heavy metal(loid)s contaminated soils – to mobilize or to immobilize? J Hazard Mater 266:141–166. https://doi.org/10.1016/j.jhazmat.2013.12.018
Brantley SL, White TS, White AF, Sparks D, Richter D, Pregitzer K et al (2006) Frontiers in exploration of the critical zone: report of a workshop sponsored by the NSF. National Science Foundation (NSF), Newark
Brevik EC, Pereg L (2017) History of soils in relation to animal and human health. In: Singh BR, McLaughlin MJ, Brevik EC (eds) The nexus of soils, plants, animals and human health. Schweizerbart Publishers, Stuttgart, pp 8–15
Byrne LB (2007) Habitat structure: a fundamental concept and framework for urban soil ecology. Urban Ecosyst 10:255–274. https://doi.org/10.1007/s11252-007-0027-6
Certini G, Scalenghe R (2011) Anthropogenic soils are the golden spikes for the Anthropocene. The Holocene 21:1269–1274. https://doi.org/10.1177/0959683611408454
Crutzen PJ (2002) Geology of mankind. Nature 415:23. https://doi.org/10.1038/415023a
Davidson DA, Dercon G, Stewart M, Watson F (2006) The legacy of past urban waste disposal on local soils. J Archaeol Sci 33:778–783. https://doi.org/10.1016/j.jas.2005.10.017
De Kimpe CR, Morel J-L (2000) Urban soil management: a growing concern. Soil Sci 165:31–40
Dubay B (2012) Urban soil genesis, weathering of waste building materials, and bioavailability of lead in a chronosequence at former demolition sites. ProQuest Dissertations Publishing, Detroit
Fang Y, Ceola S, Paik K, McGrath G, Rao PSC, Montanari A, Jawitz JW (2018) Globally universal fractal pattern of human settlements in river networks. Earth's Future 6:1134–1145. https://doi.org/10.1029/2017EF000746
Filippelli GM, Morrison D, Cicchella D (2012) Urban geochemistry and human health. Elements 8:439–444. https://doi.org/10.2113/gselements.8.6.439
Foley JA, DeFries R, Asner GP, Barford C, Bonan G, Carpenter SR et al (2005) Global consequences of land use. Science 309:570–574. https://doi.org/10.1126/science.1111772
Giusti L (2013) The chemistry and parent material of urban soils in Bristol (UK): implications for contaminated land assessment. Environ Geochem Health 35:53–67. https://doi.org/10.1007/s10653-012-9468-9
Grimm NB, Faeth SH, Golubiewski NE, Redman CL, Wu J, Bai X, Briggs JM (2008) Global change and the ecology of cities. Science 319:756–760
Hillel D (1991) Out of the earth: civilization and the life of the soil. Free Press, New York
Hillel D (2006) Human culture and soils. In: Logsdon S (ed) Encyclopedia of soil science. Taylor & Francis, pp 835–837. https://doi.org/10.1081/E-ESS-12000660

Ho SV, Sheridan PW, Athmer CJ, Heitkamp MA, Brackin JM, Weber D, Brodsky PH (1995) Integrated *in situ* soil remediation technology: the lasagna process. Environ Sci Technol 29:2528–2534

Howard JL, Olszewska D (2011) Pedogenesis, geochemical forms of heavy metals, and artifact weathering in an urban soil chronosequence, Detroit, Michigan. Environ Pollut 159:754–761. https://doi.org/10.1016/j.envpol.2010.11.028

Isbell RF (1996) In: CSIRO (ed) The Australian soil classification, Australian soil and land survey handbook, 4. CSIRO Publishing, Melbourne, 143 pp

IUSS Working Group WRB (2014) *World Reference Base for soil resources 2014*, world soil resources reports food and agriculture Organization of the United. Nations, Rome

Kummu M, de Moel H, Ward PJ, Varis O (2011) How close do we live to water? A global analysis of population distance to freshwater bodies. PLoS One 6:e20578. https://doi.org/10.1371/journal.pone.0020578

Laidlaw MAS, Filippelli GM (2008) Resuspension of urban soils as a persistent source of lead poisoning in children: a review and new directions. Appl Geochem 23:2021–2039. https://doi.org/10.1016/j.apgeochem.2008.05.009

Laidlaw MAS, Alankarage DH, Reichman SM, Taylor MP, Ball AS (2018) Assessment of soil metal concentrations in residential and community vegetable gardens in Melbourne, Australia. Chemosphere 199:303–311. https://doi.org/10.1016/j.chemosphere.2018.02.044

Lal R, Stewart BA (eds) (2017) Urban soils. Advances in soil science. CRC Press/Taylor & Francis Inc., Boca Raton, 416 pp

Liu Z, He C, Zhou Y, Wu J (2014) How much of the world's land has been urbanized, really? A hierarchical framework for avoiding confusion. Landscape Ecology 29. https://doi.org/10.1007/s10980-014-0034-y

Luo XS, Yu S, Zhu YG, Li XD (2012) Trace metal contamination in urban soils of China. Sci Total Environ 421–422:17–30

Lyons WB, Harmon RS (2012) Why urban geochemistry? Elements 8:417–422. https://doi.org/10.2113/gselements.8.6.417

Mapbox (2019) Mapbox data attribution, https://www.mapbox.com/about/maps/. Accessed 20190705

McClintock N (2015) A critical physical geography of urban soil contamination. Geoforum 65:69–85. https://doi.org/10.1016/j.geoforum.2015.07.010

McDonald RI, Mansur AV, Ascensão F, Colbert MI, Crossman K, Elmqvist T et al (2019) Research gaps in knowledge of the impact of urban growth on biodiversity. Nat Sustain. https://doi.org/10.1038/s41893-019-0436-6

McNeill JR, Winiwarter V (2004) Breaking the sod: humankind, history, and soil. Science 304:1627–1629. https://doi.org/10.1126/science.1099893

National Environment Protection Council (2013) Schedule B (1): guideline on the investigation levels for soil and groundwater, National Environment Protection (assessment of site contamination) measure (amended). Commonwealth of Australia, Canberra

National Research Council, Committee on Basic Research Opportunities in the Earth Sciences, Board on Earth Sciences and Resources (2001) Basic research opportunities in earth science. National Academy of Sciences/National Academies Press, Washington, DC

Paul MJ, Meyer JL (2001) Streams in the urban landscape. Annu Rev Ecol Syst 32:333–365

Pepper IL (2013) The soil health-human health nexus. Crit Rev Environ Sci Technol 43:2617–2652. https://doi.org/10.1080/10643389.2012.694330

Pouyat RV, Yesilonis ID, Szlavecz K, Csuzdi C, Hornung E, Korsós Z et al (2008) Response of forest soil properties to urbanization gradients in three metropolitan areas. Landsc Ecol 23:1187–1203. https://doi.org/10.1007/s10980-008-9288-6

Rate AW (2018) Multielement geochemistry identifies the spatial pattern of soil and sediment contamination in an urban parkland, Western Australia. Sci Total Environ 627:1106–1120. https://doi.org/10.1016/j.scitotenv.2018.01.332

Redman CL (2011) Social-ecological transformations in urban landscapes – a historical perspective. In: Niemelä J, Breuste JH, Guntenspergen G (eds) Urban ecology: patterns, processes, and applications. Oxford Biology. OUP Oxford, Oxford

Reis AP, Patinha C, Wragg J, Dias AC, Cave M, Sousa AJ et al (2013) Urban geochemistry of lead in gardens, playgrounds and schoolyards of Lisbon, Portugal: assessing exposure and risk to human health. Appl Geochem 44:45–53. https://doi.org/10.1016/j.apgeochem.2013.09.022

Reséndiz-Paz ML, Gutiérrez-Castorena MC, Gutiérrez-Castorena EV, Ortiz-Solorio CA, Cajuste-Bontempts L, Sánchez-Guzmán P (2013) Local soil knowledge and management of Anthrosols: a case study in Teoloyucan, Mexico. Geoderma 193–194:41–51. https://doi.org/10.1016/j.geoderma.2012.09.004

Satterthwaite D, McGranahan G, Tacoli C (2010) Urbanization and its implications for food and farming. Philos Trans R Soc B Biol Sci 365:2809–2820. https://doi.org/10.1098/rstb.2010.0136

Sauerwein M (2011) Urban soils – characterization, pollution, and relevance in urban ecosystems. In: Niemelä J, Breuste JH, Guntenspergen G (eds) Urban ecology: patterns, processes, and applications. Oxford biology. Oxford University Press, Oxford

Schleuß U, Wu Q, Blume H-P (1998) Variability of soils in urban and periurban areas in Northern Germany. Catena 33:255–270. https://doi.org/10.1016/S0341-8162(98)00070-8

Sillanpää M, Hillamo R, Saarikoski S, Frey A, Pennanen A, Makkonen U et al (2006) Chemical composition and mass closure of particulate matter at six urban sites in Europe. Atmos Environ 40:212–223. https://doi.org/10.1016/j.atmosenv.2006.01.063

Soga M, Gaston KJ, Yamaura Y (2017) Gardening is beneficial for health: a meta-analysis. Prev Med Rep 5:92–99. https://doi.org/10.1016/j.pmedr.2016.11.007

Swartjes F (2015) Human health risk assessment related to contaminated land: state of the art. Environ Geochem Health 37:651–673. https://doi.org/10.1007/s10653-015-9693-0

Teixeira da Silva R, Fleskens L, van Delden H, van der Ploeg M (2018) Incorporating soil ecosystem services into urban planning: status, challenges and opportunities. Landsc Ecol 33:1087–1102. https://doi.org/10.1007/s10980-018-0652-x

Thebo AL, Drechsel P, Lambin EF (2014) Global assessment of urban and peri-urban agriculture: irrigated and rainfed croplands. Environ Res Lett:9.: , 9pp. https://doi.org/10.1088/1748-9326/9/11/114002

UNdata (2019) Demographic statistics database. United Nations Statistics Division. http://data.un.org/. Accessed 20191128

United Nations (2017) World population prospects: the 2017 revision. Custom data acquired via https://population.un.org/

United Nations (2018) World urbanization prospects: the 2018 revision. Custom data acquired via https://population.un.org/

Voloshyn I (2004) Теоретичні засади ландшафтних еколого-геохімічних досліджень (Theoretical fundamentals of landscape ecological-geochemical researches). Серія географічна (Series Geographic) 31:72–78

Wakefield S, Yeudall F, Taron C, Reynolds J, Skinner A (2007) Growing urban health: community gardening in south-east Toronto. Health Promot Int 22:92–101. https://doi.org/10.1093/heapro/dam001

Wong CI, Sharp JM Jr, Hauwert N, Landrum J, White KM (2012) Impact of urban development on physical and chemical hydrogeology. Elements 8:429–434. https://doi.org/10.2113/gselements.8.6.429

Zhang G-L, Yang F-G, Zhao Y-G, Zhao W-J, Yang J-L, Gong Z-T (2005) Historical change of heavy metals in urban soils of Nanjing, China during the past 20 centuries. Environ Int 31:913–919. https://doi.org/10.1016/j.envint.2005.05.035

Zhuo X, Boone CG, Shock EL (2012) Soil lead distribution and environmental justice in the Phoenix metropolitan region. Environmental Justice 5:206–213. https://doi.org/10.1089/env.2011.0041

Chapter 2
Formation and Properties of Urban Soils

Andrew W. Rate

Abstract Urban soils form by the same overall set of processes that are involved in the formation of all soils. In this chapter we combine two general approaches to understanding soil formation, the first based on state factors and the second on soil fluxes. We review the soil properties used for identification and classification. The concepts of anthroposequences and urbanisation gradients are introduced, and the soil groups important for urban environments, Anthrosols and Technosols, are described. Changes in geomorphology caused by urbanisation, such as modification of hydrology and landforms created by additions of removal of material, are discussed. Specific examples of soils in modified urban environments, such as reclaimed coastal land, landfills, and constructed wetlands, are presented. Finally, this chapter explores the archaeological landforms and soil properties present in historical and contemporary cities, and the range of chemical, physical, and biological archaeological information stored in urban soils.

Keywords Soil formation · Pedogenesis · Urban soils · Geomorphology · Anthropogenic landforms · Archaeology · Soil properties · Technosols

A. W. Rate (✉)
School of Agriculture and Environment, University of Western Australia, Crawley, WA, Australia
e-mail: andrew.rate@uwa.edu.au

A. W. Rate (ed.), *Urban Soils*, Progress in Soil Science,
https://doi.org/10.1007/978-3-030-87316-5_2

Some of what you could learn from this chapter:

- The basics of soil formation processes (pedogenesis) and soil classification as it applies to urban soils
- How landforms in urban environments have been modified
- The ways in which urban soils differ from (or resemble) non-urban soils
- The types of archaeological information preserved by soils and how archaeological information can be obtained from soils

2.1 Introduction to Urban Pedology and Pedogenesis

The processes which result in formation of urban soils from their parent materials can be understood using the conceptual frameworks used for soil formation in general. As a result, this chapter will first present and discuss the basics of the main concepts involved when considering soil formation in any environment.

McKenzie et al. (2004) summarise the two main approaches to understanding *pedogenesis*, or soil formation: first, the *state factor* approach, where observable soil properties reflect the environment during soil formation (including climate, organisms, parent material, relief, and time), and second, what we will call the *soil fluxes* approach, involving additions and losses to and from soil systems and transformations and translocations of materials within the soil environment. It is most useful to have an understanding of both approaches in order to have a complete understanding of soil formation.

2.1.1 State Factors and Soil Formation

The **state factor** approach to understanding soil properties and formation was most likely first developed in the late 1800s by the Russian scientist Vasily Dokuchaev (Evtuhov 2006). Dokuchaev viewed a soil as an independent environmental compartment, which has properties reflecting the combined influence of subsoils, climate, flora and fauna, geological age, and relief in the same location. This concept was developed further by several scientists and became a foundational idea in soil science following publication of *Factors of Soil Formation* by American scientist Hans Jenny (Jenny 1941). A common way of expressing the state factor model of soil formation is in the so-called clorpt equation (Eq. 2.1):

$$\text{soil properties}, S = f\left(\text{cl,o,r,p,t,}\ldots\right) \tag{2.1}$$

where $f(\,)$ represents 'a function of'; cl = climate, o = organisms, r = relief (topography/altitude), p = parent material, t = time, … = any other factors (e.g. localised phenomena such as fire).

The scale of the *clorpt* factors is similar to the scale of observation for difference in soil properties. For example, when considering difference on soil profiles on a continental scale, we would consider large-scale differences in factors such as basin- or craton-scale differences in geological parent material, global climatic zones such as those in the Köppen-Geiger climate classification (Kottek et al. 2006), and large-scale relief such as mountain belts. In contrast, for an urban ecosystem, we would need to consider smaller-scale phenomena such as urban microclimates (e.g. an urban heat island), changes in parent material over short distances as a result of human disturbance, and smaller-scale relief such as individual hillslopes or excavations.

2.1.2 Soil Fluxes and Soil Formation

The **soil fluxes** approach to understanding soil formation and properties is derived from Simonson (1959) in which the focus is on the processes occurring in the soil itself. In this approach (called the 'process-systems' model by Schaetzl and Anderson 2005), the observable properties of a soil profile represent the balance of additions to or losses from soil, as well as translocations and transformations of material within soil. If the balance between additions, losses, translocations, and transformations differs, then the resulting soil profile will have different properties. The soil fluxes approach implies a more dynamic view of soils, since the processes involved are common to all soils, but the relative degree to which they occur affects the soil properties which can be observed at any point in time. For example, continued additions of material to a soil environment, such as net accumulation of organic matter, will ultimately result in a different soil profile.

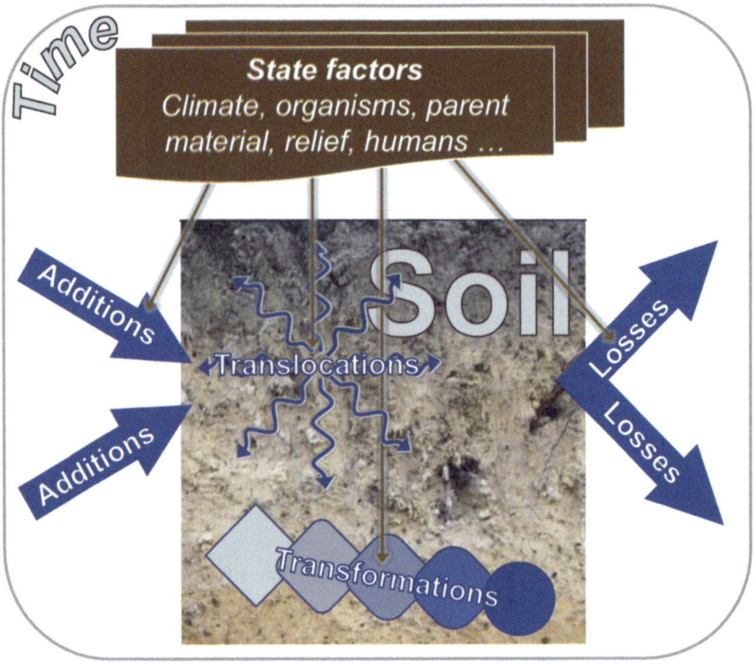

Fig. 2.1 A two-way conceptual approach to soil formation, showing the state factors in *italic text* and the soil fluxes in **bold text**. The state factors affect the relative amounts and types of soil fluxes. Time is shown as a separate overarching state factor since the other state factors are not constant but vary with time

A soil's properties reflect the relative sizes of *soil fluxes*: additions, losses, translocations, and transformations of material; however, this approach alone is not enough to fully understand soil formation and the resulting soil properties. A combination of both approaches described here is more satisfying, in that it is the *state factors* that affect the sizes and types of soil fluxes that occur. We have tried to show this combined model of soil formation graphically in Fig. 2.1, which also treats time as a special factor because it is a dimension that all other factors and fluxes operate in. The unique combination of time, state factors, and fluxes results in different observable properties in a soil profile, and a list of the types of soil properties considered important in discriminating different types of soils appears in Box 2.1. Most these properties are important for other purposes as well, since they affect the ability of soils to perform critical environmental functions such as supplying water and nutrients to plants or modifying the behaviour and toxicity of pollutants. We will address these soil functions in detail in Chaps. 4, 5, 6, 7 and 8.

Box 2.1: Soil Properties Used to Identify Different Soil Types and Classify Soils

All of the soil properties below are controlled by the effects of the state factors on soil fluxes.

Horizons – these are the approximately horizontal, layer-like features in soil caused by pedogenesis. Soils differ in the types, thicknesses, colour, etc. of horizons and other properties (see below), whether or not there is distinct contrast between horizons, whether horizons are well-developed, and so on.

Soil organic matter – concentration of soil organic carbon, depth(s) of accumulation.

Soil texture – measured by the relative amounts of sand (0.05–2 mm), silt (2 µm–0.05 mm), or clay-sized (<2 µm) grains or particles in the *fine earth* (<2 mm) fraction of soils.

Mineral types – especially type of clay but also carbonate minerals, iron oxides, silica, or presence/absence of disordered minerals, volcanic glasses, gypsum, etc.

Exchangeable cations and soil pH – some soil materials (especially clay minerals and organic matter) carry negative electrostatic charge which is balanced by the dominant cations in soil. The relative concentrations of these cations (H^+, Al^{3+}, Na^+, Mg^{2+}, K^+, Ca^{2+}) are related in part to soil pH.

Soluble salt content – a few minerals (salts) dissolve easily in water, and, if salts are abundant in soil, the result is high concentrations of salts in soil pore water. Such a soil would be considered saline.

Climate- and/or hydrology-controlled features – frozen subsoils, extremely leached horizon(s), saturation with water, arid-zone soils, desert pavements.

Presence of rock-like materials – stoniness and composition of rock fragments, cementation within or between stony components.

Degree of weathering or alteration – how different the soil material(s) are from the parent material.

Human modification – such as mixing by cultivation, presence of anthropogenic artefacts.

2.1.3 Pedogenesis of Urban Soils

A combined state factor-soil fluxes approach makes sense in urban systems if the effects of human activity are simply included in the 'organisms' state factor. Some soil scientists, however, consider anthropogenic effects as a separate state factor (e.g. Amundson and Jenny 1991). Even though inclusion of a human state factor is not restricted to urban environments, this is a useful approach to take given the great importance of human modification in cities. The impact of humans can also change some other state factors affecting soil formation, for example, by introducing new plant and animal species, creating new landforms and, more recently, modifying local and global climates.

In some environments, the state factor approach can be understood more easily by carefully choosing soils in locations which allow us to isolate the effects of a single factor on soil formation and properties. This leads to the concept of *soil sequences*, in which a series of geographically separated soils show a gradient in only one state factor, with the other state factors being approximately constant throughout the soils' development. The most commonly studied of these is probably the *toposequence*, where the changing state factor is **relief**, such as a sequence of soils from the top to bottom of a hillslope. A detailed discussion of soil sequences (e.g. toposequences, chronosequences, climosequences) is not within the scope of this book, but readers are referred to excellent discussions of this topic in Schaetzl and Anderson (2005) and White (2006). It is possible, however, to make an analogy between the more commonly studied soil sequences and soil sequences where the state factor that changes is predominantly the human factor. This type of soil sequence, an *anthroposequence*, has been studied along urban to rural gradients, showing changes in several soil properties from the rural-urban fringe to the urban core (Pouyat et al. 1995). In many cases the changes in soil properties along an urban-rural gradient represent additions of substances to soils by human activity (Figs. 2.2 and 2.3).

The existence of soil sequences along urbanisation gradients means that there is effectively a continuum of urban effects related to urbanisation. A question then emerges: do truly urban soils exist? The best answer to this question is probably

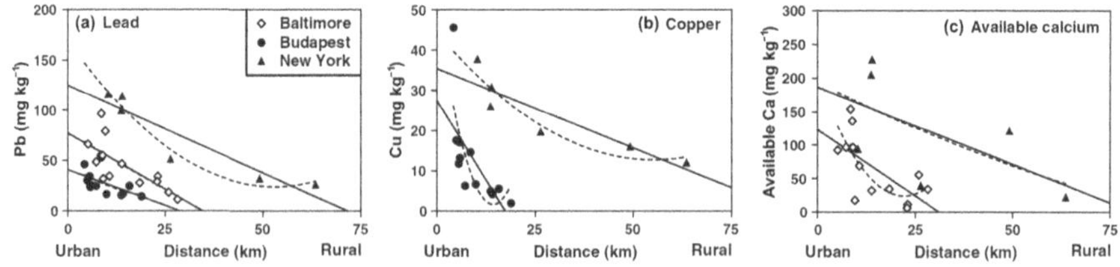

Fig. 2.2 An example of a soil anthroposequence shown by gradients in soil properties: (**a**) lead concentration, (**b**) copper concentration, and (**c**) available calcium concentration, in relation to distance from urban centres (from Pouyat et al. (2008); used with permission from Springer). 'Available calcium' refers to Ca extracted from soil using dilute acid solutions

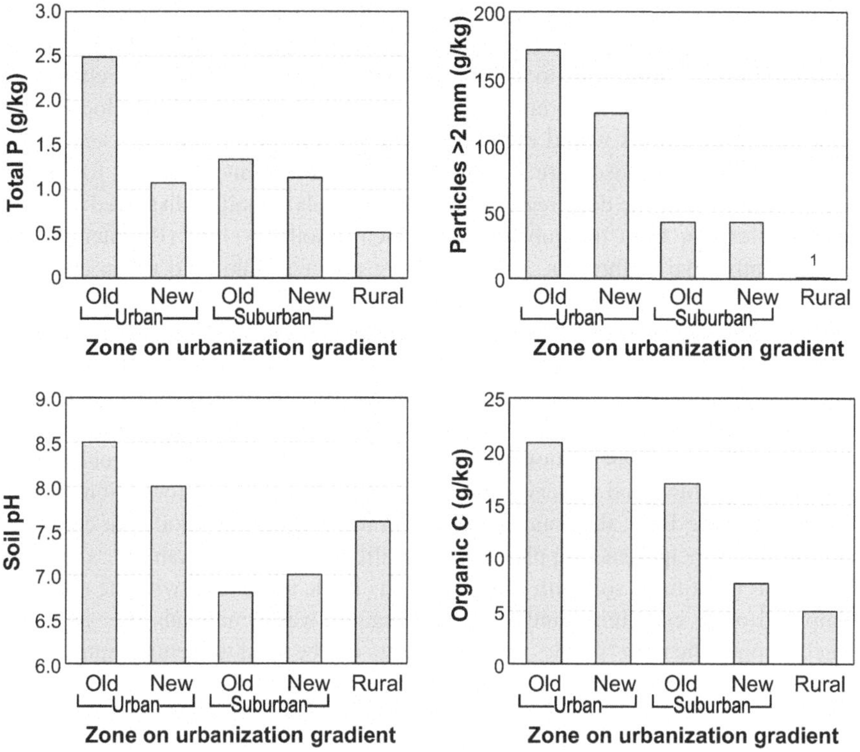

Fig. 2.3 Differences in soil properties along an urbanisation gradient. (Redrawn from data in Table 1 in Zhang (2004))

provided by soil classification schemes. There are numerous soil classification schemes in use, which can be confusing, so we will use examples from the most widely used international soil classification system, the International Union of Soil Sciences' 'World Reference Base for Soil Resources' (IUSS Working Group WRB 2014). Soil classification schemes are most often hierarchical; that is, they classify soils into broad categories based on diagnostic soil properties such as the existence of a certain type of *horizon*. If we assume that an urban soil must reflect human influence on its formation, then two groups of soils in the World Reference Base, called *Anthrosols* or *Technosols* (IUSS Working Group WRB 2014), are the most likely candidates.

Another framework for understanding urban soils is that of Soils of Urban, Industrial, Traffic, Mining, and Military Areas (SUITMAs) as proposed by Morel and Heinrich (2008). The SUITMA concept considers urban soils together with other soils having significant anthropogenic influence on their formation but which are not necessarily located in cities.

The *Anthrosol* soil group mainly relates to the effects of long-term cultivation on soil formation. The historical association of urban areas with fertile soils and food production that we covered in Chap. 1 makes it likely that many urban soils are also (former) Anthrosols. However, Anthrosols clearly also occur in non-urban environments, so they are not unique to urban areas. The *Technosol* soil group is characterised by the presence of materials which have been manufactured or relocated by humans, so Technosols would therefore seem to more obviously represent urban soils. Of course, Technosols are also present in non-urban environments; for example, in remote areas, the deep *regolith* or rock materials ('spoils') displaced by mining activities can form the substrate on which a soil develops (Rossiter 2007). Consequently, like Anthrosols, Technosols are also not unique to urban environments. Box 2.2 contains a summary of the criteria used to identify anthropogenic soils in different environments, with examples from three soil classification schemes.

The lack of uniqueness of the World Reference Base's two anthropogenic Reference Soil Groups to urban environments does not mean that soil classification schemes are inadequate tools for describing or defining urban soils. Another great advantage of a soil classification is that it provides a structured conceptual framework for describing and understanding soil properties and soil formation. In the World Reference Base, the concepts that help us to understand soils are encapsulated in the principal and supplementary qualifier terms. For example, a 'garbic Technosol' identifies a soil with ≥20% artefacts in the upper 1 m, with the artefacts composed of or containing anthropogenic organic waste materials (i.e. garbage, which inspires the term 'garbic'). The qualifiers can be used to identify human features in non-anthropogenic Reference Soil Groups as well; for example, the suffix 'transportic' is used to indicate natural soil material which has been moved (transported) by humans to another location (for more explanation see Rossiter 2007). Similarly, the presence of artefacts below the 20% threshold required for classification as a Technosol can be important, so the qualifier 'technic' can be appended to many of the Reference Soil Groups.

The concepts summarised in soil classifications such as the World Reference Base can therefore help us with the identification and description of soil environments consisting of both natural materials affected by urban phenomena and natural processes acting on urban materials. The emphasis, in many soil classifications, on horizons as the primary diagnostic criterion (i.e. macroscale phenomena) means that more subtle effects may be missed. Even when the classification can be based on soil composition rather than a diagnostic horizon, the thresholds imposed by soil classification schemes may not allow informative terms to be used within the classification. For example:

- The World Reference Base requires anthropogenic horizons to be at least 50 cm thick for the soil to be an Anthrosol.
- The World Reference Base requires artefacts to comprise ≥20% of soil volume for the classification of Technosol to be applied or 10–20% artefacts by volume to use the 'technic' qualifier.
- The Australian Soil Classification (Isbell 1996) requires additions of anthropogenic material ≥30 cm deep or that soil features reflecting natural pedogenesis have been erased by human activity, for some suborders of Anthroposols.

Box 2.2: Examples of Diagnostic Properties of Anthropogenic Soils in Soil Classifications.

Classification	Highest-level category	Criteria for highest category	Subcategories and criteria
World Reference Base of the International Union of Soil Sciences (IUSS Working Group WRB 2014)	*Anthrosols* (reference soil groups)	Various types of horizon ≥50 cm deep created by activities related to cultivation	*Hydragric* – paddy soils *Irragric* – irrigated soils *Hortic* – fertilisation and organic residues *Plaggic* – manure and sod application *Pretic* – charcoal, artefacts *Terric* – added mineral material and deep cultivation
(as above)	*Technosols*(reference soil groups)	Primarily, ≥20% by volume of artefacts in upper 1 m or with a geomembrane or hard, consolidated layer of industrial origin	*Ekranic* – hard material ≤5 cm from surface *Linic* – low permeability geomembrane ≤1 m from surface *Urbic* – ≥20 cm layer with ≥20% vol. rubble/refuse *Spolic* – ≥20 cm layer with ≥20% vol. industrial waste *Garbic* – ≥20 cm layer with ≥20% vol. organic waste *Cryic* – permanent ice in upper 2 m *Isolatic* – soil on a hard manufactured surface, e.g. roof, pots *Leptic* – hard continuous layer ≤1 m from surface *Subaquatic* – permanently submerged *Tidalic* – affected by tidal water *Reductic* – reducing conditions in top 1 m *Hyperskeletic* – < 20% vol. fine earth
Australian Soil Classification (Isbell and National Committee on Soil and Terrain 2016)	*Anthroposols* (soil orders)	'Profound modification' from human activity including truncation, burial, and creation of new parent materials. Burial depth ≥ 30 cm; removal of natural pedological features	*Fusic* – ≥0.3 m with ≥20% ash material from burnt peat *Cumulic* – ≥0.3 m human deposited material, e.g. middens, mill mud *Hortic* – incorporation of organic residues to remove natural pedological features *Garbic* – soil underlain by landfill of mainly organic refuse *Urbic* – soil underlain by landfill of mainly manufactured mineral refuse *Dredgic* – soils on mineral material dredged from marine/aquatic sediments *Spolic* – soils on material transported by humans *Scalpic* – soils on land surfaces with pre-existing soil truncated by humans
US soil taxonomy (Soil Survey Staff 2014)	(none)	(none)	Recognises multiple *anthropic epipedons* = surface soil layers derived from human alteration or transportation of soil material, which exist on landforms or disturbed areas created by humans

In many cases, human modification of soils changes their properties but not to the extent that they then meet the requirements for classification as anthropogenic soils or even anthropogenic subcategories of natural soils. Such human modifications, such as additions of new material from construction debris, street dusts, or other waste materials, can significantly change the chemical, physical, and biological properties of the soil environment (Jim 1998; Lehmann and Stahr 2007; Pouyat et al. 2007; Taylor et al. 2010; Wei and Yang 2010; Rate 2018).

2.2 Soil-Related Changes in Urban Geomorphology

Numerous changes are made to landforms in urban areas as cities evolve; many of these geomorphological changes create a more convenient environment for urban infrastructure such as buildings, roads, and below-ground pipe/cable networks. In addition, 'new' land suitable for urban use may be generated by reclamation of inland water bodies or, especially, on coasts.

2.2.1 Modification of Surface Hydrology

From a hydrological perspective, the changes in geomorphology due to urbanisation have been well-documented. Ehrenfeld (2000) reviews the changes in hydrology and wetland geomorphology caused by direct modification such as infilling or drainage but also from other changes to the urban hydrological environment including the following: covering of land surfaces with impermeable layers, stream modification, and flow regulation. Clearly the infilling of wetland basins (see the example in Fig. 2.4) represents a geomorphological change (in land elevation and slope modification), and the wetlands themselves change in form due to processes like increased erosion (Ehrenfeld 2000). Similarly, Paul and Meyer (2001) review the changes to urban stream hydrology and geomorphology; in urban environments, stream channels may be filled in or converted to surface or below-ground artificial drains (see the example in Fig. 2.5). This removal of natural drainage channels, together with large proportions of impervious land surfaces, has profound effects on urban hydrology and geomorphology (we will address some of the hydrological issues in Chap. 5). The effect of geomorphological changes on urban soils is less well documented. Soils developed on landforms created by human activity are included within various soil classification schemes (see Box 2.1 above), but there do not seem to be any systematic studies of how soil properties are affected, despite the known coupling of soil properties with hydrology (e.g. Schaetzl and Anderson 2005). The relationship between geomorphology and hydrology is not one-sided. As described above, changes in geomorphology strongly affect hydrology, and the converse is also true: hydrological changes affect geomorphology (i.e. landforms and soils), leading to complex feedbacks.

Fig. 2.4 Infill of lakes and wetlands in the central metropolitan area of Perth, Western Australia (−31.951 S, 115.86 E), shown by an overlay of a map from 1838 (State Library of Western Australia) on an aerial photograph from 2016 (Mapbox 2019)

Drainage of land, and groundwater extraction, in urban areas is also known to have caused land subsidence (Brown and Nicholls 2015), due to reduction of the effective stress of groundwater pressure (Galloway and Burbey 2011). Groundwater extraction also changes soil chemical properties, because it changes water-filled pores into air-filled pores. For example, Salmon et al. (2014) describe formation of acid sulphate soils when groundwater levels decreased by ca. 3 m between 2000 and 2010, resulting in entry of atmospheric oxygen into subsoils and consequent oxidation of sulphide minerals. For an understanding of soils in urban environments, then, a knowledge of landform changes that affect hydrology is essential.

2.2.2 Coastal Land Reclamation

Changes in geomorphology caused by human activity also include new landforms created by coastal reclamation; not all urban landform changes involve surface sealing and altered drainage. In coastal reclamation, land is 'reclaimed' from coastal fringing (e.g. saltmarsh), tidal, and even permanently submerged coastal water environments. There are two main types of coastal reclamation: (1) excluding marine and tidal water from salt-affected coastal land by constructing physical barriers such as dykes and installing artificial drainage (Li et al. 2014) and (2) creating new land by filling in submerged marine environments with imported soil- or sediment-like material (El Banna and Frihy 2009; Semmens et al. 2011).

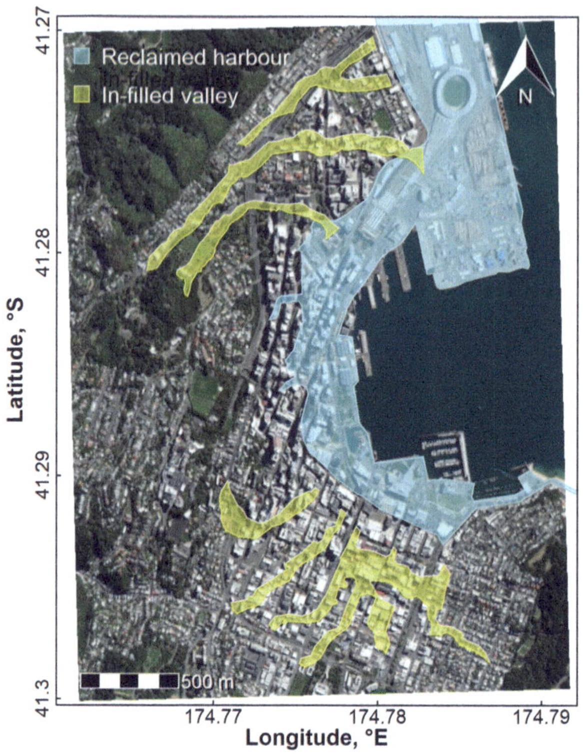

Fig. 2.5 Aerial photograph map of Wellington, New Zealand (−41.285 S, 174.775 E), showing areas of reclaimed land in former stream valleys (yellow) and harbour margins (pale blue). Base map from Mapbox (2019); overlays based on a map in Semmens et al. (2011)

From a soil science perspective, the first type of reclamation is most relevant, since the resulting soils allowed multiple land uses, including agriculture or horticulture. There is a long history of such reclaimed-land soils, which are often called 'polders', from the reclamation of low-lying coastal land in Northeastern Europe and South-East Asia. Some partially urbanised examples include Noordoostpolder 52.72 N, 5.76 E in the Netherlands and the large Bắc Hưng Hải polder near Hanoi in Vietnam 21.01 N, 105.90 E. (Coordinates are provided so the locations can be viewed in Google Earth or similar software or web mapping service; e.g. in Google Earth, remove the non-numeric information and search for '52.72, 5.76'). Pons and van der Molen (1973) investigated the properties of soils developed within 1000-year-old polders in the Netherlands. These centuries-old Netherlands polder soils have undergone a process of physical, chemical, and biological changes culminating in formation of distinct soil types at the higher order of soil classification schemes (USA), depending on the composition of the original sediment (parent material). In many cases, even several centuries of pedogenesis resulted in changes mainly in the surface soils, with deeper subsoils remaining similar to the original submerged sediments.

In many cases, construction of new land by filling in (reclamation type 2; see Fig. 2.5) has been performed because there was insufficient space, especially on flat land, for building development, so much of the new land area is ultimately occupied by infrastructure leaving minimal actual soil exposed at the surface (Jim 1998). There would appear to have been very few studies of soils developed from materials used to fill in marine or freshwater submerged environments, despite this being a common practice worldwide over time frames which are long enough for some pedogenic alteration to occur (Brown 1970; Bowler et al. 1995; Semmens et al. 2011). Some examples of infill of submerged harbour or estuarine environments are in Perth, Scotland (56.395 N, −3.43 W); Wellington, New Zealand (−41.27 S, 174.785 E, Fig. 2.5); and Rio de Janeiro, Brazil (−22.895 S, −43.19 W).

2.2.3 Other Built-up Landforms

These include the 'positive landforms' of Brown (1970) and various 'constructional anthropogenic landforms' listed in the USA soil taxonomy (Soil Survey Staff 2014), ranging in scale from middens (a few metres) to artificial islands (up to kilometres). Some examples of these are discussed separately below.

Archaeological mounds. Large mounds of soil-forming material constructed by human activity (intentional or unintentional) are known from early periods in the history of humanity. The city of Adria in Northern Italy is the site of a large mound, up to 5 m high and 20 ha in areal extent, caused by cumulative additions of urban material over an approximately 700-year period (Corrò and Mozzi 2017, 45.052 N, 12.057 E). Numerous other examples of anthropogenic mounds exist in ancient urban or pre-urban areas worldwide, for example, in Amazonia (Roosevelt 2013) and the Middle East (Faust and Katz 2015). Not only do these mounds represent new parent materials for soil formation, but they also have the potential to modify local hydrology.

Waste stockpiles and landfills. Human activities, especially those in urban environments, produce large quantities of waste material. Despite efforts to reduce the size of waste streams, considerable quantities of waste from cities require disposal (Grimm et al. 2008) and are disposed of into landfills or (temporarily) in mounded landforms, commonly on land reserved for these purposes (Cherubini et al. 2009). Soils forming on these landforms are the urbic, garbic, or spolic Technosols of the World Reference Base classification (IUSS Working Group WRB 2014). Figure 2.6 shows an example of a mounded landfill in an urban industrial zone which has been landscaped into public open space. Occasionally, stockpiled material is soil placed in temporary mounds during urban development, which is intended to be replaced or removed at project completion.

Land-disposed dredge spoils. Material removed by dredging from submerged freshwater or marine sediments, or *dredge spoil*, has historically been disposed of onto land (Almeida et al. 2001). Land disposal still occasionally occurs, despite a large body of evidence showing that the spoils commonly contain sulphide minerals

Fig. 2.6 A landfill mound in the Homebush Bay area (western Sydney, Australia −33.8454 S, 151.0559 E). (Image date April 2016; used with permission from funambulator (2016))

such as pyrite which oxidise under non-submerged conditions to form acid sulphate soils (Morse 1994; Borma et al. 2003; Clark and McConchie 2004). Land disposal of dredge spoils can form large elevated landforms; examples include South Yunderup, Western Australia (−32.59 S, 115.782 E; see Fig. 2.7). Soils developed on dredge spoil are classified as spolic Technosols in the WRB (IUSS Working Group WRB 2014) and, specifically, as dredgic Anthroposols in the Australian Soil Classification (Isbell and National Committee on Soil and Terrain 2016).

2.2.4 *Landforms Modified by Removal of Material*

The construction of urban infrastructure commonly requires levelling of land on various scales, from creating a flat base for house foundations to much larger-scale modifications such as road, railway, and canal cuttings (Fig. 2.8) – or even complete removal of hills (Brown 1970) (Fig. 2.9). In addition, the 'negative landforms' identified by Brown (1970) include excavations made for other purposes, such as quarries (Fig. 2.8) or stormwater compensation basins (Appleyard 1993). Numerous other excavated landforms are listed in the USA soil taxonomy (Soil Survey Staff

Fig. 2.7 South Yunderup, Western Australia (–32.59 S, 115.782 E), showing a large area of dredge spoil in the foreground (with diagonal fill pattern) which has acidified due to the oxidation of pyrite and other sulphide minerals contained in the original estuarine sediment. (Photograph by Chris Yanicki (2017), used with permission)

Fig. 2.8 (**a**) A large road cutting in Tongwynlais in South Wales (51.533 N, –3.253 W) in 1971 (public domain image from Gillham 2017); (**b**) Pak Shing Kok quarry landform in urban Hong Kong (22.308 N, 114.271 E). (Public domain image from 2008 by Martin Ng)

2014), in which they are called 'destructional anthropogenic landforms'. Like the positive anthropogenic landforms described in Sect. 2.2.3, excavations will affect local hydrology, and there is the potential for new soils to form on excavated surfaces. Such soils are not considered explicitly in the World Reference Base classification but would be considered scalpic Anthroposols in the Australian system (Isbell and National Committee on Soil and Terrain 2016) (Box 2.1).

Fig. 2.9 Landscape truncation by complete removal of 'Denny Hill' in Seattle, USA (47.61 N, 122.33 W), in approximately 1910; public domain image by Shakespeare at https://commons.wikimedia.org/w/index.php?curid=16477653

2.3 Characteristics of Urban and Anthropogenic Soils

It would be practically impossible to describe all the possible variations of urban soils. This section will discuss common features of soils in urban environments, with specific examples of soils in some important urban contexts.

2.3.1 Urban Soils with Minimal Modification

It is difficult, if not impossible, to identify soils in urbanised areas which have no changes resulting from human activity. Recognition of minimally modified soils is, however, important in the context of understanding the various impacts on urban soils. For example, in some regulatory frameworks (e.g. National Environment Protection Council 2013), *background concentrations* of potential contaminants are required for a full assessment of their potential environmental impacts. In many urban areas, soils having negligible human modification are rare but may exist in nature reserves or on undeveloped peri-urban land. These soils would not be defined as anthropogenic soils and may still contain traces of anthropogenic materials if deposition of airborne dusts or aerosols is locally significant.

2.3.2 Distinctive Properties of Soils in Urban Environments

We will deal with many of the soil properties encountered in urban environments in later chapters and in the sections below, but we should be aware of them at this stage as well.

Some of the diagnostic properties of Anthrosols and Technosols, which include many urban soils, are presented in Box 2.2: relatively high artefact content ('artefacts' can be imported soil-like material), *geomembranes* or industrial hard layers, rubble/refuse, industrial waste, organic waste, deep enrichment with organic matter and/or nutrients, and so on. These 'classification-based' properties encompass properties such as impermeable surface cover and many of the consequences of modifying urban geomorphology.

As mentioned, however, many of the characteristics of urban soils relate to addition of material below the thresholds required for soil classifications – for example, the threshold for artefacts in a Technosol is quite high, at 20% by volume. Of the numerous human additions, the most troubling are of contaminants, the very wide range of substances that, directly or indirectly, can have adverse effects on organisms including humans. These substances include nutrients (N, P, etc.), trace elements, asbestos, radionuclides, (micro)plastics, manufactured nanoparticles, and other inorganic contaminants like cyanide (see Chap. 6). Organic contaminants are also of great concern, including hydrocarbons, chlorinated organic compounds, pesticides, endocrine disrupting chemicals, various pharmaceuticals, and many more (see Chap. 7).

Biological contamination is also possible; urban activities may introduce pathogens into soils, and it is in the biology of soils that concerning declines may be observed rather than additions, in the form of individual species decline or loss of biodiversity. Chapter 8 discusses the biological properties which are relevant in urban soil environments.

2.3.3 Coastal Reclaimed Soils

The formation and properties of polder soils have been reviewed by Li et al. (2014), who use the term 'coastal reclaimed soils'. Some consistent properties are observed for coastal reclaimed soils; they tend to be wet soils, with finer texture and better structure than the original sediments. Chemical soil fertility generally improves with time since reclamation of coastal soils, as soil organic matter accumulates; a major constraint to their use in plant production is the residual salinity from their tidal or marine origins, with formation of acid sulphate soils occurring if the parent sediments contained pyrite or other sulphide minerals. In general, more favourable soil properties were established in coastal reclaimed soils in South-East Asia than in North America or Europe (Li et al. 2014).

2.3.4 Soils on Landfills

Soils developed on landfill materials would fall within the garbic or spolic Technosol classifications of the WRB (IUSS Working Group WRB 2014). Landfills are commonly constructed with an overlying clean soil material, however; depending on the depth of surface fill, non-anthropogenic soil groups or orders may be more relevant. The type of waste material (e.g. construction waste, organic wastes, or mixtures of different waste types) disposed of in the landfill structure affects the subsequent soil-forming processes.

Early soil development on landfill materials has been shown to be associated with compaction, as the newly deposited waste and cover materials settle, increasing the density and reducing porosity and maximum water storage (Tifafi et al. 2017). Settling on landfills containing organic waste may also reflect decreases in volume due to decomposition of putrescible organic waste material (Oakley and Jimenez 2012). The land elevation may also decrease due to the mass of overlying landfill compacting the underlying soil or sediment (El-Fadel and Khoury 2000).

One of the almost universal properties of landfills is that they contain contaminants; the actual contaminants present depend on the types of waste that have been disposed of. Soils that develop on landfills may also therefore be contaminated, depending on the properties and thickness of the clean cover material. The types of contaminants present in landfills are extremely diverse and include metals, excess nutrients, hydrocarbons, volatile organic compounds, pesticides, pathogens and other microorganisms, microplastics, and asbestos (Department of Environment and Conservation 2009; Plant et al. 2014). The details of contaminant behaviour in urban soils will be discussed from a chemical perspective in Chaps. 6 and 7 and in the context of soil biology in Chap. 8.

Landfills used to dispose of organic wastes commonly generate methane, as a product of *anoxic* decomposition of organic matter. The methane is emitted from the surface soil layers, regardless of whether clean soil overlies waste material (Blume 1989). Methane is a greenhouse gas (Bellucci et al. 2012) and also represents a safety or health hazard due to its flammability or if landfill gas enters closed buildings with poor air exchange (US EPA 2017).

2.3.5 Soils on Dredge Spoils and Coastal Acid Sulphate Soils

Soils developed on dredged materials are included in the categories of spolic Technosols in the World Reference Base (IUSS Working Group WRB 2014) or dredgic Anthroposols in the Australian Soil Classification (Isbell and National Committee on Soil and Terrain 2016). Coastal acid sulphate soils (CASS) developed from land drainage are not necessarily categorised as anthropogenic soils, being classified instead in other soil groups such as gleysols in the WRB or hydrosols in Australia.

In urban and other environments, drainage of coastal soils is common (Brady 1974). The formation of acid sulphate soils from drainage of coastal soils or land disposal of dredge spoils is a well-known phenomenon (Morse 1994; Dent and Pons 1995). The acidification process is commonly associated with increased mobility and potential bioavailability of iron, aluminium, sulphur, and trace elements, including potentially toxic metals and metalloids such as As, Cd, Co, Cr, Cu, Mn, Mo, Ni, Pb, and Zn (e.g. Sohlenius and Öborn 2004). The metal(loid) contaminants released by acid sulphate oxidation may then be lost from the soil system and enter other environments such as water and aquatic sediments (Huerta-Diaz et al. 1993; Gröger et al. 2011). Chapter 6 will address the geochemical processes and properties in acid sulphate soils in more detail.

Acidification of estuarine soils to form acid sulphate soils creates a soil environment that is not conducive to plant growth, due to toxicity of aluminium and other ions released under acidic conditions. The lack of plant cover on the soil surface commonly results in the generation of dust end export of potentially contaminated soil material into the atmosphere and other environments (Ljung et al. 2010). In addition, the salinity of acid sulphate environments, related to the estuarine or marine origin of the original sediments and to the production of soluble salts during acid sulphate oxidation, also suppresses plant growth (Fanning 1990).

2.3.6 Soil-Like Materials

Natural and constructed wetlands. Water bodies which are relatively shallow overlie sediments or *wetland soils*, which have many of the characteristics of *upland soils*. Wetlands of various types, including shallow and/or seasonal water bodies, are present in many urban environments worldwide. A key feature of wetland soils is their different oxidation-reduction chemistry, driven by the restriction of oxygen supply in water-filled pores and the consumption of oxygen and other electron acceptors by microorganisms (Gambrell 1994). The wet conditions promote storage rather than decomposition of organic matter, so that wetlands may be important for carbon storage in urban environments (Pouyat et al. 2006; Vepraskas and Vaughan 2016). The other main differences in wetland soils are related to changes in the form of iron, such that the reduced form (Fe^{2+}) predominates (giving paler 'gleyed' colours), and, if the supply of sulphur (e.g. as atmospheric sulphate) is great enough, accumulation of sulphides (mineral phases containing S^{2-} or S_2^{2-}) (Vepraskas et al. 2016). The sulphide phases formed under anoxic conditions incorporate trace elements along with iron, and so wetland soils can accumulate trace element contaminants in immobile forms. In constructed wetlands and infiltration basins, the soils most likely develop into subaquatic Technosols, according to the IUSS World Reference Base classification (IUSS Working Group WRB 2014).

Many urban wetlands have been drained and/or infilled during development. The wetland sediments may persist in drained conditions or beneath the imported fill materials, with associated risks of acid sulphate soil development, especially if the local hydrology changes towards drier conditions.

Infiltration basin sediments. Stormwater drainage networks are important infrastructure in urban environments and are very necessary due to the generally high proportion of impermeable surface cover which increases run-off. The constrained channels of stormwater drains (particularly open drains), however, have the potential to increase the risk of flooding. A flood control measure that can be applied is the inclusion of *infiltration basins* – high-volume sections of drains which are typically much deeper and wider than the drain itself – along the lines of stormwater drainage. Also called compensating basins or detention basins, these are designed to fill with storm water during flood events and thereby reduce the risk of flooding; an additional benefit is groundwater recharge while water exists above the base level of the infiltration basin.

The sediments in infiltration basins may be actual soils (e.g. public open space in natural landscape depressions), deliberately excavated basins, or natural lakes/ponds included in the stormwater drainage network. If permanently or seasonally submerged, they bear many of the properties of soils in natural and constructed wetlands. In some cases where the infiltration basin is dry for most of the year (e.g. empty basins in sandy soils with buried pipework to carry base flow), they may most resemble upland soils.

Green roofs. The use of roof spaces on urban buildings to create 'green roofs' (Fig. 2.10) is a practice which is increasing in frequency, since it offers benefits to the urban environment such as cooling through shading and evapotranspiration, or acting as storm water buffers. A green roof typically has an imported, constructed soil substrate composed of the following (from the surface to deeper layers): optional mulch, soil-like growing medium, filter membrane, drainage layer, waterproof/root-excluding membrane, thermal insulation, vapour control, and structural roof support. In the world reference base classification, they are included in isolatic Technosols (IUSS Working Group WRB 2014). Few studies have explicitly considered ongoing formation processes in soil materials on green roofs, probably as their installation in contemporary cities is a recent phenomenon, so there is not yet a consensus on soil-forming pathways. Bouzouidja et al. (2018) show that the properties of soils on green roofs evolve relatively rapidly with time. Over a 4-year period, concurrent with development of the vegetation, eluviation of fine particles, increases

Fig. 2.10 (**a**) Layers in a typical manufactured green roof construction; (**b**) aerial view of green roofs in Singapore. (Public domain image by chuttersnap on unsplash.com/photos/IfmqOuOkaOA)

Fig. 2.11 Accumulation of road dust and other debris in an urban environment in Perth, Western Australia. If left undisturbed, such deposits provide incipient parent materials for soil and associated organisms. (Photograph by Andrew W. Rate; entire ruler for scale is 15.8 cm long)

in macro- and microporosity (at the expense of mesopores), decrease in organic carbon content, and increase in nitrogen content, all occurred. In contrast, Schrader and Böning (2006) measured greater organic carbon and N contents, but lower pH, in the soil of older (compared with younger) green roofs. Regardless of constructed or ongoing soil formation, however, a green roof environment would need continuing intensive management to preserve ecosystem functions.

Street dusts. Dust accumulating on impervious urban surfaces (Fig. 2.11) is normally transient but may persist for long enough in some microenvironments that plants can establish on dust accumulations. The source of the dust may in fact be soil particles which become resuspended in air by wind (De Miguel et al. 1997). Dusts from within urban environments or from remote sources are also known to represent pedogenetic additions to urban soils. To our knowledge, researchers have not yet investigated pedogenesis where the main parent material is any type of urban dust.

2.4 Archaeological Features of Urban Soils

The soils and landforms of cities contain clues to their history; cities remain in one place for a long time, and humans leave behind many traces of their habitation which accumulate over time. These clues may take the form of evidence of past landscape modification by humans, physical soil components such as anthropogenic artefacts, chemical signatures such as accumulation of nutrients or contaminants, or particular microfossils related to human modification of ecosystems. In some case the archaeological heritage in cities is of great cultural and historical value and takes precedence over urban development. For example, 'rescue archaeology', the retrieval of artefacts accidentally excavated during development construction, was common in many municipalities until urban planning procedures began considering archaeological issues explicitly.

2.4.1 Archaeological Anthropogenic Landforms

Cities evolve most obviously by changes in lateral extent, usually to cover more land area (see Chap. 1), but urban development may also involve creation or destruction of landforms (see Sect. 2.2 above), which cause vertical changes in urban landscapes. For example, Faust and Katz (2015) studied the Bronze to early Iron Age urbanisation of Tel 'Eton in contemporary Israel, showing from archaeological strata on a large mound landform that multiple phases of urbanisation had occurred from ca. 2000 to 300 BC. Corrò and Mozzi (2017) analysed buried urban strata and showed a history of elevation change in Adria, Italy, dating back to the sixth century BC. In contrast, the presence of archaeological artefacts can help to constrain a time frame for natural soil- or landscape-forming events (see Völkel et al. 2012).

As discussed in Chap. 1, the location of ancient settlements is also related to geomorphology and the quality of soils. The location of the ancient Tel 'Eton city is thought to reflect both its proximity to fertile alluvial soils and geomorphologically controlled transport and trade routes (Faust and Katz 2015).

2.4.2 The Soil 'Cultural Layer'

The concept of a *cultural layer* in soils, an anthropogenic soil horizon which contains artefacts derived from human occupation and disposal of materials, has been used in the context of stratigraphic excavation in archaeology since the early twentieth century (Browman and Givens 1996). Cultural layers are commonly found in urban soils, commonly as anthropogenic horizon(s) superimposed above natural soils. The underlying natural soils are sometimes truncated (their upper layers removed) by excavation or erosion. Alexandrovskaya and Alexandrovskiy (2000) describe cultural layers from the fifteenth century and younger in the city of Moscow, Russia, which are typically 2–5 m and can be up to 20 m deep. Naturally, such large volumes of soil material derived from human activities contain many artefacts such as construction and food wastes, metal and ceramic objects, remains of cooking fires, and so on. Organic materials can be preserved in the anoxic conditions created by saturation of soil. Both Alexandrovskaya and Alexandrovskiy (2000) and Zhang et al. (2005) in urban soils of Nanjing, China, also found urban soil cultural layers to be enriched in organic matter, nutrients, and metals. The soil cultural layers in Nanjing spanned five Chinese dynasties across 2–6 m of anthropogenic horizons

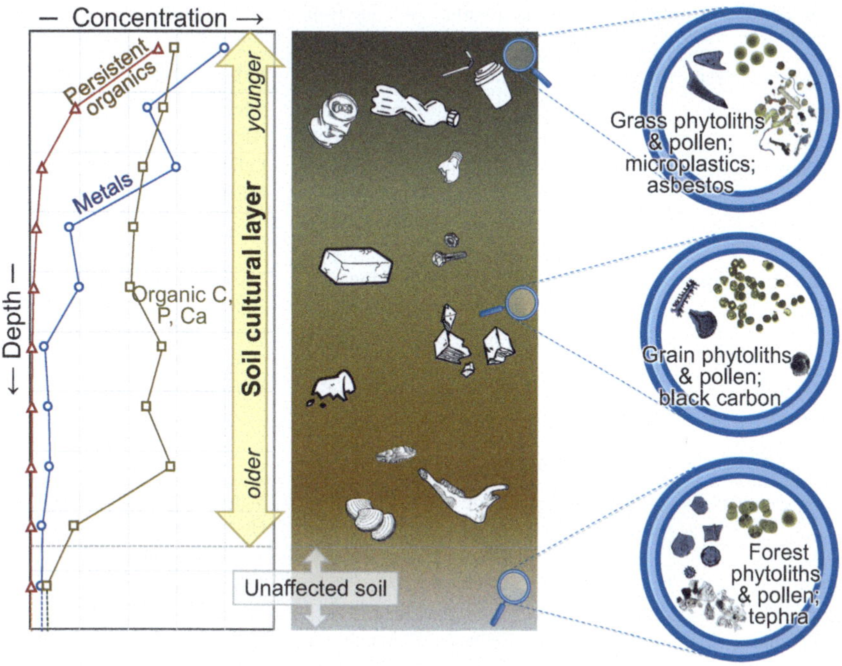

Fig. 2.12 Schematic of an idealised series of superimposed soil cultural layers. Different artefacts (anthropogenic and biological) are found in each layer depending on its age. (Graphic by Andrew W. Rate)

and had greater concentrations of Cu, Pb, and Zn than the underlying 'background' soils, similar to the Moscow soils which had also become enriched in arsenic since the seventeenth century. In general, we would expect to find different artefacts, microfossils, and chemical composition in soil cultural layers of different ages, and this is shown in idealised form in Fig. 2.12.

Evidence of patterns of human habitation and activities has also been deduced from much thinner soil layers. For example, information on human activities in ancient urban structures in both Songo Mnara in Tanzania (fourteenth to sixteenth centuries, Sulas and Madella 2012) and Brussels, Belgium (twelfth to eighteenth centuries, Devos et al. 2013), was obtained using soil micromorphology, with both chemical analyses and identification of *phytoliths*. In both of these studies, soil layers or features on the millimetre scale or smaller were identified, such as the coatings around larger soil grains, and the artefacts were the phytoliths of plant species cultivated by humans. Similar information can be generated by analysis of other microfossils in soils, such as pollen grains, or diatoms which can indicate the use of irrigation (Sánchez-Pérez et al. 2013) (Table 2.1).

Table 2.1 Types of archaeological information available from analysis of urban soils

Target of analysis	Soil material analysed	Information obtained	Reference(s)
Artefact content	Cultural layers	Age of urban habitation; types of human activities	Alexandrovskaya and Alexandrovskiy (2000)
Organic carbon content	Soil profile	Location of cultural layer; land use; activities in and around buildings, etc.	Lehmann et al. (2003) and Mazurek et al. (2016)
Nutrient (esp. P) content	Soil profile, surface soil	Type and location of human activities (e.g. cooking)	Alexandrovskaya and Alexandrovskiy (2000), Wells et al. (2000), and Mazurek et al. (2016)
Other major elements	Soil profile, surface soil	Type and location of human activities (e.g. fireplaces, buildings, roads)	Wilson et al. (2005)
Trace elements	Soil profile, surface soil	Type and location of human activities (e.g. smelting); provenance of artefacts	Wilson et al. (2005, 2007), Hellemans et al. (2014), and Sylvester et al. (2017)
Organic compounds	Cultural layers, soil profile, grave sites	Individual human behaviour and activities (e.g. diet, health); soil redox conditions	Pickering et al. (2018)
Phytoliths	Cultural layers, soil profile	Diet	Vuorela et al. (1996)
Other microfossils: pollen, diatoms, ostracods	Cultural layers, soil profile, lake sediments	Sedimentary history, irrigation, erosion, diet	Vuorela and Hiekkanen (1991), Shen et al. (2006), and Fleury et al. (2014)

2.4.3 Archaeological Information from Major Elements

Enrichment of urban soils with major elements (i.e. the more common chemical elements in the Earth's crust or in biological systems, such as C, N, P, K, S, Ca, or Fe) is a common phenomenon, since there are many human activities which can lead to increases in concentrations. Waste disposal sites such as food waste middens result in enrichment of soil with carbon, phosphorus, and calcium (from artefacts such as shells and bones; see Fig. 2.13); similar enrichments have been attributed to the use of manures as fertilisers (Entwistle et al. 1998; Davidson et al. 2006; Sánchez-Pérez et al. 2013). Greater concentrations of calcium and associated elements may also be associated with the previous locations of hearths or cooking fires (Wilson et al. 2005). Human burial sites may also become enriched in major elements such as phosphorus (Pickering et al. 2018) or calcium (Ottaway and Matthews 1988).

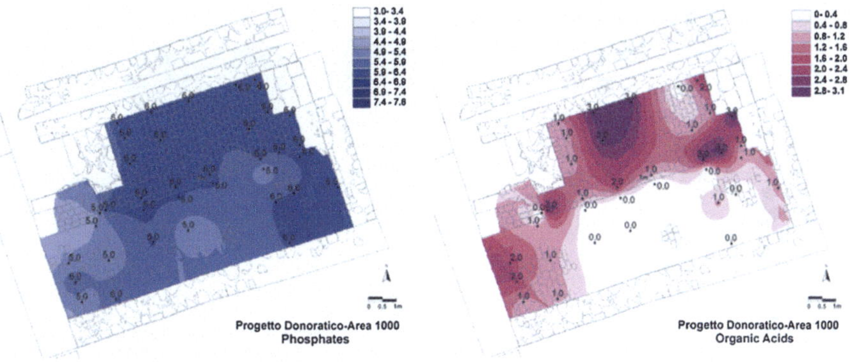

Fig. 2.13 Distributions of phosphate (left) and organic fatty acids (right) in soil materials on the former floor of Donoratico medieval castle in Central Tuscany, Italy. (From Middleton et al. 2010 and used with permission from Springer)

2.4.4 Archaeological Information from Trace Elements

The concentrations, and depth-wise and spatial distributions, of trace elements in urban soils can provide substantial information on (pre)historical human behaviour and activities. Ottaway and Matthews (1988) found that different trace elements were enriched (relative to the underlying unaffected soil) in different patterns in a soil profile, depending on the age of the anthropogenic stratum sampled (see also Fig. 2.14). They were able to relate enrichment of trace elements to the period of occupation: for example, early Neolithic samples (ca. 7000 years old) showed minimal enrichment, whereas late Neolithic and Eneolithic samples (5000–5500 years old) showed enrichment of Cu and Zn. Enriched strontium (Sr) was related to the period of human occupation in general, with greater Sr concentrations in more recent medieval samples (Ottaway and Matthews 1988).

Trace element concentrations, especially of multiple elements, can also yield information about the function of different areas at archaeological sites. Wilson et al. (2005) showed that the concentrations of Ba, Cu, Sr, and Zn in surface soils could be used to discriminate areas such as fields, gardens, middens, byres, houses, and hearths. The Roman urbanisation of Calleva Atrebatum in Hampshire, UK (first century BC to fifth century AD), was studied by Sylvester et al. (2018), who showed that the entire historical city area was enriched in multiple elements, particularly gold and silver, relative to background soils. Localised high values of individual element concentrations or multi-element indices at Calleva Atrebatum were attributed to metal extraction activities such as smelting and cupellation. In a more recent context, Rate (2018) used multi-element signatures to delineate zones on an urban site which related spatially and logically to nineteenth- and twentieth-century land uses or contamination sources such as market gardening, dumping of glass waste, road traffic, and stormwater drainage.

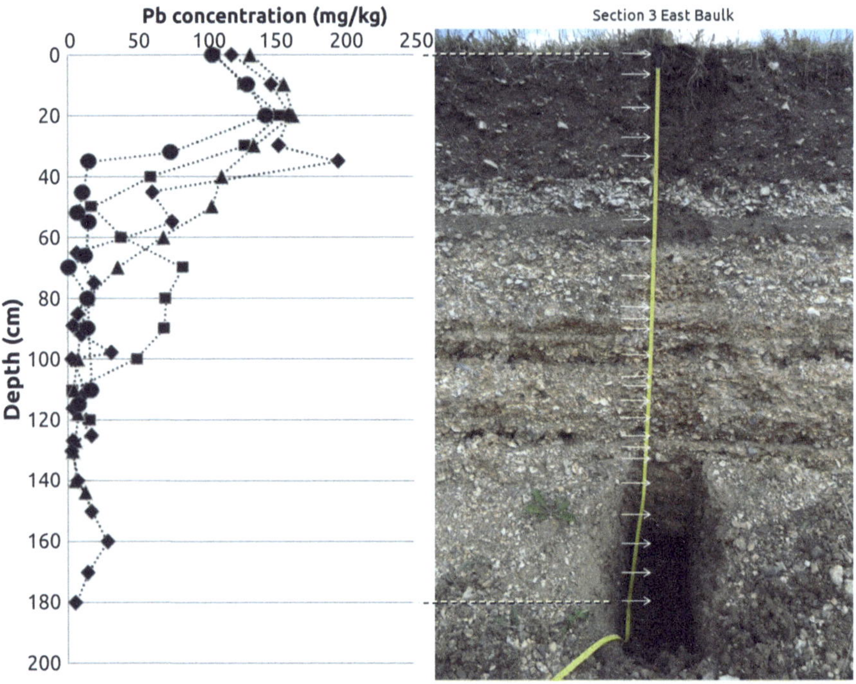

Fig. 2.14 Depth profile of soil lead (Pb) concentrations at Calleva Atrebatum in Hampshire, UK, which was occupied from the first century BC to the fifth century AD. (From Sylvester 2017, based on original work by Dr Chris Speed at the University of Reading)

The trace element signatures (comprising the concentrations of a suite of elements by which materials can be discriminated) of some anthropogenic items can yield information on the origins of these artefacts found in cultural layers of urban soils and therefore to infer patterns of commerce or migration (Hellemans et al. 2014).

2.4.5 Archaeological Information from Other Chemical Substances

More recently, organic chemical signatures have been used as archaeological tracers. For example, various organic fatty acids (see Fig. 2.13) have the potential to provide information about food residues, manures, or sacrificial rituals and distinguish location based on these activities or materials (Middleton et al. 2010). In addition, Zou et al. (2010) found that (1) polycyclic aromatic hydrocarbons (PAHs) could be used to indicate the presence and location of ancient fires; (2) PAH-like biomarkers such as the terpenoid organic compounds, cadalene and simonellite, were indicators of natural plant communities.

2.4.6 *Archaeological Information from Geophysical Techniques*

The most widely used geophysical techniques for characterising surface soils are electrical conductivity/resistivity-based techniques, measurement of variations in local magnetic field (magnetometry), and ground-penetrating radar (Herz and Garrison 1998). Such techniques, with appropriate signal processing and numerical analysis, provide potentially powerful tools for assessing subsurface soil environments in a non-invasive and non-destructive manner.

Magnetic measurements are possibly the most commonly used geophysical method in archaeology: for example, Boschi (2012) described the use of magnetic gradiometry (i.e. measurements of magnetic field gradient) to delineate various buildings in the fifth-century AD town of Classe in Northeast Italy (44.395 N, 12.219 E). Similarly, Cella and Fedi (2015 #160; see Fig. 2.15) used derivative magnetic gradiometry to obtain details of the buried ruins of buildings at the Torre Galli archaeological site in Calabria, Southern Italy (38.641 N, 15.939 E). *Magnetic susceptibility* measurements have also provided archaeological information (Fleisher and Sulas 2015).

The ability of subsurface layers to conduct (or resist) an electrical current has also been used to provide archaeological information. The simplest implementation is using an electromagnetic induction device which measures bulk soil electrical conductivity (Benech and Marmet 1999). More detailed archaeological data can be

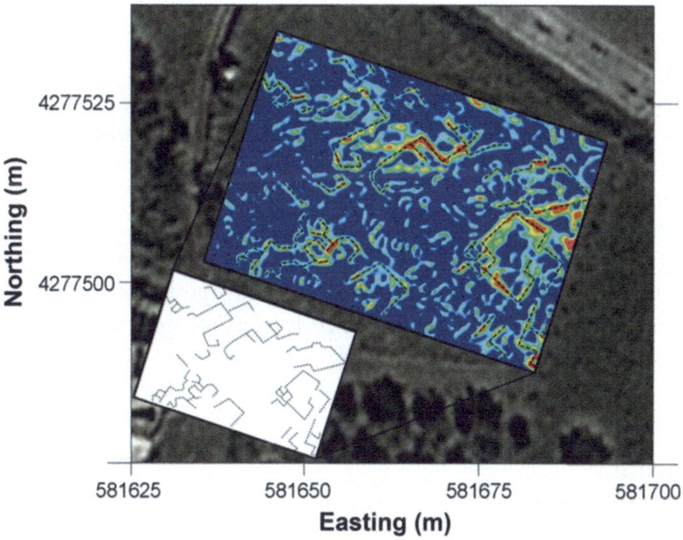

Fig. 2.15 An example of the use of soil magnetic measurements for geoarchaeology, from Cella and Fedi (2015). The large oblique rectangle on the map shows a mathematically processed magnetic field gradient of the soil, which is used to infer the historical location of walls of dwellings. (Used with permission from Springer)

acquired using electrical resistance tomography to create 2D or 3D images of the subsurface (De Giorgi and Leucci 2017).

Two- and three-dimensional images of subsurface archaeological structures can also be obtained using ground-penetrating radar (GPR). For example, Millaire and Eastaugh (2014) used GPR successfully to delineate walls and other structures in three dimensions (to a depth of 40 cm) in the pre-Hispanic city of Gallinazo, Peru (100 BC–AD 700).

2.4.7 *Archaeological Information from Soil Microbial Properties*

The microbial properties of soil such as microbial biomass, fungal biomass, and respiration rate have been found to vary between locations having historical human modification (Bronze Age; sixteenth to tenth century BC) and reference sites and within ancient anthropogenic soils themselves (Peters et al. 2014). Using more advanced DNA-based, phospholipid fatty acid profiling and substrate-based diversity techniques, Margesin et al. (2017) studied soils at Monte Iato in Western Sicily (occupied during the eighth to sixth centuries BC). The microbial analyses of the Monte Iato soils showed that the soil microbial community varied between anthropogenically modified soils in terms of both functional, physiological, and genetic diversity.

2.5 Additional Reading

Amundson R, Jenny H (1991) The place of humans in the state factor theory of ecosystems and their soils. Soil Sci. , 151:99–109

Kaye JP, Groffman PM, Grimm NB, Baker LA, Pouyat RV (2006) A distinct urban biogeochemistry? Trends Ecol Evol 21:192–199

Morel JL, Heinrich AB, (2008) SUITMA-soils in urban, industrial, traffic, mining and military areas: an interdisciplinary working group of the 'International Union of Soil Science' (IUSS) dedicated to soils strongly modified by human activities. J Soils Sediments 8:206–207

2.6 Summary

- The processes of soil formation are similar for urban soils and non-urban soils. Both can be described by a combination of a state factor model and a soil fluxes approach. In urban soil environments, however, we can define soil *anthropose-*

quences (or urbanisation gradients), where the state factor that changes is predominantly the human factor.

- Soil classification provides a structured conceptual framework for describing and understanding soil properties and soil formation. Soil classification schemes such as the World Reference Base include specific categories for soils modified by humans, which include urban soils. The most relevant broad category is that of the Technosols, soils which contain materials which have been manufactured or relocated by humans, such as impermeable surfaces, various waste materials, or geomembranes.
- The landforms in urban environments have been modified profoundly by human activity since ancient times. Wetlands, valleys, and even near-shore marine environments have been filled in to flatten or extend usable land. Streams and rivers have been straightened or forced into artificial channels, even underground, and land may subside due to groundwater extraction. Mounds have been created to store wastes, move solid-earth material to different locations, or have grown as successive layers of urbanisation are built superimposed on one another. Conversely, elevated landforms such as hills have been cut through for transportation routes or even removed completely.
- Urban soils differ from non-urban soils when there have been substantial landform or land cover changes or when significant amounts of various materials and substances have been added. Some specific examples of distinctive soils which have been modified by humans and which occur in urban environments include reclaimed coastal soils, soils on landfills, and acid sulphate soils. Urban environments may require a widening of our usual concepts of soils to include wetland and drainage basin soils, green roofs, and soils developed from anthropogenic dusts.
- In an archaeological context, we can identify a *cultural layer* in many urban soils, an anthropogenic soil horizon which contains artefacts derived from human occupation and disposal of materials. The superposition of multiple cultural layers can be derived from separate phases of historical urban development. Soil cultural layers are also commonly enriched in carbon, nutrient elements such as phosphorus, and various organic marker compounds, from use and disposal of organic materials such as foodstuffs and manures. Urban soils may also become enriched in trace elements from early times, reflecting extraction and use of metals and associated elements. The physical presence of artefacts and building ruins in soils can be ascertained using a range of geophysical techniques. Finally, the chemical and physical changes due to human activity can be reflected in changes in soil microbial abundance and diversity.

2.7 Questions

2.7.1 Checking Your Understanding

1. What are the parameters in the state factor 'c-l-o-r-p-t' equation, and how does each one affect the material fluxes in, and therefore the properties of, soils? Which parameters are affected by urbanisation?
2. What would be the requirements for a landscape transect over which to measure soil properties on an anthroposequence?
3. How would we determine if a particular urban soil was a Technosol? What are the various subcategories of Technosols?
4. Choose an urban geomorphological change (e.g. valley filling, redirection of drainage networks, or large-scale excavation) and summarise the effects you would expect the changed landforms to have on the resulting soils.
5. Make a table which lists the type of changes we see in urban environments in separate rows in the first column, with biological, chemical, and physical effects on soils identified in the next three columns for each type of change.
6. How could we use archaeological soil information to determine layers of different ages (e.g. differentiating pre-industrial from post-industrial and distinguishing pre-urban from urbanised)?

2.7.2 Thinking About the Issues

7. Is an anthroposequence a useful way of thinking about urban soils, in the same way as a more traditional toposequence or chronosequence? If so, why or, if not, why not?
8. As well as acid sulphate soils forming in excavated dredge spoils and on drained coastal land, what other urban practices might result in the formation of acid sulphate soils?
9. What processes or events might confuse, mask, or erase archaeological information in urban soils?

2.7.3 Contemplating Urban Soils Creatively

10. If we were able to travel forward through time for 5000 years, what might we find in our urban soils?

References

Alexandrovskaya EI, Alexandrovskiy AL (2000) History of the cultural layer in Moscow and accumulation of anthropogenic substances in it. Catena 41:249–259. https://doi.org/10.1016/S0341-8162(00)00107-7

Almeida MSS, Borma LS, Barbosa MC (2001) Land disposal of river and lagoon dredged sediments. Eng Geol 60:21–30. https://doi.org/10.1016/S0013-7952(00)00085-5

Amundson R, Jenny H (1991) The place of humans in the state factor theory of ecosystems and their soils. Soil Sci 151:99–109. https://doi.org/10.1097/00010694-199101000-00012

Appleyard SJ (1993) Impact of stormwater infiltration basins on groundwater quality, Perth metropolitan region, Western Australia. Environ Geol 21:227–236

Bellucci F, Bogner JE, Sturchio NC (2012) Greenhouse gas emissions at the urban scale. Elements 8:445–450. https://doi.org/10.2113/gselements.8.6.445

Benech C, Marmet E (1999) Optimum depth of investigation and conductivity response rejection of the different electromagnetic devices measuring apparent magnetic susceptibility. Archaeol Prospect 6:31–45. https://doi.org/10.1002/(sici)1099-0763(199903)6:1<31::aid-arp112>3.0.co;2-n

Blume HP (1989) Classification of soils in urban agglomerations. Catena 16:269–275. https://doi.org/10.1016/0341-8162(89)90013-1

Borma LD, Ehrlich M, Barbosa MC (2003) Acidification and release of heavy metals in dredged sediments. Can Geotech J 40:1154–1163

Boschi F (2012) Magnetic prospecting for the archaeology of Classe (Ravenna). Archaeol Prospect 19:219–227. https://doi.org/10.1002/arp.1430

Bouzouidja R, Rousseau G, Galzin V, Claverie R, Lacroix D, Séré G (2018) Green roof ageing or Isolatic Technosol's pedogenesis? J Soils Sediments 18:418–425. https://doi.org/10.1007/s11368-016-1513-3

Bowler D, Cachart R, Cheer P, Cox A, Smith C (1995) Tay Street, Perth: the excavation of an early harbour site. Proc Soc Antiqu Scotl 124:467–489

Brady NC (1974) The nature and properties of soils. Macmillan, New York

Browman DL, Givens DR (1996) Stratigraphic excavation: the first "new archaeology". Am Anthropol 98:80–95. www.jstor.org/stable/682955

Brown EH (1970) Man shapes the earth. Geogr J 136:74–85. https://doi.org/10.2307/1795683

Brown S, Nicholls RJ (2015) Subsidence and human influences in mega deltas: the case of the Ganges–Brahmaputra–Meghna. Sci Total Environ 527-528:362–374. https://doi.org/10.1016/j.scitotenv.2015.04.124

Cella F, Fedi M (2015) High-resolution geophysical 3D imaging for archaeology by magnetic and EM data: the case of the iron age settlement of Torre Galli, Southern Italy. Surv Geophys 36:831–850. https://doi.org/10.1007/s10712-015-9341-3

Cherubini F, Bargigli S, Ulgiati S (2009) Life cycle assessment (LCA) of waste management strategies: landfilling, sorting plant and incineration. Energy 34:2116–2123. https://doi.org/10.1016/j.energy.2008.08.023

Clark MW, McConchie DM (2004) Development of acid sulfate soil in sub-aerially disposed dredge spoil at Fisherman Islands, Brisbane, Australia. Aust J Soil Res 42:553–567

Corrò E, Mozzi P (2017) Water matters. Geoarchaeology of the city of Adria and palaeohydrographic variations (Po Delta, Northern Italy). J Archaeol Sci Rep 15:482–491. https://doi.org/10.1016/j.jasrep.2016.08.001

Davidson DA, Dercon G, Stewart M, Watson F (2006) The legacy of past urban waste disposal on local soils. J Archaeol Sci 33:778–783. https://doi.org/10.1016/j.jas.2005.10.017

De Giorgi L, Leucci G (2017) The archaeological site of Sagalassos (Turkey): exploring the mysteries of the invisible layers using geophysical methods. Explor Geophys 49:751–761. https://doi.org/10.1071/EG16154

De Miguel E, Llamas JF, Chacón E, Berg T, Larssen S, Røyset O, Vadset M (1997) Origin and patterns of distribution of trace elements in street dust: unleaded petrol and urban lead. Atmos Environ 31:2733–2740
Dent DL, Pons LJ (1995) A world perspective on acid sulphate soils. Geoderma 67:263–276
Department of Environment and Conservation (2009) Landfill waste classification and waste definitions 1996 (as amended December 2009). Government of Western Australia, Perth
Devos Y, Nicosia C, Vrydaghs L, Modrie S (2013) Studying urban stratigraphy: Dark Earth and a microstratified sequence on the site of the Court of Hoogstraeten (Brussels, Belgium). Integrating archaeopedology and phytolith analysis. Quat Int 315:147–166. https://doi.org/10.1016/j.quaint.2013.07.024
Ehrenfeld JG (2000) Evaluating wetlands within an urban context. Ecol Eng 15:253–265. https://doi.org/10.1016/S0925-8574(00)00080-X
El Banna MM, Frihy OE (2009) Human-induced changes in the geomorphology of the northeastern coast of the Nile delta, Egypt. Geomorphology 107:72–78. https://doi.org/10.1016/j.geomorph.2007.06.025
El-Fadel M, Khoury R (2000) Modeling settlement in MSW landfills: a critical review. Crit Rev Environ Sci Technol 30:327–361. https://doi.org/10.1080/10643380091184200
Entwistle JA, Abrahams PW, Dodgshon RA (1998) Multi-element analysis of soils from Scottish historical sites. Interpreting land-use history through the physical and geochemical analysis of soil. Journal of Archaeological Science 25:53–68. https://doi.org/10.1006/jasc.1997.0199
Evtuhov C (2006) The roots of Dokuchaev's scientific contributions: cadastral soil mapping and agro-environmental issues. In: Warkentin BP (ed) Footprints in the soil: people and ideas in soil history. Elsevier Science and Technology, Amsterdam, pp 125–148
Fanning DS (1990) Salinity problems in acid sulfate coastal soils. In: Towards the rational use of high salinity tolerant plants. Kluwer Academic Publishers, Dordrecht, pp 491–500
Faust A, Katz H (2015) A Canaanite Town, a Judahite Center, and a Persian Period Fort: excavating over two thousand years of history at Tel 'Eton. Near East Archaeol 78:88–102. https://doi.org/10.5615/neareastarch.78.2.0088
Fleisher J, Sulas F (2015) Deciphering public spaces in urban contexts: geophysical survey, multi-element soil analysis, and artifact distributions at the 15th–16th-century AD Swahili settlement of Songo Mnara, Tanzania. J Archaeol Sci 55:55–70. https://doi.org/10.1016/j.jas.2014.12.020
Fleury S, Malaizé B, Giraudeau J, Galop D, Bout-Roumazeilles V, Martinez P et al (2014) Impacts of Mayan land use on Laguna Tuspán watershed (Petén, Guatemala) as seen through clay and ostracode analysis. J Archaeol Sci 49:372–382. https://doi.org/10.1016/j.jas.2014.05.032
funambulator (2016) Homebush Bay perambulations I: a walk through 'toxicity' … munitions dumps and toxic mounds. http://www.7dayadventurer.com/2016/04/08/homebush-bay-perambulations-i-a-walk-through-a-history-of-munitions-dumps-toxic-mounds/. Accessed 11 Aug 2019
Galloway DL, Burbey TJ (2011) Review: regional land subsidence accompanying groundwater extraction. Hydrogeol J 19:1459–1486. https://doi.org/10.1007/s10040-011-0775-5
Gambrell RP (1994) Trace and toxic metals in wetlands – a review. J Environ Qual 23:883–891. https://doi.org/10.2134/jeq1994.00472425002300050005x
Gillham M (2017) View from East end of Little Garth past crushing plant further left (photograph). https://www.flickr.com/photos/marygillhamarchiveproject/30491559003. Accessed 6 Aug 2019
Grimm NB, Faeth SH, Golubiewski NE, Redman CL, Wu J, Bai X, Briggs JM (2008) Global change and the ecology of cities. Science 319:756–760
Gröger J, Proske U, Hanebuth TJJ, Hamer K (2011) Cycling of trace metals and rare earth elements (REE) in acid sulfate soils in the Plain of Reeds, Vietnam. Chem Geol 288:162–177
Hellemans K, Vincke A, Cagno S, Herremans D, De Clercq W, Janssens K (2014) Composition and state of alteration of 18th-century glass finds found at the Cistercian nunnery of Clairefontaine, Belgium. J Archaeol Sci 47:121–133. https://doi.org/10.1016/j.jas.2014.03.039
Herz N, Garrison EG (1998) Geological methods for archaeology. Oxford University Press, Oxford, 352 pp

Huerta-Diaz MA, Carignan R, Tessier A (1993) Measurement of trace metals associated with acid volatile sulfides and pyrite in organic freshwater sediments. Environ Sci Technol 27:2367–2372. https://doi.org/10.1021/es00048a009
Isbell RF (1996) In: CSIRO (ed) The Australian soil classification, Australian soil and land survey handbook, 4. CSIRO Publishing, Melbourne, 143 pp
Isbell RF, National Committee on Soil and Terrain (2016) The Australian soil classification, Australian soil and land survey handbooks series. CSIRO Publishing, Clayton South
IUSS Working Group WRB (2014) World reference base for soil resources 2014, World soil resources reports food and agriculture. Organization of the United Nations, Rome
Jenny H (1941) Factors of soil formation: a system of quantitative Pedology, McGraw-Hill publications in the agricultural sciences. McGraw-Hill Book Co, New York, 281 pp
Jim CY (1998) Urban soil characteristics and limitations for landscape planting in Hong Kong. Landsc Urban Plan 40:235–249. https://doi.org/10.1016/S0169-2046(97)00117-5
Kottek M, Grieser J, Beck C, Rudolf B, Rubel F (2006) World map of Köppen-Geiger climate classification updated. Meteorol Z 15:259–263
Lehmann A, Stahr K (2007) Nature and significance of anthropogenic urban soils. J Soils Sediments 7:247–260. https://doi.org/10.1065/jss2007.06.235
Lehmann J, Pereira da Silva J Jr, Steiner C, Nehls T, Zech W, Glaser B (2003) Nutrient availability and leaching in an archaeological Anthrosol and a Ferralsol of the Central Amazon basin: fertilizer, manure and charcoal amendments. Plant Soil 249:343–357
Li J, Pu L, Zhu M, Zhang J, Li P, Dai X et al (2014) Evolution of soil properties following reclamation in coastal areas: a review. Geoderma 226-227:130–139. https://doi.org/10.1016/j.geoderma.2014.02.003
Ljung K, Maley F, Cook A (2010) Canal estate development in an acid sulfate soil-implications for human metal exposure. Landsc Urban Plan 97:123–131
Mapbox (2019) Mapbox data attribution. https://www.mapbox.com/about/maps/. Accessed 5 July 2019
Margesin R, Siles JA, Cajthaml T, Öhlinger B, Kistler E (2017) Microbiology meets archaeology: soil microbial communities reveal different human activities at archaic Monte Iato (sixth century BC). Microb Ecol 73:925–938. https://doi.org/10.1007/s00248-016-0904-8
Mazurek R, Kowalska J, Gąsiorek M, Setlak M (2016) Micromorphological and physico-chemical analyses of cultural layers in the urban soil of a medieval city—a case study from Krakow, Poland. Catena 141:73–84. https://doi.org/10.1016/j.catena.2016.02.026
McKenzie NJ, Jacquier D, Isbell R, Brown K (2004) Australian soils and landscapes: an illustrated compendium. CSIRO Publishing, Collingwood
Middleton WD, Barba L, Pecci A, Burton JH, Ortiz A, Salvini L, Suárez RR (2010) The study of archaeological floors: methodological proposal for the analysis of anthropogenic residues by spot tests, ICP-OES, and GC-MS. J Archaeol Method Theory 17:183–208. https://doi.org/10.1007/s10816-010-9088-6
Millaire JF, Eastaugh E (2014) Geophysical survey on the coast of Peru: the early Prehispanic city of Gallinazo group in the Viru valley. Lat Am Antiq 25:239–255. https://doi.org/10.7183/1045-6635.25.3.239
Morel JL, Heinrich AB (2008) SUITMA-soils in urban, industrial, traffic, mining and military areas: an interdisciplinary working group of the 'International Union of Soil Science' (IUSS) dedicated to soils strongly modified by human activities. J Soils Sediments 8:206–207. https://doi.org/10.1007/s11368-008-0023-3
Morse JW (1994) Interactions of trace metals with authigenic sulfide minerals: implications for their bioavailability. Mar Chem 46:1–6
National Environment Protection Council (2013) Schedule B (1): guideline on the investigation levels for soil and groundwater. In: National environment protection (assessment of site contamination) measure (amended). Commonwealth of Australia, Canberra

Oakley SM, Jimenez R (2012) Sustainable sanitary landfills for neglected small cities in developing countries: the semi-mechanized trench method from Villanueva, Honduras. Waste Manag 32:2535–2551. https://doi.org/10.1016/j.wasman.2012.07.030

Ottaway JH, Matthews MR (1988) Trace element analysis of soil samples from a stratified archaeological site. Environ Geochem Health 10:105–112. https://doi.org/10.1007/bf01758678

Paul MJ, Meyer JL (2001) Streams in the urban landscape. Annu Rev Ecol Syst 32:333–365

Peters S, Borisov AV, Reinhold S, Korobov DS, Thiemeyer H (2014) Microbial characteristics of soils depending on the human impact on archaeological sites in the northern Caucasus. Quat Int 324:162–171. https://doi.org/10.1016/j.quaint.2013.11.020

Pickering MD, Ghislandi S, Usai MR, Wilson C, Connelly P, Brothwell DR, Keely BJ (2018) Signatures of degraded body tissues and environmental conditions in grave soils from a Roman and an Anglo-Scandinavian age burial from Hungate, York. J Archaeol Sci 99:87–98. https://doi.org/10.1016/j.jas.2018.08.007

Plant R, Wilmot K, Ege C (2014) Contaminated soil wastes in Australia. [Prepared for the Australian Department of the Environment]. , Institute for Sustainable Futures, University of Technology, Sydney

Pons LJ, Van Der Molen WH (1973) Soil genesis under dewatering regimes during 1000 years of polder development. Soil Sci 116:228–235. https://doi.org/10.1097/00010694-197309000-00011

Pouyat RV, McDonnell MJ, Pickett STA (1995) Soil characteristics of oak stands along an urban-rural land-use gradient. J Environ Qual 24:516–526. https://doi.org/10.2134/jeq1995.00472425002400030019x

Pouyat RV, Yesilonis ID, Nowak DJ (2006) Carbon storage by urban soils in the United States. J Environ Qual 35:1566–1575. https://doi.org/10.2134/jeq2005.0215

Pouyat RV, Yesilonis ID, Russell-Anelli J, Neerchal NK (2007) Soil chemical and physical properties that differentiate urban land-use and cover types. Soil Sci Soc Am J 71:1010–1019. https://doi.org/10.2136/sssaj2006.0164

Pouyat RV, Yesilonis ID, Szlavecz K, Csuzdi C, Hornung E, Korsós Z et al (2008) Response of forest soil properties to urbanization gradients in three metropolitan areas. Landsc Ecol 23:1187–1203. https://doi.org/10.1007/s10980-008-9288-6

Rate AW (2018) Multielement geochemistry identifies the spatial pattern of soil and sediment contamination in an urban parkland, Western Australia. Sci Total Environ 627:1106–1120. https://doi.org/10.1016/j.scitotenv.2018.01.332

Roosevelt AC (2013) The Amazon and the anthropocene: 13,000 years of human influence in a tropical rainforest. Anthropocene 4:69–87. https://doi.org/10.1016/j.ancene.2014.05.001

Rossiter DG (2007) Classification of urban and industrial soils in the world reference base for soil resources. J Soils Sediments 7:96–100. https://doi.org/10.1065/jss2007.02.208

Salmon SU, Rate AW, Rengel Z, Appleyard SJ, Prommer H, Hinz C (2014) Reactive transport controls on sandy acid sulfate soils and impacts on shallow groundwater quality. Water Resour Res 50:4924–4952. https://doi.org/10.1002/2013WR014404

Sánchez-Pérez S, Solleiro-Rebolledo E, Sedov S, de Tapia EM, Golyeva A, Prado B, Ibarra-Morales E (2013) The black San Pablo paleosol of the Teotihuacan Valley, Mexico: Pedogenesis, fertility, and use in ancient agricultural and urban systems. Geoarchaeology 28:249–267. https://doi.org/10.1002/gea.21439

Schaetzl RJ, Anderson S (2005) Soils: genesis and geomorphology. Cambridge University Press, New York

Schrader S, Böning M (2006) Soil formation on green roofs and its contribution to urban biodiversity with emphasis on collembolans. Pedobiologia 50:347–356. https://doi.org/10.1016/j.pedobi.2006.06.003

Semmens S, Perrin ND, Dellow G, Van Dissen R (2011) NZS 1170.5:2004 site subsoil classification of Wellington City. Ninth Pacific conference on earthquake engineering: building an earthquake-resilient society. New Zealand Society for Earthquake Engineering and University of Auckland, Auckland, New Zealand

Shen J, Jones RT, Yang X, Dearing JA, Wang S (2006) The Holocene vegetation history of Lake Erhai, Yunnan province southwestern China: the role of climate and human forcings. The Holocene 16:265–276. https://doi.org/10.1191/0959683606hl923rp
Simonson RW (1959) Outline of a generalized theory of soil genesis. Soil Sci Soc Am J 23:152–156. https://doi.org/10.2136/sssaj1959.03615995002300020021x
Sohlenius G, Öborn I (2004) Geochemistry and partitioning of trace metals in acid sulphate soils in Sweden and Finland before and after sulphide oxidation. Geoderma 122:167–175
Soil Survey Staff (2014) Keys to soil taxonomy. Natural Resources Conservation Service, United States Department of Agriculture, Washington, 362 pp
Sulas F, Madella M (2012) Archaeology at the micro-scale: micromorphology and phytoliths at a Swahili stonetown. Archaeol Anthropol Sci 4:145–159. https://doi.org/10.1007/s12520-012-0090-7
Sylvester GC (2017) The application of high resolution mobile metal ion (MMI) soil geochemistry to archaeological investigations: an evaluation of its effectiveness. PhD thesis, The University of Western Australia, Perth, 417 pp
Sylvester G, Mann AW, Rate AW, Wilson CA (2017) Application of high resolution soil geochemistry to archaeological investigations: an example from a Roman metal working site, Somerset, United Kingdom. Geoarchaeology 32:563–574. https://doi.org/10.1002/gea.21618
Sylvester GC, Mann AW, Cook SR, Wilson CA (2018) MMI partial extraction geochemistry for the resolution of anthropogenic activities across the archaeological Roman town of Calleva Atrebatum. Geochem Explor Environ Anal 18:58–74. https://doi.org/10.1144/geochem2017-009
Taylor MP, Mackay AK, Hudson-Edwards KA, Holz E (2010) Soil Cd, Cu, Pb and Zn contaminants around Mount Isa city, Queensland, Australia: potential sources and risks to human health. Appl Geochem 25:841–855
Tifafi M, Bouzouidja R, Leguédois S, Ouvrard S, Séré G (2017) How lysimetric monitoring of Technosols can contribute to understand the temporal dynamics of the soil porosity. Geoderma 296:60–68. https://doi.org/10.1016/j.geoderma.2017.02.027
U.S. EPA (2017) LFG energy project development handbook. United States Environmental Protection Agency (Landfill Methane Outreach Program), Washington, DC
Vepraskas MJ, Vaughan KL (2016) Morphological features of hydric and reduced soils. In: Vepraskas MJ, Craft CB (eds) Wetland soils: genesis, hydrology, landscapes, and classification, 2nd edn. CRC Press LLC, Boca Raton, pp 189–217
Vepraskas MJ, Polizzotto M, Faulkner SP (2016) Redox chemistry of hydric soils. In: Vepraskas MJ, Craft CB (eds) Wetland soils: genesis, hydrology, landscapes, and classification, 2nd edn. CRC Press LLC, Boca Raton, pp 105–131
Völkel J, Murray A, Leopold M, Hürkamp K (2012) Colluvial filling of a glacial bypass channel near the Chiemsee (Stöttham) and its function as geoarchive. Z Geomorphol 56:371–386. https://doi.org/10.1127/0372-8854/2012/0070
Vuorela I, Hiekkanen M (1991) The urban milieu of the late- and postmedieval town of Porvoo, southern Finland investigated by means of pollen analysis. Ann Bot Fenn 28:95–106
Vuorela I, Grönlund T, Lempiäinen T (1996) A reconstruction of the environment of Rettig in the city of Turku, Finland on the basis of diatom, pollen, plant macrofossil and phytolith analyses. Bull Geol Soc Finl 68:46–71. https://doi.org/10.17741/bgsf/68.2.005
Wei B, Yang L (2010) A review of heavy metal contaminations in urban soils, urban road dusts and agricultural soils from China. Microchem J 94:99–107. https://doi.org/10.1016/j.microc.2009.09.014
Wells EC, Terry RE, Parnell JJ, Hardin PJ, Jackson MW, Houston SD (2000) Chemical analyses of ancient Anthrosols in residential areas at Piedras Negras, Guatemala. J Archaeol Sci 27:449–462. https://doi.org/10.1006/jasc.1999.0490
White RE (2006) Principles and practice of soil science: the soil as a natural resource. Blackwell Science, Malden

Wilson CA, Davidson DA, Cresser MS (2005) An evaluation of multielement analysis of historic soil contamination to differentiate space use and former function in and around abandoned farms. The Holocene 15:1094–1099. https://doi.org/10.1191/0959683605hl881rr
Wilson CA, Davidson DA, Cresser MS (2007) Evaluating the use of multi-element soil analysis in archaeology: a study of a post-medieval croft (Olligarth) in Shetland. Atti della Soc Toscana di Sci Nat Mem Ser A 112:69–77
Zhang MK (2004) Phosphorus accumulation in soils along an urban-rural land use gradient in Hangzhou, Southeast China. Commun Soil Sci Plant Anal 35:819–833. https://doi.org/10.1081/CSS-120030360
Zhang G-L, Yang F-G, Zhao Y-G, Zhao W-J, Yang J-L, Gong Z-T (2005) Historical change of heavy metals in urban soils of Nanjing, China during the past 20 centuries. Environ Int 31:913–919. https://doi.org/10.1016/j.envint.2005.05.035
Zou S, Li R, Xie S, Zhu J, Wang X, Huang J (2010) Paleofire indicated by polycyclic aromatic hydrocarbons in soil of Jinluojia archaeological site, Hubei, China. J Earth Sci 21:247–256. https://doi.org/10.1007/s12583-010-0089-x

Chapter 3
Spatial Variability and Data Analysis in Urban Soils

Andrew W. Rate

Abstract Urban soils are likely to be even more variable than soils in other environments, due to the inherent heterogeneity of urban environments which intensifies natural soil variability. This chapter examines soil variability related to urbanisation at several different spatial scales, from regional phenomena to differences observed on the scale of individual soil profiles. Soil sampling strategies and designs are described, with discussion of the issues related to sampling density, sample numbers, the geometric arrangement of sampling locations, and 'hotspot' detection. We present methods for qualitative and quantitative analysis of soil spatial data using maps, spatial autocorrelation analysis, and variograms and kriging. Basic but rigorous statistical methods are described in the context of soils and compositional data, including comparisons, relationships, and multivariate techniques.

Keywords Variability · Urban soils · Spatial scales · Spatial analysis · Spatial autocorrelation · Kriging · Statistics · Data analysis

A. W. Rate (✉)
School of Agriculture and Environment, University of Western Australia, Crawley, WA, Australia
e-mail: andrew.rate@uwa.edu.au

A. W. Rate (ed.), *Urban Soils*, Progress in Soil Science,
https://doi.org/10.1007/978-3-030-87316-5_3

What you could learn from this chapter:

- That soils are naturally variable, and whether urban centres create additional sources of soil variability. You'll also consider the spatial scales over which soils vary and how these are related to urban environments.
- How to sample soils in urban environments – there are several strategies, and we will discuss the reasons why we choose certain depths, combinations of sampling locations, how many samples we should take, and how far apart.
- How we conduct spatial data analysis to investigate the relationships between soil samples from different locations and the specific spatial statistical techniques we use.
- How we apply common statistical methods to describe, explore, and assess urban soil data.

3.1 Soil Variability in Urban Environments

Urban environments are far from uniform; cities are characterised by extreme variability, reflecting the range of types, intensity, and timescales of human modification of the natural environment (Grimm et al. 2000; Pickett et al. 2001). Not surprisingly, this general heterogeneity of urban environments is also present in urban soil environments. Prior to human interference, natural soils already showed significant spatial variation due to differences in soil parent material and other soil-forming factors (see Chap. 2). This pattern of natural soil variability has, superimposed upon it, the imprint of diverse human activity, creating an even more variable soil landscape.

3.1.1 Cities and Regional Soil Variability

There is some evidence from continental-scale geochemical surveys (Cicchella et al. 2015; Mann et al. 2015) that urbanised areas can be distinguished from non-urbanised land based on the concentration of some (potential contaminant) elements such as lead (Pb), with an example shown in Fig. 3.1. This distinction between urban and non-urban areas is detectable despite the variable background from differences in the chemical composition of parent materials. The geochemical signature of urbanisation extends to peri-urban areas (Cicchella et al. 2015). The anthroposequences discussed in Chap. 1 and elaborated on in the work of Richard Pouyat and others (e.g. Pouyat and McDonnell 1991; Pouyat et al. 2007; Pouyat et al. 2008) also provide evidence that soils in cities may be distinguished from their rural counterparts on a regional scale.

3.1.2 Soil Variability at the Scale of Cities

The next largest scale of soil variability in urban environments is constrained by the dimensions of an individual city, taking into account the effects of the urban centre on the surrounding peri-urban and non-urbanised land.

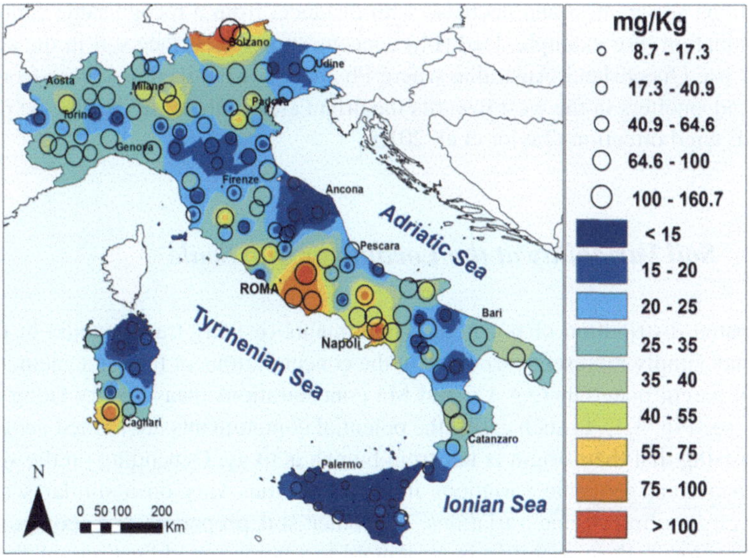

Fig. 3.1 Lead (Pb) concentrations in Italian agricultural surface soils, showing the effect of the urban centres of Roma and Napoli (from Cicchella et al. (2015) and used with permission from Elsevier)

An important parameter for the spatial distribution of soils in cities is the amount of soil remaining exposed after urban development. This can be assessed using conventional ground-based surveying techniques or by remote sensing and image analysis (Wu and Murray 2003; Wu 2004). The remote sensing techniques commonly categorise land into vegetation, impervious, or soil categories (the V–I–S model). Urban catchments can range from 5 to 100% impervious cover, depending on the density of infrastructure (Fletcher et al. 2004). Assuming that the vegetation component of the remaining pervious cover represents plants growing in soil, then urban soil cover may range from approximately 95% to 0%. Whole-city soil cover ranges from 10 to 35% for smaller cities (pop. < 100,000; Bauer et al. 2008) and up to at least 60% in larger cities (Zhu et al. 2017). The proportion of soil cover in urban environments is itself highly variable and shows gradients related to distance from urban centres and age of urban development (Powell et al. 2007). Soil cover tends to be lowest in urban centres and to decrease as the time since development increases.

Other urban soil properties also show systematic variation on whole-city scales. Johnson and Ander (2008) reviewed multiple studies of the spatial distribution of trace elements in urban environments, explaining that such studies have multiple objectives, including collection of baseline data, and identifying contaminated areas and their sources and risks. In many cities, the concentrations of some contaminants are greatest in the metropolitan centre (often the oldest area in the city), and lower concentrations occur at greater distance from the urban centre (e.g. Pb concentrations in surface soils in Pueblo, Colorado (Diawara et al. 2006); see Fig. 3.2). This implies a cumulative and ongoing input of contaminants, relating to more diffuse sources such as road traffic, construction, etc. In other urban environments, the concentrations are greatest and decrease with distance, from a recognisable source of contamination. For example, lead (Pb) concentrations in surface soil in the city of Mount Isa, Queensland, Australia, where Pb and Cu are mined, decrease from the mine and smelters in the west towards the urban area in the east, despite the dominant SE wind direction (Taylor et al. 2010).

3.1.3 Soil Variability at the Locality or Site Scale

The spatial distribution of potential contaminants (usually trace metals) in urban soils may simply represent variations in the concentrations of the same elements in the soil parent materials (e.g. Co and Mn concentrations measured by Gong et al. 2010) (see Fig. 3.3). In such cases, the potential contaminants are termed *geogenic*, emphasising that their origin is not from human activity. Depending on the size of the urban area, geogenic variations in soil properties vary on a similar scale to whole-city anthropogenic variations. For other soil properties (e.g. contaminant concentrations), the variability is attributable to human modifications of the soil environment. These modifications include excavation, dumping, or construction and point or diffuse sources of contamination such as vehicle traffic or industrial emissions. For example, the high concentrations of Pb and Zn in urban centre soils

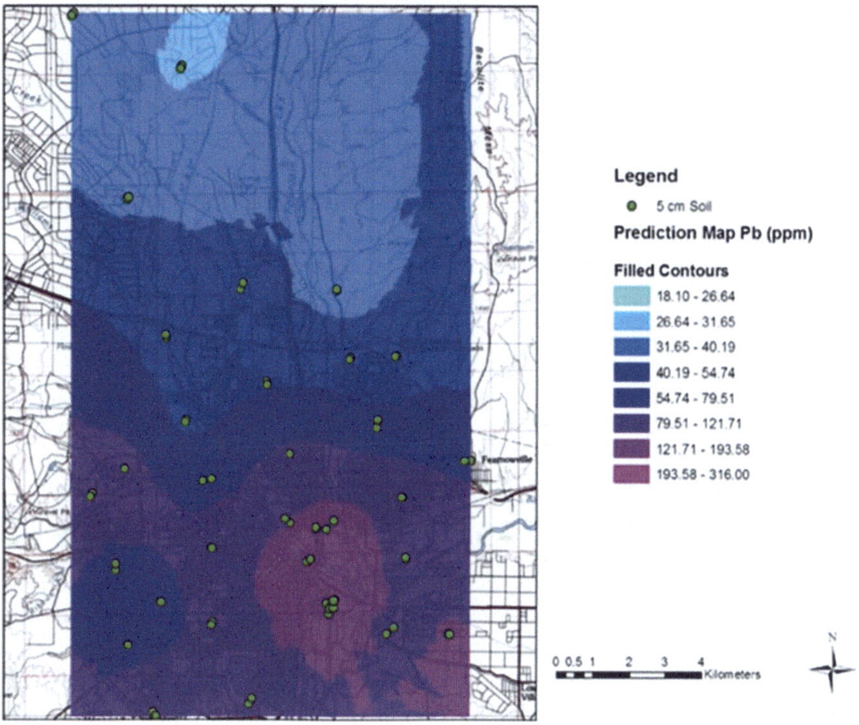

Fig. 3.2 Spatial distribution of lead (Pb) in surface soil (0–5 cm) sampled in Pueblo, Colorado, USA (from Diawara et al. (2006); used with permission from Springer)

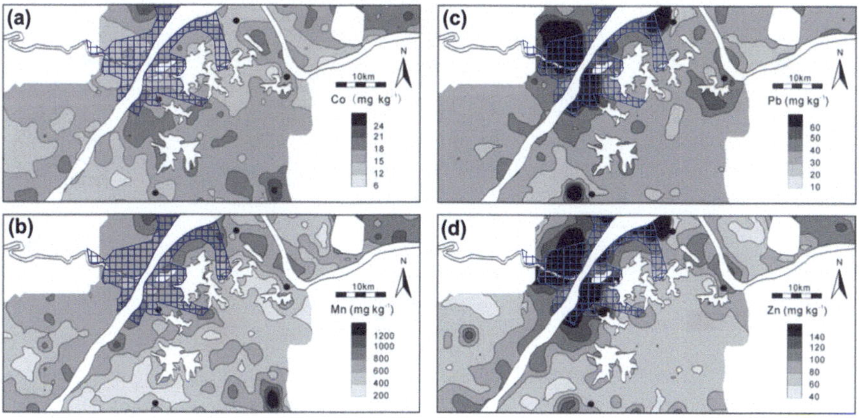

Fig. 3.3 Concentration distributions of (**a**) cobalt, Co; (**b**) manganese, Mn; (**c**) lead, Pb; and (**d**) zinc, Zn, from a gridded sampling design in the area in and around Wuhan city, Hubei, China (blue cross-hatching ▦ shows urban area). Cobalt and manganese (**a** and **b**) show a geogenic distribution, whereas lead and zinc (c and d) show accumulation in the urban area related to anthropogenic additions. Redrawn from Gong et al. (2010); used with permission from Springer

in Wuhan, China, were attributed by Gong et al. (2010) to domestic heating and road traffic.

In practice, urban soil investigations are more often conducted at the scale of specific sites, for example, to obtain site-specific data for environmental impact assessment (e.g. Carr et al. 2008). Figure 3.4 shows a typical pattern of spatial variability in soil properties for urban parkland soils, where the soil properties reflect a combination of both geogenic variation (some of the parkland occupies former lake beds) and anthropogenic modification (horticulture, landfill, construction, and demolition wastes). For example, zones of low pH in Fig. 3.4 are probably due to drainage of lake sediments and subsequent formation of acid sulphate soils; high pH probably reflects the use and/or disposal of limestone-based products such as construction cement. High EC (a measure of soluble salts in soils) could be caused by release of dissolved ions by acid sulphate oxidation or demolition of a building containing gypsum-based materials (e.g. wall and ceiling panels) in the south of the study site.

Soil variability at the site scale is also shown more simply in concentration-distance relationships, for example, along transects (which are analogous to the anthroposequences described for city- or regional-scale transects by Pouyat et al. (2008)). Figure 3.5 presents two examples of such variability, showing gradients away from likely contaminant sources (a major roadway and a coal-fired power station site). In both cases, it is likely that contaminants are carried from their sources by wind (as aerosols) and deposited on soil surfaces in decreasing amounts as distance from the source increases. Transport of material from roads to soils by surface flow of water is also possible.

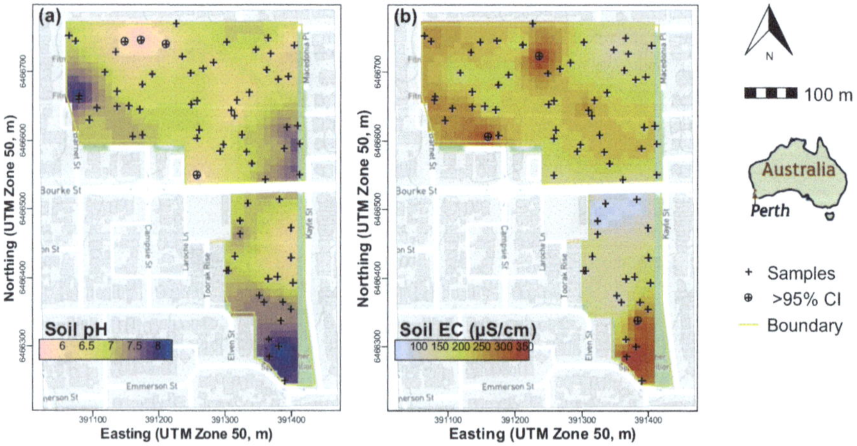

Fig. 3.4 Spatial variability of (**a**) surface soil pH and (**b**) electrical conductivity (EC) of a 1:5 soil/water suspension at Charles Veryard and Smith's Lake Reserves, urban parklands in Perth, Western Australia. Maps generated in R with the packages 'OpenStreetMap' (Fellows 2019) and 'geoR' (Ribeiro Jr. et al. 2020), and using kriging with exponential variogram models (graphic by Andrew W. Rate)

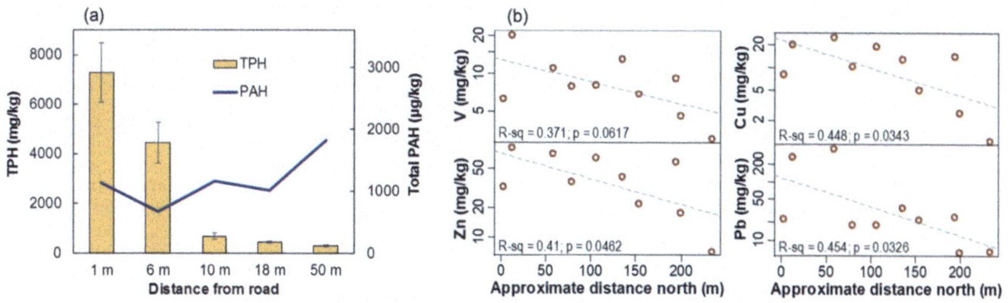

Fig. 3.5 Concentration-distance gradients for (**a**) total petroleum hydrocarbons (TPH) and polycyclic aromatic hydrocarbons (PAH) in urban roadside soils (redrawn from Nikolaeva et al. 2017, and used with permission from Springer) and (**b**) metals V, Zn, Cu, and Pb in urban parkland soils north of a decommissioned coal-fired power station (Banks Reserve, Perth, Western Australia; graphic by Andrew W. Rate)

3.1.4 Contamination Hotspots

Some urban events, such as chemical spills, can produce soil variability on a very small scale. Relatively small (metre-scale) areas of high concentration are commonly termed 'hotspots'. An example of a contamination hotspot at this scale is shown for an urban public open space in Fig. 3.6. In the example shown in Fig. 3.6, the contamination with several metals in the hotspot most likely represents disposal of contaminated sediments from the adjacent open stormwater drain onto the soil surface during drain maintenance. The hotspot was identified by sampling soils on transects along expected contamination gradients, rather than grid sampling, so in a sense the hotspot was identified 'accidentally'!

The term 'hotspot' is not reserved for metre-scale areas, however; Nriagu (1988) argues that urban areas themselves represent hotspots on regional, continental, or even global scales. Similarly, Li et al. (2004) identify district-sized hotspots within the city of Hong Kong. A better definition of a hotspot, then, is an area of high concentration of contaminants in soil that is localised, or small, relative to the scale of observation.

3.1.5 Soil Variability with Depth

Soil properties can vary substantially with depth, that is, in the vertical rather than horizontal dimension which has been the focus of the sections above. For some soil properties, such as organic carbon content, the source of the variation is natural pedogenesis. The process of soil formation results in greater organic carbon content in a soil's surface horizon(s), since most additions of organic materials (litter fall, excretion) and losses (microbial decomposition) occur in the surface soil (Coleman et al. 2017). Natural pedogenesis may also result in enrichment of some soil components at greater depth, such as clay or iron content, due to the complex

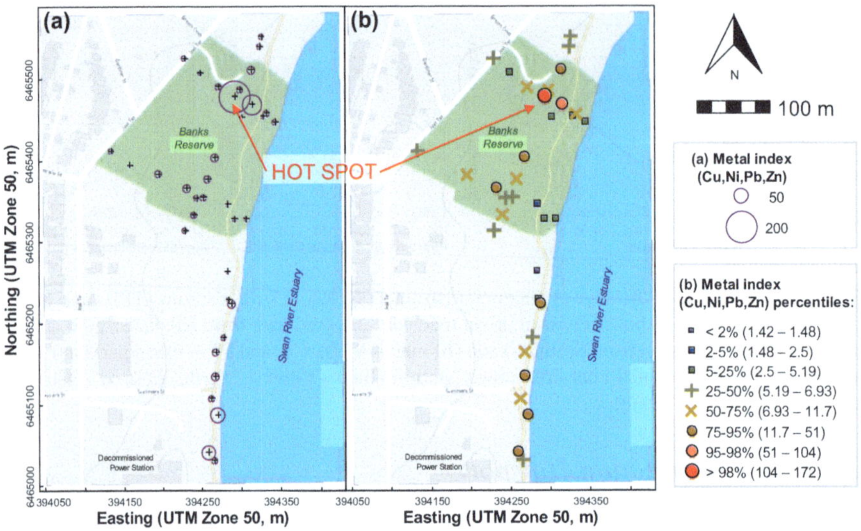

Fig. 3.6 A soil contamination hotspot (labelled) shown by the spatial distribution of a metal contamination index (an additive index of normalised Cu, Ni, Pb, and Zn concentrations – contamination indices will be explained in Chap. 6). Data are shown with two representation styles for point spatial data: (**a**) area-proportional 'bubbles' and (**b**) symbols showing percentile categories

interactions of soil and hydrological processes. General soil science texts, such as those by Schaetzl and Anderson (2005) and White (2006), discuss natural pedogenic processes resulting in depth variation in much more detail.

In urban soils, soil properties do not always vary with depth in straightforward ways. A huge variety of human modifications, such as additions of new material (and often burial of existing soils), excavation and refilling, and contamination, are commonly superimposed upon natural pedogenic processes. Figure 3.7 shows some depth profiles for soil properties in a highly disturbed urban area.

In some cases the concentration vs. depth relationships (referred to as depth profiles) in urban soils, even for contaminants, are closely related to depth profiles for pedogenic features. For example, metal contaminant concentrations in soil adjoining highways in Cincinnati, USA, decreased with increasing depth (Turer et al. 2001) but were shown to be more closely related to soil organic matter content than to other predictors. A similar control by soil organic matter was deduced for phthalate and BPA contaminants (organic compounds used as plasticisers) in urban and other soil environments (Tran et al. 2015) and for polycyclic aromatic hydrocarbons (PAH) in soils in Beijing, China (Bu et al. 2009). The relationship of contaminants to a pedogenic parameter such as soil organic matter content demonstrates the ability of organic matter to act as a sink for added contaminants. Soil clays or iron oxides may also provide contaminant sinks (e.g. Rate 2018).

For biological parameters, the rule of thumb that biological activity decreases as soil depth increases may not apply in all urban soil environments. For example, Zhao et al. (2012) found that soil microbial biomass (measured as carbon or

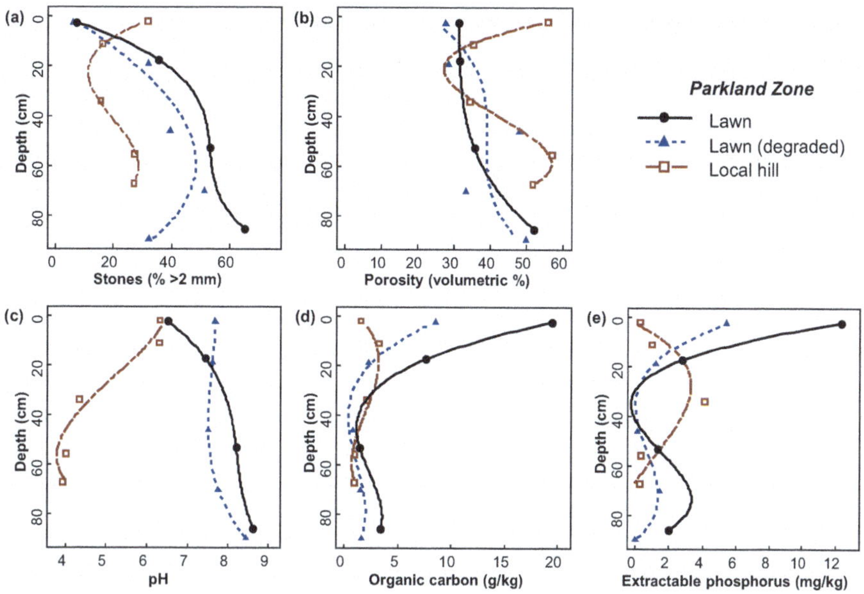

Fig. 3.7 Variation of physical and chemical soil properties with depth, in soils from an urban parkland in Hong Kong (redrawn by Andrew W. Rate from data tabulated in Jim 1998). Smooth lines are cubic regressions intended only as visual guides. The high stone contents (**a**) reflect the use of fill material, with many low porosity values (**b**) reflecting compaction by foot traffic. High pH (**c**), except in the relatively natural local hill soil, is related to use of cement-based materials. The generally low organic carbon (**d**) and extractable P (**e**) contents reflect the young age of the soil materials, with concentration peaks at depth possibly indicating burial of pre-existing soil

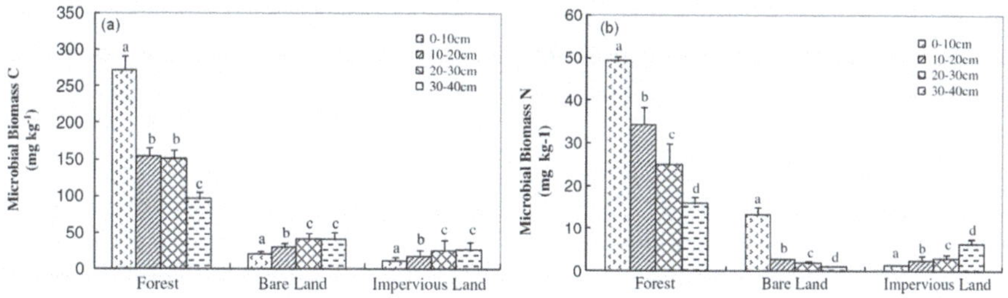

Fig. 3.8 Variation with depth of soil microbial biomass carbon (**a**) and nitrogen (**b**) in different land cover types in urban Beijing, China (from Zhao et al. 2012 and used with permission from Springer)

nitrogen) could decrease or increase with increasing depth, in urban soils of Beijing, China, depending on the measurement or land cover type (Fig. 3.8). In this study, microbial biomass C or N was greater in deeper soil samples under impervious cover (Zhao et al. 2012).

3.2 Measurement and Description of Soil Variability

The previous sections in this chapter described the scales of soil variability in urban systems and provided some explanations for this variability. Soil variability is, logically, more pronounced in urban soil environments, due to the combination of a wide range of natural and human processes which affect soil properties and formation. The following sections will therefore deal with how soils can be characterised in highly variable urban environments and how the variability can be both accounted for, and described, by suitable sampling and statistical methods. In the next chapter, we will also address some of the implications of soil heterogeneity for land utilisation in cities.

The way in which urban soils are sampled is guided by the purpose(s) of the sampling program. Johnson and Ander (2008) review multiple urban geochemical surveys and identify several objectives for mapping soil properties (especially contaminant concentrations) in urban soil environments. These objectives include determining the current (baseline) status of an urban environment; regulatory drivers, such as fulfilling the requirements for environmental impact assessment for development; identifying and locating polluted urban areas; identifying (potential) contaminant sources, including separating geogenic and anthropogenic sources; and assessing risks to other urban environmental compartments such as water bodies and human health (Johnson and Ander 2008).

3.2.1 Sampling Depth

In all of the strategies outlined in Sect. 3.2.2 for generating a two-dimensional spatial arrangement of sampling points, a decision must also be made about the depth range(s) of soil to be sampled. In some protocols (e.g. Smith 2004), this is guided by the depth of a particular soil horizon, commonly the A horizon but in other studies different horizons are targeted (e.g. Reimann et al. 2008, who describe sampling of O, B, and C horizons). In soils in urban environments, horizons are commonly poorly developed or obscured by truncation or addition of material. As a result, sampling of urban soil profiles is usually done on the basis of depth increments; 0–10 cm is common for surface soils, and narrower increments are more subject to sampling variability (Johnson and Ander 2008). Depth ranges for 'surface soils' in the published literature generally range from 0–2 cm to 0–20 cm. When assessing an environment for contamination, many protocols (e.g. Department of Environmental Protection 2001) recommend sampling at multiple depths, and examples of studies where depthwise sampling has been performed are Turer et al. (2001), Zhang (2004), Zhao et al. (2012), and Corrò and Mozzi (2017). Some studies assume that deeper samples remain relatively uncontaminated, especially for the purposes of calculating contamination indices (e.g. Gong et al. 2008; see Chap. 6).

3.2.2 Sampling Strategies and Designs for Urban Soils

3.2.2.1 Sampling Density

The number of samples required depends on the spatial scale and objectives of the study. In their review, Johnson and Ander (2008) identify 'systematic' and 'targeted' surveys. Systematic surveys are based on hundreds to thousands of samples over entire urban or regional areas at densities of 1–4 samples per km^2. In contrast, targeted surveys have tens of samples (or fewer), within a specific urban land use, at densities of 4–50 samples per km^2. The specific case of hotspot identification discussed below has more stringent constraints on the number of samples required for identifying very localised contamination. If there is a requirement for samples to be independent, then spatial statistics in the form of variogram analysis may be required. Essentially a variogram is an analysis of the variance (in some variable, such as the concentration of a contaminant) between samples as a function of distance between the samples. In theory, sufficient distance between samples maximises the between-sample variance, meaning that samples are independent (Oliver and Webster 2014). We will look into variogram analysis, and how it is used in practice, later in this chapter. It should be said, though, that the variogram analysis requires field data – but in this context, the purpose of using the variogram is to guide the sampling required to obtain that field data! The use of variograms to inform sampling therefore requires preliminary data to be acquired, which is unlikely in practice except perhaps for studies for scientific research purposes.

3.2.2.2 Sampling for Hotspots

The general aim of identifying and locating polluted urban areas can be specified as the search for contamination hotspots. There are methods for calculating the required number of samples for hotspot identification based on rigorous statistical principles. The Western Australian Department of Environmental Protection (2001) describes such a method, shown in Box 3.1. Bugdalski et al. (2014) emphasise that if the objective of sampling is to detect hotspots, then inadequate sampling can lead to type II errors or false negatives (i.e. not all hotspots on a site are detected).

Box 3.1 Calculating sampling grid size and sample numbers for hotspot detection

The grid size, G, should be calculated using Eq. 3.1:

$$\mathbf{G = R / 0.59} \tag{3.1}$$

where: **G** = grid size of the sampling plan, in metres **R** = radius of the smallest hotspot that the sampling intends to detect, in metres **0.59** = factor derived from 95% detection probability, assuming circular hotspots

The number of sampling points **n** should then be calculated from Eq. 3.2:

$$\mathbf{n = A / G^2} \tag{3.2}$$

where.: **A** = area to be sampled, in square metres
G = grid size of the sampling pattern, from Eq. 3.1, in metres.

Of course the size of unknown hotspots (**R** in Eq. 3.1) cannot be estimated in advance with complete confidence, so there is again the possibility of type II errors, unless the estimate of **R** is biased towards lower values. As with calculation of the maximum distance between independent samples, the size of contamination hotspots will, in practice, most often be measured after sampling is completed.

Grid Sampling Sampling urban and peri-urban soils on regular grids is most commonly used over larger (city-wide or regional) scales to assess the current state, or baseline, of a soil environment. This would equate to systematic sampling in Johnson and Ander's (2008) study; the other, more common, usage of the term 'systematic sampling' is a sampling design that is based on regular intervals across a landscape, such as various types of grid (Fig. 3.9). The geometry of a grid may be rectangular. For example, Lv et al. (2015) sampled soils across an entire province in China on a rectangular 2 × 2 km grid (0.25 samples/km^2; see Fig. 3.10). A grid sampling design provides a dataset which is well-suited to statistical spatial analysis (see Oliver and Webster 2014). For some targeted soil sampling, the lesser number of samples may be more suited to a transect (essentially a one-dimensional grid) across expected gradients in soil properties.

Stratified Sampling In some urban soil environments, there is sufficient pre-existing information to classify land into categories. This information includes data such as soil types, underlying geology, geomorphological zones, previous or current land use, vegetation communities, and so on. In such cases the total sampling area can be subdivided into sub-areas, or sampling strata, with samples collected within each of the strata (USEPA 2002). An example of a quite complex stratified sampling design for the city of Kumasi, Ghana, conducted by Nero and Anning (2018) is shown in Fig. 3.10. Stratified sampling may be used to ensure that sufficient sample numbers are taken in each sub-area and therefore can be used to test hypotheses, for example, about differences in soil properties between sampling strata.

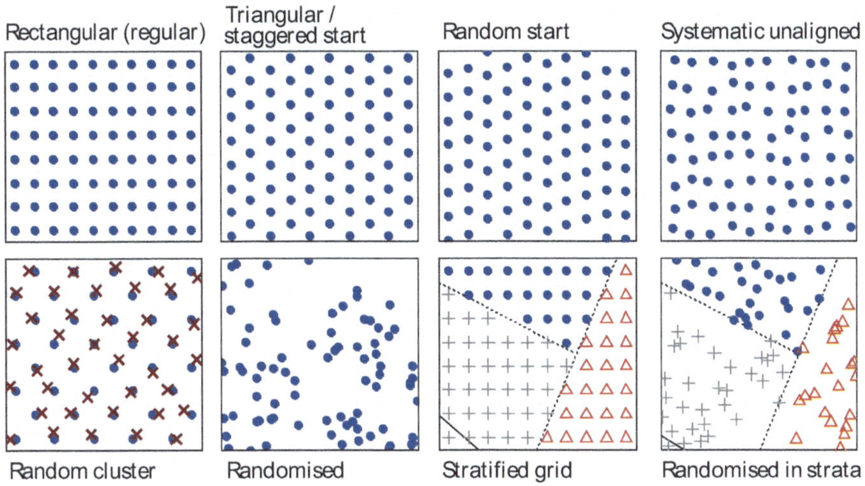

Fig. 3.9 Geometries of grid, cluster, and random sampling designs (graphic redrawn by Andrew W. Rate, based on multiple unattributed online sources)

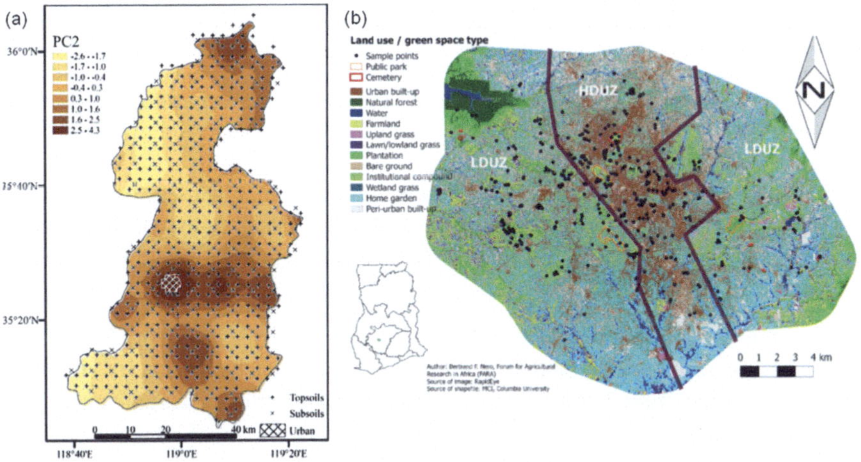

Fig. 3.10 Examples of soil sampling designs in urban and peri-urban environments
(**a**) Grid sampling Ju County, Shandong Province, China (redrawn from Lv et al. 2015; used with permission from Springer), superimposed on an interpolated map of principal component 2 (PC2, which emphasised signals from Cd, Pb, and Zn) from a multivariate statistical analysis
(**b**) Stratified sampling in the city of Kumasi, Ghana (Nero and Anning 2018; used with permission from Springer). Samples were taken from strata defined by eight unique green space types within each of two urban zones: high (HDUZ) and low density (LDUZ)

Randomised Sampling Completely randomised sampling is straightforward to implement and may be suitable for areas which have minimal variation (e.g. no hotspots USEPA 2002). Samples are theoretically statistically independent, but the

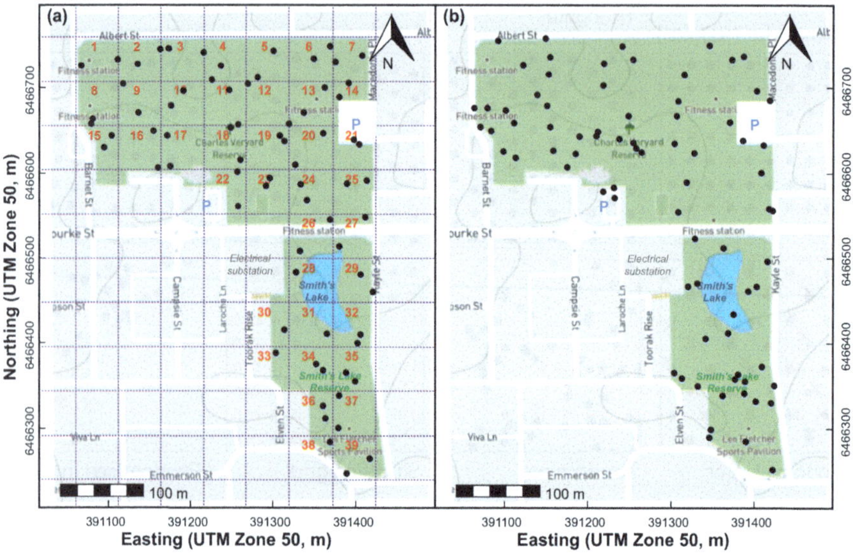

Fig. 3.11 (**a**) Random-in-grid sampling plan used to generate the data in Fig. 3.4. The grid size (52 m) was adjusted to fit the sampling area, and two samples were taken per grid square. Filled circle symbols ● show planned sampling locations with known coordinates; these were adjusted if necessary during field sampling (e.g. if a sample position was on a paved surface). (**b**) A completely randomised sampling equivalent of (a) showing large gaps in sampling area caused by uncontrolled randomisation. Other issues (e.g. samples on paved areas) can be addressed by adjustments during field sampling. Graphics by Andrew W. Rate

completely random location of sample points may mean that 'gaps' exist within the sampling area (Fig. 3.11), such that features like hotspots are missed. Since urban soils are very heterogeneous, completely randomised sampling is seldom used and is not recommended.

A variation on randomised sampling is random-in-grid sampling (Fig. 3.11), where samples are taken at randomised locations within each polygon (or stratum), or using a predetermined grid. The effect resembles cluster or systematic unaligned sampling (see Fig. 3.9) and is a compromise between the good area coverage of systematic grid sampling and the statistical independence of randomised sampling.

3.3 Analysis of Spatial Data

3.3.1 Maps

Spatial information is complex, and the most straightforward way of assessing spatial data from urban soil environments is using maps. The map itself presents the land surface as a two-dimensional representation, and other layers of information (such as land elevations, soil properties, contaminant concentrations, etc.) can be represented in different ways such as contour lines, sets of related symbols (e.g. size proportional to concentration), and so on. We have already seen Figs. 3.3 and 3.4 which show examples of contour-based information and Fig. 3.6 which has two different examples of symbol sets to show information layers on maps.

Maps are usually created in Geographic Information System (GIS) software, although more general open-source platforms such as R (R Core Team 2020) in combination with specific packages (such as those by Fellows 2019; Ribeiro Jr. et al. 2020) can provide excellent results. In addition, the QGIS package (QGIS.org 2020) is an open-source alternative to commercial GIS software and has a wide user base.

3.3.2 *Spatial Autocorrelation*

One of the objectives of spatial analysis is to investigate the effect of between-sample distance on soil variables. In effect, this is a test of Tobler's first law of geography (Tobler 2004), stating:

> *Everything is related to everything else,*
> *but near things are more related than distant things.*

Whether samples are spatially related or not, in terms of a particular soil property, is expressed by the spatial autocorrelation or Moran's I (Zhang et al. 2008). The Moran's I statistic is based on comparison of the values of a variable at one point with a specified number (or within a specified distance) of neighbouring points. A positive Moran's I autocorrelation suggests that locations with similar values of the variable considered tend to cluster together. A Moran's I close to zero suggests no autocorrelation (values of the variable considered are randomly located), and a larger negative Moran's I suggests that similar values of the variable considered tend to be further apart than expected from a random spatial distribution. The Moran's I statistic is tested against the null hypothesis of no spatial autocorrelation and will vary with the number of neighbouring points in the calculation, with more points giving weaker autocorrelations (Kalogirou 2019).

The basic Moran's I is a global autocorrelation, across the whole spatial dataset being analysed. The local Moran's I can also be calculated and shows the extent of significant spatial clustering of similar values (of the variable considered) around each observation. The two examples in Fig. 3.12 show local Moran's I as map symbols for the soil pH and EC data which we have already seen in Fig. 3.4. Interestingly, at the urban parkland location in Fig. 3.12, soil pH at most sampling points is spatially associated with similar values, but EC is not.

Spatial autocorrelation statistics, usually Moran's I, can be calculated using various GIS and statistical software, including several packages which add functionality to R. A useful expansion of the type of spatial autocorrelation information presented in Fig. 3.12 is local indicators of spatial association (LISA) analysis (Anselin 1995).

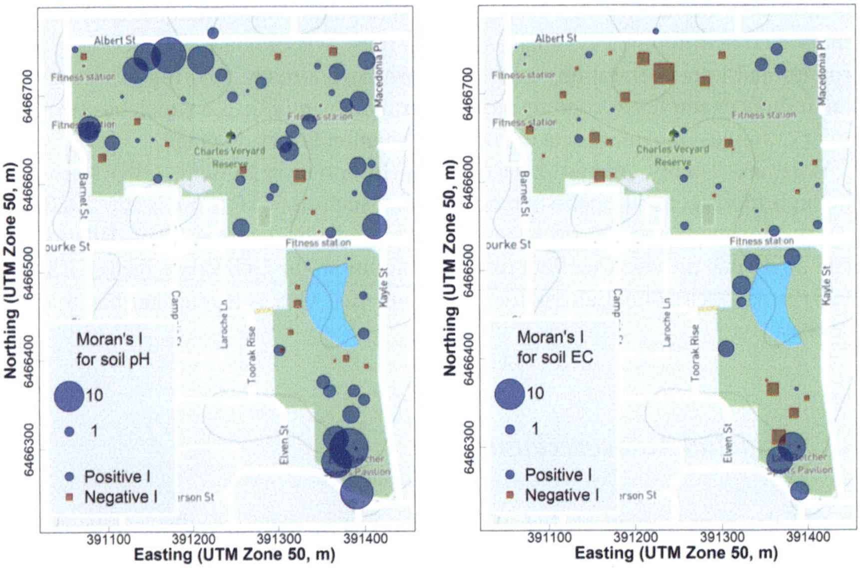

Fig. 3.12 Local Moran's I autocorrelation maps for the soil pH (left) and EC (right) data in Fig. 3.4. Larger positive values of local Moran's I imply significant spatial clustering of similar values; negative local Moran's I implies significant spatial clustering of dissimilar values. Global Moran's I (5 neighbours) for pH is positive and significant ($p \leq 0.001$,.) but for EC is near zero and $p > 0.05$ (graphics and data by Andrew W. Rate)

A LISA analysis identifies, for map points or polygons, whether significant values of local Moran's I represent association of high values of a variable with other high values, low with low, high with low, or low with high.

3.3.3 *Variograms and Kriging*

The variogram, simplistically, is the relationship between the variance between sample points and the distance separating those sample points. In many instances, it is desirable to predict a soil property at locations where samples have not been taken, and this requires some assumption(s), usually a mathematical model, about how that soil property varies with distance. This information is actually provided by the variogram. In many cases the form of the variogram relationship can be simulated adequately with a mathematical function. The variogram function can then be used to interpolate between points – a process known as kriging (after the originator of the method and pioneer of geostatistics, Professor Danie Krige). Variograms and kriging are summarised expertly by Oliver and Webster (2014) and Reimann et al. (2008). Webster and Oliver (1993) argue that at least 100 observations (and preferably more) are needed for kriging interpolation, based on variogram analysis to establish the relationship between sample points as a function of separation

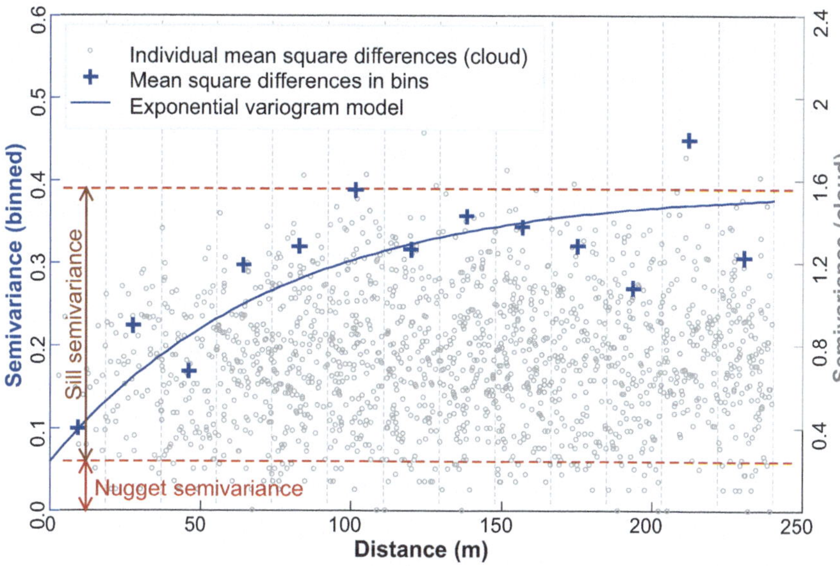

Fig. 3.13 The empirical variogram used to interpolate the soil pH in Fig. 3.4. The individual mean square differences are plotted at all possible pairwise distances between points up to the defined maximum distance (240 m, where the maximum possible pairwise distance at the site was 599 m). Vertical dashed lines represent the boundaries of the distance intervals ('bins') used to calculate the binned variogram for fitting the exponential model. The model parameters were nugget = 0.06, sill = 0.33, and range = 75 m (practical range = 225 m)

distance. The US Environmental Protection Agency (2002) reports that, for variograms and kriging, stratified sampling can have a lower sample number requirement than a simple grid but that kriging accuracy is similar for all sampling designs.

Figure 3.13 shows some of the key concepts of variogram analysis. There is some semivariance that exists even for very closely spaced samples, and this is called the 'nugget'. This semivariance increases with increasing distance between samples to a limiting value called the 'sill'. At some distance there is no increase in semivariance (which then approaches the variance of the complete dataset), and this distance is called the 'range', the value of which depends on the mathematical model used to describe the semivariance-distance relationship. The 'practical range', the distance at which samples are independent, is related to the model range by a factor dependent on the model equation.

Kriging and the associated variogram analysis can be very subjective in practice (Bohling 2005). Real soil data do not behave in an ideal fashion (see the scatter of binned points in Fig. 3.13), and there is no systematic way to make the choices of:

- The number of inter-sample distance categories or 'bins'.
- The maximum inter-sample distance to be considered in variogram analysis (Reimann et al. (2008) recommend the actual maximum inter-sample distance × approximately 0.4).

- The function used to model the variogram: exponential, spherical, Gaussian, etc.
- The mathematical algorithm for fitting the function to the variogram, e.g. using least squares, or maximum likelihood, or even heuristically.
- Whether the fitting process should be weighted (and there are several options).
- Whether to fix key variogram parameters such as the nugget or sill (see Fig. 3.13).
- Whether to assume an underlying trend in the data.
- Whether the variation with distance is the same in all directions (i.e. whether an isotropic or anisotropic variogram model should be used)
- .. . and so on.

Variogram model fitting can, therefore, appear to be more like an art than a science (Bohling 2005). Alternative forms of interpolation (e.g. splines, inverse distance) to predict soil properties between sampling points are not universally recommended, however, and can result in unusual predictions depending on the mathematical interpolation method used.

3.4 Comparison of Sampling Strata

3.4.1 Comparing Mean or Median Values

The most convenient way of comparing between strata is to use some form of statistical means comparison (Fig. 3.14). The comparisons are *univariate*, in that mean values are compared one variable at a time. The first step of this analysis would be to assess some parameters describing the distribution of the variable of interest. We do this because standard statistical means comparison methods such as a t-test and analysis of variance (ANOVA) assume that the variable is normally distributed and that the variance is approximately equal in each stratum. We use a method such as the Shapiro-Wilk test to test for normality (against the null hypothesis that the distribution is not different from a normal distribution); it may be possible to transform variables to achieve normality (see Sect. 3.4.2). We also apply the Bartlett test or equivalent to test for heteroscedasticity (against the null hypothesis that the variance in each stratum is equal, i.e. homoscedastic). The conventional (parametric) statistical tests include the t-test (commonly implemented as Welch's t, for heteroscedastic variables) for two-level comparisons and the f-test as either standard ANOVA or Welch's f for heteroscedastic variables for comparisons with three or more levels. If the assumptions of normally distributed variables are not met, then non-parametric comparisons such as the Wilcoxon (for two-level comparisons) or Kruskal-Wallis (for multiple-level comparisons) tests can be used. The null hypothesis for all means comparison tests is that means in each stratum are equal, and the result of the tests is the probability that the null hypothesis is true for the population, given the values and variability that we have measured in our sample.

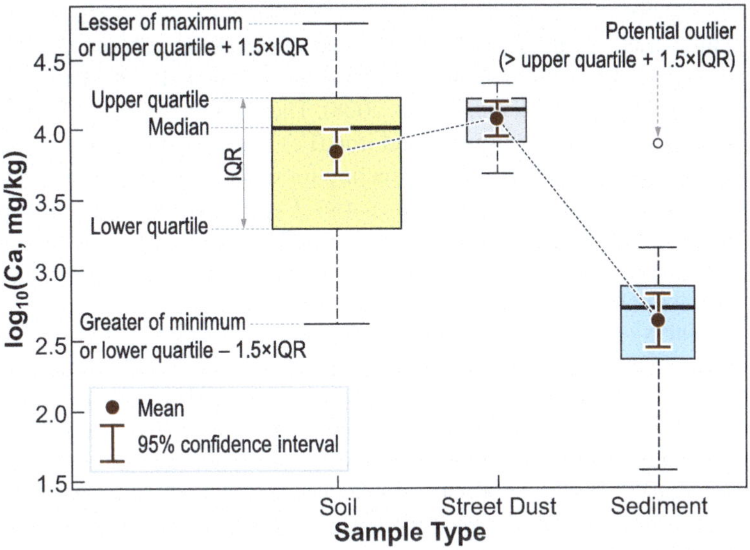

Fig. 3.14 Graphical comparisons of means and median concentrations of calcium in different sample types in an urban parkland, using an annotated and enhanced version of a standard 'Tukey' box plot. IQR = interquartile range (data and graphic by Andrew W. Rate)

3.4.2 Transforming Variables

In order to meet the assumptions of parametric statistical tests (e.g. t-tests, ANOVA/f-test), variables should be normally distributed. This is seldom the case for soil measurements (except sometimes soil pH), which commonly show positively skewed distributions. For continuous positively skewed variables, such as the concentration of a soil constituent, we tend to either use transform variables to logarithms (base 10 is most convenient and interpretable) or use a power function of the variable for transformation ($x_{\text{transformed}} = x^a$, where a is the power term). The power term which transforms a variable to have a distribution 'closest' to normal can be estimated from the Box-Cox algorithm, which is implemented in most statistical software. The power term a can be negative, which reverses the ordering of the variable (i.e. the greatest value will become the least and vice versa). In this case the ordering of the original variable can be preserved by including a factor of -1 in the power transform calculation ($x_{\text{transformed}} = -(x^a)$).

For different types of variables, particular transformations are required. For example, variables which are counts rather than continuous variables should not be transformed; instead alternative statistical models such as generalised linear models assuming a Poisson or negative binomial distribution should be used (O'Hara and Kotze 2010). Compositional variables (including concentrations or proportions of land surface coverage) are technically part of a fixed-sum closed set. For example, with data on percent land use over an urban area, all percentages add up to 100%! If uncorrected, fixed-sum closure can lead to very misleading conclusions, especially when relationships between variables are being investigated, as in correlation analyses or multivariate methods such as principal component analysis. Closed data require specialised transformations to remove closure, such as calculation of centred or additive log ratios (Reimann et al. 2008). The example in Fig. 3.15 shows the relationship between phosphorus (P) and iron (Fe) in soil/sediment materials in an acid sulphate environment. Without correcting for compositional closure, the P vs. Fe plot implies that P increases as Fe increases. Correcting for compositional closure, however, suggests the opposite, with P negatively related to Fe! In this case, if we had used conventional transformations, we might have come to a very wrong conclusion about the sediment properties affecting phosphorus.

The business of comparing means for environmental variables is possibly more complicated than we might have expected (as described above), but, to choose the correct method, the criteria are logical. The flow diagram in Fig. 3.16 shows how we can make the choice using three relatively simple questions: How many groups do we want to compare? Are our variables (transformed if necessary) normally distributed? And, does the variance of our variable depend on which group it is in? Once we have been guided in this way to the correct statistical test, we have two more questions that need to be asked. First, if we have more than two groups, which means are different from each other? We can answer this question rigorously using 'pairwise comparison' tests, as described below. Second, do we have a meaningful difference or just a statistical one? This is deciding whether we have a large or small 'effect size', and we discuss this below as well.

3.4.3 *Pairwise Comparisons*

The means comparison tests above will help us to make a decision about whether the mean value of some variable differs between strata. For sites with exactly two strata (where we would use a t-test or Wilcoxon test), the test tells us if the

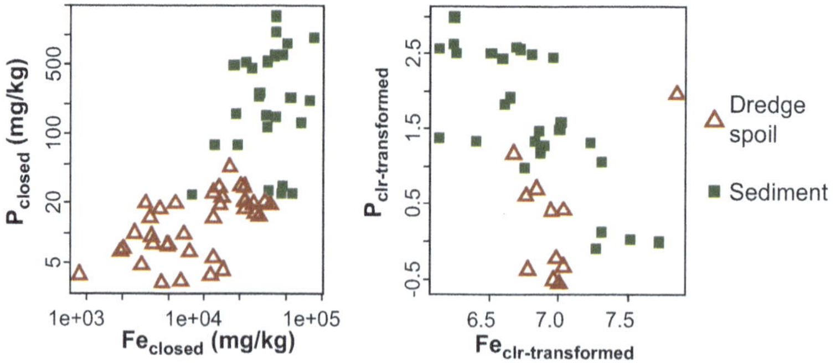

Fig. 3.15 Comparison of relationships between P and Fe for (**a**) compositionally closed concentrations showing a positive relationship and (**b**) concentration variables corrected for compositional closure using centred log ratios showing a negative relationship. Data from Xu et al. (2018); graphic by Andrew W. Rate

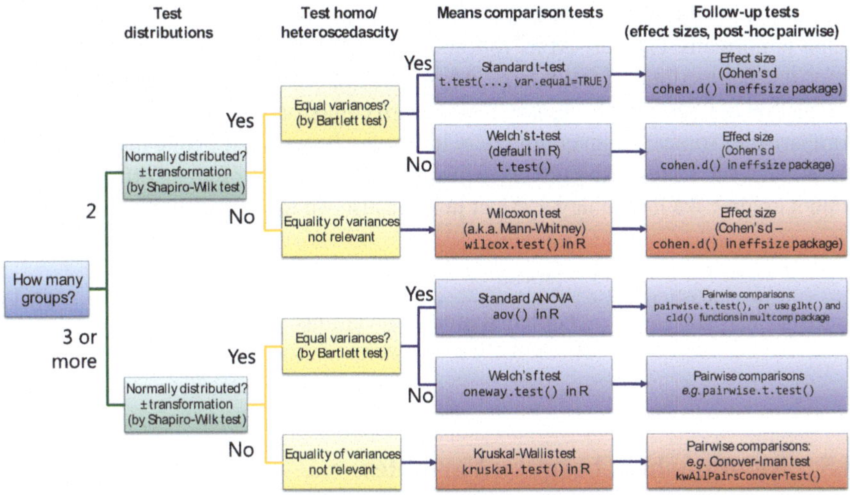

Fig. 3.16 Decision tree for choosing appropriate statistical tests for overall and pairwise comparisons of mean values of measurements between different sampling strata. Functions from the R statistical computing environment (R Core Team 2020) are shown for various tests in monospaced font (graphic by Andrew W. Rate)

difference in means is between all combinations of strata – since there is only one possible combination! In most sites where we study urban soils, we commonly have sites with three or more strata, and we want to compare mean values of our measurements between the strata. Ideally we would like to know which means are different from the others (not just if we can reject the null). So, if the f-test or Kruskal-Wallis test allows rejection of the statistical null hypothesis (i.e. equal means), we usually follow up with a pairwise test. For well-conditioned data where the assumptions of ANOVA are met, we can use the Tukey set of statistical analyses (least significant difference and rigorous pairwise p-values). For heteroscedastic variables this is not strictly allowed, but we can still apply something like a pairwise t-test with adjustment of p-values for multiple comparisons. Finally, with a non-parametric (e.g. Kruskal-Wallis) test of three or more means, we can use pairwise Wilcoxon tests with adjustment of p-values for multiple comparisons or a specialised pairwise comparison test such as Conover's test.

3.4.4 Effect Sizes for Comparing Means

We cannot rely on just rejection of the null hypothesis (of equal means), since it is mathematically possible for a statistically significant difference to exist when the practical difference is meaningless. In some cases an effect size statistic, of which the most common is Cohen's d (Eq. 3.1), can help us assess the magnitude of the difference between central values (means or medians). Cohen's d is a standardised measure of the difference between means for exactly two groups (e.g. strata), and its value is normally categorised as follows: d < 0.2 negligible; 0.2 < d < 0.5 small; 0.5 < d < 0.8 medium; d > 0.8 large. Most contemporary statistical software allows calculation of Cohen's d for binary (two group) comparisons; for multiple (pairwise) comparisons, some custom coding (e.g. in R) may be required. (Note that the size of the p-value for a t-test, ANOVA, or equivalent does not represent an effect size! We cannot assume we have a larger effect just because we have a smaller p-value.)

$$\text{Cohen's } d = \frac{\left(\text{mean}_{\text{group}\,1} - \text{mean}_{\text{group}\,2}\right)}{\text{Pooled standard deviations}} \quad (3.1)$$

3.5 Relationships between Variables

In the spatial context, we can use correlation or regression statistics to assess relationships between a soil variable and distance (e.g. distance from a potential or suspected source of contamination). We also investigate relationships between variables for other reasons, such as finding which observations do not follow the expected relationship, and we will look at an example of this in Chap. 6. For now, we will go through the basics of correct application of correlation and regression analyses.

3.5.1 *Correlation Analysis*

The most commonly used measure of correlation between two variables (bivariate) is Pearson's correlation coefficient, r, which can vary between −1 and 1, with an r value of zero meaning no correlation. The assumptions behind calculation of Pearson's r require that each variable is normally distributed; sometimes this can be achieved with an appropriate transformation (e.g. taking logarithms or a power function, bearing in mind the discussion in Sect. 3.4.2). Variables which are unable to be transformed to a normally distributed variable are unsuitable for Pearson's correlation analysis, but we can use a non-parametric method, Spearman's correlation, in such cases. Spearman's correlation is based on comparison of ranks within each ordered variable and is therefore independent of transformation. The p-value for correlation tests is for the null hypothesis of no relationship between the pair of variables, against the existence of a true correlation in the population from which the sample is taken.

Very often, it is useful for exploratory data analysis to generate a correlation matrix, which calculates a correlation coefficient (e.g. Pearson's or Spearman's) for all possible pairwise relationships between the variables selected. These are subject to the same requirements in terms of the distribution of variables as bivariate correlations, with the additional precaution that p-values should be adjusted upwards to account for the increased likelihood of type 1 errors (false positives) when multiple comparisons are made. Most statistical software will calculate these corrected p-values, for example, using Holm's method.

With any correlation analyses, it is essential to check the relationships graphically. It is easy to misinterpret r values if the data behave unexpectedly. For example, outliers may still exist in transformed variables, which have a large influence on the value of Pearson's r. The variables may show grouping or bimodality, so that the true relationships are masked by considering the data as whole or a strong relationship may exist which is non-linear as assumed in the Pearson correlation. Inspection of (appropriately transformed) bivariate plots can identify these types of issues, and most statistical software will allow plotting of scatterplot matrices to streamline this task.

3.5.2 Regression Analysis

If a relationship between variables exists, it should be possible to estimate, or predict, one variable from another. This prediction is the goal of regression models; in their simplest form of bivariate linear regression, they are conceptually similar to Pearson's correlation, but the focus is on the ability of the regression model (commonly a mathematical equation) to predict one variable, the 'dependent variable', from one or more 'predictor variables'. A thorough discussion of regression models would itself take a whole book, so we will not do that here! Instead we will look at a sequence of steps we can take to generate and assess different types of linear regression models, foreshadowing the example in Chap. 6, Fig. 6.10; the procedures in all of these steps should be available in any up-to-date statistical software. The aim of our regression model is to predict arsenic (As) concentration in soil, from the other soil measurements we have made. The dataset includes soil EC and pH, plus concentrations of numerous major and trace elements.

The general form of the multiple or simple linear regression models we will discuss is:

$$y = a + \sum_{i=1}^{n} (b_i x_i) + e \tag{3.2}$$

where y is the dependent variable to be predicted, n is the number of predictor variables (which can equal 1), a is the constant 'intercept' term, x_i are the predictor variables, b_i are the coefficients for each predictor, and e is the error term or 'residual'.

Initial Assumptions and Transformations For a valid linear regression model, the residuals need to be normally distributed. In practice we can increase the likelihood that this will be the case by transforming our variables to remove skewness, commonly with a $\log_{10}$ transformation. Since the goal is prediction, rather than analysis of the relationship itself, we can argue that issues like compositional closure can be ignored. In our example, we $\log_{10}$ transform all variables except soil pH which is used untransformed.

Choosing Predictors The next decision that needs to be made is what the predictor variable(s) should be. Realistically, we would normally try to predict a variable that is difficult, unreliable, or expensive to measure, since we would usually rather have an actual measurement than an estimate from prediction. The predictor variables, then, would logically be those which are more easily, reliably, and/or inexpensively measured. **More importantly**, we should try to choose predictors that make sense in the real world. For example, the concentrations of trace elements in soils are often closely related to one another due to similar geochemistry or common sources. In reality, though, it's unlikely that one trace element would have an effect on another in soils since the concentrations of both are too low. In soils, then, we tend to use 'bulk soil' properties, such as pH, clay content, organic carbon and other major element content, EC, redox potential, and so on, as predictors, since they fulfil both the 'easily measured' and 'realistic effect' criteria. For our example, we want to predict arsenic (As) concentrations, and our initial list of predictors is pH, EC, Al, Ca, Fe, K, Mn, Na, P, and S.

Collinearity of Predictors The predictors we select should not be linearly related to one another (collinear). The criteria we use to assess this are the Pearson correlations (which should be ≤ 0.8 between any pair of predictors) and variance inflation factors (VIF), which estimate how much greater the variance of a regression coefficient (the b_i values in Eq. 3.2) is, due to collinearity. There are various rules of thumb for selecting predictors on the basis of VIF: a value above 10 suggests that a predictor should be removed; $4 < \text{VIF} < 10$ should be noted. In our example, the following pairs of predictor variables have Pearson's r greater than 0.8: Al-Fe, Fe-Mn, and Mn-P. The variance inflation factors are listed in Table 3.1. We will choose to remove Al and Mn from the 'maximal' regression model, but different variables could have been removed.

Refinement of Predictors Not all of the possible predictors that we select will have a significant influence on the value of our dependent variable. The output of statistical software (e.g. Table 3.2) usually has a p-value from a test of significance for each predictor (using the null hypothesis that the predictor has no effect on the dependent variable), as well as an analogous null hypothesis significance test for the model as a whole. Inspection of this output may imply that that some predictors have no

Table 3.1 Variance inflation factors (VIF) for a multiple regression model predicting As_{log} from pH, EC_{log}, Al_{log}, Ca_{log}, Fe_{log}, K_{log}, Mn_{log}, Na_{log}, P_{log}, and S_{log} (subscript $_{log}$ denotes $\log_{10}$ transformation)

Predictor	pH	EC_{log}	Al_{log}	Ca_{log}	Fe_{log}	K_{log}	Mn_{log}	Na_{log}	P_{log}	S_{log}
VIF	3.163	1.911	8.123	5.571	7.982	3.298	9.575	7.997	7.768	5.774

effect. To remove non-significant predictors, we use a stepwise regression algorithm, which systematically adds and removes predictors from a set of models, using an 'information criterion' to select the best subset of predictors which all contribute 'information' or predictive ability to the model. The stepwise algorithm can be configured to add predictors from a list to a basic model (forward selection) or to remove predictors from a maximal model or both. Ideally, different implementations of stepwise procedures, using the same data, should arrive at the same final answer.

We can see in Table 3.2 that the null hypothesis of no prediction ability is rejected at $p \leq 0.05$ for all predictors in the final model except S_{log} (this relates to the different selection criteria for predictors in the stepwise procedure). The model is good at predicting As concentration; the multiple R^2 (r-squared) value is 0.8155, so nearly 82% of the variance in $\log_{10}$As is explained by the four predictors. We can also reject (p-value = 2.6×10^{-16}, so $p \leq 0.05$) the null hypothesis of no prediction ability for the overall model. The VIF values are all close to 1, meaning negligible collinearity.

Model Checking Many of the assumptions for regression relate to the residuals, and we use a number of diagnostic tests and/or plots (Fig. 3.17) to assess these assumptions. First, the residuals have a median value close to zero (−0.0058, Table 3.2), and the mean residual value is 1.8×10^{-18}. By applying a Shapiro-Wilk test to the residuals from the model, we find that the null hypothesis (that the distribution is not different from a normal distribution) cannot be rejected, satisfying the assumption of normally distributed residuals.

Table 3.2 Summary of final regression model predicting As_{log} from initial predictors pH, EC_{log}, Al_{log}, Ca_{log}, Fe_{log}, K_{log}, Mn_{log}, Na_{log}, P_{log}, and S_{log} (subscript $_{log}$ denotes $\log_{10}$ transformation). Values explained in the text are in shaded cells with ***bold italic*** text. The same final set of predictors was obtained by either forward or backward stepwise selection of predictors in R (R Core Team 2020)

Residuals:	Min	1Q	Median	3Q	Max	
	−0.300155	−0.082543	***−0.00583***	0.049328	0.296157	
Coefficients:		Estimate	Std. error	t value	Pr(>|t|)	
	(intercept)	−1.08038	0.31638	−3.415	***0.001343***	**
	Fe_{log}	0.57608	0.05875	9.805	***7.58E-13***	***
	pH	−0.06438	0.01821	−3.535	***0.000944***	***
	EC_{log}	−0.16794	0.07197	−2.333	***0.024051***	*
	S_{log}	0.15322	0.07629	2.008	***0.050487***	.
Signif. Codes:	*** ≤ 0.001	** ≤ 0.01	* ≤ 0.05	. ≤ 0.1		
Statistics:	Residual standard error		0.1245 on 46 degrees of freedom			
	Multiple R-squared		***0.8155***	Adjusted R-squared		0.7994
	F-statistic	50.82 on 4 & 46 DF		p-value		***2.61E-16***
Predictor:	**Fe_{log}**	**pH**	**EC_{log}**	**S_{log}**		
VIF:	***1.374***	***1.192***	***1.028***	***1.562***		

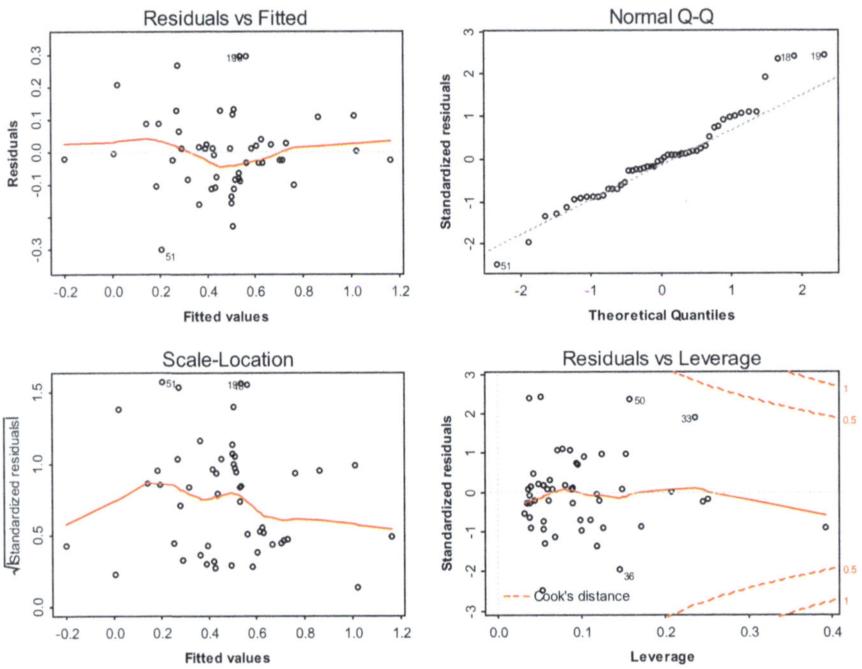

Fig. 3.17 Diagnostic plots for the final regression model predicting As_{log} from predictors pH, EC_{log}, Fe_{log}, and S_{log} (subscript log denotes $\log_{10}$ transformation)

The standard set of diagnostic plots (Fig. 3.17) allows us to assess, visually, some further regression assumptions. The residuals vs. fitted plot checks that the mean residual is close to zero and that there is no systematic trend in the residuals (this can be assessed separately by calculating residual autocorrelation; the autocorrelation coefficients should be close to zero). The normal Q-Q plot is a visual assessment of whether the residuals are normally distributed; the dotted straight line represents a perfect normal distribution with the same mean and standard deviation as the residuals, and the points lie approximately along this line, confirming the Shapiro-Wilk test result above. The scale-location plot assesses whether the residuals show homoscedasticity (i.e. the size of the residuals should be independent of the value of the dependent variable, measured or predicted). In our case there seems to be a 'bulge' of greater residuals in the middle of the plot, suggesting that this assumption may not be fulfilled for our model (again, we can test for this in more detail separately, e.g. using the Breusch-Pagan test (Hothorn et al. 2020)). Finally, the residuals vs. leverage plot is one way of testing if any individual observation has an unexpectedly large influence on the model parameters. Cook's distance is a measure of the change in regression parameters when a point is removed; ideally its value should be zero. There are a number of rules of thumb defining excessively large Cook's distance values, e.g. 4/n, where n is the number of observations (points).

The final regression model can be used in different ways. We can never use correlation or regression to make conclusions about whether one measurement causes another; …*correlation is not causation*. We could certainly use the soil pH, EC, and Fe and S contents, however, to predict As concentration with some accuracy. Actually, though, total As concentration is not so difficult to measure! We may choose to include regression models as part of more complex environmental simulation models where many parameters are required and we do not have access to data for all possible locations where prediction is required. One of the more powerful ways we can use regression models in urban environments is to make use of the deviations from the model – with well-chosen predictors, these can provide a good indication of truly unusual samples, and we look at an example of doing this in Chap. 6.

Of course, multiple regression is not the only variation on simple linear regression that we can make use of when studying urban soils. If we have different sampling strata (see Sect. 3.2.2 above), we can make use of our stratified sampling design in regression. We would not necessarily expect the same linear relationships between variables in different strata (which might, for instance, include both industrial land and undisturbed nature reserves). In this case we can use grouped linear regression (see Fig. 3.18 and Table 3.3), which effectively includes a separate intercept and coefficient(s) for each stratum within our data.

The general form of the grouped or simple linear regression models is similar to that for multiple regression:

$$y = \sum_{i=1}^{n} (a_i + b_i x) + e \tag{3.3}$$

Fig. 3.18 Regressions predicting $\log_{10}$As from $\log_{10}$S, showing grouped (separate symbols, solid lines) and ungrouped (dashed line with shaded 95% confidence interval) regression models

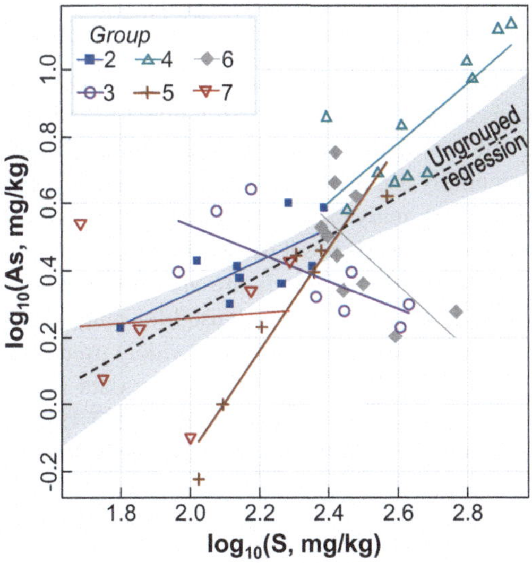

Table 3.3 Summary and interpretation of ungrouped and grouped regression statistics

Ungrouped regression					
$\log As = -0.905 + 0.587 \cdot \log S$					
$R^2 = 0.371$, F statistic = 28.86 on 1 and 49 DF, p-value: 2.13×10^{-6}					
Grouped regression					
$\log As = -0.621 + 0.478 \cdot \log S$ (group 2)					
$\log As = 1.38 - 0.419 \cdot \log S$ (group 3)					
$\log As = -1.54 + 0.894 \cdot \log S$ (group 4)					
$\log As = 2.85 + 1.53 \cdot \log S$ (group 5)					
$\log As = 2.85 - 0.959 \cdot \log S$ (group 6)					
$\log As = 0.11 + 0.0742 \cdot \log S$ (group 7)					
$R^2 = 0.802$, F statistic = 14.3 on 11 and 39 DF, p-value = 1.72×10^{-10}					
Comparison of models					
Res.Df RSS	Df	sum of Sq F		Pr(>F)	
Ungrouped	49	2.43273			
Grouped 39	0.76705	10	1.6657	8.469	3.928e-07

(P-values is ≤0.05 so the null hypothesis, that the more complex model makes no improvement in prediction, can be rejected.)

where the terminology is as for Eq. 3.3, except that now we have a single predictor x with different intercepts (a_i) and slopes (b_i) for each group of observations.

We should always check if the more complex model is actually better at prediction or whether it is simply 'over parameterised'. We can compare linear regression models by using analysis of variance (Table 3.3) if they are nested, that is, a simpler model is a subset of a more complex model.

3.5.3 Multivariate Analysis

It is quite common to measure many variables in studies of urban soils. In the sections above, we have discussed how to analyse a dataset to interpret one variable at a time using univariate methods (although use of multiple linear regression does potentially use many variables to explain one other variable). Using various types of ordination analysis, we can use the information contained in multiple variables to create a reduced subset of variables containing nearly the same amount of information. Ordination methods are also referred to, for this reason, as 'data reduction' methods and are commonly used for multivariate analysis.

One of the earliest and most widely used ordination methods for exploration and dimension-reduction of multivariate data is principal components analysis (PCA; see the explanation in Box 3.2 and Fig. 3.19). Imagine a dataset with many samples (rows) and n continuous numeric variables (columns) which contain quantitative

Box 3.2 Principal Components Analysis

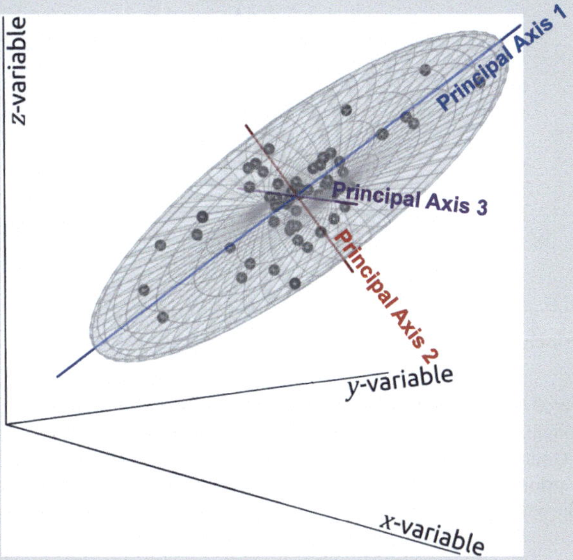

Fig. 3.19 Visualisation of principal components analysis for three variables/dimensions x, y, and z: we can conceptualise an ellipsoid encapsulating the 'cloud' of points (i.e. samples). The longest dimension (Principal Axis 1) of the ellipsoid, which accounts for the greatest proportion of multiple variance, is in the direction of its major axis and is a function of the variables x, y, and z. Principal Axis 2 must be orthogonal to Principal Axis 1 and is a different function of the variables x, y, and z which accounts for the next highest-possible proportion of multiple variance. Principal Axis 3 must be orthogonal to both Principal Axes 1 and 2; is a unique function of x, y, and z; and accounts for the remainder of multiple variance. For n variables/dimensions (n > 3), the analogy is an n-dimensional hyper-ellipsoid which has n orthogonal axes, which is very difficult to visualise

information about each sample such as concentrations, heights, velocities, etc. For these n variables/dimensions, the principal component calculation generates n new variables, or principal components, which are each a function of the set of all the original variables (so each principal component is defined by a weighting or coefficient for each of the original variables). We may choose to omit some variables from the analysis if they contain too many missing observations or if there is another valid reason to doubt their integrity. Since each principal component is selected to account for successively smaller proportions of the multiple variance, it is usually the first few principal components which explain most of the variance and therefore contain the most useful information. We conventionally visualise this in a 'scree plot' (Fig. 3.20a), a kind of bar graph showing the decrease in variance accounted for by each component (the 'eigenvalue').

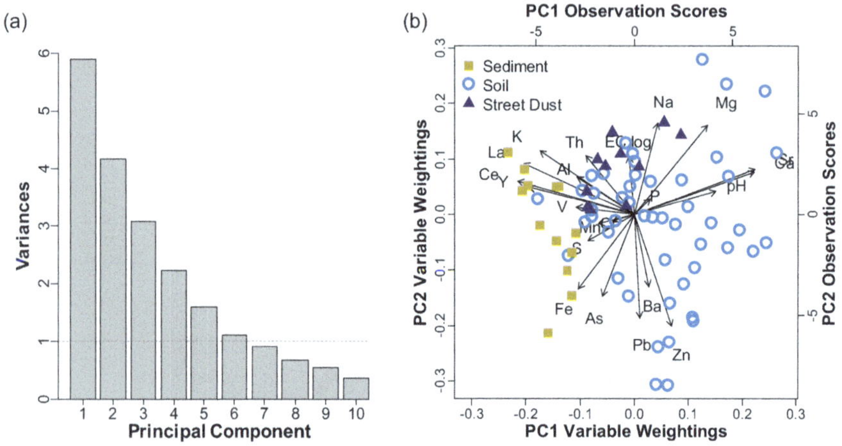

Fig. 3.20 (**a**) Scree plot of component variances and (**b**) biplot of the first two components, for principal components analysis on variables related to chemical properties of samples from an urban parkland. Concentration variables were transformed to centred log ratios to remove compositional closure prior to calculation of principal components on scaled, zero-centred variables (graphic by Andrew W. Rate)

As well as the component variances, the useful results of principal components analysis include the variable weightings or 'rotations' for each principal component. In addition, every individual observation (sample) is a multivariate point, the observation scores for all samples in each principal component based on the values of the variables used in PCA for that sample. It is conventional to plot both of these two types of output in a principal component 'biplot', as shown in Fig. 3.20b. Before discussing the biplot, we should note that the sign (positive or negative) of variable weightings and observation scores (i.e. the direction of the axes) is arbitrary and should not affect our interpretations.

The principal component biplot is useful because the variable weightings group together for variables (measurements) that are related to one another. For example, in the biplot in Fig. 3.20b, the variables are mainly concentrations of elements (which have been corrected for compositional closure before PCA using a log ratio transformation). These variables are shown as vectors (arrows) in the biplot of principal components PC1 and PC2, and elements which are geochemically related have vectors of similar length and/or direction. For example, the elements La, Ce, and Y are all geochemically similar rare-earth elements and plot closely together on the biplot, and the same is true for Ca and Sr which commonly co-occur in carbonate minerals. The other main information we obtain from principal components biplots is from the observation scores. These will plot at locations similar to their dominant variables: for example, in Fig. 3.20b, the sediment samples all plot towards the left of the biplot in the same direction as the K, La, Ce, Y, S, and Fe variable weighting vectors. This suggests that they are characterised by greater values of these variables (e.g. wetland sediments may contain higher concentrations of

Fe, S, and rare-earth elements due to formation of sulphides and Fe and S cycling (Morgan et al. 2012)).

We have used an example based on soil chemical data, but many other types of numerical data can be used in principal components analysis. These types of datasets could include soil physical data, composition of vegetation or microbial communities, and so on. For example, a dataset with variables which measure different plant species composition might show, in a PCA biplot, grouping of wetland plant species in riparian sampling strata, weedy species in disturbed urban land, native species in reserves, and so on. (Note that variables such as percent species compositions would comprise a fixed-sum closed set and would require a transformation to remove closure before rigorous multivariate data analysis such as PCA!) The provision of information of this type makes ordination methods such as principal components analysis powerful tools for exploratory data analysis. Different algorithms for ordination of multivariate data are based on different criteria than the maximisation of multiple variance used in PCA, for example, similarity or dissimilarity between samples. We will discuss different multivariate methods for analysis of data related to soil microbiology in Chap. 8.

3.6 Further Reading

Oliver MA, Webster R (2014) A tutorial guide to geostatistics: computing and modelling variograms and kriging. Catena 113:56–69. https://doi.org/10.1016/j.catena.2013.09.006

Reimann C, Filzmoser P, Garrett RG, Dutter R (2008) Statistical data analysis explained: applied environmental statistics with R. John Wiley & Sons, Chichester, England, 343 pp

3.7 Summary

- Cities affect soil variability on a subcontinental or regional scale, reflecting the concentration of human populations and resources in urban environments. Cities themselves contain variable soils on the scale of whole metropolitan areas, localities, sites, and individual soil profiles, and these are related to the age of human habitation and the types of activities conducted. Variation in soil properties with depth can inform us about site history or the extent of anthropogenic additions to soil.
- Sampling of soil needs to match the objective, that is, the type of information required. To capture regional- or metropolitan-scale soil variability, large systematic sampling exercises including hundreds or even thousands of samples are conducted. Studies of smaller spatial scales requires tens of samples at higher density, and depth variability is assessed with 2–30 vertical increments. Specific

approaches are required to detect discrete hotspots or sample pre-existing spatial strata.

- Analysis of spatial data begins with visual analysis in the form of maps with layer(s) of soil property data or scatterplots vs. distance along transects. More rigorous spatial data analysis techniques include variations of spatial autocorrelations or construction of variograms which allow spatial prediction using kriging.
- Other trends in spatial data, such as differences in soil parameters between strata, can be assessed with rigorously applied standard statistical techniques, both parametric and non-parametric, for comparison of central tendencies, assessing relationships, regression models, or multivariate ordination. For credible interpretation of urban soil data, care must be taken to ensure that the assumptions of each statistical method are met.

3.8 Questions

3.8.1 Checking your Understanding

1. What are the spatial scales of soil variability that we have considered? Are there any other scales that might be important in urban soil environments (your answer might differ depending on whether your focus is on ecosystem services, soil management, or soil research)?
2. How does the concept of an anthroposequence relate to the analysis (e.g. qualitatively or using quantitative measures like autocorrelation) of soil variability?
3. How many samples would you need to collect to have a 95% chance of detection of a circular contamination hotspot, 5 m in radius, over an area of 4 hectares? How far apart would the samples need to be?
4. List the advantages and disadvantages of the following four sampling designs: grid, stratified, random in grid, and completely randomised.
5. What is spatial autocorrelation measured with global Moran's I? What does the value of local Moran's I tell us about the relationships between samples which are close to one another?
6. Summarise the situations (i.e. properties of the variables and factors in a dataset) in which you would use the following mean/median comparison tests: Student's t, Wilcoxon, analysis of variance, and Kruskal-Wallis. When would you apply a pairwise comparison test?

3.8.2 *Thinking about the Issues*

7. Would it be reasonable to expect steep (sudden) gradients in soil properties between adjacent sampling strata or would soil properties show a more gradual change? Explain your answer, and try to think of an example of where the opposite to your initial answer might be true.
8. Figure 3.4 presents spatial data mapped as a continuous surface, whereas Fig. 3.6 presents similar data as point symbols containing the soil property information but without interpolation. Comment on the pros and cons of each approach.
9. How many reasons can you think of for using multivariate data analysis methods instead of multiple applications of conventional (uni- or bivariate) methods?

3.8.3 *Using Your Creative Brain*

10. Imagine that you are an expert witness, and a land developer has used a mean comparison to show that the average concentration (based on 50 soil samples) of polycyclic aromatic hydrocarbons (PAHs) on the land for the proposed development is not significantly different from 10 identical measurements on 'background' soil. They further argue that this means that there is no need for concern. Tell the hearing what is wrong with the land developer's reasoning, and suggest what a more appropriate analysis of the data would be.

References

Anselin L (1995) Local indicators of spatial association—LISA. Geogr Anal 27:93–115. https://doi.org/10.1111/j.1538-4632.1995.tb00338.x

Bauer ME, Loffelholz BC, Wilson B (2008) Estimating and mapping impervious surface area by regression analysis of Landsat imagery. In: Weng Q (ed) Remote sensing of impervious surfaces. CRC Press / Taylor & Francis, Boca Raton, pp 3–20

Bohling G (2005) Introduction to Geostatistics and Variogram analysis, Kansas Geological Survey

Bu QW, Zhang ZH, Lu S, He FP (2009) Vertical distribution and environmental significance of PAHs in soil profiles in Beijing, China. Environ Geochem Health 31:119–131. https://doi.org/10.1007/s10653-008-9171-z

Bugdalski L, Lemke LD, McElmurry SP (2014) Spatial variation of soil lead in an urban community garden: implications for risk-based sampling. Risk Anal 34:17–27. https://doi.org/10.1111/risa.12053

Carr R, Zhang C, Moles N, Harder M (2008) Identification and mapping of heavy metal pollution in soils of a sports ground in Galway City, Ireland, using a portable XRF analyser and GIS. Environ Geochem Health 30:45–52. https://doi.org/10.1007/s10653-007-9106-0

Cicchella D, Giaccio L, Dinelli E, Albanese S, Lima A, Zuzolo D et al (2015) GEMAS: spatial distribution of chemical elements in agricultural and grazing land soil of Italy. J Geochem Explor 154:129–142. https://doi.org/10.1016/j.gexplo.2014.11.009

Coleman DC, Crossley DA Jr, Callaham MA (2017) Fundamentals of soil ecology. Elsevier Science & Technology, Saint Louis

Corrò E, Mozzi P (2017) Water matters. Geoarchaeology of the city of Adria and palaeohydrographic variations (Po Delta, Northern Italy). J Archaeol Sci Rep 15:482–491. https://doi.org/10.1016/j.jasrep.2016.08.001

Department of Environmental Protection (2001) Development of sampling and analysis programs. Contaminated Sites Management Series. Government of Western Australia, Perth, Western Australia, 69 pp.

Diawara DM, Litt JS, Unis D, Alfonso N, Martinez LA, Crock JG et al (2006) Arsenic, cadmium, lead, and mercury in surface soils, Pueblo, Colorado: implications for population health risk. Environ Geochem Health 28:297–315. https://doi.org/10.1007/s10653-005-9000-6

Fellows I (2019) OpenStreetMap: Access to open street map raster images, using the JMapViewer library by Jan Peter Stotz.. (R Package Version 0.3.4) http://CRAN.R-project.org/package=OpenStreetMap http://blog.fellstat.com/?cat=15. Accessed 20210616

Fletcher T, Duncan H, Poelsma P, Lloyd S (2004) Stormwater flow and quality, and the effectiveness of non-proprietary Stormwater treatment measures : a review and gap analysis. Technical Report 04/8, Melbourne, Australia

Gong Q, Deng J, Xiang Y, Wang Q, Yang L (2008) Calculating pollution indices by heavy metals in ecological geochemistry assessment and a case study in parks of Beijing. J China Univ Geosci 19:230–241. https://doi.org/10.1016/S1002-0705(08)60042-4

Gong M, Wu L, Bi X-Y, Ren L-M, Wang L, Ma Z-D, Li Z-G (2010) Assessing heavy-metal contamination and sources by GIS-based approach and multivariate analysis of urban–rural topsoils in Wuhan, Central China. Environ Geochem Health 32:59–72. https://doi.org/10.1007/s10653-009-9265-2

Grimm NB, Grove JG, Pickett STA, Redman CL (2000) Integrated approaches to long-term studies of urban ecological SystemsUrban ecological systems present multiple challenges to ecologists—pervasive human impact and extreme heterogeneity of cities, and the need to integrate social and ecological approaches, concepts, and theory. Bioscience 50:571–584. https://doi.org/10.1641/0006-3568(2000)050[0571:IATLTO]2.0.CO;2

Hothorn T, Zeileis A, Farebrother RW, Cummins C (2020) lmtest: Testing Linear Regression Models. R package version 0.9-38 https://CRAN.R-project.org/package=lmtest. Accessed 2021-06-03

Jim CY (1998) Soil characteristics and management in an urban park in Hong Kong. Environ Manag 22:683–695. https://doi.org/10.1007/s002679900139

Johnson CC, Ander EL (2008) Urban geochemical mapping studies: how and why we do them. Environ Geochem Health 30:511. https://doi.org/10.1007/s10653-008-9189-2

Kalogirou S (2019) Spatial autocorrelation, Vignette for R package 'lctools', https://cran.r-project.org/web/packages/lctools/vignettes/SpatialAutocorrelation.pdf. Accessed 20191204

Li X, Lee SL, Wong SC, Shi W, Thornton I (2004) The study of metal contamination in urban soils of Hong Kong using a GIS-based approach. Environ Pollut 129:113–124. https://doi.org/10.1016/j.envpol.2003.09.030

Lv J, Liu Y, Zhang Z, Dai J, Dai B, Zhu Y (2015) Identifying the origins and spatial distributions of heavy metals in soils of Ju country (Eastern China) using multivariate and geostatistical approach. J Soils Sediments 15:163–178. https://doi.org/10.1007/s11368-014-0937-x

Mann A, Reimann C, de Caritat P, Turner N, Birke M, Team GP (2015) Mobile Metal Ion® analysis of European agricultural soils: bioavailability, weathering, geogenic patterns and anthropogenic anomalies. Geochem Exploration Environ Analysis 15:99–112. https://doi.org/10.1144/geochem2014-279

Morgan B, Rate AW, Burton ED (2012) Trace element reactivity in FeS-rich eutrophic estuarine sediments: influence of formation environment and acid sulfate soil drainage. Sci Total Environ 438:463–476. https://doi.org/10.1016/j.scitotenv.2012.08.088

Nero BF, Anning AK (2018) Variations in soil characteristics among urban green spaces in Kumasi, Ghana. Environ Earth Sci 77:317. https://doi.org/10.1007/s12665-018-7441-3

Nikolaeva O, Rozanova M, Karpukhin M (2017) Distribution of traffic-related contaminants in urban topsoils across a highway in Moscow. J Soils Sediments 17:1045–1053. https://doi.org/10.1007/s11368-016-1587-y

Nriagu JO (1988) A silent epidemic of environmental metal poisoning? Environ Pollut 50:139–161. https://doi.org/10.1016/0269-7491(88)90189-3

O'Hara RB, Kotze DJ (2010) Do not log-transform count data. Methods Ecol Evol 1:118–122. https://doi.org/10.1111/j.2041-210X.2010.00021.x

Oliver MA, Webster R (2014) A tutorial guide to geostatistics: computing and modelling variograms and kriging. Catena 113:56–69. https://doi.org/10.1016/j.catena.2013.09.006

Pickett STA, Cadenasso ML, Grove JM, Nilon CH, Pouyat RV, Zipperer WC, Costanza R (2001) Urban ecological systems: linking terrestrial ecological, physical, and socioeconomic components of metropolitan areas. Annu Rev Ecol Syst 32:127–157. https://doi.org/10.1146/annurev.ecolsys.32.081501.114012

Pouyat RV, McDonnell MJ (1991) Heavy metal accumulations in forest soils along an urban- rural gradient in Southeastern New York, USA. Water Air Soil Pollut 57-58:797–807. https://doi.org/10.1007/BF00282943

Pouyat RV, Yesilonis ID, Russell-Anelli J, Neerchal NK (2007) Soil chemical and physical properties that differentiate urban land-use and cover types. Soil Sci Soc Am J 71:1010–1019. https://doi.org/10.2136/sssaj2006.0164

Pouyat RV, Yesilonis ID, Szlavecz K, Csuzdi C, Hornung E, Korsós Z et al (2008) Response of forest soil properties to urbanization gradients in three metropolitan areas. Landsc Ecol 23:1187–1203. https://doi.org/10.1007/s10980-008-9288-6

Powell RL, Roberts DA, Dennison PE, Hess LL (2007) Sub-pixel mapping of urban land cover using multiple endmember spectral mixture analysis: Manaus, Brazil. Remote Sens Environ 106:253–267. https://doi.org/10.1016/j.rse.2006.09.005

QGIS.org (2020) QGIS geographic information system. Open source geospatial foundation project https://qgis.org. Accessed 20210616

R Core Team (2020) R: A language and environment for statistical computing (Version 4.0.3 'Bunny-Wunnies Freak Out'). R Foundation for Statistical Computing, Vienna, Austria, https://www.R-project.org http://www.R-project.org. Accessed 20210616

Rate AW (2018) Multielement geochemistry identifies the spatial pattern of soil and sediment contamination in an urban parkland, Western Australia. Sci Total Environ 627:1106–1120. https://doi.org/10.1016/j.scitotenv.2018.01.332

Reimann C, Filzmoser P, Garrett RG, Dutter R (2008) Statistical data analysis explained: applied environmental statistics with R. Wiley, Chichester, 343 pp

Ribeiro Jr. PJ, Diggle PJ, Schlather M, Bivand R, Ripley B (2020) geoR: Analysis of Geostatistical Data. {R package version 1.8.1}, http://CRAN.R-project.org/package=geoR. Accessed 20201215

Schaetzl RJ, Anderson S (2005) Soils: genesis and geomorphology. Cambridge University Press, New York

Smith DB (2004) Appendix 1: soil sampling protocols for geochemical landscapes pilot studies. U.S. Geological Survey, Denver

Taylor MP, Mackay AK, Hudson-Edwards KA, Holz E (2010) Soil Cd, Cu, Pb and Zn contaminants around Mount Isa city, Queensland, Australia: potential sources and risks to human health. Appl Geochem 25:841–855

Tobler W (2004) On the first law of geography: a reply. Ann Assoc Am Geogr 94:304–310

Tran BC, Teil M-J, Blanchard M, Alliot F, Chevreuil M (2015) Fate of phthalates and BPA in agricultural and non-agricultural soils of the Paris area (France). Environ Sci Pollut Res 22:11118–11126. https://doi.org/10.1007/s11356-015-4178-3

Turer D, Maynard JB, Sansalone JJ (2001) Heavy metal contamination in soils of urban highways: comparison between runoff and soil concentrations at Cincinnati, Ohio. Water Air Soil Pollut 132:293–314. https://doi.org/10.1023/a:1013290130089

USEPA, 2002. Guidance on choosing a sampling Design for Environmental Data Collection for use in developing a quality assurance project plan. EPA QA/G-5S, United States Environmental Protection Agency, Washington, DC
Webster R, Oliver MA (1993) How large a sample is needed to estimate the regional variogram adequately? Geo 1:155–166
White RE (2006) Principles and practice of soil science: the soil as a natural resource. Blackwell Science Ltd, Malden
Wu C (2004) Normalized spectral mixture analysis for monitoring urban composition using ETM+ imagery. Remote Sens Environ 93:480–492. https://doi.org/10.1016/j.rse.2004.08.003
Wu C, Murray AT (2003) Estimating impervious surface distribution by spectral mixture analysis. Remote Sens Environ 84:493–505. https://doi.org/10.1016/s0034-4257(02)00136-0
Xu N, Rate AW, Morgan B (2018) From source to sink: rare-earth elements trace the legacy of sulfuric dredge spoils on estuarine sediments. Sci Total Environ 637-638:1537–1549. https://doi.org/10.1016/j.scitotenv.2018.04.398
Zhang MK (2004) Phosphorus accumulation in soils along an urban-rural land use gradient in Hangzhou, Southeast China. Commun Soil Sci Plant Anal 35:819–833. https://doi.org/10.1081/CSS-120030360
Zhang C, Luo L, Xu W, Ledwith V (2008) Use of local Moran's I and GIS to identify pollution hotspots of Pb in urban soils of Galway, Ireland. Sci Total Environ 398:212–221. https://doi.org/10.1016/j.scitotenv.2008.03.011
Zhao D, Li F, Wang R, Yang Q, Ni H (2012) Effect of soil sealing on the microbial biomass, N transformation and related enzyme activities at various depths of soils in urban area of Beijing, China. J Soils Sediments 12:519–530. https://doi.org/10.1007/s11368-012-0472-6
Zhu W, Egitto BA, Yesilonis ID, Pouyat R (2017) Chapter 5: soil carbon and nitrogen cycling and ecosystem service in cities. In: Lal R, Stewart BA (eds) Urban Soils. Advances in Soil Science. CRC Press/Taylor & Francis Inc., Boca Raton, pp 121–135

Chapter 4
Urban Soil Functions

Andrew W. Rate

Abstract In a textbook on urban soils, there is a need to present and explain some of the fundamental concepts in soil science. This chapter is framed around the ecosystem functions performed by soils and covers physical, chemical, and biological functions. The soil physics topics covered include volume relationships of solids and fluids, soil water retention, soil texture and structure, soil water potentials, water flow through soils, soil water balance, and soil temperature and the soil energy balance. The soil biological topics covered include discussions of how biological processes affect soil physical properties, nutrient cycling, and soil food webs. An explanation of bioavailability leads into examination of the chemical reactions and properties in soils which underpin ecosystem functions. The chapter concludes with a brief discussion of soil functions affecting human health and well-being.

Keywords Soil functions · Urban soils · Ecosystem services · Physical properties · Soil biology · Nutrient cycling · Chemical properties · Soil physics · Soil chemistry

A. W. Rate (✉)
School of Agriculture and Environment, University of Western Australia,
Crawley, WA, Australia
e-mail: andrew.rate@uwa.edu.au

A. W. Rate (ed.), *Urban Soils*, Progress in Soil Science,
https://doi.org/10.1007/978-3-030-87316-5_4

What you could learn from this chapter:

- What the key physical properties of soils are and how they allow soils to function, especially considering the behaviour of water in soil
- How the biological components (organisms) in soils control important soil functions and gain an understanding of nutrient cycling and bioavailability
- The chemical composition of soils: minerals, compounds, and ions – and what the important types of chemical reaction are
- Why and how soils are important for humans – in addition to the ecosystem services provided by soils.

4.1 Roles and Ecosystem Services of Soils in Urban Systems

In many urban environments, the roles and function of soils are no different from non-urban soils. Morel et al. (2015) recognise four general categories of *ecosystem services* provided by soils in any environment: supporting plant growth, maintaining biodiversity, removal and cycling of materials, and storage of ecosystem resources (see Table 4.1, which also summarises the mechanisms behind each function, i.e. how each function works). In addition to the four types of ecosystem services, Morel et al. (2015) identify three further ecosystem services that emphasise human uses of soils: a source of raw materials, physical and cultural support of human activities, and preserving a geological and archaeological record. This chapter will address primarily the functions and ecosystem services provided by soils in general, without necessarily being restricted to urban environments. The constraints imposed by the properties, features, and composition of urban soils will be addressed in following chapters.

Table 4.1 Functions of soil in urban systems

Soil function or property	Mechanism
Support primary productivity	Regulation of water and nutrient supply for plants; physical support for plants via roots
Nutrient and other element cycling	Multiple biotic and physicochemical mechanisms, e.g.: • Mineralisation of N, P, and S from organic matter (pedo→hydrosphere) • Biological immobilisation of inorganic forms of N and P • Plant uptake of subsoil elements (K, Ca) and litterfall to surface (litho → bio → pedosphere) • Provision of a sink for metal pollutants (atmo and hydro→pedosphere) • …and so on
Habitat for microflora and micro- and mesofauna	Provide food for heterotrophic organisms (soil organic matter); soil porosity allows shelter and movement
Filter between surface water and groundwater	Physical filtration of particulate material in soil pore network; chemical immobilisation of dissolved or gas-phase substances
Source/sink for atmospheric carbon (CO_2, CH_4) + NO_x	Conversion of biological detritus to soil organic matter with long residence time. In some soils, abiotic conversion of atmospheric CO_2 to HCO_3^-
Hydrological 'buffering'	Allowing infiltration to decrease run-off. Storage of water in soil pores
Chemical buffering and attenuation	Reversible sorption of dissolved or gas-phase substances on clays, oxide minerals, and organic matter. Biological degradation. Precipitation
Retain signatures of human land use, underlying geology, pollution, and so on	Reversible sorption and formation of longer-term sinks such as new mineral phases formed by (co) precipitation (see Chap. 6)

4.2 Soil Functions Related to Soil Physical Properties and Processes

The physical properties of soil are essential to the functioning of most terrestrial ecosystems, and some of the key phenomena are illustrated in Fig. 4.1. In this textbook we provide a summary of our view of the important concepts in soil physics, with particular reference where necessary to urban soils. In doing so we have relied on general soil science textbooks such as McKenzie et al. (2004), White (2006), and Schaetzl and Anderson (2005). For more detail about soil physical functions and concepts, readers should refer to a specialised soil physics textbook such as Hillel (2014).

The most basic of a soil's physical properties, treated first in many soil science textbooks, are the *volume relationships*. The soil volume is commonly subdivided into phases: solids, liquids, and gases. To complicate soil volume relationships a little further, soil scientists may separate the solid phase on the basis of size: the

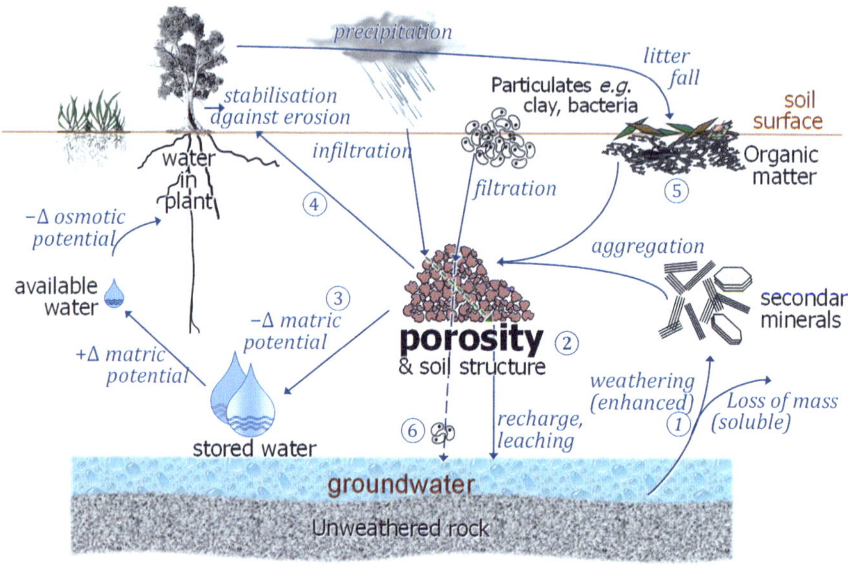

Fig. 4.1 Physical processes involved in maintaining ecosystem services in soils and some of the controls and interactions between them. (Graphic by Andrew W. Rate):
1. The mass loss and formation of secondary minerals allows formation of soil porosity and aggregation
2. Soil porosity allows recharge of, and leaching to, groundwater (especially with macropores from soil structure)
3. Soil porosity allows water storage, with some water available for biota, such as photosynthetic plants
4. Both soil structure and plant cover protect the soil surface against erosion
5. The biological community returns organic matter to the soil, further favouring aggregation and porosity
6. The soil pore network acts to filter particulates, including contaminants such as pathogenic bacteria

fractional volume in each of the three phases is measured (or estimated) for the soil's fine earth fraction, having grain sizes less than 2 mm, with the coarse material (≥ 2 mm) comprising a separate volume element. This grain size distinction is particularly relevant in urban soils, which often contain large proportions of rubble or artefacts (see the definition of *Technosols* in Chap. 2). The obvious question at this stage is whether these volume relationships are important aspects of soil functioning. The answer lies in the *porosity*; the volume of soil which is occupied by fluids, the liquids (mostly water containing dissolved material) and the gases (mostly air, but compositionally different from the above-ground atmosphere). The pore space in soil is extremely, and surprisingly, important. The proportion of pores relative to the total soil volume affects the ability of the soil to store and conduct water, the ease with which plant roots can penetrate, and the soil's suitability as a habitat for microorganisms and larger creatures from nematodes to earthworms, in terms of both living space safe from predation and allowing the diffusion of life-giving oxygen and carbon substrates.

Since the total soil volume is the sum of the three dominant phases, solids, liquids, and gases, then an increase in the volume fraction of any of these phases must be at the expense of the others; soil volume is a fixed-sum closed set. The presence of large, abundant solids having low porosity (such as natural or anthropogenic stone) in urban or other soils therefore results in lower pore volume, as porosity only exists between the finer grains. By a similar argument, the density of a soil, which is controlled by the packing of solid materials, also controls porosity. High density results from close packing, that is, more solids per volume than for a lower density soil. The higher volumetric solid content of dense soils leaves less space for pores and the vital functions that such pores allow.

The total volume of pores in a soil is not yet sufficient information to understand soil functioning completely. The size and connectivity of pores are critical; pores which are large (and by large, we mean $\gtrsim 0.03$ mm in diameter) hold water only weakly by capillary forces and are unable to hold water against gravitational drainage. Conversely, pores which are very small ($\lesssim 0.2$ μm or 0.0002 mm in diameter) retain water very tightly, such that plant roots cannot exert sufficient osmotic pressure to extract it. To maximise the ecosystem service provided by soil porosity, a soil needs to have a range of pore sizes: some larger, to drain excess water, and some smaller, to store water in a state in which it can be gradually released to plants and other soil organisms. An ideal range of pore sizes is not observed in all soils, even natural ones!

There are two main controls on soil pore size distribution. First, it makes sense that the spaces or *pores* between larger grains (sand, between 0.02 mm and 2 mm in diameter) will be larger than the pores between small grains (silt, between 0.002 mm and 0.02 mm in diameter, or clay ≤ 2 μm). We now need to introduce another soil science concept that of soil *texture*. The texture of a soil is a category, defined using a *texture triangle*, based on the relative proportions of mineral grains of different sizes in a soil (i.e. after organic matter is removed). In the international (IUSS) classification (Fig. 4.2), three ranges of grain sizes are used; sand, silt, and clay as defined above. The texture triangle defines thresholds between the texture categories; for example, a soil with 50% sand, 30% silt, and 20% clay by weight in the fine earth fraction would be classified as a clay loam in the IUSS system. There are several grain size and texture classifications in use in different countries (Minasny and McBratney 2001), which can confuse the issue.

The second control on the pore size distribution in soils is also related to grain size – the aggregation of individual grains into *aggregates*, or *peds*. Soil aggregates are arrangements of many individual soil grains held together by a range of forces and separated by pores and planes of weakness. The existence of such aggregates, and the size and geometry of the aggregates and their surrounding and internal pore spaces, is called the soil *structure*. Soil aggregates can range in size from dimensions similar to individual sand grains, up to several centimetres. The pore dimensions can therefore span a similar range; in addition, the aggregates themselves contain smaller 'micropores'.

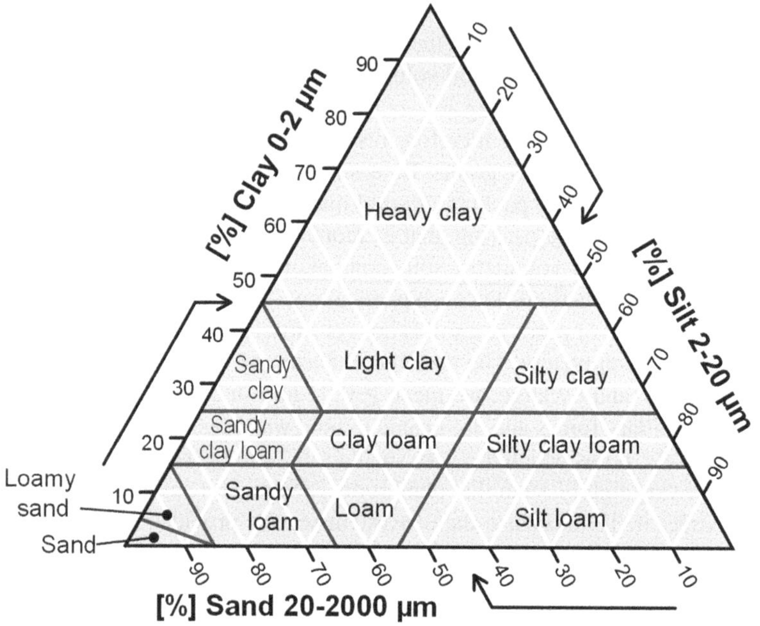

Fig. 4.2 Soil texture triangle based on mineral grain size fractions and categories from the International Union of Soil Sciences (IUSS) (drawn using the soiltexture R package by Moeys 2018)

Soil structure forms when the attractive forces between soil particles exceeds the tendency to exist as single grains. Although it is beyond the scope of this book to explain the mechanisms of soil aggregation in detail, we can identify a few qualitative soil properties that favour the development of structure: moderate to high clay content; low proportions (e.g. ≤6%) of exchangeable sodium (Na^+) as a fraction of total exchangeable cations (see Section 4.4.2 below); conditions favouring precipitation of cementing agents (e.g. iron oxides, carbonate minerals); specific types of organic matter, such as polysaccharides and other plant root or microbial exudates that possess adhesive qualities; plant roots themselves; and fungal hyphae. External forces such as shrink-swell cycles caused by wetting and drying, or freeze-thaw cycles, also facilitate formation of soil structure if favourable soil properties exist. The microscopic basis of soil structure is the 'clay domain' (Fig. 4.3).

So, with the pore network in soils controlled by soil texture and structure, we are better informed to understand how soil physical properties affect the storage and movement of water. Due to its adhesive and cohesive properties (e.g. surface tension), water in soils will tend to minimise its potential energy (symbolised as the matric potential, ψ_M) when present in the smallest possible pores. This tendency is summarised in the relationship between volumetric water content, θ_V, and soil matric potential, ψ_M (Fig. 4.4b). Simplistically, ψ_M is the pressure required to remove water from the largest water-filled pore (explaining why it's hard to get water from a dry soil!).

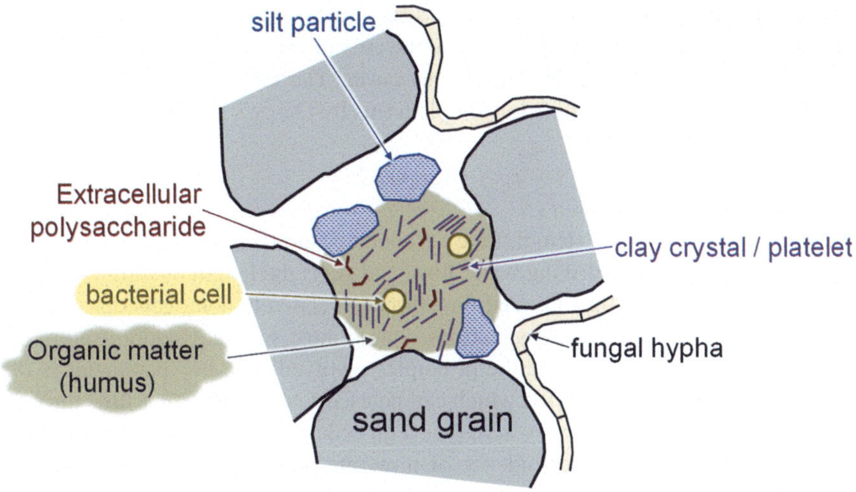

Fig. 4.3 Complexity of soil aggregation which results in soil structure. The structure-forming agents, clays, organic materials such as humus and polysaccharides, and enmeshing components such as fungal hyphae and plant roots, allow the formation of 'clay domains' which are the nuclei of soil structure that allow the larger grains (silt, sand) to be incorporated into aggregates

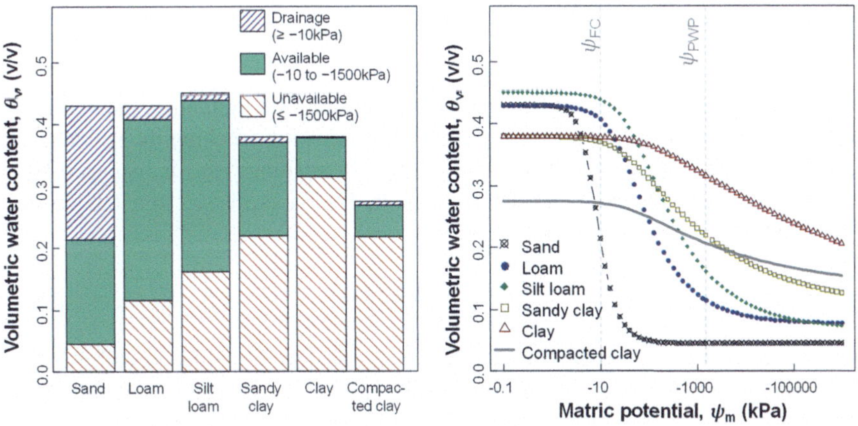

Fig. 4.4 (**a**) Comparison of available and unavailable soil water for a range of textures; (**b**) soil water retention curves for different textures using data from Carsel and Parrish (1988) and for compacted clays adapted from Tinjum et al. (1997). The vertical lines are labelled at the potentials representing field capacity (ψ_{FC}) and permanent wilting point (ψ_{PWP}, −1500 kPa) .Water fractions were calculated, and θ-ψ relationships were plotted, from the van Genuchten equation implemented using the 'soilphysics' R package (de Lima et al. 2016)

Similarly, pore size distribution also affects the ability of a soil to conduct water, an important property in urban soils since the area of exposed soil into which water can infiltrate is diminished by impervious cover. The rate with which water can travel through a soil is expressed in simple terms by Darcy's law (Eq. 4.1):

$$q_w = -K_s\left(\Delta h / \Delta z\right) \quad (4.1)$$

where q_w is the water flux (length/time), K_s is the saturated hydraulic conductivity, Δh is the hydraulic head driving water flow, and Δz is the length of soil.

Darcy's law is valid for saturated flow through relatively homogeneous soils, so it often does not apply in field situations, where most soils are seldom saturated; nevertheless, it does summarise the principles involved. The key parameter is the saturated hydraulic conductivity, which can be used to compare water flow through different soils. Values of K_s vary across several orders of magnitude, reflecting the theoretical inverse square dependence of flow rate on pore radius summarised in Poiseuille's law (McLaren and Cameron 1990). For example, while a sand may have $K_s = 1000$ mm/h, K_s for a clay-textured soil may be <0.1–5 mm/h (White 2006).

The pore structure of soils is also important in other contexts that we will not cover in this chapter: transport of gases though soils and the ability of the soil pores to provide habitats for soil (micro)organisms, as described in Chap. 8.

4.2.1 Soil Energy

A further important physical role of soils in supporting ecosystems is the ability of soil to act as a heat sink or source. The net solar radiation reaching the soil surface (affected by planetary albedo, slope, slope aspect, shading, etc.) is balanced by a number of heat fluxes from and within the soil. This is expressed in a soil surface energy balance equation (Eq. 4.2):

$$J_n = \mathrm{S} + \mathrm{A} + LE \quad (4.2)$$

where J_n is the net solar radiation flux, S is heat transferred within the soil, A is the 'sensible' (radiative) heat flux from the soil, and LE is the latent heat flux from evaporation, and transpiration by plants, of soil water.

Soil can therefore have a cooling effect on its local atmosphere, as it exports energy as water vapour to the atmosphere via the evapotranspiration (latent) heat flux (Hillel 2014). This cooling effect has also been recognised to be important for global climate regulation (Ban-Weiss et al. 2011). Soil also has a dampening effect, within the soil environment, on the extremes of atmospheric temperature, an effect which is more pronounced in subsoils compared with surface soils. The rate of heat conductance in soils also means that subsoil temperature changes lag behind those at the surface (see White 2006; Hillel 2008 for details). Both the dampening effect of depth and the lag caused by heat conductance are illustrated in Fig. 4.5. At the surface (5 cm in Fig. 4.5), soil temperatures are similar to the air temperature – the

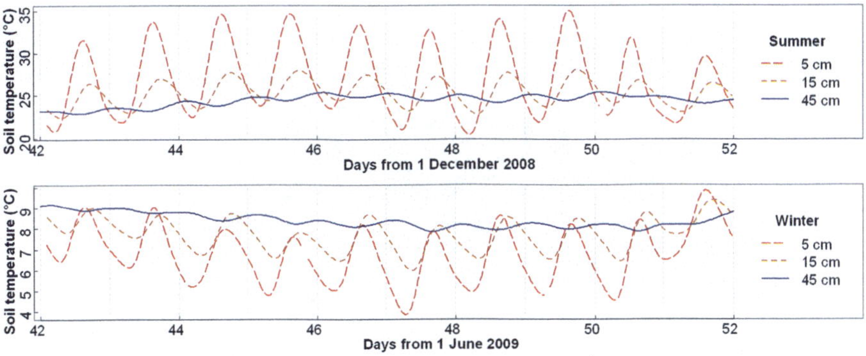

Fig. 4.5 Trends in soil temperature showing diurnal fluctuations at different depths over 10-day periods in summer and winter (note different temperature scales!) for an urban site near Canberra Airport, ACT, Australia. Data are from the OzNet data archive at www.oznet.org.au (Smith et al. 2012); curves are slightly smoothed to reduce discretisation effects. (Graphic by Andrew W. Rate)

'weather' temperature. At greater soil depth, soil temperatures do not increase as much during the day as at the surface, nor do temperatures decrease by as much as the surface at night. Not only is the soil temperature range less at depth, though, but the maxima and minima occur later than at the surface. The final feature shown by Fig. 4.5 is seasonal; in summer, soil temperature is cooler at depth than at the surface, but in winter the deeper soil is warmer, on average, than the surface soil.

In urban soil environments such as raised garden beds or green roofs, there is greater contact between soil and atmosphere. This means that even deeper soils can gain and lose heat more quickly, so that subsoil temperatures show a greater range which is more similar to surface soils.

4.2.2 Soil Functions Related to Hydrological Properties and Processes

The pore network of soils means that soils have a role in controlling the balance between run-off (overland flow) and infiltration (water penetrating into the soil) in the hydrological cycle (Fig. 4.6). Run-off contributes directly to stream flow and is possible when the maximum infiltration rate of the soils (essentially the hydraulic conductivity at the atmosphere-soil interface) is exceeded by the flux of liquid water added by precipitation (rainfall or melting snow or hail). Run-off contributes water to natural (streams, rivers) and anthropogenic drainage networks but, in excess, results in soil erosion (Hillel 2008).

Infiltration of water into soil has multiple consequences. Most directly, infiltration of precipitation water increases soil water content, where it is stored in the soil pore system. Water held at matric potentials less negative than about −10 kPa will drain under the influence of Earth's gravity; strictly vertical drainage will replenish groundwater, and drainage with a significant lateral movement contributes to

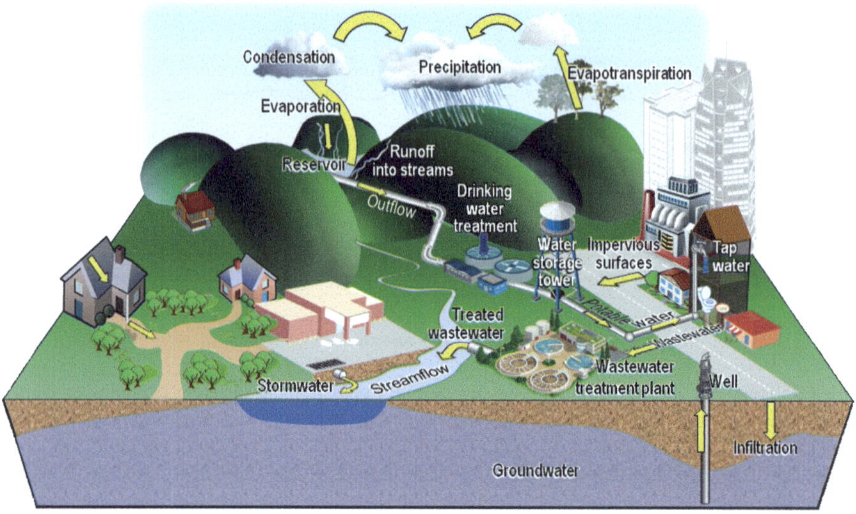

Fig. 4.6 The hydrological cycle including urban contributions. (Graphic by Andrew W. Rate inspired by a diagram by the Cary Institute of Ecosystem Studies (2020))

surface water bodies (drainage such as streams, or static bodies such as lakes). On hillslopes, the mass of soil with high water content can result in mass movement phenomena such as soil slippage or landslides, depending on the slope steepness and soil composition (Schaetzl and Anderson 2005).

The conservative nature of water cycle is expressed by a soil water balance (Eq. 4.3). The closed nature of this balance means that if one component changes, at least one other component will also change so that the balance is maintained. In this regard, run-off from, and infiltration into, soils are commonly a complementary pair; if run-off increases, infiltration decreases, and *vice versa* (White 2006).

$$\begin{array}{ll} \Sigma(\text{inputs}) & = \Sigma(\text{changes}) + \Sigma(\text{outputs}) \\ (\mathrm{P}+\mathrm{I}+\mathrm{U}) & (\Delta S + \Delta V) + (\mathrm{R}+\mathrm{D}+\mathrm{E}+\mathrm{T}) \end{array} \quad (4.3)$$

where P is precipitation, I is irrigation, U is upward capillary flow from below the root zone, ΔS is the change in soil water content, ΔV is the water content change in plants, R is run-off, D is (vertical) drainage, E is direct evaporation from soil, and T is transpiration by plants (Hillel 2008).

Both directly and indirectly, soil porosity also contributes to atmospheric compartments of the hydrological cycle. Water can evaporate from moist soil and be transpired by plants; both processes add water vapour to the air and (as observed above) exert a cooling effect on local and global atmospheres.

4.3 Soil Functions Related to Soil Biological Properties and Processes

Soils are distinguished from other solid Earth materials by the presence of an active biological community; soils are astonishingly rich in life (Fortuna 2012). After plants, the largest mass of living organisms on Earth is in soil fungi and bacteria! (Bar-On et al. 2018). In Sect. 4.2, we started to discuss how the soil's physical properties affected some biological processes, by providing a growth medium and habitat for plants and microorganisms and allowing storage and release of the water required for their survival. Soil systems contain complex networks of effects and feedbacks, in that plants and microorganisms can mutually promote one another's survival through food webs, trophic transfers, and nutrient cycling (Coleman et al. 2017). The biological components of soils in the form of plants and microorganisms can also affect soil physical and chemical properties, in many cases accelerating soil processes and promoting the establishment of soil conditions that are beneficial for their own survival.

4.3.1 Biological Effects on Soil Physical Properties

Soil Aggregation

Soil aggregates which are the building blocks for soil structure may be created by biological activity by enmeshment by fungal hyphae or plant roots, exudation of organic binding agents by plant roots and microbiota, and faecal pellets, or casts, of larger soil animals such as earthworms or mites (Coleman et al. 2017).

Soil Macroporosity

The penetration of plant roots and the burrowing of soil fauna create large continuous soil pores or *macropores* (McLaren and Cameron 1990; Schaetzl and Anderson 2005). Soil macropores are important, especially in soils with otherwise low hydraulic conductivity, for increasing overall infiltration rates.

Soil Mixing

The mixing of soils by organisms, termed *bioturbation*, is very important in some soil environments, increasing porosity and reducing soil density (Schaetzl and Anderson 2005). For example, the burrowing activity of earthworms and ants may be able to modify the entire upper metre of soil on timescales of centuries (de Bruyn and Conacher 1990; Feller et al. 2003).

4.3.2 Nutrient Cycling

Soils are, for many elements, the engines of terrestrial *biogeochemical cycling* (Chorover et al. 2007). The elements of greatest relevance are those required to sustain life: the elements C, H, and O primarily captured by photosynthesis; the *macronutrients* N, P, K, Ca, Mg, and S; and a range of *micronutrients*, of which Fe, Mn, Zn, Cu, B, Mo, Cl, and Ni are essential for plants. Additional trace elements (Co, Cr, I, Se) are required for organisms other than plants, including humans. Except for carbon, hydrogen, and oxygen which plant photosynthesis captures from the Earth's atmosphere, all of the other elements are supplied by the soil or are reliant on soil-based processes. For any one element, the biogeochemical cycle is a conceptual model which represents the various transformations and fluxes by which an element moves within and between environmental compartments. For example, Fig. 4.7 represents a combined carbon-nitrogen cycle for soils (Yang et al. 2009). The nitrogen and carbon in soils are derived from the atmosphere by primary

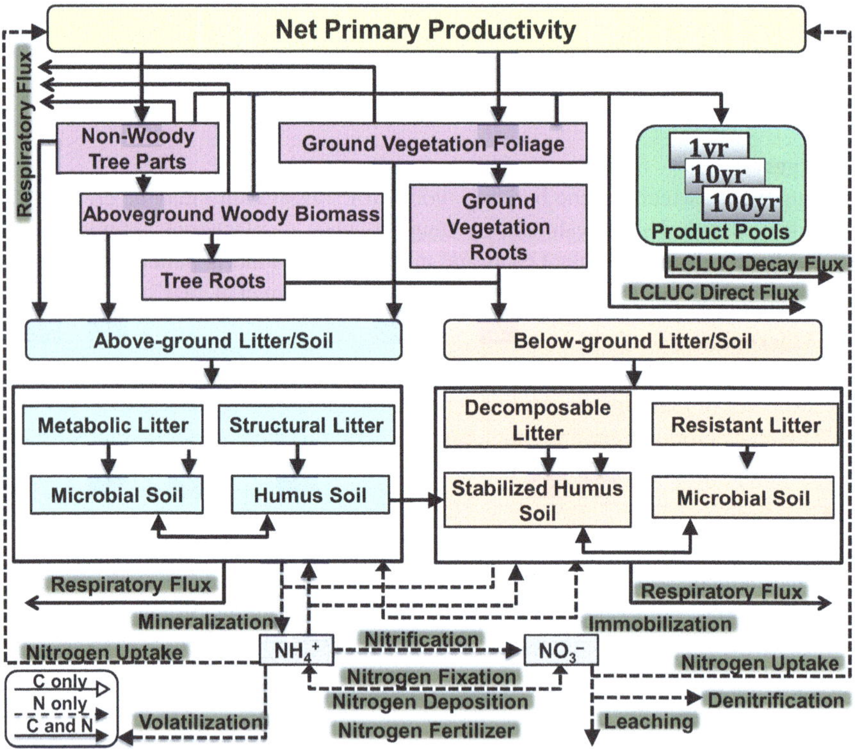

Fig. 4.7 Combined conceptual carbon and nitrogen cycle for soils. (From Yang et al. 2009 and used with permission from John Wiley and Sons). The boxes represent different compartments (forms) of C and/or N, and the arrows (solid lines for the C cycle, dashed lines for the N cycle) represent fluxes (transfers, transformations) of C and N. LCLUC is land cover and land use change

productivity – the growth, life, and death of photosynthetic plants. Plant litter becomes soil organic matter or 'humus', which in turn provides a substrate for the growth of microorganisms (mainly fungi and bacteria). Respiration by soil organisms, either directly from soil organic matter or through food webs (Sect. 4.3.3), completes the cycle by returning carbon and nitrogen to the atmosphere. Other soil biological processes contribute carbon and nitrogen to other environmental compartments such as water or retain carbon and nitrogen in the soil. We can use the same general concepts for cycling of any element like the essential elements listed above or even pollutant elements (or stable materials such as water); the cycling of many elements in soils is important, and we discuss some aspects of nutrient cycling below. It should become apparent that, to fully understand a biogeochemical cycle, we also need to know how the separate biological, chemical, and physical processes work.

Nutrient Cycling Involving Plants

Terrestrial plants and other photosynthesising organisms such as algae capture atmospheric carbon to synthesise a vast array of organic compounds, from nucleic acids to lignin. A significant proportion of these carbon compounds, also containing elements derived from soils (mainly, but not limited to, N, P, and S; see Table 4.2), enter the soil environment as the organisms, or parts of them, die and begin to decompose. In addition some organic molecules are released from living organisms into the soil environment as exudates from plant roots (and microbial exudates). The transfer of organic material derived from plants into the soil environments provides the substrate for microorganisms that utilise it as a source of energy, carbon, and nutrients (Chorover et al. 2007). Plants also redistribute nutrients (and other elements, since non-essential elements are not excluded from uptake) from deeper soil or regolith into the surface soil *via* uptake by deep root translocation within the plant and redeposition as dead material (litter fall) at or near the soil surface (Brantley et al. 2007). Finally, certain types of plants (most notably the legumes) have the ability, in symbiosis with specific groups of bacteria, to capture nitrogen from the atmosphere without relying on soil sources of N (Haynes 1986).

Microbial Nutrient Cycling

Much nutrient cycling is associated with the heterotrophic metabolism of organic compounds in soils by soil microorganisms (bacteria, fungi, and archaea). This is important in two contexts. First, elements contained in organic molecules (including carbon (C) and also N, P, and S) are released as simple inorganic forms during soil organic matter decomposition, or *mineralisation*, as the soil microbiota utilise carbon compounds to produce energy and metabolites in the process of *microbial respiration*. The simple inorganic ions of molecules can also be reincorporated into microbial organic matter in the process of microbial *immobilisation*. Second, respiration is an oxidative process, and, in the absence of sufficient oxygen in the soil atmosphere (or dissolved in soil pore water), other chemical substances can act as terminal electron acceptors, or oxidants, during oxidation of carbon compounds. These include compounds or ions of several essential elements, such as NO_3^-, Fe^{III} [hydr]oxides, Mn^{IV} [hydr]oxides, and SO_4^{2-} (Chorover et al. 2007; Coleman et al.

Table 4.2 Essential elements in soils and their *bioavailable* (e.g. plant-available) forms, forms present in soils, and key biogeochemical cycling processes

Essential elements	Form(s) taken up by plants	Main forms present in soils (~order of abundance)[a]	Key biological cycling processes (including interactions with non-living soil organic matter)
N (nitrogen)	NH_4^+, NO_3^- (ammonium, nitrate)	N in organic molecules Exchangeable NH_4^+ Dissolved NH_4^+, NO_3^-	Ammonification: release of NH_4^+ during heterotrophic metabolism of organic compounds Immobilisation: microbial reuptake of NH_4^+, NO_3^- to form organic forms of N Denitrification: use of NO_3^- as an electron acceptor for heterotrophic microbial respiration under anoxic conditions; forms $NO_{(g)}$, $N_2O_{(g)}$, $N_{2(g)}$
P (phosphorus)	$H_2PO_4^-$, HPO_4^{2-} (orthophosphate)	P in insoluble minerals P in organic molecules Adsorbed[b] $H_2PO_4^-$, HPO_4^{2-} Dissolved $H_2PO_4^-$, HPO_4^{2-}	Mineralisation: release of $H_2PO_4^-$, HPO_4^{2-} during heterotrophic metabolism of organic compounds Immobilisation: microbial reuptake of $H_2PO_4^-$, HPO_4^{2-} to form organic forms of P Reductive microbial dissolution of Fe^{III}[hydr]oxides to release adsorbed $H_2PO_4^-$, HPO_4^{2-} Organic acids exuded by roots, microbiota dissolve insoluble minerals, release $H_2PO_4^-$, HPO_4^{2-}
K, Ca, Mg (potassium, calcium, magnesium)	K^+, Ca^{2+}, Mg^{2+}	In insoluble minerals 'Non-exchangeable' K^+ Exchangeable ions Dissolved K^+, Ca^{2+}, Mg^{2+}	Organic acids exuded by roots, microbiota dissolves insoluble minerals, releases K^+, Ca^{2+}, Mg^{2+} Deep root uptake followed by translocation and litter fall recycles K^+, Ca^{2+}, Mg^{2+} to soil surface
S (sulphur)	SO_4^{2-} (sulphate)	S in organic molecules Insoluble sulphide minerals Adsorbed[b] SO_4^{2-} Dissolved SO_4^{2-} (HS^-, $S_2O_3^{2-}$)	Mineralisation: release of SO_4^{2-}, HS^- during heterotrophic metabolism of organic compounds Sulphate reduction: microbial metabolism of carbon using SO_4^{2-} as the electron acceptor to form HS^- Microbially catalysed oxidation of sulphide minerals (e.g. pyrite) to release SO_4^{2-}, Fe^{2+}, and H^+

Fe (iron)	Fe^{2+} (Fe^{3+})	Fe^{3+} in insoluble minerals ([hydr]oxides, phosphates) Fe^{2+} in insoluble minerals (sulphides, carbonates) Adsorbed[b] Fe^{2+}, Fe^{3+} Dissolved Fe^{2+} (Fe^{3+})	Reductive microbial dissolution of Fe^{III}[hydr]oxides (i.e. Fe^{III}[hydr]oxides are electron acceptors for organic carbon oxidation) to release Fe^{2+} plus ions adsorbed on, or substituted in, the oxide mineral Organic acids exuded by roots, microbiota dissolve insoluble Fe minerals, release Fe^{2+} (Fe^{3+}) Non-living soil organic matter can adsorb[b] Fe^{2+}, Fe^{3+} ions strongly
Mn (manganese)	Mn^{2+}	Mn^{4+} in insoluble minerals(e.g. [hydr]oxides) Adsorbed[b] Mn^{2+}, Mn^{4+} Dissolved Mn^{2+} (MnO_4^-)	Reductive microbial dissolution of Mn^{IV}[hydr]oxides (i.e. Mn^{IV}[hydr]oxides are electron acceptors for organic carbon oxidation) to release Mn^{2+} plus ions adsorbed on, or substituted in, the oxide mineral Organic acids exuded by roots, microbiota dissolve insoluble Mn minerals, release Mn^{2+} Non-living soil organic matter can adsorb[b] Mn^{2+} ions strongly
Cu, Zn, Ni (copper, zinc, nickel)	Cu^{2+}, Zn^{2+}, Ni^{2+}	In insoluble minerals Ions adsorbed[b] to colloids[c] exchangeable ions Dissolved ions and complexes	Reductive microbial dissolution of Fe^{III}[hydr]oxides or Mn^{IV}[hydr]oxides, to release adsorbed or co-precipitated [d] Cu^{2+}, Zn^{2+}, Ni^{2+} Organic acids exuded by roots, microbiota dissolve insoluble minerals, to release adsorbed or co-precipitated [d] Cu^{2+}, Zn^{2+}, Ni^{2+} Organic acids exuded by roots, microbiota form stable and soluble complexes with Cu^{2+}, Zn^{2+}, Ni^{2+}, increasing total dissolved Concentrations and suppressing adsorption, precipitation, and ion exchange Non-living soil organic matter can adsorb[b] Cu^{2+}, Zn^{2+}, Ni^{2+} ions strongly

(continued)

Table 4.2 (continued)

Essential elements	Form(s) taken up by plants	Main forms present in soils (~order of abundance)[a]	Key biological cycling processes (including interactions with non-living soil organic matter)
B (boron)	$B(OH)_3$, $B(OH)_4^-$ (borate) $B_4O_7^{2-}$ (tetraborate)	In insoluble minerals Adsorbed[b] to colloids[c] As soluble salts Dissolved ions/molecules	Some non-living soil organic matter can adsorb[b] $B(OH)_3$, $B(OH)_4^-$ strongly
Mo (molybdenum)	$HMoO_4^-$, MoO_4^{2-} (molybdate)	Mo in insoluble minerals Adsorbed[b] $HMoO_4^-$, MoO_4^{2-} Dissolved $HMoO_4^-$, MoO_4^{2-}	Reductive microbial dissolution of FeIII[hydr]oxides to release adsorbed $HMoO_4^-$, MoO_4^{2-} Organic acids exuded by roots, microbiota dissolve insoluble minerals, release $HMoO_4^-$, MoO_4^{2-}
Cl(chlorine)	Cl^- (chloride)	Dissolved Cl^- As soluble salts	–

[a]Approximate order of abundance, omitting gas-phase forms
[b]Adsorbed forms are ions or molecules that form chemical or physical bonds with the surface of a colloid, so that they accumulate at that surface (see Sect. 4.4.3)
[c]colloids are small particles able to form stable suspensions – in soils these are mainly clays, oxide minerals, and organic matter (see Sect. 0)
[d]Co-precipitated ions are those that take the place of the major structural ion in a mineral (see Sect. 4.4.4)

2017). The metabolism of organic carbon under anoxic (oxygen-limited) conditions in soils also affects some elements indirectly. For example, if an iron(III) oxide mineral acts as the terminal electron acceptor in the anoxic oxidation of organic carbon, phosphate or trace metal ions adsorbed to the surface of that iron oxide particle, or incorporated as impurities into its structure, will be released into solution (see Table 4.2).

Nutrient cycles in soils are very complex, and in this textbook, we can only discuss a few of the principal issues involved. Some of the more important nutrient cycling processes are presented in Table 4.2, which also contains information on chemical forms of elements, since the topics of element cycling and bioavailability are where soil biology and soil chemistry overlap extensively.

4.3.3 Soil Food Webs

Stored energy, in the form of metabolisable carbon compounds, is transferred from organisms upwards through trophic levels to provide them with the energy and substrate (food) for survival. The photosynthetic organisms or primary producers capture solar energy; the herbivores and decomposers that feed on plant material pass energy, carbon, and nutrients to the next level, and so on (see Fig. 4.8). The incomplete efficiency of these transfers means that the highest proportions of stored

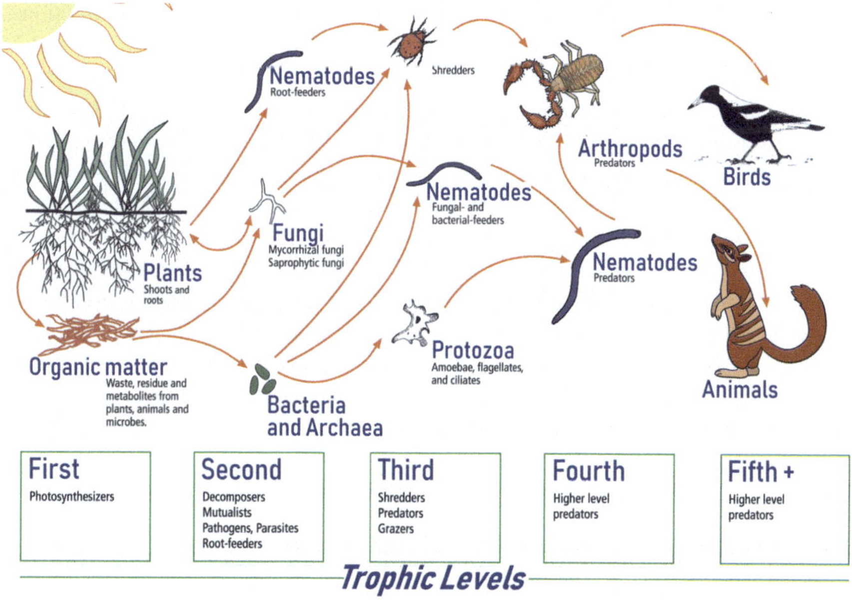

Fig. 4.8 Simplified soil food web showing trophic transfers (energy, carbon, and nutrients) and some of the diversity of soil organisms (based on Ingham n.d.). The arrows show the direction of flow of energy, carbon, and nutrients

energy, carbon, and nutrients exist in the organisms at lower trophic levels. By far the greatest proportion of terrestrial biomass on Earth exists in plants, which contain more than 35 times more biomass than the next most abundant group, soil fungi, followed by soil bacteria (Bar-On et al. 2018). The most abundant animals (by number of individuals, not mass) on Earth's terrestrial near-surface environments are nematodes, which exist at multiple trophic levels (van den Hoogen et al. 2019). In terms of biomass, the most abundant soil animals are probably arthropods or annelids (Bar-On et al. 2018).

4.4 Soil Functions Related to Soil Chemical Properties and Processes

As with the soil physics section earlier in this chapter, we present here a summary of our view of the important concepts in soil chemistry, again making particular reference where necessary to urban soils. The presentation of this material also relies on the general soil science textbooks by McKenzie et al. (2004), White (2006), and Schaetzl and Anderson (2005). To study soil chemical functions and concepts in more depth, readers should refer to a specialised soil or environmental chemistry textbook such as Sposito (2008) or Ryan (2014).

Numerous soil chemical properties and processes are important for the delivery of ecosystem processes, and these are summarised in Fig. 4.9. This section is intended to give an overview of most of the more important chemical soil functions. We will go into some more detail about soil chemical processes in Chaps. 6 and 7. A central concept is that of chemical *speciation* – the existence of multiple chemical forms of an element or compound, the identity of which affects not only its chemical reactions but also biological and physical processes in the soil environment as well. Table 4.3 explains and summarises chemical speciation as it applies to soils. In addition, soil chemists recognise two 'master variables' – *soil pH* and soil *redox potential*. The soil pH has a profound effect on nearly every chemical process, from mineral dissolution or precipitation to the *bioavailability* of nutrient elements or the degradation of organic contaminants (Sposito 2008). Likewise, the soil redox potential, or E_h, can significantly affect soil chemical reactions; E_h is nearly constant in aerobic soils, but once available oxygen in a soil decreases, usually due to saturation with water and microbial respiration, a cascade of chemical reactions is initiated (McBride 1994), some of which are outlined in Table 4.2.

4.4.1 Formation of Secondary Minerals and Organic Matter

A fundamental process in the conversion of soil parent materials (mainly rocks and the minerals contained in them) is chemical weathering (Brantley et al. 2007). In general, the solid minerals in rocks weather by reactions such as dissolution or

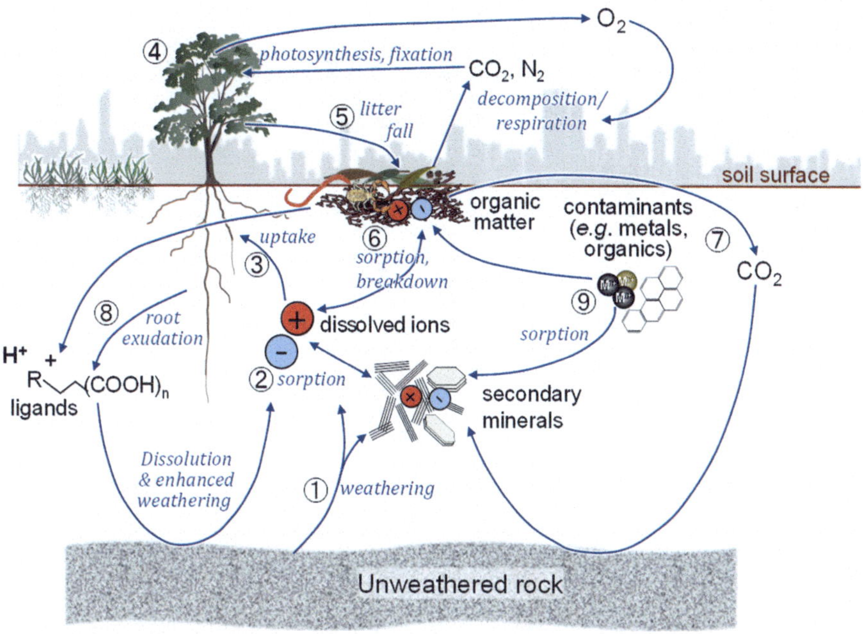

Fig. 4.9 Chemical processes involved in maintaining ecosystem services in soils, with controls/interactions (graphic by Andrew W. Rate):

1. Weathering of soil parent materials to give secondary minerals with small grain size, plus dissolved ions
2. The ions can interact with the secondary minerals by cation exchange or chemisorption mechanisms
3. The ions released by weathering, and held in potentially bioavailable reserve by sorption, comprise essential elements for plant growth (K^+, Ca^{2+}, HPO_4^{2-}, etc.)
4. Primary production by photosynthesis (and for some producers, nitrogen fixation) is facilitated by the uptake of nutrient ions from soil
5. Eventually, some of the carbon fixed by photosynthesis is added to the soil via litter fall, to form organic matter
6. The soil organic matter provides additional substrate for the adsorption of ions (including those essential for biota); some organic matter degrades to give soluble organic matter.
7. The utilisation of organic matter by heterotrophic soil organisms and the metabolic processes in plant roots result in respiration, releasing CO_2 which further enhances weathering of soil parent materials
8. Soil organic matter decomposition and exudation of organic compounds by roots and microbiota provide organic acids/anions which act as metal complexing ligands to further enhance dissolution of minerals
9. Contaminants such as metals, nutrients, and organic molecules are also chemically immobilised by secondary minerals and soil organic matter

Table 4.3 Chemical speciation explained – some common nomenclature with examples and explanations

Type of species and synonym(s)	Examples	Diagram(s)	Definition and notes
Hydrated cation (aquo-complex)	K^+, Ca^{2+}, $Cu(H_2O)_6^{2+}$	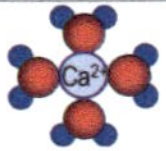	Aqueous (dissolved) phase. The hydration water is often omitted since, in many reactions, it does not participate
Anion	F^-, Cl^-	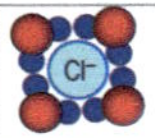	Aqueous phase. Will also be variably hydrated
Oxyanion	NO_3^-, SO_4^{2-}, $H_2PO_4^-$, $HAsO_4^{2-}$	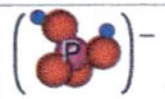 $(H_2PO_4^-)$$(H_2PO_4^-)$	Aqueous phase. Will also be variably hydrated. A subset of anions
Ion pair	$CaSO_4^0$, $MgHCO_3^+$	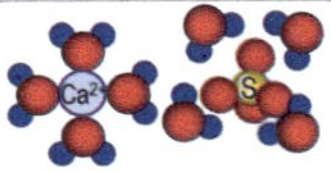	Aqueous phase. All or part of the hydration sphere retained. Weak bonding
Hydrolysis product (hydroxy complex, hydroxide complex)	$Al(OH)^{2+}$, $Fe^{III}(OH)_4^-$, $Cu(OH)_4^{2-}$	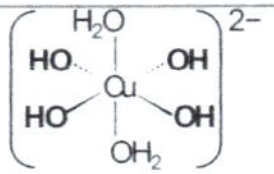	Aqueous phase. Will also be variably hydrated, e.g. $Al(OH)^{2+} = Al(H_2O)_5(OH)^{2+}$
Ligand (complexing agent)	H_2O, OH^-, F^-, Cl^-, SO_4^{2-}, $H_2PO_4^-/HPO_4^{2-}$, $[oxalate]^{2-}$, $EDTA^{4-}$, fulvic acid, DOM* *DOM = dissolved organic matter	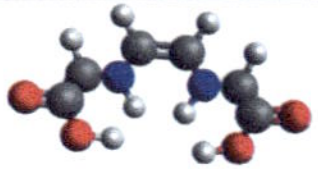 (H_4EDTA)	Ion or molecule surrounding the central ion in a complex. Aqueous phase. *Chelating ligands* are able to form more than one bond with the central ion
Complex (coordination complex)	$[Cu(Cl)_4]^-$, $[Pb(EDTA)]^{2-}$	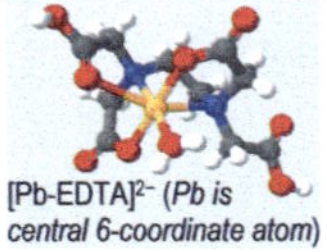 $[Pb\text{-}EDTA]^{2-}$(*Pb is central 6-coordinate atom*)	Aqueous phase. Hydration sphere lost. Strong covalent bonding between ligand(s) and central ion. *Chelates* are complexes of a central ion with a chelating ligand (see above)

(continued)

Cation exchange (electrostatic adsorption)	Accumulation of Na^+, K^+, Mg^{2+}, Ca^{2+}, or NH_4^+ around negatively charged surfaces of solid/colloidal particles of clay or oxide minerals, or organic matter (e.g. see Fig. 4.11)	
Chemisorption (inner-sphere adsorption, chemical adsorption, specific adsorption, surface complexation, ligand exchange)	Accumulation of ions or molecules at the surface of soil solid/colloidal particles by formation of strong electron-sharing chemical bonds (e.g. see Fig. 4.11)	
Precipitation	$CaCO_3$ (calcite) $Al(OH)_3$ (gibbsite) $Al_2Si_2O_5(OH)_4$ (kaolinite) FeS (mackinawite)	The formation of new solid phases, by reaction of individual chemical species in solution. The reacting species are removed from the solution and included in the chemical/mineralogical structure of the solid or precipitate. Requires *oversaturation* of the solution with the reacting ions, defined by the *solubility product*
Co-precipitation	$[Ce, La, Nd, Th]PO_4$ (rhabdophane) $Na_{0.5}Al_2(Si_{3.5}Al_{0.5})O_{10}(OH)_2 \cdot n(H_2O)$ (beidellite, bentonite)	Inclusion of small amounts of 'foreign' ions in the structure of a precipitating solid, normally limited to a few mole percent of the foreign ion(s). The examples show minerals for which co-precipitation is the rule; other minerals such as iron and aluminium (hydr)oxides also include trace elements such as Cr, Cu, Mn, Ni, Zn, etc.

oxidation to produce two main products: dissolved ions and molecules in the aqueous (liquid water) phase and secondary minerals in the solid phase. For example, the weathering of the common silicate mineral orthoclase feldspar can be expressed by:

$$\underset{\text{(K feldspar)}}{2KAlSi_3O_{8(s)}} + \underset{\text{(carbonicacid)}}{2H_2CO_3} + \underset{\text{(water)}}{9H_2O} \rightarrow \underset{\text{(kaolinite)}}{Al_2Si_2O_5(OH)_{4(s)}} + \underset{\text{(dissolved silica, potassium ions and bicarbonate ions)}}{4H_4SiO_{4(aq)} + 2K^+_{(aq)} + 2HCO_3^-} \tag{4.4}$$

where the subscripts $_{(s)}$ indicate a solid phase, and $_{(aq)}$ indicate an aqueous phase (dissolved in water).

The chemical reaction symbolised in Eq. 4.4 emphasises the role of atmospheric CO_2 dissolved in water to give carbonic acid, providing a weathering fluid which dissolves silicate minerals by acid hydrolysis. The consumption of CO_2 by weathering represents a long-term control on Earth's atmospheric composition and therefore the greenhouse effect (Kump et al. 2000). From a soil ecosystem service perspective, however, the products of this type of reaction are more immediately important: the secondary solid(s) (the *phyllosilicate clay mineral* kaolinite in this example) and the dissolved material (Fig. 4.9). The phyllosilicate clay minerals are important in a range of chemical reactions, as we will see below. In addition, the clay content of soils exerts significant control on pore size distribution and soil structure (see Sect. 4.2 above). The release of dissolved components is also extremely important. Loss of rock mass as dissolved ions is a key contributor to the development of porosity (Brimhall and Dietrich, 1987), and, crucially, the ions themselves (e.g. the K^+ in Eq. 4.4) provide a source of available nutrients for plants and other soil biota.

Secondary minerals also form by direct precipitation from solution, given favourable conditions. For example, if a weathering reaction generates dissolved Fe^{3+} ions, the following reaction (Eq. 4.5) can occur.

$$Fe^{3+}_{(aq)} + 3H_2O \rightarrow Fe(OH)_{3(s)} + 3H^+_{(aq)} \tag{4.5}$$

The new *secondary mineral* is the ferric hydroxide $Fe(OH)_3$ (similar to the documented oxyhydroxide mineral ferrihydrite). The existence of H^+ as a product of Eq. 4.5 means that this reaction will be favoured at higher pH, since at high pH the concentration of H^+ ions is low, and part of chemical equilibrium theory (Le Chatelier's principle, or the 'equilibrium law') means that if a product of any reaction decreases in concentration, the reaction will tend to progress further towards those products.

Finally, the stable organic matter in soils, although not a mineral phase, can be considered to be generated from 'weathering' (i.e. fragmentation and chemical alteration) of its parent materials – the organic detritus that enters soils during natural cycles of life and death of organisms.

4.4.2 *Cation and Anion Exchange Reactions*

The phyllosilicate clay minerals form from a solution generated during chemical weathering, and since parent rocks usually contain many different types of minerals, this solution contains a diverse range of dissolved ions. As a result, the clay minerals which form are seldom 'pure' aluminosilicates; their structures contain other cations in the structural positions of Al^{3+} or Si^{4+}. For example, the structure of the common 2:1 clay mineral montmorillonite (having the nominal chemical formula $(Na,Ca)_{0.3}(Al,Mg)_2Si_4O_{10}(OH)_2 \cdot n(H_2O))$ is shown in Fig. 4.10, showing the positions of the Al^{3+} or Si^{4+} ions which, at mineral formation, are nearly always partially occupied by 'impurity' ions such as Mg^{2+} or Fe^{2+}. Since the substituting ions (Mg^{2+} or Fe^{2+}) have a lower positive charge than the dominant Al^{3+} or Si^{4+}, the entire crystal structure forms with a deficit of positive charge – the clay crystals are negatively charged! An electrostatically charged particle cannot exist independently in the soil environment, and the negative charge is **always** balanced by *exchangeable cations* which are electrostatically attracted to the external surfaces and interlayer spaces of the clay crystal. In addition, since clay minerals mainly have grain sizes in the clay-size range (see Sect. 4.2), their surface area to mass ratio is high, another factor which favours chemical reactions. The exchangeable cations, most often Na^+, K^+, Mg^{2+}, Ca^{2+}, or NH_4^+, are of great importance for ecosystem functioning in soils (Fig. 4.9). First, the fact that they are held on a solid phase means not all the cations released during chemical weathering will be lost. Second, electrostatic interactions are relatively weak, so the cations are held in rapid and dynamic equilibrium between the mineral surface(s) and the pore water and therefore represent an

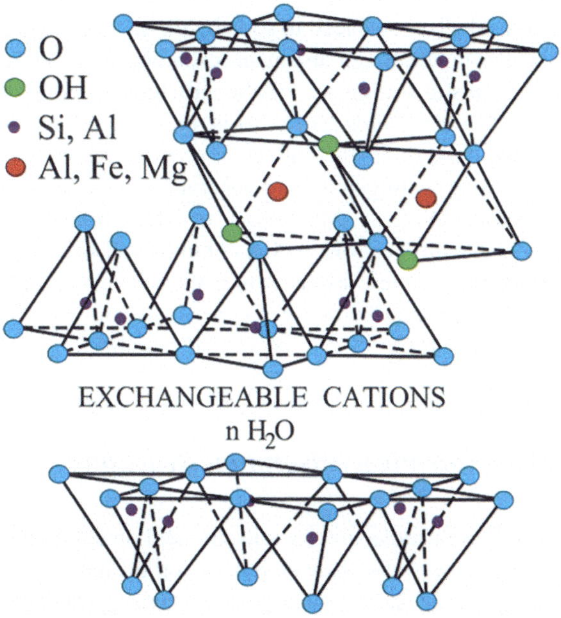

Fig. 4.10 Crystal (chemical) structure of montmorillonite, a 2:1 phyllosilicate clay mineral in the smectite group. Image from Poppe (2004)

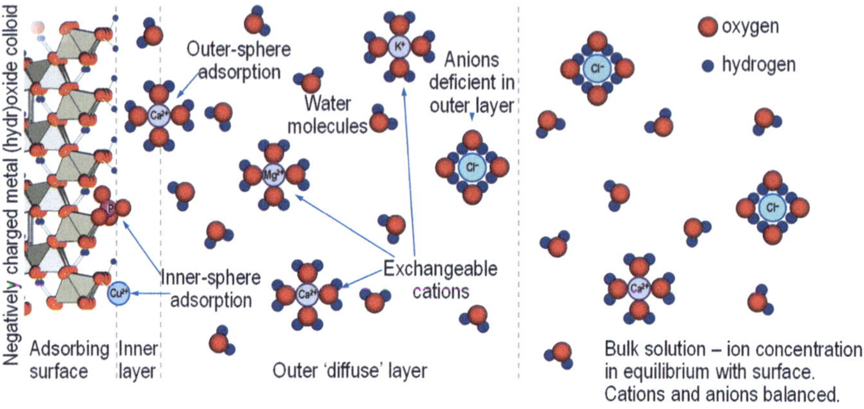

Fig. 4.11 Distribution of ions around a negatively charged oxide surface (graphic by Andrew W. Rate using ideas from Thompson and Goyne (2012) and Yang et al. (2006))

important reserve of elements for plants and other soil biota which are potentially bioavailable – both essential elements and potentially toxic elements (see Chap. 6).

Other negatively charged materials also exist in soils. Soil organic matter forms from the detritus of living organisms by a wide combination of reactions including hydrolysis, oxidation, and condensation, to create high molecular weight, highly variable structures. Soil organic matter consistently contains structural elements (functional groups) such as *carboxylates* as part of its structure. Carboxylates are anions of weak acids, so if the soil pH is high enough ($\gtrsim$ pH 3.5, which is most often is), then the carboxylates are in dissociated (anionic) form. Most soil organic matter exists in the solid phase, so organic particles also carry a negative electrostatic charge which is balanced in soils by the same exchangeable cations as clays.

Other minerals can also manifest a negative electrostatic charge: the hydroxide (OH^-) ions at the surface of the structure of hydroxides of aluminium, iron, and manganese are *amphoteric*; that is, they can behave as weak acids or bases. Under alkaline conditions, the collective surface OH on hydroxide minerals behave as a base, losing hydrogen ions to generate negative charge which is again balanced by cations like Na^+, K^+, Mg^{2+}, Ca^{2+}, or NH_4^+. Iron(III) hydroxides, in particular, can also **gain** hydrogen ions within common soil pH ranges ($\lesssim$ pH 8) to generate **positive** charge. In a few instances (e.g. the subsoils of highly weathered and acidic soils with low concentrations of charged clays or organic matter), then, exchange of anions such as Cl^-, SO_4^{2-}, or NO_3^- is possible.

4.4.3 Chemical Adsorption Reactions

Some ions or molecules in soils are held on solid phases much more strongly than the relatively weak associations of ion exchange. This type of reaction is called *adsorption* or in the usual case, where chemical bonding by electron-sharing between atoms is involved, *chemisorption*. Adsorption reactions still hold adsorbed ions in equilibrium with dissolved ions in the soil solution, but the equilibrium position is much further towards the surface species, the reactions are much slower to reach equilibrium, and generally rapid exchange is not observed.

Cations such as those of the trace metals Cu^{2+}, Ni^{2+}, and Zn^{2+} form strong chemical bonding with oxygen-containing structural components on soil solid phases. For example, the carboxylate groups on soil organic matter (and in some cases N-containing amino- functional groups) contain non-bonded electron pairs which can be shared to form covalent bonds with trace metal cations. Similarly, the surface hydroxides on oxide minerals and clays also contain non-bonded electron pairs allowing bonding with trace metal cations. Cation adsorption is more favourable at higher pH; this pH dependence is because bonding of a metal ion displaces hydrogen ions from the mineral or organic matter surface (Fig. 4.12). Since the O (or organic N) atoms contribute all of the electrons to the chemical bonds, the bonding is analogous to *coordination complexes* of metals, and the cation adsorption mechanism is sometimes called *surface complexation*.

Anion adsorption is somewhat different from cation adsorption. It generally only occurs significantly on inorganic materials: the phyllosilicate clays and hydroxide minerals. Again, the reaction sites are the surface hydroxides on mineral surfaces; however, bonding of an oxyanion like phosphate ($H_2PO_4^-$ or HPO_4^{2-}) occurs between the phosphate ion and the structural cations (e.g. Fe^{3+} or Al^{3+}) in the clay or hydroxide, and the previously bonded OH^- anions are displaced into solution. This is an example of *ligand exchange*, where both the incoming phosphate ions, and the outgoing hydroxide ions or water molecules, are ligands which exist in a coordination complex with metal ions near the surface of the adsorbing mineral structure. Anion adsorption is therefore more favourable at lower pH, since under acidic conditions the low concentration of OH^- anions tends to push the anion adsorption equilibrium towards reaction products (Fig. 4.12).

Adsorption of organic compounds is dependent on whether the molecule can be ionised in water and, especially for non-polar molecules, soil organic matter content. The behaviour of organic molecules in soils will be covered in more detail in Chap. 7.

4.4.4 Precipitation and Co-precipitation

Precipitation is the formation of new solid phases, by reaction of individual chemical species in solution. The reacting species are removed from the solution and included in the chemical/mineralogical structure of the new solid, or 'precipitate'. A very large range of minerals can precipitate in soils, including phyllosilicate clays, oxides, hydroxides, carbonates, sulphides, sulphates, and phosphates (Lindsay 1979), and a few examples are listed in Box 4.1.

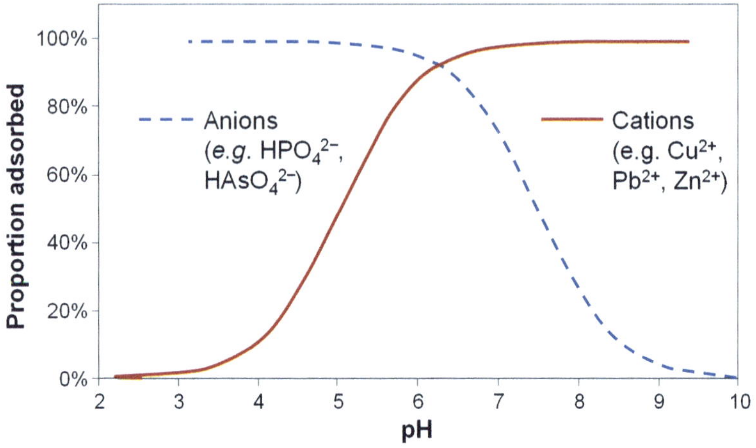

Fig. 4.12 Idealised adsorption of cations and anions on soils as a function of pH. The curves may be shifted horizontally or have different gradients, depending on the identity and concentration of the ion involved and the soil materials involved in the adsorption reaction. (Graphic by Andre W. Rate)

Box 4.1 Examples of Precipitation Reactions in Soils

1. $$Ca^{2+}{}_{(aq)} + HCO_3^-{}_{(aq)} \rightleftharpoons \mathrm{CaCO}_{3(s)} + \mathrm{H}^+$$
calcium + bicarbonate ⇌ calcite + hydrogenion

2. $$Fe^{3+}{}_{(aq)} + 3\mathrm{H_2O}_{(l)} \rightleftharpoons Fe(OH)_{3(s)} + 3\mathrm{H}^+$$
ferricion + water ⇌ ferrihydrite + hydrogenion

3. $$Fe^{2+}{}_{(aq)} + HS^-{}_{(aq)} \rightleftharpoons FeS_{(s)} + \mathrm{H}^+$$
ferrousion + hydrogensulfide ⇌ mackinawite + hydrogenion

4. $$\mathrm{K}^+{}_{(aq)} + 3Fe^{3+}{}_{(aq)} + 6OH^-{}_{(aq)} + 2\mathrm{SO_4}^{2-}{}_{(aq)} \rightleftharpoons KFe_3(OH)_6(\mathrm{SO_4})_{6(s)}$$
potassium, ferric, hydroxide, and sulfate ions ⇌ jarosite

Precipitation requires oversaturation of the solution with the reacting ions, defined by the *solubility product* K_{sp}. The solubility product is a simplified equilibrium constant for a solid substance dissolving in an aqueous solution: the product of the component ion concentrations each raised to the power of their stoichiometric coefficient (Eqs. 4.6 and 4.7).

For the reaction of a moles of ion A and b moles of ion B to give the solid $A_aB_{b(s)}$:

$$aA_{(aq)} + bB_{(aq)} \rightleftharpoons A_aB_{b(s)} \tag{4.6}$$

$$K_{sp} = [A]_{eq}^{a} \times [B]_{eq}^{b} \tag{4.7}$$

Ionic charges have been omitted for clarity (subscripts (aq) show aqueous-phase (dissolved) species and (s) solid-phase species, and square brackets with subscript 'eq' ($[\ ]_{eq}$) means concentration of the enclosed species in equilibrium with the solid).

In any solution, not necessarily at equilibrium with any solid, we can calculate the ion product $Q_{sp} = [A]^a \times [B]^b$ and its relationship to the equilibrium constant K_{sp}. If $Q_{sp} = K_{sp}$ for any solid phase, the solution is in equilibrium with that solid phase, and so the solution is said to be saturated. If $Q_{sp} < K_{sp}$ for any solid phase, the solution is undersaturated and that solid phase cannot form. If $Q_{sp} > K_{sp}$ for any solid phase, the solution is oversaturated with respect to that solid, and that solid phase will (usually) form; sometimes the solution needs to reach a certain value of Q_{sp}/K_{sp} for the solid to start forming. In fact, some solids will almost never form immediately; the Ostwald step rule (Threlfall 2003) proposes that the least stable mineral variant will form first, followed by progressive recrystallisation into more stable (and less soluble) mineral phases, a process known as Ostwald ripening.

4.4.4.1 Dependence of Precipitation on pH and Redox

Many precipitation reactions are dependent on soil pH, as implied by some of the reactions in Box 4.1 which include H^+ or OH^- as reactants or products. For example, a reaction producing H^+ (on the right side of the chemical equation) will be more favourable at high soil pH; soils with greater pH will react to consume the H^+ produced, forcing the equilibrium further towards products. Similarly, redox potential will control the precipitation of some minerals; for example, sulphide minerals will form much more favourable under reducing conditions. This is because under chemically reduced conditions, the reduced forms of both iron (Fe^{2+}) and sulphur (HS^-/S^{2-}) will be present in greater concentration, and these are both required reactants for sulphide mineral formation.

4.4.4.2 Co-precipitation

Unless a solid precipitates from a solution containing *only* its component ions, other ions of similar charge and ionic radius may be included in the structure of the precipitating mineral. This process is known as co-precipitation and involves inclusion of small amounts of 'foreign' ions in the structure of a precipitating solid, normally limited to a few mole percent of the foreign ion(s). We already know that the phyllosilicate clay minerals co-precipitate with cations in their structure other than Al^{3+}

and Si^{4+} (Sect. 4.4.2), and this is a very important phenomenon in soils and sediments. Other types of mineral are also known to form co-precipitates; for example, iron oxides may include trace or greater amounts of Al, As, Cd, Co, Cr, Cu, Mn, Ni, Pb, Sn, V, and Zn (e.g. see Singh and Gilkes 1992).

4.5 Soil Functions Related to Human Concerns

Direct use of soil by human communities affects issues related to physical and emotional health and well-being, food production, community cohesion, and construction. Soil may also have indirect effects on health through differences in land use or moderation of climate or may represent an archaeological or historical archive. We have addressed soil functions related to human concerns in the other chapters in this book.

In our discussion of soil formation in urban environments (Chap. 2), we examined the archaeological function of urban soils, that is, their ability to store and preserve different types of record of human habitation and activities. Chapters 2 and 3 also discuss the geological information stored in some urban soils in terms of their original parent material. The manufactured items that we find in soil help us learn about human history, and urban soils also offer us many opportunities to learn about the natural world through urban agriculture and environmental education (Gregory et al. 2016), a topic we address in Chap. 10.

The main 'use' of soil in urban environments is to provide living space; we construct our dwellings on (and in some cases, *from*) soil. We cover some of the engineering properties of soil as a building material and foundation in Chaps. 5 and 10. Construction of buildings, however, almost always insulates us from the soil beneath, a constraint which makes access to gardens and public open spaces crucial. In contrast, the use of soil for construction or as an artisanal raw material puts humans in very direct contact with soil. Some of the most obvious activities where humans use soil as a raw material are in pottery fabrication (Oladimeji et al. 2015) and where bricks or other items for building construction are made from soil, such as rammed earth which has been used since ancient times (Spence 1975; Ghavami et al. 1999; Liu et al. 2010). The soils in cities can provide a basis for addressing a range of concerns related to individual and community health and well-being; the health risks are discussed in Chaps. 6, 7, 8, and 9, and health benefits are addressed for multiple contexts in Chap. 10.

4.6 Summary

- The functions of soils can be expressed in terms of the services that soil provide to ecosystems, in general including support of plant growth, maintenance of biodiversity, removal and cycling of materials, and storage of ecosystem resources.

- The sizes and arrangements of solid materials in soils, known as soil texture and structure, control soil porosity and pore size distribution. Understanding soil pores allows us to understand the storage and movement of water in soils.
- We also need to account for soil water in the context of other Earth subsystems, and this is expressed in the soil water balance. Similarly, we can construct an energy balance for soils and use this to understand trends in soil temperature or even soil's effect on climates.
- The organisms that live in or on soils perform a vast range of ecosystem services, ultimately related to the ability of soils to capture and store carbon fixed by photosynthetic organisms – mainly, but not restricted to, plants. The transfers of carbon, nutrients, and energy through biological food webs in soils encapsulate some of the most important aspects of biogeochemical cycles of carbon and macro- and micronutrient elements.
- The interactions of soil biological and physical processes control crucial soil properties such as aggregation and porosity. Soil biology and soil chemistry interact in their effects on bioavailability of the chemical elements.
- Humans depend on all these natural soil functions and have also developed other uses for soils: we construct things from soil; we use soil to store, deliberately or otherwise, our diverse array of waste materials; the waste materials themselves preserve a record of human existence.

4.7 Further Reading

Hazelton P, Murphy B (2011) Understanding soils in urban environments. CSIRO Publishing, Collingwood, 160 pp.

Hillel D (2008) Soil in the environment: crucible of terrestrial life. Elsevier/Academic Press, Amsterdam/Boston, 307 pp.

McBride MB (1994) Environmental chemistry of soils. Oxford University Press, Oxford

Paul EA (ed) (2015) Soil microbiology, ecology and biochemistry. Academic Press, San Diego

4.8 Questions

4.8.1 Checking Your Understanding

1. Explain how soil porosity and pore size distribution affects the behaviour of water in soil.
2. If finer-textured soils hold on to water more tightly, how is it that loamy soils can have more plant-available water than sandy soils?

3. What are the soil microbial processes involved in biogeochemical cycling of carbon and nutrients?
4. What are the forms of macro- and micronutrients in soils which are available to plants? What do these have in common, and how do they differ?
5. What are 'secondary minerals' in soils, and why are they important?
6. List multiple ways by which soil can immobilise various contaminants.

4.8.2 Thinking About the Issues

7. What do humans do in urban environments that damages soil structure? Conversely, what can we do to improve urban soil structure?
8. Do you think it would be possible for plants or soil organisms to degrade or remove contaminant substances? If so, which ones, and how does the degradation of removal occur? If not, why not?
9. Humans have added many substances to urban soils which degrade soil quality, so is there anything we can add to soils to *improve* their ability to supply health ecosystem functions?

4.8.3 Contemplating Soils Creatively

10. Would people place more value on soil in general if they 'got their hands dirty', that is, came into more frequent direct contact with soil?

References

Ban-Weiss GA, Bala G, Cao L, Pongratz J, Caldeira K (2011) Climate forcing and response to idealized changes in surface latent and sensible heat. Environ Res Lett:6. https://doi.org/10.1088/1748-9326/6/3/034032

Bar-On YM, Phillips R, Milo R (2018) The biomass distribution on Earth. Proc Natl Acad Sci 115:6506–6511. https://doi.org/10.1073/pnas.1711842115

Brantley SL, Goldhaber MB, Ragnarsdottir KV (2007) Crossing disciplines and scales to understand the critical zone. Elements 3:307–314. https://doi.org/10.2113/gselements.3.5.307

Brimhall GH, Dietrich WE (1987) Constitutive mass balance relations between chemical composition, volume, density, porosity, and strain in metasomatic hydrochemical systems: results on weathering and pedogenesis. Geochim Cosmochim Acta 51:567–587. https://doi.org/10.1016/0016-7037(87)90070-6

Carsel RF, Parrish RS (1988) Developing joint probability distributions of soil water retention characteristics. Water Resour Res 24:755–769. https://doi.org/10.1029/WR024i005p00755

Cary Institute of Ecosystem Studies (2020) Water & watersheds, Millbrook, New York. https://www.caryinstitute.org/eco-inquiry/teaching-materials/water-watersheds (accessed 20210615)

Chorover J, Kretzcschmar R, Garcia-Pichel F, Sparks DL (2007) Soil biogeochemical processes within the critical zone. Elements 3:321–326. https://doi.org/10.2113/gselements.3.5.321
Coleman DC, Crossley DA Jr, Callaham MA (2017) Fundamentals of soil ecology. Elsevier Science & Technology, Saint Louis
de Bruyn LAL, Conacher AJ (1990) The role of termites and ants in soil modification: a review. Aust J Soil Res 28:55–93. https://doi.org/10.1071/SR9900055
de Lima RP, da Silva AR, da Silva AP, Leao TP, Mosaddeghi MR (2016) Soilphysics: an R package for calculating soil water availability to plants by different soil physical indices. Comput Electron Agric 120:63–71. https://doi.org/10.1016/j.compag.2015.11.003
Feller C, Brown GG, Blanchart E, Deleporte P, Chernyanskii SS (2003) Charles Darwin, earthworms and the natural sciences: various lessons from past to future. Agric Ecosyst Environ 99:29–49. https://doi.org/10.1016/S0167-8809(03)00143-9
Fortuna A (2012) The soil biota. Nat Educ Knowledge 3:1. https://www.nature.com/scitable/knowledge/library/the-soil-biota-84078125
Ghavami K, Toledo Filho RD, Barbosa NP (1999) Behaviour of composite soil reinforced with natural fibres. Cem Concr Compos 21:39–48. https://doi.org/10.1016/S0958-9465(98)00033-X
Gregory MM, Leslie TW, Drinkwater LE (2016) Agroecological and social characteristics of New York city community gardens: contributions to urban food security, ecosystem services, and environmental education. Urban Ecosyst 19:763–794. https://doi.org/10.1007/s11252-015-0505-1
Haynes RJ (1986) Mineral nitrogen in the plant-soil system. Academic Press, Orlando
Hillel D (2008) Soil in the environment: crucible of terrestrial life. Elsevier/Academic Press, Amsterdam/Boston, 307 pp
Hillel D (2014) Environmental soil physics fundamentals, applications, and environmental considerations. Academic Press, San Diego
Ingham ER (n.d.) Soil food web. Natural Resources Conservation Service (Soils), United States Department of Agriculture. https://www.nrcs.usda.gov/wps/portal/nrcs/detailfull/soils/health/biology/?cid=nrcs142p2_053868 (accessed 20190802)
Kump LR, Brantley SL, Arthur MA (2000) Chemical weathering, atmospheric CO_2, and climate. Annu Rev Earth Planet Sci 28:611–667
Lindsay WL (1979) Chemical equilibria in soils. Wiley, New York
Liu J, Hu R, Wang R, Yang L (2010) Regeneration of vernacular architecture: new rammed earth houses on the upper reaches of the Yangtze River. Front Energy Power Eng China 4:93–99. https://doi.org/10.1007/s11708-010-0002-4
McBride MB (1994) Environmental chemistry of soils. Oxford University Press, New York
McKenzie NJ, Jacquier D, Isbell R, Brown K (2004) Australian soils and landscapes: an illustrated compendium. CSIRO Publishing, Collingwood
McLaren RG, Cameron KC (1990) Soil science - an introduction to the properties and Management of new Zealand Soils. Oxford University Press, Auckland
Minasny B, McBratney AB (2001) The Australian soil texture boomerang: a comparison of the Australian and USDA/FAO soil particle-size classification systems. Aust J Soil Res 39. https://doi.org/10.1071/SR00065
Moeys J (2018) soiltexture: functions for soil texture plot, classification and transformation. R package version 1.5.1. The Comprehensive R Archive Network. https://CRAN.R-project.org/package=soiltexture (accessed 20210116)
Morel JL, Chenu C, Lorenz K (2015) Ecosystem services provided by soils of urban, industrial, traffic, mining, and military areas (SUITMAs). J Soils Sediments 15:1659–1666. https://doi.org/10.1007/s11368-014-0926-0
Oladimeji Y, Adepoju SA, Abdulsalam Z (2015) Reviving pottery enterprise: an impetus to poverty alleviation and self-reliance among women folks in Ilorin, Kwara state, Nigeria. Int J Dev Sustain 4:145–160
Poppe LJ, Paskevich VF, Hathaway JC, Blackwood DS (2004) A laboratory manual for X-Ray powder diffraction. U. S. Geological Survey Open-File Report, U.S. Department of the Interior,

U.S. Geological Survey, Woods Hole, MA, USA. http://pubs.usgs.gov/openfile/of01-041/index.htm
Ryan P (2014) Environmental and low temperature geochemistry. Wiley, Chichester
Schaetzl RJ, Anderson S (2005) Soils: genesis and geomorphology. Cambridge University Press, New York
Singh B, Gilkes RJ (1992) Properties and distribution of iron oxides and their association with minor elements in the soils of South-Western Australia. J Soil Sci 43:77–98. https://doi.org/10.1111/j.1365-2389.1992.tb00121.x
Smith AB, Walker JP, Western AW, Young RI, Ellett KM, Pipunic RC et al (2012) The Murrumbidgee soil moisture monitoring network data set. Water Resour Res 48:W07701. https://doi.org/10.1029/2012WR011976
Spence RJS (1975) Predicting the performance of soil-cement as a building material in tropical countries. Build Sci 10:155–159. https://doi.org/10.1016/0007-3628(75)90031-6
Sposito G (2008) The chemistry of soils, 2nd edn. Oxford University Press, New York, 329 pp
Thompson A, Goyne KW (2012) Introduction to the sorption of chemical constituents in soils. Nat Educ Knowledge 4:7. https://www.nature.com/scitable/knowledge/library/introduction-to-the-sorption-of-chemical-constituents-94841002/
Threlfall T (2003) Structural and thermodynamic explanations of Ostwald's rule. Org Process Res Dev 7:1017–1027. https://doi.org/10.1021/op030026l
Tinjum JM, Benson CH, Blotz LR (1997) Soil-water characteristic curves for compacted clays. J Geotech Geoenviron 123:1060–1069. https://doi.org/10.1061/(ASCE)1090-0241(1997)123:11(1060)
van den Hoogen J, Geisen S, Routh D, Ferris H, Traunspurger W, Wardle DA et al (2019) Soil nematode abundance and functional group composition at a global scale. Nature 572:194–198. https://doi.org/10.1038/s41586-019-1418-6
White RE (2006) Principles and practice of soil science: the soil as a natural resource. Blackwell Science, Malden
Yang H, Lu R, Downs RT, Costin G (2006) Goethite, α-FeO(OH), from single-crystal data. Acta Crystallogr E Struct Rep Online E62:i250–i252. https://doi.org/10.1107/S1600536806047258/wm2061Isup2.hkl
Yang X, Wittig VE, Jain AK, Post WM (2009) Integration of nitrogen cycle dynamics into the integrated science assessment model for the study of terrestrial ecosystem responses to global change. Glob Biogeochem Cycles 23:GB4029. https://doi.org/10.1029/2009GB003474

Chapter 5
Urban Soil Physics

Andrew W. Rate

Abstract Urban soils commonly have physical properties which reflect human activity in cities and which affect the functioning of urban soils and the ecosystem services they provide. This chapter starts with presenting some of the physical constraints present in urban soils, such as surface sealing, artificial layering, loss of structure, increased density, and the common presence of coarse fragments. Soil strength is addressed in the context of construction and also soil erosion and slope failure. We also cover the urban heat island phenomenon as it applies to soils in urban environments. The consequences of the sometimes adverse physical properties of urban soils are examined in the context of water-sensitive urban design, soil heating and its consequences, the 'urban karst' effect, plant growth, and bearing capacities for buildings and other infrastructure. The final sections cover methods for soil physical measurement in the context of urban environments, starting with standard soil physical measurement techniques and progressing to geophysical and remote sensing methods.

Keywords Soil physics · Urban soils · Physical properties · Physical constraints · Impervious surfaces · Soil erosion · Urban heat island · Methods · Geophysical methods

A. W. Rate (✉)
School of Agriculture and Environment, University of Western Australia, Crawley, WA, Australia
e-mail: andrew.rate@uwa.edu.au

A. W. Rate (ed.), *Urban Soils*, Progress in Soil Science,
https://doi.org/10.1007/978-3-030-87316-5_5

What you could learn from this chapter:

- The various physical constraints which exist in many urban soils.
- How physical constraints affect the delivery of ecosystem services by urban soils and how the same constraints affect suitability of land for engineered structures.
- The range of soil physical parameters that are important for understanding and assessing urban soil environments. An overview of field, laboratory, geophysical, and remotely sensed methods for obtaining soil physical data.

5.1 Introduction

In Chap. 4, we covered some of the fundamental physical properties and processes in soils, as they relate to the ecosystem services provided by soils in general. This chapter will discuss common physical constraints on urban soil functions, given the particular physical properties of urban soils introduced in Chaps. 1 and 4. We should remember that urban soils are highly variable, and the physical properties are affected by factors such as the original soil properties and parent material(s), any anthropogenic material added to the soil, soil disturbance, the time elapsed since human modification, climate, and land use.

5.2 Physical Constraints Observed in Urban Soils

5.2.1 Surface Sealing and Subsoil Layering

The existence of high proportions of land having impervious surfaces, sealed as a result of buildings or paving on the soil surface, is one of the most influential physical features of urban soils (Paul and Meyer 2001; Wong et al. 2012). Since impervious surfaces may cover 60% of cities (Zhu and Carreiro 2004) and may locally be 100% of the land surface (Fletcher et al. 2004), then there are substantial implications for urban soil functioning and ecosystem services. A large proportion of impervious surfaces reduces both shallow and deep infiltration (Fig. 5.1), which is expected to result in lower soil water contents (Coutts et al. 2013) and greatly increase run-off and potentially soil erosion.

A separate but related issue is that of artificial soil layering, which is common in urban soils and is the practice of creating a soil profile with specific properties designed to achieve a desired function. Some examples of artificially layered or *engineered soil profiles* are those in green roofs (Morel et al. 2015) or putting greens on golf courses (USGA 2018). Artificial soil layering may cause impeded water flow and/or shallow *perched water tables* (Jim 1998b). Artificial soil layering may involve abrupt changes in texture and/or density with depth, which will control the downward and upward movement of water and solutes by formation of permeability or capillary barriers (Li et al. 2013).

5.2.2 Soil Density and Porosity

The density of soil, measured as the dry bulk density, is commonly greater in urban soils than in natural soils. Dry bulk density greater than 1.6 Mg/m^3 is generally considered to be less suitable for ecosystem functioning, as root growth is restricted (McKenzie et al. 2004). For example, Short et al. (1986a), in urban soils in Washington DC (USA), measured mean soil densities of 1.61 Mg/m^3 in surface soils and 1.74 Mg/m^3 in subsoils, with several values clearly exceeding the 1.6 Mg/m^3 threshold (Table 5.1). Soil bulk density was higher in young urban residential

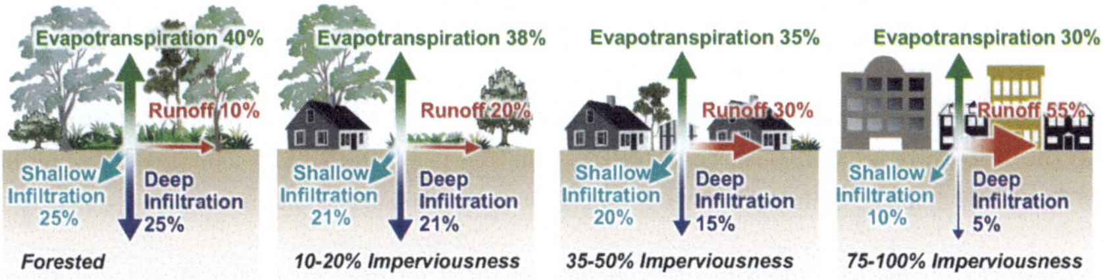

Fig. 5.1 Simplified city-wide water balance at different proportions of impervious surface cover. (Redrawn from USEPA 1993)

Table 5.1 Bulk density and porosity measurements in urban soils

City	Soil depth (cm)	Dry bulk density (Mg/m^3)		Porosity (vol. %)[a]	References
		Mean	Range		
Hong Kong	10–100 (range of depths)	1.65	1.15–2.63	18.5–57.0	Jim (1998b)
Washington DC, USA	Clods from 0 cm	1.61	1.25–1.85	36.6	Short et al. (1986b)
	Clods from 30 cm	1.74	1.40–2.03	32.8	
Moscow, ID, USA, and Pullman, WA, USA	0–15	–	1.39–1.75	34.0–47.5[a]	Scharenbroch et al. (2005)

[a]Estimated by the authors, assuming a particle density of 2.65 Mg/m^3

soils in two urban centres in the USA, with lower densities in older soils; the differences were attributed to pedogenesis (Scharenbroch et al. 2005) but potentially related to confounding differences in texture. The decreased bulk density was not reflected in any changed mean gravimetric soil water content. The porosity of soil is inversely proportional to the bulk density, given similar density of solids (typically ca. 2.6 Mg/m^3, similar to many silicate minerals (Cresswell et al. 2002)). Urban soils therefore often have relatively low porosity (e.g. some measurements of <20% porosity in Table 5.1), which can affect not only root growth but the movement of water, other liquids, and gases. In many instances, the high bulk density of urban soils may be related to deliberate compaction of soil materials which underlie building or road construction.

5.2.3 *Coarse Fragments and Artefacts*

Many urban soils have large proportions of their volume occupied by coarse (> 2 mm) fragments of anthropogenic origin (as might be expected from the IUSS definitions of some Technosols in Chap. 2, which specify ≥20% coarse fragments of human origin by volume). The presence of large, low-porosity solids in soils (such as natural or anthropogenic stone) results in lower pore volume, as significant porosity only exists between the finer grains. Coarse fragments include components of urban refuse such as construction rubble and household waste. An example of a Technosol with obvious artefact fragments is shown in Fig. 5.2(a); fragments may also include ceramics and, more recently, plastics (El Khalil et al. 2016; Hulisz et al. 2018) (Table 5.2).

Fig. 5.2 Examples of urban soil profiles showing (**a**) the content of coarse fragments in an 'urbic Technosol' (an urban soil with significant artefact content) and (**b**) surface sealing in an 'ekranic Technosol' (an urban soil having hard material at the surface) (from Hulisz et al. 2018; used with permission from Springer)

Table 5.2 Types of coarse fragments and artefacts found in urban soils

Study location	Coarse fragments and artefacts	References
Washington, DC, USA	Brick, concrete, nails (iron), glass	Short et al. (1986b)
Hong Kong, China	Stones (building rubble, other foreign substances)	Jim (1998b)
Hong Kong, China	Stones (granitic fill, shells)	Jim (1998a)
Moscow, Russia	Construction wastes, stones, brick debris, cement, concrete, metallic materials, wood, nutshells, leather	Alexandrovskaya and Alexandrovskiy (2000)
Marrakech, Morocco	Bones, fabrics, glass, metal, plastic, wood	El Khalil et al. (2016)
Toruń and Zielona Góra, Poland	Gravel, concrete, bricks, cinder, metals, ceramics, plastic, bones, glass, charcoals	Hulisz et al. (2018)

5.2.4 Soil Structure

Urban soils are typically poorly structured in their early stages of development. Soil aggregation to form structural elements (peds) occurs progressively along with other soil formation processes. Many urban soils are developed on soil materials modified or created by human activity and so are relatively young, with minimal change due to pedogenesis – including minimal development of soil structure (Jim 1998b; Chen et al. 2014). The development of soil structure is one of the mechanisms that increases soil porosity, provided that the soil materials have suitable properties (e.g. sufficient clay content – see White 2006). The high bulk density and corresponding low porosity of many urban soils (Table 5.1) can be another consequence of the limited development of structure, or processes such as compaction and disturbance may cause both the lack of structure and the low porosity (high density).

Improvements in urban soil structure may result from amendment of soils with composts made from urban waste materials. Fourvel et al. (2019) studied the effect of green waste compost on soil and dredged dam sediment, finding that compost improved the structural parameters of the soil materials. Increases in mean weight diameter of soil aggregates (i.e. better structure), decreases in bulk density, and increases in macroporosity persisted for up to 18 months following compost addition to the soil materials. Structural improvements can also be achieved using organic materials from other waste streams, such as biosolids (digested sewage sludge) (Kumar and Hundal 2016). Dredged dam sediments can provide a potentially fertile material to offset soil loss in urban environments, and their amendment with organic waste materials therefore represents beneficial reuse of both materials (Almeida et al. 2001), although dredged materials may contain potentially acidifying sulphides.

5.2.5 Soil Strength

Soil strength is important in different ways depending on the context. For maintenance of a biological community, high soil strength is undesirable, but in the context of preventing erosion or slope failure or supporting built infrastructure, high soil strength is advantageous.

Soils with high strength are common in urban environments, and this is frequently a consequence of deliberate or accidental soil compaction. Only a few studies have measured soil strength, for example, in terms of penetration resistance, in urban environments. For example, Chatterjea (2007) found significantly higher penetration resistance on and around walking trails in an urban park in Singapore, with on-trail penetration resistance frequently $\geq$1 MPa. The increased resistance to penetration in Chatterjea's (2007) study was related to the compaction caused by human foot traffic. Rocha et al. (2015) also measured high penetration resistance in soils being rehabilitated to forest in a peri-urban environment in Brazil (Fig. 5.3).

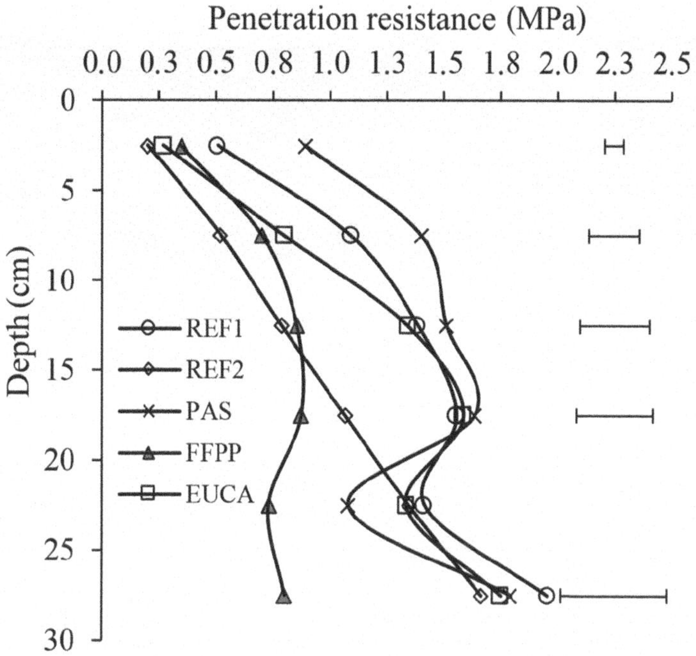

Fig. 5.3 Variation of penetration resistance with depth in soils being rehabilitated to forest in Garça city, Sao Paulo, Brazil. REF1 and REF2 refer to sites reforested with Brazilian native tree species; PAS is a site under pasture; FFPP is a remnant forest site (control); and EUCA is planted with *Eucalyptus* species. Translated version redrawn from Rocha et al. (2015) and used under the conditions of a CC-BY-4.0 licence

5.2.6 Soil Erosion and Erodibility

It has been known for some time that urbanisation causes local increases in soil erosion, for example, due to construction of buildings and roads exposing bare soil. Erosion is exacerbated by the increased volume and velocity of run-off from impermeable surfaces. The increased sediment yield from erosion of urban soils generally has the consequence of increased sediment load of urban streams (Wolman and Schick 1967). The short-term rates of erosion in terms of soil depth with time can be up to 18 cm/year on soil materials exposed or deposited by construction practices. For individual projects, therefore, soil erosion is a crucial consideration (Rowlands 2018).

Less severe water erosion is usually in the form of sheet (or sheetwash) and rill erosion (Fig. 5.4), caused by water flowing over unconsolidated land surfaces such as bare soil or with minimal vegetation cover (Knox et al. 2000). Soil loss by sheet and rill erosion is dependent on rainfall intensity and landscape factors such as slope steepness, slope length, and vegetation or impermeable cover. Soil factors also affect soil loss by water erosion; a soil's intrinsic erodibility depends on properties

Fig. 5.4 Sheet and rill erosion of bare soil on a slope exposed by urban construction activities. (Photograph by Andrew W. Rate)

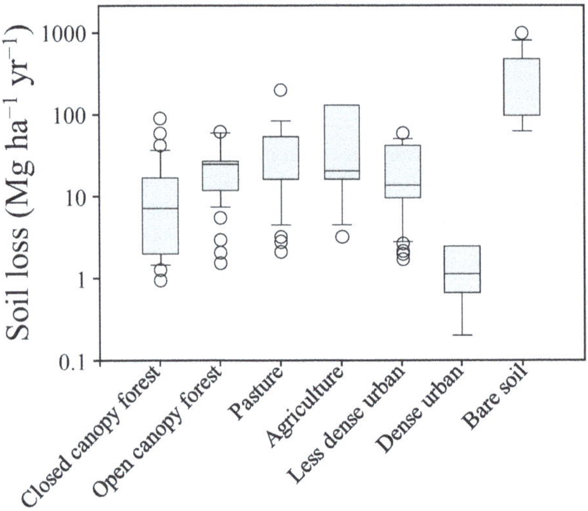

Fig. 5.5 Soil loss by erosion for different land use types in Puerto Rico. Redrawn from Del Mar López et al. (1998); used with permission from the College of Arts and Sciences, University of Puerto Rico, Mayagüez

like soil structure and texture, organic matter content, hydraulic conductivity, and soil strength (White 2006).

On a larger scale, therefore, urban soil erosion may be lower than for other land uses, because on average the land surface is covered with either vegetation (e.g. lawns) or impermeable surfaces that protect the soil surface from rainwater impact (Del Mar López et al. 1998; Knox et al. 2000) (Fig. 5.5). Del Mar López et al. (1998) modelled erosion with the Revised Universal Soil Loss Equation (RUSLE2; see Box 5.1) and assumed that the 'crop factor' (i.e. protection of land by vegetation cover) was more protective for densely developed urban land than for any other land use category, including closed-canopy forest. Even though vulnerable soil in urban environments is far more erodible than in rural settings (Wolman and Schick 1967; USDA 2000), the overall effect of surface sealing in urban areas can be to decrease erosion relative to natural environments on a whole-city scale (Fig. 5.5).

The erosion of soil from urban environments may still exceed erosion from other land use types such as forest or agriculture, especially when rapid urbanisation is occurring (e.g. see Erskine et al. 2003; Martin et al. 2003; Ozsoy and Aksoy 2015).

Wind Erosion

Erosion of particles from soils by wind in arid and semi-arid environments can be a significant pathway for soil loss. For example, Khresat et al. (1998) recognised urban expansion as a contributor to desertification in Jordan, with some of the most important mechanisms being erosion by water and wind. Urban soil erosion is also a significant source of airborne particulate matter (Eliasson et al. 2008) (Fig. 5.6), which may have adverse effects on human health. Although the source of some soil particles in air is external to urban centres, some studies show greater particulate concentrations in city centres compared with peri-urban areas (Eliasson et al. 2008).

Urban development has been a contributing factor in catastrophic landslides. For example, the 1979 Abbotsford landslide in New Zealand (Fig. 5.7), involving ca. 5×10^6 m^3 of soil and underlying unconsolidated sediment, was caused by multiple natural and urban factors. Natural factors included the slope of 7–10° along sediment bedding planes and soil and underlying material containing smectite clay with very low shear strength. The factors related to urbanisation which were identified included excavation of material on the lower slope, a leaking water main pipe that increased pore water pressure, and minor contributions from the increased mass of buildings and paved areas and removal of vegetation (Hancox 2008).

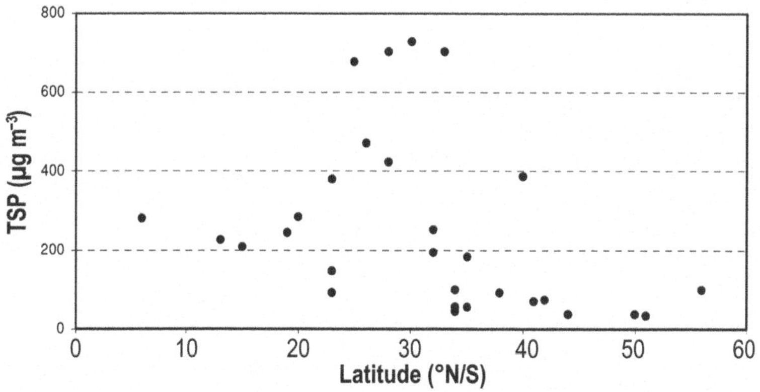

Fig. 5.6 Daily mean total suspended particulates (TSP, μg m^{-3}) in urban air for different cities, as a function of latitude (°N or °S). (From Eliasson et al. 2008; used with permission from Springer)

Fig. 5.7 Catastrophic slope failure in suburban Abbotsford, New Zealand, 1979. (Image from Hancox 2008; used with permission from Springer)

Box 5.1 The Revised Universal Soil Loss Equation

The Revised Universal Soil Loss Equation (RUSLE) predicts soil loss by water erosion and has been modified several times since the original USLE was developed by Walt Wischmeier at the US Department of Agriculture in the 1960s. The USLE model was originally intended for predicting soil losses from croplands in the USA and was updated for prediction in other environments, such as constructed areas, in 1978 (Renard et al. 1997). The latest version, RUSLE2, is based on the following equation (Foster et al. 2003):

$$A = S\sum\left(r_i k_i l_i c_i p_i\right)$$

where:

A is the average annual erosion
r is the rainfall/run-off erosivity factor
k is the soil erodibility factor
l is the topographic factor (slope steepness, roughness, etc.)
c is the cover-management factor
p is the support practices factor
and the subscript i denotes the daily index value for each factor (r, k, l, c, p)

RUSLE2 multiplies the daily factor values to estimate daily erosion values, which are summed (indicated by the $\sum$ symbol in the equation) for all days in a year to estimate average annual erosion. Previous versions of RUSLE included the k and c factors varying with time but not all factors as in RUSLE2.

There are fewer applications of the USLE or RUSLE to soil loss by erosion of urban soils than for agricultural soils, possibly as the model assumptions are not fulfilled. More advanced modelling approaches such as the USDA's Water Erosion Prediction Project (WEPP) may offer more reliable prediction (Laflen and Flanagan 2013).

5.2.7 *Soil Temperature and Heat Fluxes*

The *urban heat island* effect is the tendency of urban areas to have greater air and land surface temperatures than the surrounding peri-urban and rural areas. Urban heat islands have been known to exist since the early 1800s and have been confirmed by numerous studies using micrometeorological and remote sensing techniques over the last 40–50 years (Landsberg and Maisel 1972; Hafner and Kidder 1999; Small 2006; Min et al. 2019). There are multiple potential causes of urban heating, including the abundance of impermeable surfaces with large heat capacities and low reflectivity, intentional release of heat from combustion, the low proportions of vegetated land and open water which would otherwise confer an evaporative cooling effect, and lower soil water contents requiring less heat input for evaporation from soil. The increased surface area and heat capacity of tall buildings create the so-called canyon effect which is known to cause additional urban heating (Landsberg and Maisel 1972; Changnon 1999; Grimm et al. 2008; Tang et al. 2011). Examples of surface temperature gradients in and around several urban areas are shown in Fig. 5.8, showing that urban heat island effects are widespread, but they do not occur in all cities.

The greater air and land surface temperatures in urban areas would logically lead to greater soil temperatures, and this is generally supported by measurements. Changnon (1999) used a 60-year time series of air and soil temperature data in Champaign, Illinois, USA, to show an increasing trend attributed to urbanisation. A similar soil temperature increase was confirmed by Savva et al. (2010) who measured increases in soil temperature under both turf grass and urban forest in Baltimore, USA. Based on depth profiles from geothermal boreholes in Gateshead, UK, Banks et al. (2009) presented evidence that soil heat fluxes from an urban heat island had warmed surface soil and underlying sedimentary rocks to a depth of at least 55 m. Finally, in an extensive study in Nanjing, China, Tang et al. (2011) measured differences in urban and rural soil temperatures of between 1 and 3 °C (Fig. 5.9a). The city of Shanghai, China, however has an urban heat island only in terms of air temperature, whereas soils in the urban centre of Shanghai are cooler than soils in surrounding suburbs, a phenomenon which is likely to be due to increased shading (Fig. 5.9b).

The greater soil temperature in urban systems, caused by greater heat fluxes from atmosphere to soil, would need to be at least partly balanced by greater outgoing heat fluxes. Some of this heat loss flux would be soil heat flux to deeper soil, including to considerable depths, as suggested by Banks et al. (2009). Soil evaporation (latent heat flux), however, would also be likely to increase, resulting in drier urban

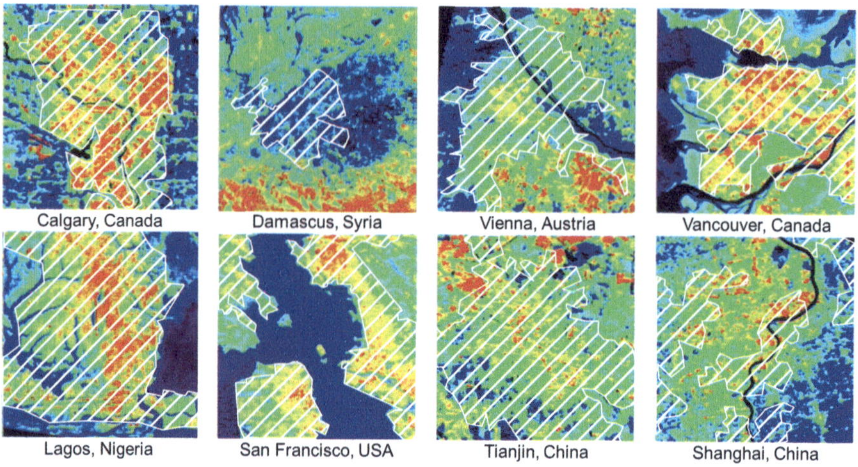

Fig. 5.8 Urban heat island effects shown by images derived from analyses of Landsat ETM+ spectra. Approximate urban extents (from Google Maps and shown as white cross-hatched polygons) are superimposed on surface temperature images selected from Small (2006) and used with permission from Elsevier. Colour scale in temperature maps ranges from black and blue tones (cooler) to red tones (warmer)

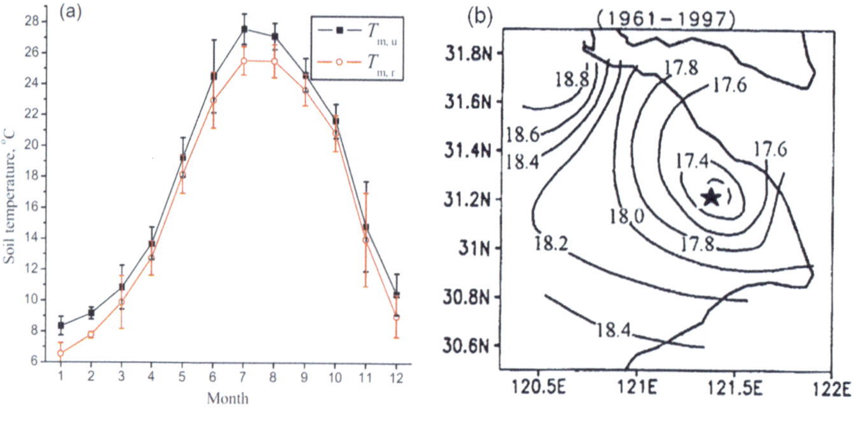

Fig. 5.9 (**a**) Yearly trend of urban ((—■— $T_{m,u}$)$T_{m,u}$) and rural ((—○— $T_{m,r}$)$T_{m,r}$) mean soil temperatures (from Tang et al. 2011) and used with permission from Elsevier; (**b**) urban 'cool island' of surface soil temperature in Shanghai, China (from Chen et al. 2003), with contours in °C and the ★ symbol representing the urban centre (used with permission from Springer)

soils. Greater soil temperature would also be expected to affect soil chemical and biological processes (such as faster chemical reactions or greater microbial activity when soil water content is sufficient; for more detail see Xiao et al. (2005) and Peters and McFadden (2010)).

Soils are important for regulation of urban microclimates (Mao et al. 2014). The ability of urban soil to cool the overlying atmosphere by way of the latent heat flux will clearly be lower for drier soils, however, since the latent heat flux represents the heat content of evaporating water vapour (Coutts et al. 2013). Urban vegetation, especially trees, promotes cooler soils by providing shade and by allowing greater evapotranspiration (i.e. latent heat flux) from soil (indirectly) to the atmosphere (Lin and Lin 2010).

5.2.8 Heterogeneity of Soil Physical Properties

Although this issue has already been addressed in Chap. 3, it is worth remembering that substantial short-scale heterogeneity in soil physical properties can exist in both anthropogenic and 'natural' soils. One example is the artificial layering discussed briefly in Sect. 5.2.1. There have not been many researchers who have studied short-range variability of soil physical properties, but it is an important issue to consider for engineering properties of soils and may require an additional margin of safety to be applied, for example, in the case of load-bearing soils for construction (Uzielli et al. 2006).

5.3 Effects of Urban Soil Physical Constraints on Ecosystem Functioning

5.3.1 Effect of Impervious Surfaces

Lower infiltration of water into a landscape with high impervious surface cover (Paul and Meyer 2001) can be assumed to result in less soil water being available for plants and soil biota (Coutts et al. 2013). Some studies, however, have found that changing the permeability of surface cover has little to no effect on the growth of urban trees (Morgenroth and Buchan 2009; Volder et al. 2009). The transfer of water to soils, and consequent availability of water to plants, in urban environments can be increased with *water-sensitive urban design* (WSUD) features such as swales and buffer strips or rain gardens (Fig. 5.10).

Impervious surfaces also differ from pervious surfaces or uncovered soil in their thermal characteristics. For example, Montague and Kjelgren (2004) showed that the albedo of different surface materials decreased in the order: concrete > gravel rock mulch > turf > asphalt > pine bark mulch > lava rock mulch. In the same study, thermal conductivity decreased in a somewhat different order: asphalt > concrete > turf > gravel rock mulch > pine bark mulch > lava rock mulch. Both low albedo and high thermal conductivity would be expected to result in greater temperatures in the

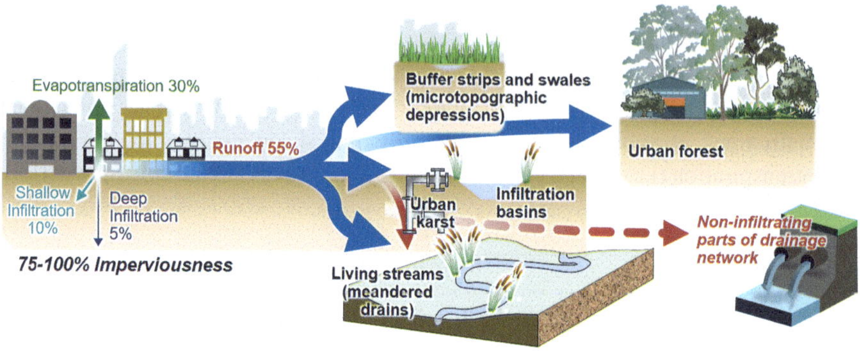

Fig. 5.10 Modified version of part of the 'classic' urban water balance diagram in Fig. 5.1, showing some possible pathways based on water-sensitive urban design for the increased run-off generated by high proportions of impervious surfaces. (Graphic by Andrew W. Rate)

underlying soils, but actual observations were only partly consistent with this expectation. Under all net solar radiation scenarios, the greatest soil temperatures were under asphalt and concrete, with the lowest soil temperatures under pine bark mulch (Montague and Kjelgren 2004).

Deliberately buried infrastructure, such as pipework for urban utilities, has the same effect as coarse fragments in reducing the effective soil volume for ecosystem services, like water storage and water and solute movement. Some authors call this the *urban karst* effect, and the phenomenon is illustrated in Fig. 5.11). This is because of the combined effects of limited infiltration areas from impervious surface cover, underground cavities, buried infrastructure, and tree roots creating *preferential flow* pathways in the same ways as natural 'karst' landscapes formed by dissolution-dominated weathering of limestones (Gwenzi and Nyamadzawo 2014; Bonneau et al. 2017). Urban impervious surfaces create more focused areas for water infiltration (sometimes in intentionally constructed basins) which increases percolation of water into discrete smaller areas, even leading to local mounding of groundwater. Installation of buried infrastructure such as pipework with smooth surfaces, and infilling of infrastructure trenches with high-permeability materials such as coarse sands and gravels, creates preferential flow pathways for water and solutes within urban soils (Bonneau et al. 2017). An important combined outcome of localised infiltration and preferential water flow in urban soils is therefore to, in some cases, increase leaching of substances dissolved in water. These substances may be contaminants such as nutrients, metals, organic compounds, or pathogens; the preferential flow means less interaction with the solid materials in the soil, consequent greater concentrations in pore water, and therefore possible groundwater contamination.

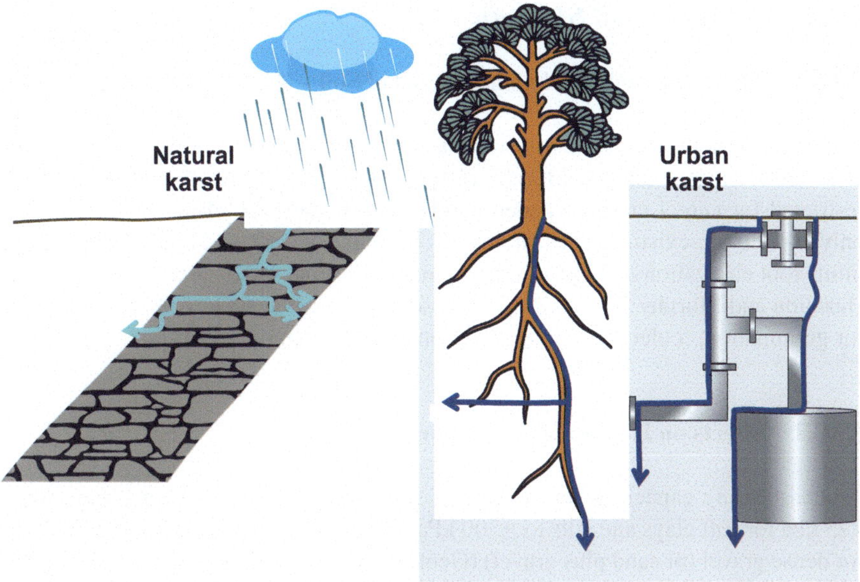

Fig. 5.11 The urban karst phenomenon, showing a schematic of preferential water flow (arrows) through natural fractured limestone karst at left and at right preferential flow along the boundary of soil with tree roots or buried infrastructure in an urban karst. (Graphic by Andrew W. Rate, inspired by US EPA (1993))

5.3.2 Effects of Soil Density and Porosity

Bulk density greater than ca. 1.6 Mg/m^3 is usually considered to be undesirable (McKenzie et al. 2004). The actual threshold bulk density value is dependent on texture (see US EPA 2011), and the upper threshold value may be as high as 1.8 Mg/m^3 before plant growth is severely restricted on sandy soils. The total porosity of soil is best understood in the context of air- and water-filled pores; air-filled porosity needs to be ca. 10% of total soil volume at field capacity water content for adequate aeration for plants and aerobic microorganisms (Hazelton and Murphy 2011). Low porosity confers a greater risk of inadequate air-filled porosity in wet soils, with consequent risks of waterlogging and anoxia (White (2006) suggests a minimum porosity of 23% by volume – have another look at Table 5.1). High water-filled pore space can decrease soil strength; low porosity, especially the absence of macropores, causes low infiltration rates resulting in run-off and potentially soil erosion.

5.3.3 *Effects of Soil Strength*

5.3.3.1 Effects on Biological Components of Soil

Plant root growth decreases with increasing soil strength (see Fig. 5.12 and Zou et al. (2001)). Hazelton and Murphy (2011) state that root growth will be severely restricted for cereal crops at penetration resistance ≥2–2.4 MPa, since roots can only explore pre-existing pores and planes of weakness in a soil. Soil shear strength limits root elongation at 70 kPa in sandy loam soils, and up to 290 kPa in clays (see Hazelton and Murphy 2011, who also present limiting values of soil shear strength for germination, 'coleoptile elongation', and seedling emergence).

5.3.3.2 Effects on Human Construction

Typical bearing capacities for a range of soils and soil-like materials range from <75 kPa for soft clays and silts to ≥300 kPa for compacted sand and up to >600 kPa for dense gravel (or sand plus gravel) (Geotechdata.info 2015). Some typical values of bearing limits for urban soils and related materials are presented in Table 5.3. These values have a safety factor applied; if soil has insufficient bearing capacity for the weight of structure, shear failure of the soil beneath and adjacent to foundations can compromise the built structure. The dependence of shear strength on grain size and morphology means that soil strength can be increased by mixing with a coarse-grained material such as rock chips (Rahardjo et al. 2008).

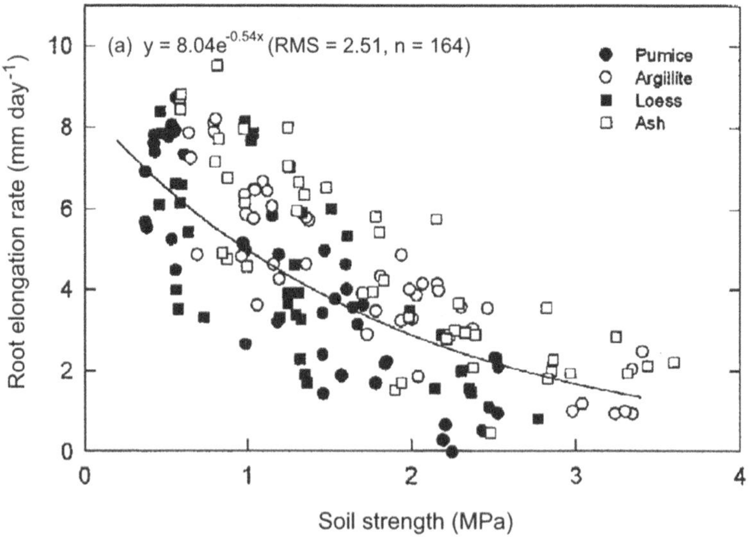

Fig. 5.12 Effect of soil strength (MPa) measured by penetration resistance on root growth (mm day^{-1}) of *Pinus radiata* seedlings (Zou et al. 2001; used with permission from Springer)

Table 5.3 Bearing limits of earth materials based on shear strength. (British Standards Institution 1986)

Material	Bearing limit (kN/m²)
Rocks	
Strong unweathered igneous or gneissic rock	10,000
Strong unweathered limestones or sandstones	4000
Unweathered schists and slates	3000
Strong unweathered shales, mudstones, or siltstones	2000
Soils	
Dense gravel/dense sand plus gravel	> 600
Medium dense gravel/medium dense sand plus gravel	< 200–600
Loose gravel/loose sand plus gravel	< 200
Compact sand	> 300
Medium dense sand	100–300
Loose sand[a]	< 100
Very stiff boulder clays and hard clays [b]	300–600
Stiff clays[b]	150–300
Firm clays[b]	75–150
Soft clays and silts[b]	< 75

[a]Depends on degree of looseness; [b]susceptible to long-term settling/compaction

Soil compressibility is also an issue, measured by a range of parameters (e.g. bulk modulus, volumetric compressibility) depending on the context (e.g. whether the soil is laterally constrained; see Terzaghi et al. 1996; Liu and Evett 2008). Coarser materials such as gravels and sands tend to have lower compressibility and are therefore more suitable for construction than more compressible silts, clays, and organic-rich soils for which settlement can be a severe problem.

In extreme cases, urban development may lead to potentially dangerous or even catastrophic events such as large sinkholes (Fig. 5.13) or landslides. The sinkhole which formed in Harbor, Oregon, USA, shown in Fig. 5.13 was along the line of a stream gully which had been infilled to allow construction. Preferential flow of water down the path of the former stream during heavy rain resulted in tunnel erosion, with the sinkhole forming as the material overlying the tunnel collapsed. Julian and Anthony (1996) discuss the increased incidence of landslides related to coastal urban development in south-eastern France, noting that human activities such as mechanical compaction, road construction, and removal of vegetation are factors contributing to slope failures.

Fig. 5.13 A large sinkhole which formed in Harbor, Oregon, USA, in 2016, probably caused by preferential subsurface water flow down an infilled former valley (from upper right to lower left of image (c)) during heavy rainfall (Photo credits: (**a**, **b**), Oregon Department of Transportation, used under terms of Creative Commons Attribution 2.0 Generic license; (**c**) still from YouTube video by Kyle Rice, Triook Technology, Harbor, Oregon, USA (used with permission))

5.3.4 Effects of Soil Erosion

Erosion of urban soils, especially during the construction phases of urban development, is a significant source of sediment to streams and rivers (Paul and Meyer 2001) and ultimately to the marine environment (USEPA 1993). The ecology of urban streams can be affected significantly by increased sediment load, resulting in effects such as eutrophication, reduced biodiversity of plants and invertebrates, and reduced diversity and population declines for fish species (Paul and Meyer 2001). Export of sediment to streams by soil erosion is also associated with stream sediment contamination (Sutherland 2000). In stormwater drainage systems, excessive sedimentation from soil erosion may necessitate drain maintenance by excavation of drain sediments (Department of Environment 2004).

Wind erosion of soil (including urban soil) increases the concentration of fine particles suspended in the atmosphere (Chan et al. 2008; Eliasson et al. 2008;

Athanasopoulou et al. 2010). Erosion by wind is particularly relevant in drier soil environments, which in urban environments may result from higher soil temperatures and reduced deep and shallow infiltration due to impervious surfaces. The combined influences of both the urban heat island effect, and the increasing temperature trend due to climate change, may result in increasing severity of wind (*aeolian*) erosion of urban soils.

Erosion of surface soils by water and wind also represents a loss of fertility due to the vertical stratification of nutrients and soil organic matter, in that the greatest concentrations of nutrients and organic matter are at or near the soil surface. Establishment of vegetation may therefore be more difficult (e.g. in rehabilitation of urban soils) if erosion occurs, unless soil amendments are used to manage fertility (US EPA 2011).

5.3.5 Effects of Warmer Soils

Since (with some exceptions, such as Damascus, Lebanon (Fig. 5.8), or Shanghai, China (Fig. 5.9)) urban soils are likely to be warmer than their natural soil counterparts, it is worthwhile considering the effects of greater soil temperature on soil processes and functions.

Warmer soils will tend to be drier; the lower water content is driven by external energy inputs which are balanced by the latent heat flux of the soil (Hillel 2014; see Chap. 4). The theory is consistent with measurements in urban soil systems; for example, Wang et al. (2011) measured greater evapotranspiration by urban trees at higher soil temperatures. Evaporation of water directly from soil also requires there to be a relative humidity gradient between soil and the atmosphere, however, so warm soils will not always dry out.

Combined warming and drying of soil will generally cause decreases in biological activity. Plants will experience water stress (Hillel 2014), and soil microorganisms from urban soils may not survive extreme drying (Gleason et al. 2004). In some cases, though, soil microorganisms may adapt to the selection pressure applied by higher temperatures so that they can better survive warming and/or drying (McLean et al. 2005).

At greater soil temperatures in the absence of water limitation, microbial processes occur more quickly. The most obvious example is that of soil organic matter decomposition, commonly measured as respiration of CO_2 by soils. In urban soils in Auckland, New Zealand, Weissert et al. (2016) showed that soil temperature and soil water content were the best predictors of soil CO_2 emissions across a range of land uses and soil types, including urban parks and areas of remnant and planted forest. Other microbially driven processes have also been shown to respond to temperature changes in urban soils. Methane emissions from urban wetland soils (which are driven by microbial processes) in Ohio, USA, increased with increasing soil temperature (Morin et al. 2014). Similarly, the net mineralisation of nitrogen, another process dominated by soil microbial and mesofaunal activity, was greater in urban than rural environments, an effect attributed to an urban heat island (Pavao-Zuckerman and Coleman 2005).

Plant growth can also be affected by increased soil temperature in urban environments. Ziska et al. (2004) found that plant productivity was more closely related to soil temperature than to daytime air temperature or atmospheric CO_2 concentration, along a rural-urban gradient in Maryland, USA (the study ensured that soil water content or nutrients were not limiting factors to plant growth).

5.4 Soil Physical Measurements

5.4.1 Standard Soil Physical Methods

There are numerous standard field and laboratory-based methods for determining soil physical properties, and it is not our intention to review these comprehensively in this textbook. We will include a brief discussion below and refer readers to excellent compilations of soil physical methods in Cresswell et al. (2002) and Dane and Clarke Topp (2002).

There are numerous soil physical parameters which are important to be measured or estimated in urban soils. It is useful to have information on basic soil properties such as density, porosity, coarse fragments, water content, soil texture, soil structure, soil temperature, and electrical conductivity.

Parameterised models can be used to estimate or predict soil physical parameters which are difficult to measure. These may be *mechanistic*, meaning that the model is based on a theoretical understanding of the processes involved, such as the equation describing a soil water retention curve using the van Genuchten equation (e.g. see de Lima et al. 2016). Alternatively it is sometimes possible to predict the values of soil properties because they are statistically related to other (more easily measurable) soil properties, without assuming any physical mechanism. The statistical relationships are usually regression models (see Chap. 3), giving rise to the so-called pedotransfer functions, which are used to predict soil physical parameters that are difficult to measure (Cresswell et al. 2002, Chap. 22). (Table 5.4)

5.4.2 Geophysical Methods

A number of 'geophysical' techniques have the potential to generate two- or three-dimensional representations of the below-ground soil environment and therefore to provide information which include soil variability across a landscape and/or with depth. They can be especially useful in combination with each other, to cross-validate detection of subsurface soil features and properties. The soil properties accessible by geophysical methods are, not surprisingly, mainly soil physical

Table 5.4 Examples and availability of methods for determining soil physical properties in the field and laboratory

Soil physical property	In situ/field method	Laboratory method
Soil depth	Excavation; push-probe	–
Water table height	Piezometers, bores	–
Bulk density	Water replacement and related methods, neutron probe	Volumetric/gravimetric on undisturbed field cores
Coarse fragments	Sieving/gravimetric	Sieving/gravimetric
Water content	Time domain reflectometry (TDR), neutron probe	Thermogravimetric
Water potential	Tensiometers	Filter paper method
Water repellence	–	Molarity of ethanol droplet (MED) test
Texture	Manual deformation	Gravitational sedimentation
Structure	Field observation	–
Aggregate stability	Emerson test (qualitative)	Wet sieving
Penetration resistance	Field penetrometer	Penetrometer
Shear strength	–	Triaxial cell; direct shear box
Compressibility	–	Consolidometer; triaxial compression cell
Water retention curve	Simultaneous TDR or neutron probe + tensiometers	Pressure plates/tables, plus gravimetric
Infiltration rate	Double-ring or tension infiltrometer, well permeameter	–
Hydraulic conductivity	Depthwise soil pore water samplers or thermistors	Constant- or falling-head permeameters
Soil temperature	Thermocouples; thermistors	–
Heat fluxes: Latent and sensible	Eddy covariance; Bowen ratio method	–
Electrical conductivity/ resistivity	Suspension and field conductivity meter; EM survey; electrical resistance tomography	Suspension and conductivity cell
Particle density	–	Volumetric/gravimetric
Specific surface area	–	BET sorptimetry
Atterberg limits	–	Drop-cone penetrometer + thermogravimetric
Evapo(transpi)ration	Lysimeters; plant-based techniques	–
Air permeability	Surface-chamber method	Manometer/flow meter
Heat capacity, soil heat flux	Dual-probe heat-pulse method	Dual-probe heat-pulse method
Thermal conductivity	Time series of depthwise temperature gradient	Time series of depthwise temperature gradient
Solute transport (esp. models)	Lysimeters; soil pore water samplers; groundwater tracers	Hydrogeochemical simulation models; lysimeters

Source: Cresswell et al. (2002), Dane and Clarke Topp (2002)

properties (sometimes modified by chemical composition parameters such as salinity). Chemical and biological soil properties are not generally able to be determined by above-ground probes such as those described below and are dependent on physical sampling and subsequent laboratory analyses.

Ground-penetrating radar (GPR) is based on detection of the rate of propagation and strength of reflection of a pulse of radiofrequency electromagnetic radiation applied at the soil surface. It has been used to measure soil depth to bedrock and detect voids and buried infrastructure in urban contexts (Saarenketo and Scullion 2000), in situ urban tree root morphology (Stokes et al. 2002), as well as in archaeological exploration of ancient cities (Leopold et al. 2011).

Magnetic methods include magnetic gradiometry (i.e. measurements of magnetic field gradient) and magnetic susceptibility. For example, Magiera et al. (2006) measured magnetic susceptibility on a ca. 10 km grid spacing in surface soils across Poland, the Czech Republic, and Germany, finding that urban soils had a distinct magnetic signature. Similarly, magnetic gradiometry was used in Montréal, Canada, to assess the subsurface of landfill soils, in combination with other geophysical techniques (Boudreault et al. 2010). Like other methods, magnetic gradiometry is also used in urban archaeology (e.g. see Boschi (2012), and the examples in Chap. 2).

Electrical resistivity/conductivity of the subsurface is the basis for geophysical techniques such as electromagnetic (EM) mapping and electrical resistivity tomography (ERT). The simplest and most portable method is EM mapping, with handheld and vehicle-mounted instruments available. Examples of the use of EM mapping in urban soils include detection of buried infrastructure at a decommissioned coal mine in Lünen, Germany (Bell and Failey 1991) and mapping of various underground utilities (e.g. pipes, tanks) in Jeddah City, Saudi Arabia, in combination with GPR (Rashed and Atef 2015). The investigation by Rashed and Atef (2015) also utilised the ability of EM methods to provide magnetic susceptibility data, allowing discrimination of different materials (plastics, metals).

Electrical resistivity tomography is a more complex technique, logistically, in terms of numerical processing requiring inverse modelling of raw data but importantly with respect to the additional information provided. It relies on acquisition of data, over typical time frames of 0.5 to a few hours, from linear electrode arrays inserted into surface soil. The type of information provided by ERT typically relates to soil water content and soil texture, since both affect the electrical conductivity of the subsurface. An example of the use of ERT in an urban soil environment was to assess land suitability for potential urban or tourism development near the urban area of As Siliyin, El-Fayoum, Egypt {Metwaly, 2010 #5646`; see also Fig. 5.14}. The ERT information collected by Metwaly et al. (2010) was used to discriminate sediment textures and water content, to help understand the groundwater resource at the site (Fig. 5.14). Boudreault et al. (2010) showed that ERT also has the capability to detect solid, high-resistivity zones and objects in urban soil-like materials. An archaeological application of EM and ERT in cities, to locate previously unknown bronze production sites in Athens, Greece, is described by Leopold et al. (2011).

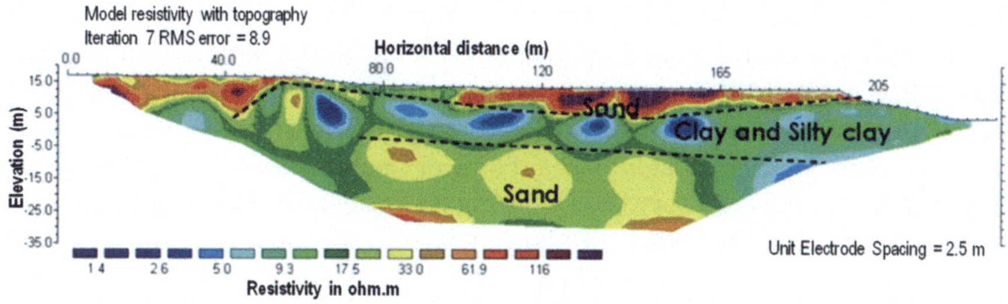

Fig. 5.14 Electrical resistivity tomography inversion image for a cross section in the peri-urban area of As Siliyin, Egypt (30.793 E, 29.381 N). (Image from Metwaly et al. (2010); used with permission from Springer)

It is worth mentioning *seismic geophysical techniques*, because as well as providing in situ information about the subsurface environment, they can provide estimates of the risks posed by earthquakes, as well as minor seismic phenomena such as vibrations from heavy vehicles and construction activities. An application of surface-based, non-destructive seismic methods for soils and volcanic sediments with a range of consolidation is described for the urban environment of Napoli, Italy, by Nunziata et al. (2004).

Other geophysical techniques include (but are not limited to) *induced polarisation*, which is related to ERT but measures the ability of different subsurface materials to hold an electrical charge (e.g. Cardarelli and Di Filippo 2004), and magnetic resonance sounding, which can yield information about free water content and hydraulic conductivity (e.g. Lubczynski and Roy 2003). Both techniques are used down pre-existing boreholes through soil and underlying material, and induced polarisation can also be used in linear electrode arrays in the same way as ERT.

5.4.3 Remote Sensing Methods

Remotely sensed data, using imagery from satellites or aircraft, have been used for many years to assess urban environments, including collection of data on urban soils. Many landform or land use parameters, and soil physical properties, are accessible by remote sensing. Only a few soil chemical and biological properties are accessible using remote sensing methods.

Satellite data. In an early example, Ormsby (1982) used Landsat 3 visible, near-infrared, and thermal data to discriminate selected pairs of land covers including urban vs. agricultural or urban vs. unvegetated/extractive industry land. More recent investigations using Landsat data are based on fitting remotely sensed imagery to the vegetation – impervious surface – soil (V-I-S) model (e.g. Phinn et al. 2002; Wu and Murray 2003), with an important outcome being reliable estimation of impervious surface cover in cities. Other satellite-derived data, such as MODIS or ATLAS, have been used to map various parameters relevant to urban soil environments, including soil temperatures and urban heat island effects (Schneider et al. 2012), detecting the sources of urban air pollution (Xu et al. 2005), inferred soil organic carbon (Bae and Ryu 2015), evapotranspiration and plant water requirements (Nouri et al. 2016), and simulation of urban soil water content (Chiesi et al. 2019). *Synthetic aperture radar* (SAR) is an additional airborne or satellite-based sensing technique that effectively simulates a very large radar antenna ('aperture') by virtue of the different positions along the flight path, allowing high-resolution images and measurement of the height of objects or land elevations. SAR has been used for some years to map urban expansion and urban land use (e.g. Henderson and Xia 1997). Additional applications include estimation of soil water content (Moeremans and Dautrebande 2000), assessment of land subsidence (Tosi et al. 2009), building height (Colin-Koeniguer and Trouvé 2014), vegetation indices (Kim et al. 2014), impervious surface cover (Zhang et al. 2018), and flood water monitoring (Chini et al. 2019).

Airborne data. Aerial photography was probably the first remote sensing data layer to be used and is still commonly used in the mapping of urban land use or land surface cover (e.g. Grant and Finlayson 1978; Fox et al. 2012). High-resolution aerial photography has also been used as calibration data for analysis of satellite images (Ormsby 1982). Airborne radiometric data have been used widely in mineral exploration for several decades and have also provided information useful to urban areas. Relying on the natural radioactivity caused by low but detectable concentrations of radioactive potassium, thorium, and uranium isotopes, airborne radiometric measurements provide remotely sensed information to allow discrimination of solid earth-surface materials such as different types of rocks and soils. Airborne radiometric measurements have been used in an urban context by Beamish and Busby (2016) to assess peri-urban geological structures for geothermal potential. Another potentially fruitful application was described by Bierwirth and Brodie (2005), who found that the radiometric thorium (Th) signal was depleted in acid sulphate soil environments. Although Bierwirth and Brodie's (2005) study was not in an urban environment, the incidence of acid sulphate soil processes in new urban developments (especially in coastal areas) could make airborne radiometrics a useful monitoring tool.

5.5 Additional Reading

Hillel D (2014) Environmental soil physics: fundamentals, applications, and environmental considerations. Academic Press, San Diego, USA

Verma SK, Sharma SP (Eds.) (2011) Urban Geophysics (Special Issue with 21 articles). *Physics and Chemistry of the Earth, Parts A/B/C*, Vol. 36, Issue 16, pp. 1209–1436. Elsevier, Amsterdam, The Netherlands

5.6 Summary

Urban soils can have numerous physical constraints which affect their ability either to perform ecosystem services or to support urban infrastructure

The potentially adverse physical properties include:

- Surface or subsurface sealing
- High density and associated low porosity
- The existence of substantial proportions of coarse materials, artefacts, or buried infrastructure
- Weak on non-existent soil structure (for ecosystem services)
- Soil strength which is inappropriate to the desired soil functions
- High soil erodibility
- High soil temperatures

Adverse soil physical properties in urban environments may have undesirable effects on the ecosystem services or engineering functions of soils. The undesired effects may differ or even oppose one another depending on the soil management objectives (e.g. supporting vegetation or use as a structural foundation)

The physical properties of soils can be either static or dynamic. There are many methods for determining these properties, based on combinations of field and/or laboratory measurements. Some measurements must be made in situ in the field (e.g. infiltration rate or soil temperature), while others must be made in a laboratory (e.g. shear strength, particle density).

Soil physical morphology and properties are amenable to geophysical and remote sensing techniques to a greater degree than are soil biological or chemical measurements. Geophysical methods such as ground-penetrating radar, magnetic methods, or techniques based on electrical properties can provide useful information over larger spatial extents in urban soils. In addition remotely sensed (airborne or satellite) data has been used to estimate urban soil physical conditions.

5.7 Questions

5.7.1 Checking Your Understanding

1. Which soil physical constraints are likely to affect soil water storage and movement in urban soils? What is the nature of the effects that you have identified?
2. Some urban soil physical constraints are likely to affect soil biological functioning – what are these constraints? Which soil organisms might be affected and why?

3. Identify the soil physical processes which (a) have adverse effects on both ecosystem services and engineered structures and (b) have opposite effects on ecosystem services and engineered structures. Try to explain the differences!
4. Draw an annotated diagram which shows the water and heat fluxes (and any changes in these fluxes) involved in cooling of urban atmosphere and land surfaces by trees.
5. List the soil physical methods which directly or indirectly relate to measurement of soil water storage or movement. Describe which aspect of the behaviour of water in soils is being measured in each case.
6. Using examples, describe the differences between ground-based and remote sensing geophysical techniques for measurement or estimation of physical properties of urban soils.

   (For the following sets of questions, you might need to do some additional reading.)

5.7.2 *Thinking About the Issues*

7. Explain why the analogy of a karst landscape is helpful for understanding water and solute flow in urban soils (or why it is not!).
8. Discuss the possible advantages and disadvantages of field versus laboratory measurement of the following soil physical properties: bulk density, water content, texture, penetration resistance, and electrical conductivity.
9. Is the often-cited water balance graphic (e.g. Fig. 5.1) still valid for urban ecosystems? Why (or why not)?

5.7.3 *Contemplating Soils Creatively*

10. Discuss whether it would be possible to remove the urban heat island effect in part or all of an urban area. What would be the best strategies for cooling soils in an urban environment?

References

Alexandrovskaya EI, Alexandrovskiy AL (2000) History of the cultural layer in Moscow and accumulation of anthropogenic substances in it. Catena 41:249–259. https://doi.org/10.1016/S0341-8162(00)00107-7

Almeida MSS, Borma LS, Barbosa MC (2001) Land disposal of river and lagoon dredged sediments. Eng Geol 60:21–30. https://doi.org/10.1016/S0013-7952(00)00085-5

Athanasopoulou E, Tombrou M, Russell AG, Karanasiou A, Eleftheriadis K, Dandou A (2010) Implementation of road and soil dust emission parameterizations in the aerosol model CAMx: Applications over the greater Athens urban area affected by natural sources. J Geophys Res Atmos:115. https://doi.org/10.1029/2009JD013207

Bae J, Ryu Y (2015) Land use and land cover changes explain spatial and temporal variations of the soil organic carbon stocks in a constructed urban park. Landsc Urban Plan 136:57–67. https://doi.org/10.1016/j.landurbplan.2014.11.015

Banks D, Gandy CJ, Younger PL, Withers J, Underwood C (2009) Anthropogenic thermogeological 'anomaly' in Gateshead, Tyne and Wear, UK. Q J Eng Geol Hydrogeol 42:307–312. https://doi.org/10.1144/1470-9236/08-024

Beamish D, Busby J (2016) The Cornubian geothermal province: heat production and flow in SW England: estimates from boreholes and airborne gamma-ray measurements. Geothermal Energy 4. https://doi.org/10.1186/s40517-016-0046-8

Bell RM, Failey RA (1991) Plant uptake of organic pollutants. In: Jones KC (ed) Organic contaminants in the environment: environmental pathways & effects, Elsevier Applied Science, pp 189–206. https://doi.org/10.1007/978-94-009-4329-2_6

Bierwirth PN, Brodie RS (2005) Identifying acid sulfate soil hotspots from airborne gamma-radiometric data and GIS analysis. Australian Government Bureau of Rural Sciences, Canberra

Bonneau J, Fletcher TD, Costelloe JF, Burns MJ (2017) Stormwater infiltration and the 'urban karst' – a review. J Hydrol 552:141–150. https://doi.org/10.1016/j.jhydrol.2017.06.043

Boschi F (2012) Magnetic prospecting for the archaeology of Classe (Ravenna). Archaeol Prospect 19:219–227. https://doi.org/10.1002/arp.1430

Boudreault J-P, Dubé J-S, Chouteau M, Winiarski T, Hardy É (2010) Geophysical characterization of contaminated urban fills. Eng Geol 116:196–206. https://doi.org/10.1016/j.enggeo.2010.09.002

British Standards Institution (1986) BS 8004: code of practice for foundations. UK, London

Cardarelli E, Di Filippo G (2004) Integrated geophysical surveys on waste dumps: evaluation of physical parameters to characterize an urban waste dump (four case studies in Italy). Waste Manag Res 22:390–402. https://doi.org/10.1177/0734242X04046042

Chan YC, Cohen DD, Hawas O, Stelcer E, Simpson R, Denison L et al (2008) Apportionment of sources of fine and coarse particles in four major Australian cities by positive matrix factorisation. Atmos Environ 42:374–389. https://doi.org/10.1016/j.atmosenv.2007.09.030

Changnon SA (1999) A rare long record of deep soil temperatures defines temporal temperature changes and an urban heat island. Clim Change 42:531–538. https://doi.org/10.1023/A:1005453217967

Chatterjea K (2007) Assessment and demarcation of trail degradation in a nature reserve, using GIS: case of Bukit Timah nature reserve. Land Degrad Dev 18:500–518. https://doi.org/10.1002/ldr.793

Chen L, Zhu W, Zhou X, Zhou Z (2003) Characteristics of the heat island effect in Shanghai and its possible mechanism. Adv Atmos Sci 20:991–1001

Chen Y, Day SD, Wick AF, McGuire KJ (2014) Influence of urban land development and subsequent soil rehabilitation on soil aggregates, carbon, and hydraulic conductivity. Sci Total Environ 494-495:329–336. https://doi.org/10.1016/j.scitotenv.2014.06.099

Chiesi M, Battista P, Fibbi L, Gardin L, Pieri M, Rapi B et al (2019) Spatio-temporal fusion of NDVI data for simulating soil water content in heterogeneous Mediterranean areas. Europ J Remote Sensing 52:88–95. https://doi.org/10.1080/22797254.2018.1557501

Chini M, Pelich R, Pulvirenti L, Pierdicca N, Hostache R, Matgen P (2019) Sentinel-1 InSAR coherence to detect floodwater in urban areas: Houston and hurricane Harvey as a test case. Remote Sens (Basel) 11. https://doi.org/10.3390/rs11020107

Colin-Koeniguer E, Trouvé N (2014) Performance of building height estimation using high-resolution PolInSAR images. IEEE Trans Geosci Remote Sens 52:5870–5879. https://doi.org/10.1109/TGRS.2013.2293605

Coutts AM, Tapper NJ, Beringer J, Loughnan M, Demuzere M (2013) Watering our cities: the capacity for water sensitive Urban Design to support urban cooling and improve human thermal comfort in the Australian context. Prog Phys Geograph: Earth & Environ 37:2–28. https://doi.org/10.1177/0309133312461032

Cresswell H, Coughlan K, McKenzie N (2002) Soil physical measurement and interpretation for land evaluation. CSIRO Publishing, Victoria, Australia, ix, 379 pp.
Dane JH, Clarke Topp G (Eds.) (2002) Methods of soil analysis. Part 4. Physical methods. SSSA Book Series, Number 5. Soil Science Society of America, Inc., Madison, Wisconsin, USA
de Lima RP, da Silva AR, da Silva AP, Leao TP, Mosaddeghi MR (2016) Soilphysics: an R package for calculating soil water availability to plants by different soil physical indices. Comput Electron Agric 120:63–71. https://doi.org/10.1016/j.compag.2015.11.003
Del Mar López T, Aide TM, Scatena FN (1998) The effect of land use on soil erosion in the Guadiana watershed in Puerto Rico. Caribb J Sci 34:298–307
Department of Environment (2004) Chapter 7 Non-structural controls. Stormwater Management Manual for Western Australia, Chalmers, L., Gray, S. (Eds.). Government of Western Australia, Perth, Australia
El Khalil H, Schwartz C, El Hamiani O, Sirguey C, Kubiniok J, Boularbah A (2016) How physical alteration of technic materials affects mobility and phytoavailabilty of metals in urban soils? Chemosphere 152:407–414. https://doi.org/10.1016/j.chemosphere.2016.02.116
Eliasson I, Jonsson P, Holmer B (2008) Diurnal and intra-urban particle concentrations in relation to windspeed and stability during the dry season in three African cities. Environ Monit Assess 154:309. https://doi.org/10.1007/s10661-008-0399-y
Erskine WD, Mahmoudzadeh A, Browning CM, Myers C (2003) Sediment yields and soil loss rates from different land uses on Triassic shales in western Sydney, NSW. Soil Res 41:127–140. https://doi.org/10.1071/SR01078
Fletcher T, Duncan H, Poelsma P, Lloyd S (2004) Stormwater flow and quality, and the effectiveness of non-proprietary Stormwater treatment measures : a review and gap analysis. Technical Report 04/8, Melbourne, Australia
Foster GR, Toy TE, Renard KG (2003) Comparison of the USLE, RUSLE1.06c, and RUSLE2 for application to highly disturbed lands, Proc. Interagency Conf. on Research in the Watersheds, 1st, Benson, AZ, USA, pp. 154–160
Fourvel GJ, Vidal-Beaudet L, Le Bocq A, Thery F, Brochier V, Cannavo P (2019) Fertility of Technosols constructed with dam sediments for urban greening and land reclamation. J Soil Sediment 19:3178–3192. https://doi.org/10.1007/s11368-018-2077-1
Fox DM, Witz E, Blanc V, Soulié C, Penalver-Navarro M, Dervieux A (2012) A case study of land cover change (1950-2003) and runoff in a Mediterranean catchment. Appl Geogr 32:810–821. https://doi.org/10.1016/j.apgeog.2011.07.007
Geotechdata.info (2015) Soil bearing capacity, http://geotechdata.info/parameter/bearing-capacity.html (accessed 20190808)
Gleason FH, Letcher PM, McGee PA (2004) Some Chytridiomycota in soil recover from drying and high temperatures. Mycol Res 108:583–589. https://doi.org/10.1017/S0953756204009736
Grant K, Finlayson AA (1978) The assessment and evaluation of geotechnical resources in urban or regional environments. Eng Geol 12:219–293. https://doi.org/10.1016/0013-7952(78)90014-5
Grimm NB, Faeth SH, Golubiewski NE, Redman CL, Wu J, Bai X, Briggs JM (2008) Global change and the ecology of cities. Science 319:756–760
Gwenzi W, Nyamadzawo G (2014) Hydrological impacts of urbanization and urban roof water harvesting in water-limited catchments: a review. Environ Process 1:573–593. https://doi.org/10.1007/s40710-014-0037-3
Hafner J, Kidder SQ (1999) Urban heat island modeling in conjunction with satellite-derived surface/soil parameters. J Appl Meteorol 38:448–465. https://doi.org/10.1175/1520-0450(1999)038<0448:UHIMIC>2.0.CO;2
Hancox GT (2008) The 1979 Abbotsford landslide, Dunedin, New Zealand: a retrospective look at its nature and causes. Landslides 5:177–188. https://doi.org/10.1007/s10346-007-0097-9
Hazelton P, Murphy B (2011) Understanding Soils in Urban Environments. CSIRO Publishing, Collingwood, Australia, 160 pp

Henderson FM, Xia ZG (1997) SAR applications in human settlement detection, population estimation and urban land use pattern analysis: a status report. IEEE Trans Geosci Remote Sens 35:79–85. https://doi.org/10.1109/36.551936

Hillel D (2014) Environmental soil physics fundamentals, applications, and environmental considerations. Academic Press, San Diego

Hulisz P, Charzyński P, Greinert A (2018) Urban soil resources of medium-sized cities in Poland: a comparative case study of Toruń and Zielona Góra. J Soil Sediment 18:358–372. https://doi.org/10.1007/s11368-016-1596-x

Jim CY (1998a) Soil characteristics and management in an urban park in Hong Kong. Environ Manag 22:683–695. https://doi.org/10.1007/s002679900139

Jim CY (1998b) Urban soil characteristics and limitations for landscape planting in Hong Kong. Landsc Urban Plan 40:235–249. https://doi.org/10.1016/S0169-2046(97)00117-5

Julian M, Anthony E (1996) Aspects of landslide activity in the Mercantour Massif and the French Riviera, southeastern France. Geomorphology 15:275–289. https://doi.org/10.1016/0169-555X(95)00075-G

Khresat SA, Rawajfih Z, Mohammad M (1998) Land degradation in North-Western Jordan: causes and processes. J Arid Environ 39:623–629. https://doi.org/10.1006/jare.1998.0385

Kim YH, Oh JH, Kim YI (2014) Comparative analysis of the multispectral vegetation indices and the radar vegetation index. J Korean Soc Surv Geod Photogramm Cartogr 32:607–615. https://doi.org/10.7848/ksgpc.2014.32.6.607

Knox EG, Bouchard CE, Barrett JG (2000) Erosion and sedimentation in urban areas (chapter 7). In: Brown RB, Huddleston JH, Anderson JL (eds) Managing soils in an urban environment. Agronomy Monograph. American Society of Agronomy, Crop Science Society of America, Soil Science Society of America, Madison, WI, pp 179–197. https://doi.org/10.2134/agronmonogr39.c7

Kumar K, Hundal L (2016) Soil in the city: sustainably improving urban soils. J Environ Qual 45:2–8. https://doi.org/10.2134/jeq2015.11.0589

Laflen JM, Flanagan DC (2013) The development of U. S. soil erosion prediction and modeling. Int Soil & Water Conserv Res 1:1–11. https://doi.org/10.1016/S2095-6339(15)30034-4

Landsberg HE, Maisel TN (1972) Micrometeorological observations in an area of urban growth. Bound-Lay Meteorol 2:365–370. https://doi.org/10.1007/bf02184776

Leopold M, Gannaway E, Volkel J, Haas F, Becht M, Heckmann T et al (2011) Geophysical prospection of a bronze foundry on the southern slope of the acropolis at Athens, Greece. Archaeol Prospect 18:27–41. https://doi.org/10.1002/arp.402

Li X, Chang SX, Salifu KF (2013) Soil texture and layering effects on water and salt dynamics in the presence of a water table: a review. Environ Rev 22:41–50. https://doi.org/10.1139/er-2013-0035

Lin BS, Lin YJ (2010) Cooling effect of shade trees with different characteristics in a subtropical urban park. HortScience 45:83–86

Liu C, Evett JB (2008) Soils and foundations, 7th edn. Pearson/Prentice Hall, Upper Saddle River, NJ

Lubczynski M, Roy J (2003) Hydrogeological interpretation and potential of the new magnetic resonance sounding (MRS) method. J Hydrol 283:19–40. https://doi.org/10.1016/S0022-1694(03)00170-7

Magiera T, Strzyszcz Z, Kapicka A, Petrovsky E (2006) Discrimination of lithogenic and anthropogenic influences on topsoil magnetic susceptibility in Central Europe. Geoderma 130:299–311. https://doi.org/10.1016/j.geoderma.2005.02.002

Mao Q, Huang G, Buyantuev A, Wu J, Luo S, Ma K (2014) Spatial heterogeneity of urban soils: the case of the Beijing metropolitan region, China. Ecological Process 3. https://doi.org/10.1186/s13717-014-0023-8

Martin A, Gunter JT, Regens JL (2003) Estimating erosion in a riverine watershed: Bayou Liberty-Tchefuncta river in Louisiana. Environ Sci Pollut Res 10:245–250. https://doi.org/10.1065/espr2003.05.153

McKenzie NJ, Jacquier D, Isbell R, Brown K (2004) Australian soils and landscapes: an illustrated compendium. CSIRO Publishing, Collingwood, VIC
McLean MA, Angilletta MJ Jr, Williams KS (2005) If you can't stand the heat, stay out of the city: thermal reaction norms of chitinolytic fungi in an urban heat island. J Therm Biol 30:384–391. https://doi.org/10.1016/j.jtherbio.2005.03.002
Metwaly M, El-Qady G, Massoud U, El-Kenawy A, Matsushima J, Al-Arifi N (2010) Integrated geoelectrical survey for groundwater and shallow subsurface evaluation: case study at Siliyin spring, El-Fayoum, Egypt. Int J Earth Sci 99:1427–1436. https://doi.org/10.1007/s00531-009-0458-9
Min M, Lin C, Duan X, Jin Z, Zhang L (2019) Spatial distribution and driving force analysis of urban heat island effect based on raster data: a case study of the Nanjing metropolitan area, China. Sustain Cities Soc 50:101637. https://doi.org/10.1016/j.scs.2019.101637
Moeremans B, Dautrebande S (2000) Soil moisture evaluation by means of multi-temporal ERS SAR PRI images and interferometric coherence. J Hydrol 234:162–169. https://doi.org/10.1016/S0022-1694(00)00251-1
Montague T, Kjelgren R (2004) Energy balance of six common landscape surfaces and the influence of surface properties on gas exchange of four containerized tree species. Sci Hortic 100:229–249. https://doi.org/10.1016/j.scienta.2003.08.010
Morel JL, Chenu C, Lorenz K (2015) Ecosystem services provided by soils of urban, industrial, traffic, mining, and military areas (SUITMAs). J Soil Sediment 15:1659–1666. https://doi.org/10.1007/s11368-014-0926-0
Morgenroth J, Buchan GD (2009) Soil moisture and aeration beneath pervious and impervious pavements. Arboriculture and Urban Forestry 35:135–141
Morin TH, Bohrer G, Frasson RPDM, Naor-Azreli L, Mesi S, Stefanik KC, Schäfer KVR (2014) Environmental drivers of methane fluxes from an urban temperate wetland park. J Geophys Res G: Biogeosci 119:2188–2208. https://doi.org/10.1002/2014JG002750
Nouri H, Glenn EP, Beecham S, Boroujeni SC, Sutton P, Alaghmand S et al (2016) Comparing three approaches of evapotranspiration estimation in mixed urban vegetation: field-based, remote sensing-based and observational-based methods. Remote Sens (Basel) **8**. https://doi.org/10.3390/rs8060492
Nunziata C, Natale M, Panza GF (2004) Seismic characterization of Neapolitan soils. Pure Appl Geophys 161:1285–1300. https://doi.org/10.1007/s00024-003-2492-z
Ormsby JP (1982) The use of Landsat-3 thermal data to help differentiate land covers. Remote Sens Environ 12:97–105. https://doi.org/10.1016/0034-4257(82)90029-3
Ozsoy G, Aksoy E (2015) Estimation of soil erosion risk within an important agricultural sub-watershed in Bursa, Turkey, in relation to rapid urbanization. Environ Monit Assess 187:419. https://doi.org/10.1007/s10661-015-4653-9
Paul MJ, Meyer JL (2001) Streams in the urban landscape. Annu Rev Ecol Syst 32:333–365
Pavao-Zuckerman MA, Coleman DC (2005) Decomposition of chestnut oak (Quercus prinus) leaves and nitrogen mineralization in an urban environment. Biol Fertil Soils 41:343–349. https://doi.org/10.1007/s00374-005-0841-z
Peters EB, McFadden JP (2010) Influence of seasonality and vegetation type on suburban microclimates. Urban Ecosyst 13:443–460. https://doi.org/10.1007/s11252-010-0128-5
Phinn S, Stanford M, Scarth P, Murray AT, Shyy PT (2002) Monitoring the composition of urban environments based on the vegetation-impervious surface-soil (VIS) model by subpixel analysis techniques. Int J Remote Sens 23:4131–4153. https://doi.org/10.1080/01431160110114998
Rahardjo H, Indrawan IGB, Leong EC, Yong WK (2008) Effects of coarse-grained material on hydraulic properties and shear strength of top soil. Eng Geol 101:165–173. https://doi.org/10.1016/j.enggeo.2008.05.001
Rashed M, Atef A (2015) Mapping underground utilities within conductive soil using multi-frequency electromagnetic induction and ground penetrating radar. Arab J Geosci 8:2341–2346. https://doi.org/10.1007/s12517-014-1358-2

Renard KG, Foster GR, Weesies GA, McCool DK, Yoder DC (1997) Predicting soil erosion by water: a guide to conservation planning with the revised universal soil loss equation (RUSLE). Agriculture handbook no. 703, U.S. Department of Agriculture, Washington DC, USA. https://www.ars.usda.gov/ARSUserFiles/64080530/RUSLE/AH_703.pdf

Rocha JHT, Santos AJM, Diogo FA, Backes C, De Melo AGC, Borelli K, De Oliveira Godinho T (2015) Reforestation and recovery of soil chemical and physical attributes [in Portugese]. Floresta e Ambiente 22:299–306. https://doi.org/10.1590/2179-8087.041613

Rowlands L (2018) Erosion and sediment control - WSUD during the construction phase of land development. In: Sharma AK, Gardner T, Begbie D (eds) Approaches to water sensitive urban design. Elsevier Science & Technology, Amsterdam, pp 163–176

Saarenketo T, Scullion T (2000) Road evaluation with ground penetrating radar. 43: 119–138. https://doi.org/10.1016/S0926-9851(99)00052-X

Savva Y, Szlavecz K, Pouyat RV, Groffman PM, Heisler G (2010) Effects of land use and vegetation cover on soil temperature in an urban ecosystem. Soil Sci Soc Am J 74:469–480. https://doi.org/10.2136/sssaj2009.0107

Scharenbroch BC, Lloyd JE, Johnson-Maynard JL (2005) Distinguishing urban soils with physical, chemical, and biological properties. Pedobiologia 49:283–296. https://doi.org/10.1016/j.pedobi.2004.12.002

Schneider A, Logan KE, Kucharik CJ (2012) Impacts of urbanization on ecosystem goods and services in the U.S. corn belt. Ecosystems 15:519–541. https://doi.org/10.1007/s10021-012-9519-1

Short JR, Fanning DS, Foss JE, Patterson JC (1986a) Soils of the Mall in Washington, DC: II. Genesis, classification, and mapping. Soil Sci Soc Am J 50:705–710. https://doi.org/10.2136/sssaj1986.03615995005000030031x

Short JR, Fanning DS, McIntosh MS, Foss JE, Patterson JC (1986b) Soils of the Mall in Washington, DC: I. statistical summary of Properties1. Soil Sci Soc Am J 50:699–705. https://doi.org/10.2136/sssaj1986.03615995005000030030x

Small C (2006) Comparative analysis of urban reflectance and surface temperature. Remote Sens Environ 104:168–189. https://doi.org/10.1016/j.rse.2005.10.029

Stokes A, Fourcaud T, Hruska J, Cermak J, Nadyezdhina N, Nadyezhdin V, Praus L (2002) An evaluation of different methods to investigate root system architecture of urban trees in situ: I. ground-penetrating radar. J Arboric 28:2–10

Sutherland RA (2000) Bed sediment-associated trace metals in an urban stream, Oahu, Hawaii. Environ Geol 39:611–627

Tang CS, Shi B, Gao L, Daniels JL, Jiang HT, Liu C (2011) Urbanization effect on soil temperature in Nanjing, China. Energ Buildings 43:3090–3098. https://doi.org/10.1016/j.enbuild.2011.08.003

Terzaghi K, Peck RB, Mesri G (1996) Soil mechanics in engineering practice, third edition. Wiley, New York, 549 pp, doi:0471086584

Tosi L, Teatini P, Carbognin L, Brancolini G (2009) Using high resolution data to reveal depth-dependent mechanisms that drive land subsidence: the Venice coast, Italy. Tectonophysics 474:271–284. https://doi.org/10.1016/j.tecto.2009.02.026

U.S. EPA (2011) Evaluation of urban soils: suitability for green infrastructure or urban agriculture, United States Environmental Protection Agency, Washington, DC, USA. https://ucanr.edu/sites/UrbanAg/files/263151.pdf

USDA (2000) Erosion and sedimentation on construction sites, urban technical note. No. 1, United States Department of Agriculture, Natural Resources Conservation Service, Auburn, AL, USA

USEPA (1993) Guidance specifying management measures for sources of nonpoint source pollution in coastal waters. 840-B-92-002, United States Environmental Protection Agency, Washington DC, USA. https://www.epa.gov/nps/guidance-specifying-management-measures-sources-nonpoint-pollution-coastal-waters

USGA (2018) USGA recommendations for a method of putting green construction, 2018 revision. United States Golf Association, Liberty Corner

Uzielli M, Lacasse S, Nadim F, Phoon K-K (2006) Soil variability analysis for geotechnical practice. Proceedings of the 2nd International Workshop on Characterisation and Engineering Properties of Natural Soils. Taylor & Francis, Singapore, pp. 103. https://doi.org/10.1201/NOE0415426916.ch3

Volder A, Watson T, Viswanathan B (2009) Potential use of pervious concrete for maintaining existing mature trees during and after urban development. Urban Forest & Urban Green 8:249–256. https://doi.org/10.1016/j.ufug.2009.08.006

Wang H, Ouyang Z, Chen W, Wang X, Zheng H, Ren Y (2011) Water, heat, and airborne pollutants effects on transpiration of urban trees. Environ Pollut 159:2127–2137. https://doi.org/10.1016/j.envpol.2011.02.031

Weissert LF, Salmond JA, Schwendenmann L (2016) Variability of soil organic carbon stocks and soil CO2 efflux across urban land use and soil cover types. Geoderma 271:80–90. https://doi.org/10.1016/j.geoderma.2016.02.014

White RE (2006) Principles and practice of soil science: the soil as a natural resource. Blackwell Science Ltd., Malden, MA

Wolman MG, Schick AP (1967) Effects of construction on fluvial sediment, urban and suburban areas of Maryland. Water Resour Res 3:451–464. https://doi.org/10.1029/WR003i002p00451

Wong CI, Sharp JM Jr, Hauwert N, Landrum J, White KM (2012) Impact of urban development on physical and chemical hydrogeology. Elements 8:429–434. https://doi.org/10.2113/gselements.8.6.429

Wu C, Murray AT (2003) Estimating impervious surface distribution by spectral mixture analysis. Remote Sens Environ 84:493–505. https://doi.org/10.1016/s0034-4257(02)00136-0

Xiao RB, Ouyang ZY, Li WF, Zhang ZM, Tarver G Jr, Wang XK, Miao H (2005) A review of the eco-environmental consequences of urban heat islands. Acta Ecol Sin 25:2055–2060

Xu X, Zhou X, Shi X (2005) Spatial structure and scale feature of the atmospheric pollution source impact of city agglomeration. Sci China Ser D Earth Sci 48:1–24. https://doi.org/10.1360/03YD0125

Zhang H, Lin H, Wang Y (2018) A new scheme for urban impervious surface classification from SAR images. ISPRS J Photogramm Remote Sens 139:103–118. https://doi.org/10.1016/j.isprsjprs.2018.03.007

Zhu WX, Carreiro MM (2004) Variations of soluble organic nitrogen and microbial nitrogen in deciduous forest soils along an urban-rural gradient. Soil Biol Biochem 36:279–288. https://doi.org/10.1016/j.soilbio.2003.09.011

Ziska LH, Bunce JA, Goins EW (2004) Characterization of an urban-rural CO2/temperature gradient and associated changes in initial plant productivity during secondary succession. Oecologia 139:454–458. https://doi.org/10.1007/s00442-004-1526-2

Zou C, Penfold C, Sands R, Misra RK, Hudson I (2001) Effects of soil air-filled porosity, soil matric potential and soil strength on primary root growth of radiata pine seedlings. Plant and Soil 236:105–115. https://doi.org/10.1023/A:1011994615014

Chapter 6
Inorganic Contaminants in Urban Soils

Andrew W. Rate

Abstract If, in the absence of prior knowledge, we were to guess one feature of urban soils, we might assume that pollution with hazardous inorganic substances, such as 'heavy metals' or asbestos, would be widespread. In many cases this is true, and so this chapter deals with the topic of inorganic contaminants in urban soils. We identify nutrients, metals and metalloids, asbestos, plastics, manufactured nanoparticles, radionuclides, and cyanide as contaminants of concern and discuss their sources in urban soils. In addition, we cover the formation and properties of acid sulphate soils. Following this, we explain the behaviour of inorganic contaminants, going into some more detail on soil chemical reactions than was presented in Chap. 4. The discussion addresses the effects of transport processes, time, and soil properties on the relevant chemical reactions. Some further discussion of bioavailability is coupled with a discussion of the effects of urban soil contamination on ecosystem services. The later sections present methods for soil chemical analyses and concepts for understanding soil chemical data. The analytical method component covers total and partial elemental analyses and field measurements. Following this, we examine the concepts of background concentrations, regulatory contamination thresholds, separation of geogenic and anthropogenic sources of contaminants, and contamination indices.

Keywords Soil chemistry · Urban soils · Chemical properties · Nutrients · Metals · Acid sulphate soils · Speciation · Bioavailability · Transport

---

A. W. Rate (✉)
School of Agriculture and Environment, University of Western Australia, Crawley, WA, Australia
e-mail: andrew.rate@uwa.edu.au

A. W. Rate (ed.), *Urban Soils*, Progress in Soil Science,
https://doi.org/10.1007/978-3-030-87316-5_6

What you could learn from this chapter:

- The inorganic contaminants of concern in urban soils and where they come from.
- An introduction to acid sulphate soils in urban environments.
- The processes controlling immobilisation, mobility, and bioavailability of inorganic contaminants.
- Why there are many ways of analysing soil for inorganic contaminants and what information the analyses give us.
- Some ways we can use soil analysis data to distinguish natural and anthropogenic sources of inorganic contaminants.

6.1 Sources of Inorganic Contaminants in Urban Soils

An understanding of the potential sources of contaminants is essential for assessment of urban soil environments. In a formal environmental impact procedure, the likely contaminant sources form part of a conceptual site model (CSM), which will be discussed further in Chap. 9. Substances which are potential contaminants may be subdivided into two categories: first, those elements and substances which are from recognised sources and have usually been added to soils during the *Anthropocene* era (i.e. since about 1950, when accelerating human activity began to significantly change environments on a global scale). Second, contaminants (often the same substances) have also been added to urban soils by humans, in historical times (subjectively, before the twentieth century), which provide indicators of urban/anthropogenic impact (e.g. Strauch et al. 2008).

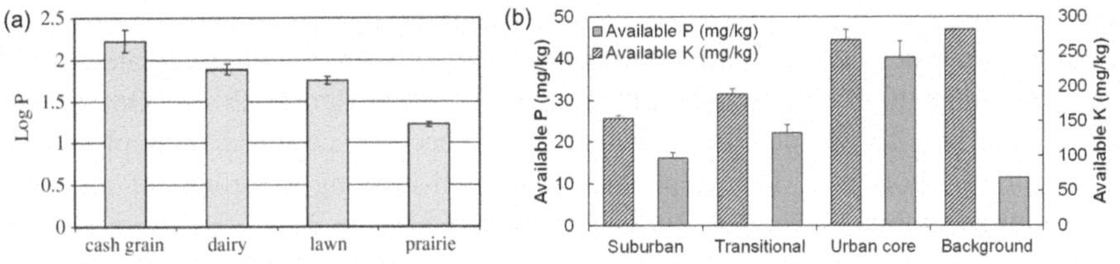

Fig. 6.1 (**a**) Phosphorus concentration ($\log_{10}$(mg/kg); extracted from soil using the Bray-1 'available' P method) in soils under different land uses in and around the Madison urban area, Wisconsin, USA (from Bennett et al. 2005; used with permission from Springer). 'Cash grain' and 'dairy' are agricultural land uses; 'lawn' is residential urban; 'prairie' is remnant natural vegetation. Urban land has less P than agricultural and greater P than natural (significance $p < 0.01$). (**b**) Available P and K in urban soils of Beijing, China (redrawn from data in Mao et al. 2014)

6.1.1 *Sources of Nutrients*

Since early history (at least as early as the Bronze age; Golyeva et al. 2018), soils in urban areas commonly have become more concentrated in nutrients. This nutrient signature is predominantly manifested as increased phosphorus and carbon contents (both elements are more persistent in soils than nitrogen; see Deevey et al. 1979; Davidson et al. 2006). Urban soils which have received recent fertiliser or organic amendments may also be enriched in nitrogen (Gregory et al. 2016). An excess of nutrients can occur in urban soils on land currently or historically used for horticulture and other forms of plant production, since these soils commonly receive large amounts of synthetic and organic fertilisers (Mann et al. 2002; Sangare et al. 2012; Mao et al. 2014). Soils used primarily for plant growth also include urban lawns (Bennett et al. 2005 – see Fig. 6.1) and those reclaimed from coastal saline environments (Li et al. 2014). The export of nutrients from urban soil (e.g. residential lawns) is strongly dependent on soil properties. Export of nitrogen and phosphorus has been measured to be much higher if soil properties favoured run-off rather than infiltration of rainwater (Easton and Petrovic 2008b; Easton and Petrovic 2008a). Another important source of excess soil nutrients in some urban environments is the discharge of untreated or minimally treated wastewater or solid wastes onto soil, even if the nutrient content of the wastewater represents beneficial reuse of an otherwise discarded material (Anikwe and Nwobodo 2002; Nyenje et al. 2010).

6.1.2 *Sources of Metals*

Natural inputs. Trace elements occur widely in the parent materials of soils, at concentrations which are broadly dependent on lithology. Global summaries of average concentrations of trace elements in different rock types have been presented by Kabata-Pendias (2011), Alloway (1995), and Adriano (2001), but to present these data here is beyond the scope of this text. Some soils have naturally high concentrations of one or more trace elements, derived from their parent rocks. For example, soils developed on ultramafic or serpentine rocks commonly contain high concentrations of chromium and/or nickel (Gonnelli and Renella 2012). During the process of soil formation, trace elements may become enriched or depleted in different soil horizons. This depends on the soil-forming environment and the predominant chemical reactions for each element (Adriano 2001). For example, Cu and Pb may become enriched relative to parent material in the surface organic horizons, following plant uptake from depth. The metals remain in surface soils because of their strong associations with pedogenic minerals (such as iron oxides and clays) and soil organic matter (Davies 1988; Baker and Senft 1995). External inputs of trace elements include atmospheric deposition, which is minimal in many natural environments (e.g. Ljung et al. 2010) but may be locally important, especially in urban areas (Nriagu 1988).

Anthropogenic Inputs Elevated concentrations of metals in soils are commonly derived from human activity, and (not surprisingly) most research into metal ion *bioavailability* (see Box 6.1) has been concerned with contaminated environments. Anthropogenic sources of trace metals have been reviewed exhaustively (e.g. see Adriano 2001; Kabata-Pendias 2011; Alloway 2012). Industrial inputs include mining and ore processing, smelting, and other metallurgical processes and a wide range of metal-utilising industries (Chen et al. 1997; Wong et al. 2002; Laidlaw et al. 2018). Several agricultural and horticultural activities, such as market gardening, contribute metals, particularly in (peri)urban systems. These sources of trace elements in horticulture include use of fertilisers with trace element supplements or impurities (Mann et al. 2002), application of metal-contaminated organic wastes including sewage sludge or effluents from intensive animal industries (Sterrett et al. 1996), and use of metal-containing pesticides, especially Cu and Zn in fungicides. Enhanced atmospheric deposition of some metals has also resulted from use and disposal of fossil fuels and their residues. Specifically urban sources include road traffic (Adachi and Tainosho 2004; Mielke et al. 2010), construction and weathering of buildings and other infrastructure (Davis et al. 2001), and indirect contributions from road dust (Harrison et al. 1981; Kelly et al. 1996; De Miguel et al. 1997; Banerjee 2003). Urban environments are not always predictably enriched in trace elements on a whole-city basis; for example, Acosta et al. (2015) found that population density does not affect metal concentrations in soil in some Spanish cities.

6.1.3 Sources of Other Contaminants

Asbestos The asbestos group of minerals (such as chrysotile, crocidolite, or tremolite, having fibrous crystal forms) is naturally occurring and has been mined and used widely in commercial products such as insulation, fireproof building materials, some concrete products, and vehicle brake linings (Alloway 2004; USEPA 2008; Department of Health (WA) 2009). In addition, naturally occurring asbestos may exist in soils near natural asbestos deposits (Lee et al. 2008). Asbestos may occur in urban soils following abrasion or disposal of any of these manufactured items or weathering of rocks containing minerals in the asbestos group. Human exposure to asbestos presents well-established health risks, which are discussed in Chap. 10. The risk is dependent on the form of asbestos in soil; the Department of Health (WA) (2009) identifies, in increasing order of risk, asbestos-containing material (intact products or larger fragments present in soil via disposal or spillage), fibrous asbestos (friable and smaller fragments of asbestos), and asbestos fines (free fibres and any fragments or aggregates of asbestos passing a 7 mm sieve). In many countries the use of asbestos, and management of land with asbestos contamination, is heavily regulated.

Plastics Contamination of global environments with plastics, particularly microplastics, is extremely widespread, with significant attention being given to the presence of plastic pollution in marine environments (Derraik 2002; Browne et al. 2011, and see Fig. 6.2a) due to their perceived and actual effects on marine organisms (Gregory 2009). Microplastics are mainly generated in terrestrial environments (Jambeck et al. 2015) and are known to occur in soils (Rillig 2012), though less is known about the occurrence and ecological effects of plastics in soils (Horton et al. 2017; He et al. 2018). An important secondary source of plastic pollution in urban environments is road dust (Abbasi et al. 2017); transport of microplastics occurs through natural drainage networks such as streams (Nizzetto et al. 2016). It is likely that constructed stormwater systems are also conduits for transport of microplastics derived from road dust and other terrestrial sources, but this issue has not yet been addressed significantly in published research. Even waste products such as cigarette butts may represent a source of persistent plastic pollution in urban soils (Green et al. 2019).

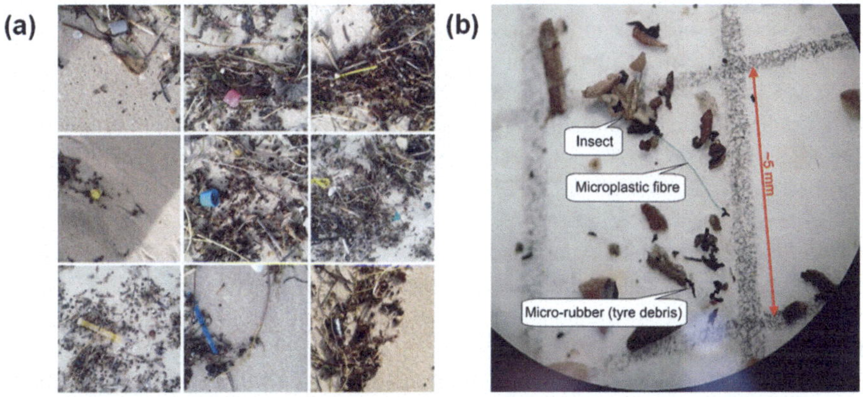

Fig. 6.2 (**a**) Plastic pollution in the wrack zone of a beach; (**b**) optical microscope image of microplastic and micro-rubber particles extracted from urban soil (images by Andrew W. Rate)

Plastics in soils, in particular microplastics (Fig. 6.2b), are of concern because they are known to interact with soil fauna (Huerta Lwanga et al. 2017; Maaß et al. 2017) and because of their capacity to adsorb and transport persistent organic pollutants (Da Costa et al. 2019) and potentially toxic metals (Zhou et al. 2019). The interactions of fauna or pollutants with plastics are not understood sufficiently well for their effects on soil ecosystem services to be assessed.

Manufactured Nanoparticles Nanomaterials, including a wide range of different type of nanoparticles, are materials manufactured with particle sizes in the range 1–100 nanometres (1 nm = 10^{-9} m). The field of nanotechnology is rapidly expanding, and nanoparticles have a wide range of applications including electronics, medical diagnosis and treatment, advanced solar power generation and storage, 'smart' fabrics, catalysis, and environmental remediation (Independent Working Group for PMSEIC 2005; National Nanotechnology Initiative 2019). The compositions of nanomaterials considered in soil research include silver (Ag), zerovalent iron (Fe), titanium dioxide, (TiO_2), magnetite (Fe_3O_4), stannic oxide (SnO_2), cerium oxide (CeO_2), antimony oxide (Sb_2O_3), zinc oxide (ZnO), carbon nanotubes, and a range of nanoparticulate plastics. Given the wide range of uses of nanomaterials and the high mobility of nanoparticles, a range of nanoparticulate substances are likely to be found in soils. To date, however, there are no measurements of manufactured nanoparticles in soils, except those that have been added experimentally to soils in a research context (Bundschuh et al. 2018). Manufactured nanoparticles need to be distinguished from the diversity of natural nanoparticles in soils (e.g. many Fe oxides, clay particles), meaning that current techniques are extremely complex (e.g. Praetorius et al. 2017 – future methods are also likely to be similarly elaborate).

Radionuclides Natural radionuclides (radioactive isotopes of some elements such as Cs, Ra, Ru, Th, U, etc.) in soils, sediments, and rocks generate the ambient radiation background which can be enhanced by urban construction using these materials (e.g. Medeiros and Yoshimura 2005). Potentially harmful concentrations of radioactive isotopes can be derived from human activities which may occur in urban areas. The most obvious of these are nuclear power generation accidents, such as Chernobyl and Fukushima (Andersson and Roed 1994; Endo et al. 2012). Other radionuclide contamination may be derived from mineral sand mining which can adjoin urban areas (Arogunjo et al. 2009) or from application of phosphogypsum fertilisers (Fernández-Caliani 2012).

Cyanide Simple cyanide compounds, mainly hydrogen cyanide (HCN) and related salts such as NaCN containing the cyanide anion CN^-, are extremely toxic with a very low fatal dose for humans. More complex cyanide compounds (such as the relatively simple $Fe^{II}(CN)_6^{4-}$) have substantially lower toxicity (Kjeldsen 1999). The most likely sources of cyanide in urban soils are current or former gas works sites, where extraction of coal gas produces HCN as one of the by-products (Mival et al. 2006). Cyanide is also used in electroplating and extraction of gold from ores, both activities which could conceivably occur in cities. Meeussen et al. (1994)

measured concentrations of free cyanide in groundwater and deduced that the concentrations were controlled by the low solubility of $Fe^{III}_4(Fe^{II}(CN)_6)_3$ (commonly called 'Prussian blue') which is formed during purification of coal gas. In a quite different context, Stec et al. (2019) found soil to be contaminated with free cyanide following the tragic Grenfell Tower fire in London, UK, in 2018. The source of cyanide in this instance was the partial combustion of, or volatilisation from, nitrogen-containing organic (e.g. polymer) materials in the fire.

6.1.4 Acid Sulphate Soils

Acid sulphate soils (often abbreviated to ASS) form from the oxidation of sulphide minerals (commonly pyrite, FeS_2) contained naturally in soils, which have previously remained in anoxic conditions due to saturation with water (Moormann 1963; Pons and Van Der Molen 1973). The exposure of such soil materials to atmospheric oxygen results in the oxidation of sulphides and the formation of sulphuric acid; if insufficient acid neutralisation capacity (e.g. carbonate minerals) is present, the soils become very acidic with $pH \leq 4$. The most common landscape category for acid sulphate soils to form in are coastal soils formed from parent materials which were previously inundated with seawater (e.g. during periods of higher sea level), such as estuaries. Drainage of these soils for urban development, such as canal construction, housing, or growing crops, reduces the water-filled pore space in the soil, allowing entry of oxygen and initiating sulphide oxidation (Figs. 6.3, 6.4). Formation of acid sulphate soils also occurs on sulphidic material dredged from drains and natural waterways (e.g. Fig. 2.7 and Fig. 6.4; Fanning 1990; Xu et al. 2018). Prior to oxidation by drainage or dredge-spoil deposition, the sulphidic soils are known as potential acid sulphate soil, or PASS, and these are distinguished from actual acid sulphate soils (sometimes abbreviated AASS) (Ahern et al. 2004).

Acid sulphate soils are known to occur in urban environments. For example, Fanning and Rabenhorst (1990) and Evans et al. (2000) describe formation of acid sulphate soils from material dredged from Baltimore Harbour, USA, within a timescale of a few weeks. Formation of acid sulphate soils on dredged sediment has also been described in the city of Brisbane, Australia (Clark and McConchie 2004). Appleyard et al. (2004) measured acidification of groundwater attributed to excavation and drainage of peat soils in Perth, Western Australia; the acidification also caused increased concentrations of dissolved arsenic, aluminium, and iron. The threats from acid sulphate soils are likely to remain an important issue, since they occur in highly populated areas in developing countries (Ljung et al. 2009). In addition, climate change and water abstraction in urban areas may accelerate their formation, if the groundwater level declines so that sulphidic materials are exposed to oxygen in soil or sediment pores (Salmon et al. 2014).

Acid sulphate soils can also be detrimental to urban infrastructure. For example, Plumlee et al. (2016) describe how the use of dredged materials in dune reconstruction caused corrosion of steel structures used for stabilisation of landforms. Concrete

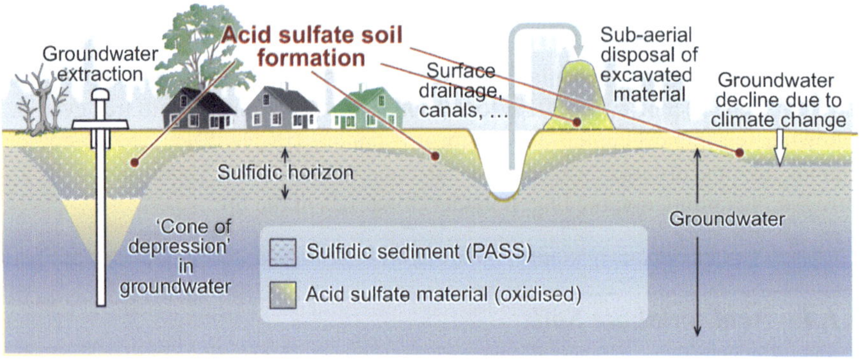

Fig. 6.3 Formation of acid sulphate soils in urban landscapes by processes which expose sulphidic material to atmospheric oxygen: groundwater extraction, excavations such as land drainage, subaerial spoil disposal, and climate change or other processes which lower groundwater such as altered hydrology (graphic by Andrew W. Rate)

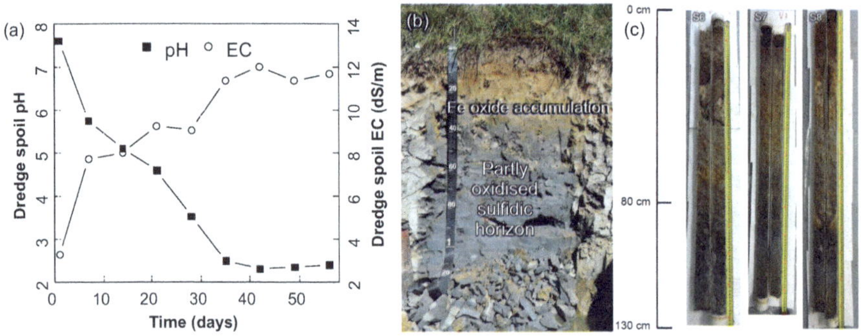

Fig. 6.4 (**a**) Change in pH and EC of dredge spoil after exposure to air (redrawn from data in from Evans et al. 2000); (**b**) acid sulphate soil profile in a peri-urban environment near Fredricksburg, Virginia, USA (coordinates 38.397 N, −77.456 W) (from Rabenhorst 2016 with annotations by Andrew W. Rate; used with permission from John Wiley and Sons); (**c**) soil cores from an acidified dredge spoil in Western Australia (−32.59 S, 115.782 E) (from Xu 2018, used with permission)

and limestone materials are also known to be damaged when exposed to acid sulphate materials (Golab and Indraratna 2009).

6.1.4.1 Other Acidic Urban Soils

Not all acidic urban soils have acidified by acid sulphate processes. Du et al. (2015) describe areas of acidic urban soils in southern China, centred on large urban areas, caused by atmospheric deposition of sulphur dioxide (SO_2), nitrogen oxides (NO_x), and ammonia (NH_3) in rainfall. In this case the SO_2, NO_x, and NH_3 are derived from emissions from industries and other urban sources.

6.2 Controls on Contaminant Behaviour in Urban Soils

6.2.1 Reactions Involving Soil Solid Phases

Many trace elements predominantly exist as cations in natural systems (metallic forms such as $Cu^{0}_{(s)}$ may exist under some redox conditions, but are uncommon). The aqueous speciation of cationic metals is mostly as stable complexes or ion pairs, as well as aquo complexes and their hydrolysis products. The cationic metals commonly form insoluble oxides, hydroxides, carbonates, phosphates and sulphides. In general, most ionic forms of nutrient and trace elements are associated in soils with solid phases as described below and in Chap. 4.

6.2.1.1 Metals in Mineral Phases

Many trace elements occur as discrete mineral phases containing the element of interest as the primary structural cation (such as hydroxides, carbonates, phosphates, and sulphides) or as ions co-precipitated in variable proportions in the structure of more common minerals such as primary and secondary silicates and oxyhydroxides of Fe, Al, and Mn. While trace metals within primary or pedogenic mineral structures provide useful data on geochemical origins and soil-forming processes, these forms of elements are not usually considered to contribute to biological uptake due to their slow cycling within terrestrial ecosystems. Exceptions to this generalisation include trace metal ions co-precipitated with secondary iron or manganese oxyhydroxides or present in sulphide minerals. Changes in soil redox potential can result in transient fluxes of dissolved metal ions from these minerals via processes such as reductive dissolution of oxides or sulphide oxidation (Gambrell 1994; Singh et al. 1996; Sullivan et al. 2013). There is some evidence to show that over a wide geographical range, several trace elements including As, Cr, Co, Cu, Ni, Pb, and Zn co-occur with iron minerals in soils (Hamon et al. 2004). It should be noted, however, that despite the predominance of metals in mineral phases and their purported lack of bioavailability (Box 6.1), correlations between total metal concentrations in soils and bioavailability as determined by plant uptake have been observed (e.g. McGrath et al. 2000).

Nanoparticulate minerals. In some cases, metals in nanoparticulate phases in soils may contribute to enhanced bioavailability (Aruoja et al. 2009; Unrine et al. 2010). It is not yet clear whether this bioavailability reflects differences in solubility of, or adsorption onto, nanoparticulate solid phases (Theng and Yuan 2008) or whether nanoparticles themselves can be assimilated by organisms. Some studies in aquatic environments have found that toxicity may be attributed to nanoparticle dissolution, rather than a direct nanoparticle effect (Franklin et al. 2007; Mortimer et al. 2010). In some cases, however, it appears that nanoparticulate Cu and Zn oxides are themselves bioavailable or induce a physiological response (Ivask et al. 2010; Johnston et al. 2010).

6.2.1.2 Metals Retained by Surface Reactions: Chemisorption, Ion Exchange

It is important to make the chemical distinction between metals contained in mineral structures in soils and metal species which are retained at the solid-solution interface by a sorption mechanism.

Chemisorption (adsorption by formation of electron-sharing bonds; see Chap. 4) is likely to be very important in terms of bioavailability, for two main reasons. First, chemisorption removes metal ions from solution and therefore controls the concentration of free ions in that solution. A theory of biological uptake called the *free ion activity model* argues that the amount of uptake of an element is related mainly to the concentration of its free (uncomplexed) ions in solution (assuming that concentration is equivalent to chemical activity, which is an acceptable approximation in dilute solutions). Second, adsorption is a reversible chemical reaction, and desorption (the reverse reaction; release from adsorption) occurs in response to depletion of the dissolved ion. Some desorption reactions are fast enough that release of ions occurs over timescales which allow significant uptake by plants and other organisms. Chemisorption is commonly considered to show poor 'reversibility', deduced from observations of desorption hysteresis, that is, an apparently different equilibrium concentration of ions for adsorption compared with desorption (McLaren et al. 1981). Desorption hysteresis is most likely an effect of slow reaction rates (Barrow 1998).

Ion exchange. The occurrence of many metals as cations (e.g. Al^{3+}, Cd^{2+}, Cu^{2+}, Ni^{2+} and Zn^{2+}) means that they can be retained by electrostatic sorption (i.e. ion exchange) on negatively charged soil colloids such as phyllosilicates and organic matter. Since variable charge is more negative at high pH, the capacity of soils to hold cations electrostatically increases as pH increases. The exchangeable fraction of metals usually represents a low proportion of total metal content in soils. Lower concentrations of exchangeable trace metals reflect the unfavourable thermodynamics of electrostatic sorption compared with other mechanisms and competition at charged particle-water interfaces by high concentrations of di- or trivalent major cations (Ca^{2+}, Mg^{2+}, Al^{3+}).

6.2.1.3 Metals in Solution and Aqueous Speciation

Concentrations of metals in soil solution are usually very low compared with those in the solid phase and reflect a combination of equilibrium with sparingly soluble and adsorbed or exchangeable forms of metals and formation of soluble inorganic and organic complexes (Wolt 1994).

6.2.2 Concentrations of Contaminants

Sorption reactions in soils can be considered to occur at 'sites' which effectively form a continuum in terms of their affinity for metal cations. At high concentrations of metals in soils, the high-affinity 'strong' chemisorption sites are fully occupied by ions. It therefore follows that the lower-affinity 'weak' sites then become more important for ion or molecule retention, since these will be the only sites left vacant for further adsorption. As a result, in contaminated soils, compared with uncontaminated, there may be a higher proportion of weakly bound ions or molecules, some of which may be held in ion-exchangeable form.

6.2.3 Transport of Metals and Nutrients

Transport phenomena for trace elements in soils have been reviewed thoroughly by Carrillo-González et al. (2006). Generally, mobility of metals is very low relative to the rate of water movement in soils. This low mobility is a consequence of reactions which retain metal ions in solid phases: (co)precipitation, adsorption, and ion exchange. The thermodynamics of these reactions favour the existence of metals in solid phases, and the low observed mobilities also suggest that rates of metal ion release to the aqueous phase are slow. Factors which affect these equilibria (mainly metal ion identity and concentration, amount and type of adsorbing phases, pH, and reduction-oxidation potential) consequently affect metal transport, as do formation of soluble complexes or adsorption on mobile colloids. The mobility of cationic metals in soils increases in the approximate order Fe^{3+} < Pb^{2+}, Hg^{2+} < Al^{3+}, Cu^{2+}, Cr^{3+} < Co^{2+}, Mn^{2+}, Ni^{2+}, Zn^{2+}, Fe^{2+} < Cd^{2+} Ca^{2+} Sr^{2+} < Na^{+}, K^{+}, NH_4^{+} (trace element order from Kabata-Pendias 2011). The mobility of common anions in soils increases in the approximate order $H_2PO_4^-$, $H_2AsO_4^-$, MoO_4^{2-}, HS^- < $H_2BO_3^-$ < SO_4^{2-} < HCO_3^-, NO_3^-, Cl^-, Br^- (Bohn et al. 2002). These rankings are only a guide, however, and the order is not replicated in all studies. For example, McQueen (2009) categorises the mobility of elements differently depending on pH (acidic, alkaline) and redox (oxidising, reducing) conditions, acknowledging that different reactions predominate depending on soil or regolith conditions.

The strong retention of many elements in soils results in small transport distances over observable timescales, which are commonly shown by the retention of contaminant-derived elements (e.g. metals; metalloids like As, Mo, or Se; or phosphate) in surface soil horizons, except in very acidic or sandy soils (Adriano 2001). Low pH, such as that encountered in acid sulphate soils or acid mine drainage, allows relatively large concentrations of metals to leach through soils (Boman et al. 2010). Mass transport of metal cations or oxyanions is likely to be dominated by preferential water flow in large, continuous soil pores (McLaren et al. 2004), and the amounts of metals transported are small reflecting the low concentrations in solution. In addition, there is increasing recognition of colloid transport (transport of sub-microscopic particles with metals or other elements attached) as a mechanism resulting in faster transport of contaminants (Sen and Khilar 2006). Metal ion supply to plants occurs more by diffusion than mass transport in uncontaminated soils (Hinsinger 2001; Degryse et al. 2009); shorter diffusion distances may remove diffusion limitations in contaminated soils. Transport by diffusion depends on the concentration gradient, ionic diffusion coefficient, and physical factors, such as soil water content and water film thickness (Sparks, 1989; Carrillo-González et al. 2006).

6.2.4 Time

Nutrients or trace metals may enter soil as ions or as components of amendments such as fertiliser or sewage sludge. After addition, short-term pedological processes act to change their speciation and bioavailability (Lu et al. 2005; Smolders et al. 2009). There are quite different timescales for urban additions, however, and commonly considered pedogenic timescales. In many cases, urban phenomena are very recent, and this represents a key difference for urban soils in which pedogenesis has had limited time to occur. Laboratory studies (Parfitt et al. 1989; Backes et al. 1995) commonly demonstrate that if nutrient or metal ions react longer with soil components they desorb less, and the desorbed fraction is released more slowly than for shorter reaction times. A contrasting concept is the ‘time bomb’ hypothesis, where it is thought that metals applied in organic residues such as sewage sludges will increase in bioavailability in the medium to long term, following mineralisation of organic adsorbing phases and associated acidification. McGrath et al. (2000), however, found that Zn and Cd extractability from soil did not change significantly more than 20 years following sewage sludge application, results which do not support either increased or decreased bioavailability in the long-term. Some steps towards resolution of this issue have been made by Bergkvist and Jarvis (2004), who show by a modelling approach that long-term changes in metal bioavailability vary according to soil and sewage sludge properties and metal content and that no universal outcomes exist. The general outcome for metal bioavailability, however, shown by Smolders et al. (2009), is for aging of metal-contaminated soils to decrease metal bioavailability (Fig. 6.5). Multiple mechanisms explain decreased bioavailability with aging, including leaching-induced changes in soil solution properties and adaptation of organisms (Smolders et al. 2009) as well as the changes in solid-state speciation implied by decreased desorption rates. Regulatory guidelines, such as the Australian National Environment Protection Measure (National Environment Protection Council 2013b), distinguish between aged and fresh contamination, with acceptable concentrations of some elements in aged contaminated soils being lower than for freshly contaminated soils.

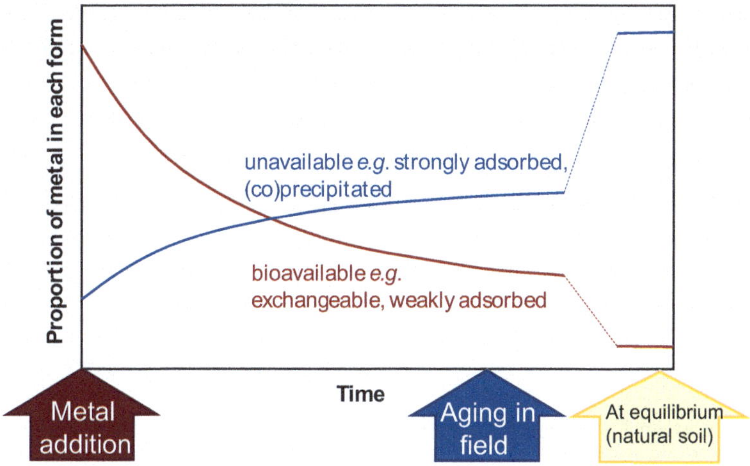

Fig. 6.5 Simplified effect of time on forms of metals in soils

6.2.5 Soil Properties

Numerous studies have shown that a range of bulk soil properties have observable effects on the behaviour of elements in soils.

6.2.5.1 Soil pH

Soil pH affects elements in different ways depending on their dominant forms and reactions in soils.

Nitrogen Low pH reduces overall decomposition of soil organic matter which is the largest store of nitrogen in soils (Haynes 1986). Nitrification (microbial oxidation of ammonium to nitrate) is more sensitive than ammonification to low pH, so ammonium tends to accumulate in acidic soils. Since there is a large range of soil biota contributing to organic matter decomposition and release of soluble nitrogen species, overall nitrogen availability is relatively insensitive to soil pH (McLaren and Cameron 1990).

Phosphorus Low pH reduces the availability of phosphorus, since dissolved concentrations of Al^{3+} and/or Fe^{3+} are greater in acidic soils, and the solubility of aluminium and iron phosphate minerals (e.g. strengite, variscite) is low. Similarly, at high pH, the dominant exchangeable cations in soils are usually Ca^{2+} and Mg^{2+} (Lindsay et al. 1989), which maintain greater equilibrium concentrations of Ca^{2+} and Mg^{2+} in soil solution. Consequently, at high soil pH, calcium and/or magnesium phosphates may exceed their solubility products and precipitate. For these reasons, the bioavailability of phosphorus is at a maximum at near-neutral pH (approximately pH 6–7).

Cationic Metals The combined effects of cation exchange, adsorption, and co-precipitation reactions in soils, as described in Chap. 4, mean that metal bioavailability (Box 6.1) decreases as pH increases (McBride 1994; Sauvé et al. 2000). Since soil pH exerts such an important control, especially on metal behaviour, it is used to adjust guideline concentrations in contaminant regulations (e.g. the Australian National Environment Protection Measure·, National Environment Protection Council 2013b).

6.2.5.2 Cation Exchange Capacity

Cation exchange capacity (CEC) in soils is a key property affecting contaminant behaviour. The CEC itself is dependent on (and probably integrates the effects of) other soil properties such as pH (see above) and the content of soil components bearing negative charge (such as organic matter and clays; see below). Smolders et al. (2009) showed that effective CEC (i.e. CEC measured at the soil's unadjusted pH) was able to explain a significant proportion of the variation in toxicity to organisms at any given total metal concentration. Like soil pH, CEC is an important control on metal behaviour and is also used to adjust regulatory guideline concentrations of contaminants (e.g. National Environment Protection Council 2013b).

6.2.5.3 Soil Organic Matter or Organic Amendments

Effects on Nutrients Soil organic matter is a dominant contributor to soil cation exchange capacity and therefore has a large influence on the behaviour of exchangeable cations in soils. Greater soil organic matter content has been shown to decrease leaching of K^+ and NH_4^+ (Griffioen 2001; Vogeler et al. 2011). In addition, the greater CEC of organic materials retains a reserve of plant-available nutrients, and organic matter added to soils may itself contain useful nutrient contents (Gallardo-Lara and Nogales 1987; Weber et al. 2007). The addition of *organic amendments* to soils, especially those containing larger proportions of easily mineralisable ('labile') carbon compounds, can lead to greater than normal decomposition of pre-existing soil organic matter. This is known as a *priming effect* and is caused by the ability of organic amendments to increase the activity and/or growth of the soil microbial population. In addition, some other microbially driven processes in soils related to nutrient cycling (e.g. sulphate reduction, denitrification) are promoted by an external source of labile carbon compounds (Thangarajan et al. 2013).

Effects on Cationic Metals It is well known that metal cations adsorb to organic matter in soils (Young 2012) and that this adsorption renders the metals both immobile and unavailable to organisms. Some external sources of organic matter, especially those derived from urban waste materials such as sewage sludges (biosolids) or municipal solid waste composts, may contain relatively concentrations of potentially toxic trace elements including cationic metals (Alloway 2012). In many cases, the metal content of added organic materials is great enough that application to soils to meet nutrient requirements may result in exceeding guideline concentrations for potentially toxic trace elements. Some research has evaluated the so-called time bomb hypothesis which predicts that the organic matter in metal-containing soil amendments will eventually decompose, reducing the ability of the soil systems to adsorb metals and creating a sudden increase in metal bioavailability (Chang et al. 1997). The 'time bomb' increase in metal bioavailability may or may not occur, however, depending on how much the organic amendment increases the natural adsorption capacity of the soil, whether the organic amendment also contains an inorganic fraction which can immobilise metals, and the actual metal content of the organic amendment (Bergkvist and Jarvis 2004). Organic matter in soils may also exist in dissolved form and provide ligands which form soluble complexes with trace element cations. In such cases the mobility of metals may actually increase as a result of addition of an organic amendment (Kabata-Pendias 2004).

6.2.5.4 Soil Redox Potential

Effects on Nutrients In aerobic soils (i.e. with an adequate supply of oxygen via air-filled soil pores), nitrogen and sulphur are most stable as the oxyanions NO_3^- and SO_4^{2-}. For nitrate (NO_3^-) in particular, this can result in leaching of N into groundwater due to the minimal retention of NO_3^- by most soils. Under reducing conditions (e.g. in saturated soils), nitrate is depleted by denitrification, and the conversion of ammonium (NH_4^+) to NO_3^- is suppressed. Prolonged reducing conditions allow microbial sulphate reduction to occur in soils to form sulphide species (HS^-/S^{2-}; see Table 4.2). The other macronutrient elements (e.g. P, K) are not directly affected by redox processes but can be affected indirectly; the main indirect effect is release into solution of adsorbed phosphate ($H_2PO_4^-$/HPO_4^{2-}) from iron oxides under reducing conditions, since anoxia favours reductive dissolution of Fe^{III} (hydr)oxides, with simultaneous release of many adsorbed ions.

Effects on Trace Elements Iron (Fe) and several minor or trace elements (including As, Cr, Co, Cu, Mn, Ni, and V and the rare-earth elements Ce and Eu) are able to exist in more than one oxidation state, depending on soil redox potential, over the range of redox potentials typically encountered in soils (McBride 1994). In aerobic soils, elements normally existing as cations remain in cationic form; trace element oxyanions are also stable, and both are predominantly adsorbed to soil colloids or present in the mineral structure of various solid phases. Changes in redox species are important reactions for some elements, and we mainly consider chromium (Cr) and arsenic (As) due to the high potential toxicity of both elements.

Chromium generally exists in the trivalent (Cr^{3+}) oxidation state in both oxidised and reduced soils in species such as mineral forms like $Cr(OH)_3$ or adsorbed $Cr(OH)_2^+$ (Gonnelli and Renella 2012). In some oxidised soils, however, including soils contaminated with chromium, the more mobile, toxic, and carcinogenic Cr^{6+} species (mainly chromate CrO_4^{2-} or dichromate $Cr_2O_7^{2-}$) can persist – these are usually called 'chromium six'and abbreviated to Cr^{VI}. Persistence of Cr^{VI} is greater if Mn oxides are abundant or if minimal concentrations of typical electron acceptors for the reduction of Cr^{VI} species exist. Under natural soil conditions, it is generally soil organic matter which reduces Cr^{VI} to non-toxic Cr^{III} species (Gonnelli and Renella 2012).

Arsenic is mainly present in aerobic soils as the oxyanionic As^V species AsO_4^{3-} (arsenate) as a structural ion in mineral solid phases or as adsorbed $H_2AsO_4^-$ or $HAsO_4^{2-}$ (still arsenate; these are the protonated forms, which are predominant when arsenate is dissolved in water). Under reducing conditions, there are two pathways for transformation of arsenate, depending on the soil organic matter content. In most soils, the product of chemical reduction is the As^{III} species arsenite $H_3AsO_3^0$; in highly organic soils, methylated forms of arsenic(III), such as methylarsonic acid or trimethylarsine oxide, can form (Wenzel 2012; Young 2012).

Reductive dissolution of iron(III) (hydr)oxides (and in some soils, Mn(IV) oxides), as for phosphate above, releases trace element cations or oxyanions which were adsorbed onto or co-precipitated with the oxide mineral (Stone and Morgan 1987). Typically, reduction of Fe and Mn oxides occurs prior to sulphate reduction in the 'redox cascade' (McBride 1994). The increase in dissolved metal concentrations may be transient, since precipitation of trace element cations may occur by formation of hydroxides at the higher pH commonly associated with reducing conditions or as sulphate reduction proceeds (see below). Organic ligands present in saturated soil environments may also prolong or enhance dissolved metal concentrations (Grybos et al. 2007).

Formation of sulphide minerals occurs under reducing conditions, with the onset of sulphate reduction, which produces free sulphide as HS^- (or H_2S). The sulphide produced by sulphate reduction reacts with cationic metals and arsenic, resulting in precipitation of discrete trace element sulphides or co-precipitation of trace elements within iron(II) sulphide minerals. Sulphide minerals maintain trace elements in forms with extremely low bioavailability, so long as reducing conditions persist (e.g. by continued saturation/submergence of soil) (Gambrell 1994). The risk of subsequent oxidation is affected by recrystallisation during early diagenesis; the initial minerals to form such as amorphous ferrous monosulphide (FeS), or mackinawite, have extremely rapid rates of oxidation, whereas more stable (recrystallised) minerals such as pyrite (FeS_2) oxidise more slowly (Morgan et al. 2012).

6.2.5.5 Soil Mineral Phases/Major Elements

The existence of mineral phases containing trace elements, or associations of trace elements with specific minerals or major elements, is difficult to prove experimentally. We cannot simply rely on correlation analyses showing a positive relationship between a trace element and a mineral or major element, since these never imply causation. In addition, the very widely used sequential selective extraction methods largely categorise trace elements into operational fractions, which are not completely specific or which overlap, so sequential extraction does not provide sufficient evidence either. More compelling data is obtained from microchemical techniques, such as energy dispersive X-ray microanalysis (EDX or EDS) which is conducted in combination with electron microscopy or micro-X-ray fluorescence (μ-XRF) which requires a synchrotron radiation source. The 'gold standard' method is obtaining X-ray or electron diffraction patterns on samples of a few microns or less in diameter, which can be used to identify individual mineral phases and which requires a transmission electron microscope. In this textbook we have tried to cite research using at least one microchemical method to support their conclusions.

Clay Content In strictly chemical terms, the role of phyllosilicate (layered aluminosilicate) clays should be distinguished from the effects of 'clay' as a soil textural category, since Fe, Al, and Mn oxides and other reactive minerals may be present in the $\leq$0.002 mm grain size fraction. Using a range of advanced techniques, phyllosilicate clay minerals have been shown to host trace element ions released by chemical weathering and redistributed in soils by formation of secondary minerals during pedogenesis (e.g. Batista et al. 2018). Trace element cations are also retained by clay in soils by surface reaction mechanisms such as adsorption (Schulthess and Huang 1990) and ion exchange (Majone et al. 1998; Abanda and Hannigan 2007).

If the simple grain size definition of clay is accepted, numerous studies show a positive correlation between metal content and clay content. Some weaker evidence that this relationship is due to the presence of aluminosilicate clays would be a similar positive correlation between metal content and aluminium content (e.g. Pardue et al. 1992) or information that other mineral phases are not present in the $\leq$0.002 mm fraction from an X-ray diffraction technique.

Iron/Fe Oxides and Hydroxides Very large databases of soil properties demonstrate relationships between trace element and iron concentrations across multiple soil environments worldwide (Hamon et al. 2004); see Fig. 6.6. The most likely explanation for these globally consistent relationships is the association, by adsorption or co-precipitation, of trace metal ions with iron(III) oxides and/or hydroxides (e.g. Singh and Gilkes 1992; Matera et al. 2003; Dąmbkowska-Naskręt 2004).

Calcium/Carbonate Minerals A number of cationic trace elements (including Cu, Ni, Pb, and Zn) are known to form discrete carbonate mineral phases or associations with carbonate minerals (Yarlagadda et al. 1995; Adamo et al. 1996; Dermatas et al.

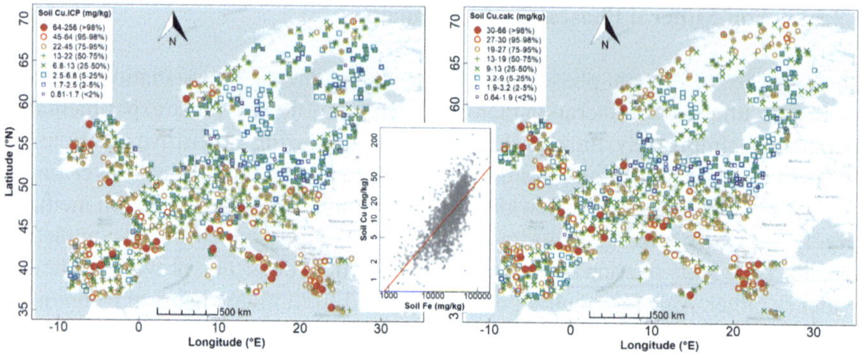

Fig. 6.6 An example of continental-scale trace element-iron relationship: copper concentration ranges in European soils (left), relationship between copper and iron concentrations (inset), and copper predicted from iron concentration (right). Data from the FOREGS survey (Salminen et al. 2005); graphic by Andrew W. Rate

2006). The existence of cadmium carbonate ($CdCO_3$, octavite) has been inferred from soil pore water concentrations of Cd^{2+} and CO_3^{2-} which appear to have an ion product close to the K_{sp} for octavite and therefore to be controlled by octavite solubility (e.g. Street et al. 1977).

Sulphides Sulphide precipitation is important for many trace elements in reduced soils, as discussed above.

Other Materials Numerous other mineral phases containing trace elements, including oxides, phosphates, sulphates, and silicates, have been detected in soil environments (e.g. Adamo et al. 1996; Buatier et al. 2001; Ettler et al. 2008).

6.3 Effects of Inorganic Soil Contaminants on Ecosystem Services

6.3.1 Nutrients

6.3.1.1 Soil Environments

The most important effects of high concentrations of nutrients in soil are related to the likelihood that the soil's retention capacity (e.g. by adsorption or ion exchange) may be exceeded. In many cases this results in export or loss of nutrient(s) from the soil system into a *receiving environment* (this issue is discussed in the following subsection below).

Soils having excessive concentrations of nutrients, or *eutrophic* soils, may show different ecosystem biodiversity and functioning than natural soils (Schindelbeck et al. 2008). For example, in some environments, nutrient toxicity to native plant species from soil eutrophication, such as the sensitivity of some Proteaceae to phosphorus, may limit land rehabilitation efforts (Nichols and Beardsell 1981; Wolff et al. 2017). The consequences of soil eutrophication for the ecology of soil microorganisms and fauna will be addressed in Chap. 8.

6.3.1.2 Receiving Environments

Concentrations of nutrient elements in soil in excess of the capacity of organisms to absorb them will often result in transfers out of the soil system or losses. Nutrients can be lost from soils by leaching (mass transport with water), soil erosion, and for some elements by soil to atmosphere fluxes (such as volatilisation of gas-phase ammonia or hydrogen sulphide or denitrification to produce nitrous oxide) (Carpenter et al. 1998; Groffman et al. 2002).

The effects of loss of nutrients into waterways is common to cause excess concentrations of nutrients in water, or *eutrophication*. The high concentrations of nutrients remove limitations to biomass production for aquatic photosynthetic organisms such as algae, causing phenomena such as algal blooms. In aquatic systems (such as streams, rivers, lakes, and estuaries), productivity is most commonly limited by the supply of phosphorus or nitrogen (Carpenter et al. 1998; Peters and Donohue 2001). Eutrophication and algal blooms have their own adverse environmental effects, such as oxygen depletion, toxicity, and formation of sedimentary organic ooze, which are beyond the scope of this book.

Gas phase losses of nutrients from soils have different types of environmental consequences. Volatilisation of ammonia or hydrogen sulphide can cause toxicity to plants, but, especially at higher concentrations, they are more commonly a nuisance odour to humans (Iglesias Jiménez and Perez Garcia 1989; Muezzinoglu 2003). Denitrification is more insidious; in theory, the end product of denitrification is the environmentally benign nitrogen gas (N_2). In reality, reaction intermediates such as oxides of nitrogen are released during denitrification in soils, in particular nitrous oxide (N_2O) which is a powerful greenhouse gas (on a mass basis, 298 times more effective at trapping solar radiation than CO_2 (IPCC 2007)). Nitrous oxide is also an ozone-depleting substance and is the most important ozone-depleting gas following the widespread ban on halocarbons in the late twentieth century (Ravishankara et al. 2009).

6.3.2 Trace Elements

6.3.2.1 Plants

It is well known that high concentrations of trace elements in soils can lead to toxic effects on plants (Nagajyoti et al. 2010). Metals have a range of biochemical effects on plants (e.g. interfering with essential enzymatic systems) which result in reduced growth related to higher concentrations of metals in plant tissues. The uptake of potentially toxic metallic or metalloid elements by plants confers the possibility of transfer of metals through food chains to higher trophic levels, including consumption of metal-contaminated produce by humans (McBride 1995; Abrahams 2002; Bolan et al. 2013).

6.3.2.2 Soil Microbiota, Meso- and Macrofauna

Soil pollution with potentially toxic trace elements is well-known to affect the growth, functioning, and diversity of soil microorganisms, mesofauna, and macrofauna (He et al. 2005; Smolders et al. 2009; Pauget et al. 2013). These issues will be discussed in much more detail in Chap. 8.

6.3.2.3 Humans

Humans can be affected by soil contamination with potentially toxic trace elements by consumption of plant produce grown on contaminated soils, as discussed previously (Abrahams 2002; Laidlaw et al. 2018). This pathway for ingestion of contaminants is of particular concern due to the growing importance of both community gardening and urban agriculture for food production (Thebo et al. 2014; Egendorf et al. 2018). A number of other mechanisms also result in soil to human transfers of trace elements. Dust entering the atmosphere from contaminated soil (Ljung et al. 2009) potentially results in a decline in air quality and the potential for adverse health effects due to inhalation of fine (< 10 μm) dust. Children are known to accidentally or deliberately consume soil, and this is a known vector for lead poisoning and ingestion of other potentially toxic elements (Mielke et al. 1999; Ren et al. 2006). Finally, trace elements may leach from contaminated soils as dissolved or colloidal forms, particularly at low pH or in very sandy soils (Imperato et al. 2003; Appleyard et al. 2004), posing a threat to water used for human consumption.

6.4 Measurements and Data

Total analyses are those which measure the total concentration of an element regardless of chemical species or physical location in the soil. The methods used are based on complete dissolution of a soil sample using a mixture of concentrated acids, fusion of the soil sample with a flux (e.g. lithium metaborate/tetraborate) followed by dilute acid dissolution, or using a spectroscopic instrumental technique such as X-ray fluorescence or neutron activation analysis (Sparks et al. 1996). Total elemental analyses do not generally correlate well with biological uptake (McLaughlin et al. 2000; Nolan et al. 2003; Rayment and Lyons 2010) and include background concentrations of contaminants as well as anthropogenic additions to soils.

Partial analyses aim to measure an actual species, such as exchangeable metals, or a conceptual chemical form, such as bioavailable phosphorus, of a nutrient or metal in soils. Despite the importance of chemical speciation in soils, however, methods for accurately determining speciation require further refinement. For example, the existence of some elements in specific mineral phases in soils may be established using electron microscopy in conjunction with microprobe techniques such as EDS (having relatively high detection limits, e.g. Essington and Mattigod 1991), synchrotron X-ray spectroscopic techniques (Xia et al. 1997), or SIMS (Chardon et al. 2008). Determination of truly dissolved forms of elements (e.g. 'free' ions) remains complicated by the difficulties in separating truly aqueous species from dispersed colloids. Techniques which target a conceptual fraction (such as 'bioavailable', 'mobile', or 'plant available') rather than a discrete species may, in fact, be more useful in predicting biological uptake than true speciation methods (see Box 6.1; McLaughlin et al. 2000; Zhang et al. 2001).

Box 6.1: Concepts for Bioavailability in Soils

Bioavailability and bioaccessibility. Semple et al. (2004) discuss two related concepts, bioaccessibility and bioavailability. A substance such as a potentially toxic trace element is bioavailable if a transfer from soil into an organism can occur (Fig. 6.7). A substance is bioaccessible if it is able to resupply the bioavailable pool, but organisms are not presently in the same space in the soil, or if a slow process limits release of the bioavailable forms.

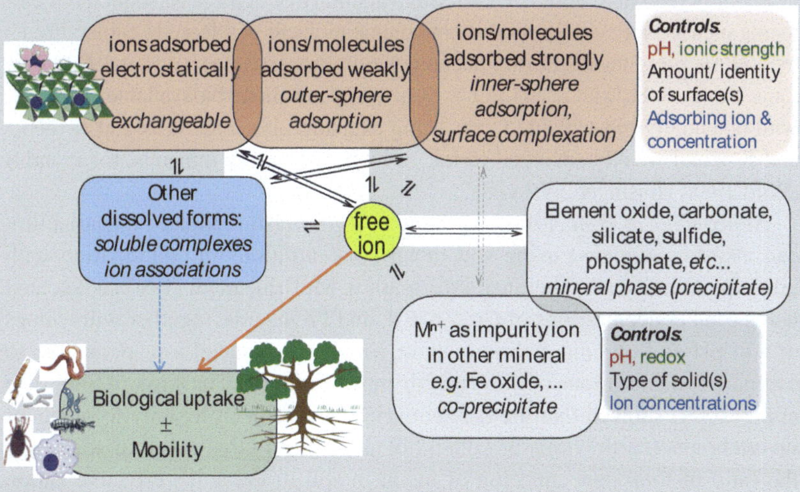

Fig. 6.7 Interactions between chemical species in soils and their relationships with bioavailability – example for metal ions (graphic by Andrew W. Rate)

(continued)

Box 6.1 (continued)

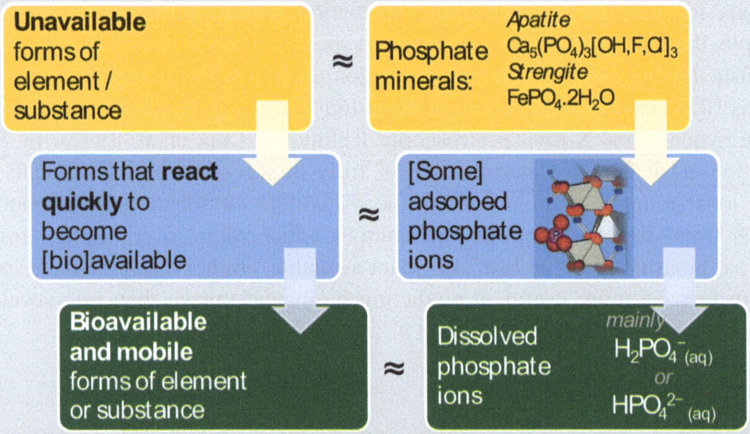

Fig. 6.8 Simplified relationship between chemical speciation and bioavailability for phosphorus (P) in soils (graphic by Andrew W. Rate)

Relationships of bioavailability to speciation. The chemical form of trace elements has a very large effect on those elements' availability or accessibility to organisms (Fig. 6.8). It is commonly assumed that 'bioavailability' continuously decreases from free ions in solution to exchangeable ions, chemisorbed ions, with ions within pedogenic, then primary, mineral structures having the lowest bioavailability (Kabata-Pendias 2004). Bioavailability of nutrients also includes elements such as N, P, and S contained in organic molecules, released ('mineralised') by soil microorganisms during decomposition processes (Stevenson 1994). The ionic forms (NH_4^+, NO_3^-, $H_2PO_4^-$, SO_4^{2-}, K^+) are those bioavailable to plants, with potentially bioavailable forms being exchangeable ions for N and K (NH_4^+ and K^+), and weakly adsorbed forms, or higher-solubility minerals, for P and S (Wild 1988; Hinsinger 2001).

Adjustment of concentrations using soil properties. Metal bioavailability can also be estimated using soil metal concentrations in combination with other quantitative soil chemical information. McBride et al. (1997) advocated using total concentrations of Cu, Zn, Cd, and Pb in soils, together with values of soil pH and organic carbon content which affect metal ion speciation, to estimate concentrations of these metals in soil solution. The work of Smolders et al. (2009) showed that the best empirical predictor of toxicity to plants or invertebrates (earthworms or collembolans), over a large range of soils, was the ratio of total concentration of metal in soil to the soil's effective cation exchange capacity (ECEC). More accurate predictions were obtained if this ratio was corrected for leaching and/or aging of contaminated soils in the

(continued)

Box 6.1 (continued)

field. Just because the metal to ECEC ratio is useful, however, does not prove that exchangeable metal ions are the actual bioavailable species. Considerable uncertainty still surrounds the ability to predict bioavailability from soil measurements (Smolders et al. 2009).

Free ion activity model(s). It is usually assumed that plants or soil microorganisms can only take up the free, uncomplexed metal ion from solution (Parker and Pedler 1997). This is the basis for the 'free ion activity model' of bioavailability. Many studies have shown a positive relationship between free metal ion activity and plant uptake (e.g. Zhang et al. 2001). The free ion activity model has conceptual shortcomings, however, since it does not account for resupply of metals from soil solid phases and also does not allow for chemical reactions on organism surfaces. The free ion activity model would probably be too simplistic for a complex environment such as a soil (Parker and Pedler 1997).

Models based on metal resupply. We now understand that metal ion uptake by plants is controlled by both the free ion concentration **and** the rate of resupply (e.g. desorption) from soil solid phases. In particular, Zhang et al. (2001) showed that the 'effective concentration' of Cu (C_E) measured by a technique called 'diffusive gradients in thin films' (DGT) was the best predictor of copper uptake by plants. DGT analysis measures the concentration of ions released by diffusion over a fixed time interval, so it closely resembles the mechanism for biological uptake. Oporto et al. (2009) and Degryse et al. (2009), however, found that DGT is not a good predictor of plant metal uptake at high concentrations of metals in soil, as would be the case in contaminated soils. In contaminated soil environments, biological uptake may not be limited by diffusion.

Bioavailability models which consider reactions at organism surfaces. The biotic ligand model (BLM) considers competitive, pH-dependent, chemical interactions between ions in soil solution, reactive solid phases, and the biological receptors on organisms (Di Toro et al. 2001). Thakali et al. (2006) showed that a BLM could predict toxicity of Cu^{2+} and Ni^{2+} to plants (measured as decreased root growth), from the amount of Cu^{2+} or Ni^{2+} bound to a biotic ligand (e.g. the root surface). Predictions by the soil BLM were made only for relatively large total metal ion concentrations, since the emphasis was on toxicity caused by contaminated soils. A simplified biotic ligand model has been shown to predict Cu toxicity for a wide range of organisms, particularly plants and invertebrates, with toxicity effects commencing at uncontaminated soil Cu concentrations (<10 mg/kg) (Qiu et al. 2013).

Practical use of bioavailability concepts. In practice, the most useful approach for incorporating the principles of bioavailability into managing soil contamination is to adjust the maximum permissible concentrations using soil

(continued)

Box 6.1 (continued)

properties. The critical soil properties are those known to affect bioavailability – usually pH, cation exchange capacity, clay content, and organic carbon content. Whether the contamination is recent or 'aged' is also important. These concepts have entered regulatory frameworks in the USA and Europe (Naidu et al. 2015), Australia (National Environment Protection Council 2013b), and in the UK (Martin et al. 2017).

There are a wide range of methods for both total and partial analyses of nutrients and metals in soils. Some of the more commonly used methods, or types of methods, are outlined in the following subsections.

6.4.1 Total Elemental Analyses

The analysis of soil for inorganic contaminants is done for numerous reasons. Some measurements, such as pH, redox potential, or clay content are used as supporting information, to characterise the soil environment for a better understanding of the behaviour of the contaminants of interest. For the measurement of nutrients or contaminants in soils, there are two main options.

Nitrogen The most convenient method for total nitrogen (and carbon) in soil is by high-temperature combustion in a specialised instrument. The nitrogen is converted to N_2 gas which is measured with a thermal conductivity detector. Total nitrogen can also be measured by 'Kjeldhal' digestion in concentrated sulphuric acid, which converts all nitrogen into ammonium which can be measured by automated colourimetric analysis (Rayment and Lyons 2010).

Phosphorus, Sulphur, and Potassium The simultaneous measurement of P, S, and K, and most other major elements, in soils is most conveniently achieved using X-ray fluorescence (XRF). Total P and K can also be measured by a fusion/dissolution method followed by separate analyses for the resulting dissolved phosphate and potassium. Total S can be measured by high-temperature combustion and detection of the evolved SO_2 gas by infrared spectrometry (Rayment and Lyons 2010).

Major Elements – Na, K, Mg, Ca, Al, Fe, Si The 'gold standard' method for major element analyses in soils is X-ray fluorescence (XRF) (Karathanasis and Hajek 1996). Comparisons in the author's laboratory indicate that, for major and minor elements in soil and regolith samples, results comparable with XRF are obtained using fusion in a lithium metaborate/lithium tetraborate flux, followed by dissolution in dilute hydrochloric acid with analysis of the solution using ICP-OES (e.g. see Du et al. 2012).

Minor and Trace Elements Most metal and metalloid contaminants are present in soils at concentrations $\leq 0.1\%$ (1000 mg/kg) by weight and so are categorised as minor or trace elements. Metal contaminants which are of interest in urban soils (not a complete list of possibilities!) include Ag, As, B, Cd, Cr, Co, Cu, Hg, Li, Mn, Mo, Ni, Pb, V, and Zn, with rising interest in the rare-earth elements such as Ce, Gd, La, and Nd due to their increasing use in electronics, medical imaging, and other technologies. The detection limits of XRF are too high ($\geq$ 5 mg/kg for some elements), and so acid dissolution or fusion methods are more commonly used, followed by a more sensitive technique such as inductively coupled plasma-optical emission spectrometry (ICP-OES) or inductively coupled plasma-mass spectrometry (ICP-MS) (Geboy and Engle 2011).

6.4.2 Partial Analyses

Nitrogen The most common method for 'available' nitrogen (mainly dissolved and exchangeable NH_4^+ and NO_3^-, with trace concentrations of NO_2^-) in soil is by extraction in 2 mol/L KCl solution. The NH_4^+ and $NO_3^- \pm NO_2^-$ in the resulting extract are usually determined using colourimetric techniques, often automated (Mulvaney 1996; Rayment and Lyons 2010).

Phosphorus There are numerous methods for measuring 'available' phosphorus, typically using solutions able to displace weakly adsorbed phosphate such as sodium bicarbonate (Rayment and Lyons 2010) or dilute acids (e.g. acetic acid, sulphuric acid – see Kuo (1996) for more detail). The extracting solution contains dissolved $H_2PO_4^-$ and/or HPO_4^{2-} which can be measured colourimetrically.

Sulphur The analysis of 'available' sulphur (dissolved and adsorbed sulphate) is conceptually similar to that for phosphate. A phosphate solution is used to displace sulphate ions into solution, and ICP-OES can be used to measure dissolved sulphur (assuming this is mostly SO_4^{2-}) (Rayment and Lyons 2010).

Potassium Plant-available potassium is often considered to be equivalent to exchangeable potassium; soil is extracted with a solution containing a competing cation (commonly ammonium or calcium salts, although the phosphate-extracting solution, sodium bicarbonate, is also used), following which K can be determined by ICP-OES (Sparks et al. 1996; Rayment and Lyons 2010).

6.4.2.1 Partial Analyses for Metals and Metalloids

Trace elements. There are probably even more choices of extracting solution for metals and metalloids than for nutrients, reflecting multiple analytical objectives (e.g. plant nutrition, environmental assessment, or mineral exploration) and the influence of soil properties. Some wet chemical extraction methods (e.g. 0.01 M $CaCl_2$; Whitten and Ritchie 1991) appear to selectively determine trace metal cations in the ion-exchangeable fraction, although this may not have been the original intention of the method. The success of some of these methods, in predicting plant uptake, may reflect the importance of weakly adsorbed and/or exchangeable ions as a buffer to replace removal of ions from the soil solution. A simple assumption we can make is that the fractions of metals in soils measured by partial chemical extraction represent the bioavailable and/or bioaccessible component, without the need to explain how or why the analysis 'works'. This is the rationale behind the numerous chemical extraction techniques for measuring 'bioavailable' or 'mobile' metals in soils. These methods extract soil with solutions containing (singly or in combination) electrolytes, dilute acids and/or buffers, complexing agents, oxidising agents, or reducing agents (Chao 1984); some examples appear in Table 6.1. Another theory is that bioavailability should be assessed by simulating the chemical conditions within an organism. For example, Ma et al. (2009) and Smith et al. (2010) assumed that using an extractant which simulates the composition of earthworm gut fluids (e.g. relevant enzymes, anoxia) will more accurately reflect uptake and toxicity than total metal and other extractable metal concentrations.

6.4.3 Field Measurements

Measurements of soil chemical properties in the field offer the advantages of immediacy of obtaining data, and there may also be benefits if samples change between sampling and laboratory analysis, for example, following exposure to CO_2 or O_2 in the atmosphere. Simple field analyses of soil properties such as pH and electrical conductivity (EC) are routine, using appropriate electrodes and battery-powered meter (calibrated with standard or buffer solutions) on a standard suspension of soil in water or electrolyte solution (e.g. 0.01 mol/L $CaCl_2$) (Rayment and Lyons 2010). The electrical conductivity measurement can be converted to an approximate content of soluble salts in soil. Measurement of soil pH in the field after oxidation with 30% hydrogen peroxide (H_2O_2) solution can also allow semi-quantitative estimation of the presence of potential acid sulphate soil materials, using the difference between pH in water and pH after peroxide oxidation (Ahern et al. 2004). Less commonly, soil reduction-oxidation potential may also be measured in field soils using a platinum electrode, but the analysis suffers from poor electrical contact in dry soils and variability due to the presence of redox microenvironments.

Even simpler field tests are based on use of colour indicators, but these are seldom used due to the reliability and availability of electrode-based pH and EC

Table 6.1 Examples of partial analysis methods, based on extracting solutions, for metals in soils

Solution	Chemical basis	Species or fraction targeted	Reference	Extracting solutions targeting similar fractions
0.005 M[a] DTPA [b], 0.1 M triethanolamine, 0.01 M $CaCl_2$, pH 7.3	DTPA forms stable, soluble complexes with metal ions; triethanolamine buffers pH, Ca^{2+} suppresses $CaCO_3$ dissolution	Plant available micronutrient metals – Also used for potentially toxic metals, probably dissolved, exchangeable and weakly adsorbed metal ions	Lindsay and Norvell (1978)	EDTA [c] (Snowdon and Birch 2004)
0.01 M $CaCl_2$	Ca^{2+} in solution can displace exchangeable metal ions	Dissolved and exchangeable metal ions	Whitten and Ritchie (1991)	Ammonium acetate, NH_4NO_3, $CaCl_2$, $MgCl_2$, $BaCl_2$, $Mg(NO_3)_2$ (Chao 1984; Shuman 1985; McLaughlin et al. 2000)
4 M HNO_3	Dilute acid will not dissolve primary minerals	Metals in some secondary minerals, plus dissolved, exchangeable, and adsorbed metal ions (unless removed by prior extraction(s))	Chao (1984)	1 M HCl (Morgan et al. 2012)
1 M sodium acetate adjusted to pH 5.0 with acetic acid	Acidity will dissolve carbonate minerals, but pH is too high to dissolve oxides	Metals in carbonate minerals, plus dissolved, exchangeable, and weakly adsorbed metal ions (unless removed by prior extraction(s))	Tessier et al. (1979)	Acetic acid (Chao 1984)
0.25 M $NH_2OH{\cdot}HCl$ in 0.25 M HCl, pH <2, 60 °C	$NH_2OH{\cdot}HCl$ is a strong reducing agent able to dissolve amorphous Fe^{III} and Mn^{IV} (hydr) oxides	Metals in reducible minerals, plus dissolved, exchangeable, and adsorbed metal ions (unless removed by prior extraction(s))	Hall et al. (1996)	Ammonium oxalate (pH 3) (Shuman 1985)
0.1 M sodium pyrophosphate, pH 10	Pyrophosphate forms very stable complexes with Fe^{3+} and Al^{3+} which would otherwise precipitate organic matter. Organic matter is ionised at pH 10	Metals associated with soil organic matter, plus dissolved, exchangeable, and adsorbed metal ions (unless removed by prior extraction(s))	Chao (1984)	H_2O_2, NaOCl (Chao 1984; Shuman 1985)

[a]M = moles per litre of solution (mol/L); [b]diethylenetriaminepentaacetic acid; [c]ethylenediaminetetraacetic acid

measurements. For example, Rayment and Lyons (2010) describe the use of a universal pH indicator in which a small volume of indicator solution is added to soil, followed by a white $BaSO_4$ powder to more easily visualise the colour. Less-commonly used field colour tests have been developed as well, for example, for dissolved aluminium for use in acidic soils or ferrous iron (Fe^{2+}) to show whether reducing conditions exist, but neither are used routinely.

Instrumental Methods Field-based instrumental methods generally have lower measurement precision and accuracy than laboratory-based methods but may allow rapid preliminary assessment of field soils without destructive sampling. The most widely used portable analytical instrument in field analysis of soils is the portable X-Ray fluorescence spectrometer (pXRF) available in preliminary forms since about the 1980s. The use of pXRF in the field can allow more rapid and cost-effective screening of metal contaminants, with acceptable accuracy and precision of measurement (Taylor et al. 2004). Field-based pXRF measurements have been used to assess the spatial distribution of contaminants such as As, Cu, Mn, Pb, V, and Zn (as well as non-contaminant elements) in urban soils (Carr et al. 2008; Schwarz et al. 2012; Paulette et al. 2015). Portable XRF has also been used for field measurements in archaeology to locate features of historical human settlements (Save et al. 2020).

Other field-portable instrumental techniques are under development. For example, Wang et al. (2020) used a handheld Fourier-transform infrared (FTIR) spectrometer to map the distribution of petroleum hydrocarbons at a site contaminated with fuels and lubricants. Similarly, visible-infrared spectroscopic methods are able to determine mineral abundances in geological applications such as core logging (Schodlok et al. 2016), but so far there have been no applications of this technique for urban soils. It remains to be seen whether or not field-based instrumental methods (even the relatively common pXRF technique) will achieve widespread use in the chemical assessment of urban soils.

Remote Sensing In Chap. 5 we discussed the use of airborne radiometrics, a remote sensing method that allows estimation of some soil chemical and parameters based on the signals from radioisotopes of potassium, thorium, and uranium. The most promising application was probably the use of depletions in the thorium signal to infer the distribution of acid sulphate soils (Bierwirth and Brodie 2005).

6.4.4 Background Concentrations

If a soil is contaminated, the contamination is understood in the context of naturally occurring concentration of that contaminant – the background or baseline concentration. This a more complex issue than it may appear initially; most recent understanding of background concentrations reflects the idea that 'background concentration' is a variable quantity rather than a constant value (Reimann and Garrett 2005). Background concentrations of nutrients and metals in soils vary with time, scale of investigation, soil depth sampled, specific location, soil properties, and type of sample preparation and analysis – factors which are all, to some extent, interdependent (Hamon et al. 2004; Reimann and Garrett 2005; Johnson and Ander 2008). In soils across several continents, many elements in soils show a positive correlation with iron content, leading Hamon et al. (2004) to recommend that background concentrations should be variable and based on soil iron concentrations. Reimann and Garrett (2005) noted that single subcontinental-scale concentrations of elements were not useful for establishing site-specific backgrounds. They also showed that a common practice, of assuming that deeper soil horizons could be analysed to determine the local background concentrations, was invalid as it failed to account for differences in concentrations resulting from pedogenesis. In some cases, background concentrations (single values or variable functions) can be estimated following analysis, mapping, and modelling of large datasets (Reimann and Garrett 2005; Johnson and Ander 2008).

6.4.5 Regulatory Contamination Thresholds

Using the current Australian 'NEPM' guidelines (National Environment Protection Council 2013b) as an example, there are a number of issues that need to be considered:

- The actual or intended land use of the area from which the soil was sampled. The NEPM considers four land use categories, in decreasing order of risk to humans, from susceptible residential to commercial/industrial.
- The age of contamination.
- Soil properties which control the bioavailability of contaminants. Depending on the contaminant involved, the NEPM considers soil pH, cation exchange capacity, and soil texture.
- The background concentration of contaminants, valid for the site or area under investigation. The NEPM considers background concentrations only for the inorganic contaminants As, Cu, Cr(III), Ni, Pb, and Zn.

The concepts and methodology for establishing guideline concentrations are addressed in Chaps. 9 and 11.

6.4.6 Distinguishing Geogenic from Anthropogenic Contamination

In the preceding sections, we have already discussed two methods which attempt to separate natural concentrations of contaminants from those added by humans. Background concentrations (Sect. 6.4.4) are considered to represent the geogenic contributions to total nutrient or metal concentrations in soils, and the difference between measured concentration and background represents the anthropogenic addition (National Environment Protection Council 2013a). Background concentrations can be based on a statistical threshold based on multiple measurements. For example, Biasioli et al. (2012) calculated background concentrations for soil data from peri-urban and rural areas as the 90th percentile of observations, after outliers above the Tukey threshold were removed (see Fig. 3.14; this followed ISO 19258). In contrast, the National Environment Protection Council (2013a) recommends that the ambient background concentration for metals be set at the 25th percentile from a dataset derived from analysis of urban and rural soils. An alternative, valid for some trace elements, is that widely consistent regression relationships between trace elements and Fe or Mn can be used to estimate background concentrations for a site if Fe and/or Mn concentrations are known and do not themselves represent contamination (Hamon et al. 2004; National Environment Protection Council 2013a).

The final approach we will discuss for establishing background concentrations is based on analysis of a comparable but uncontaminated soil material. The US EPA (2002) recommends taking physical samples at a reference area which has the same physical, chemical, geological, and biological characteristics as the site being investigated but has not received contamination from activities on the site. Sufficient numbers of background samples need to be collected to make statistical comparisons with sufficient statistical power. In some instances, it is assumed that deeper soil horizons are uncontaminated (e.g. if the only known source of potential contamination is aeolian diffuse-source pollution). In this case the subsoil material has been used to represent geogenic background concentrations (Biasioli et al. 2012).

The calculation of enrichment factors (Sect. 6.4.7, e.g. Fabietti et al. (2010)) also relies on valid background concentrations being known and is the ratio of sample concentration of a contaminant element to the concentration of the same element in the reference material, both normalised using the concentration of a reference element. Ideally the reference element is one that behaves conservatively (e.g. not added anthropogenically, nor enriched or depleted by pedogenic processes); examples of reference element for urban soil environments include Al (Liu et al. 2014), Mn (Yongming et al. 2006), Sr (Yaylalı-Abanuz 2011), and Ti (Szolnoki et al. 2013), and Sc, Fe, and Zr have also been used.

The use of regression model analysis builds on concepts outlined by Rose et al. (1979) and Hamon et al. (2004). For example, Rate (2018) developed separate multiple regression models for a range of trace elements in soil in an urban parkland. The positive residuals (deviations from model predictions) were attributed to addition of potential contaminant elements, above a variable background which depended on soil properties. An example illustrating the use of regression analysis to identify unusual values is shown in Fig. 6.9.

6.4.7 Contamination Indices

A complement to using contaminant concentrations alone to characterise urban soil is to use a form of contamination index. The simplest forms of contamination index are for individual elements and are calculated as ratios between sample and

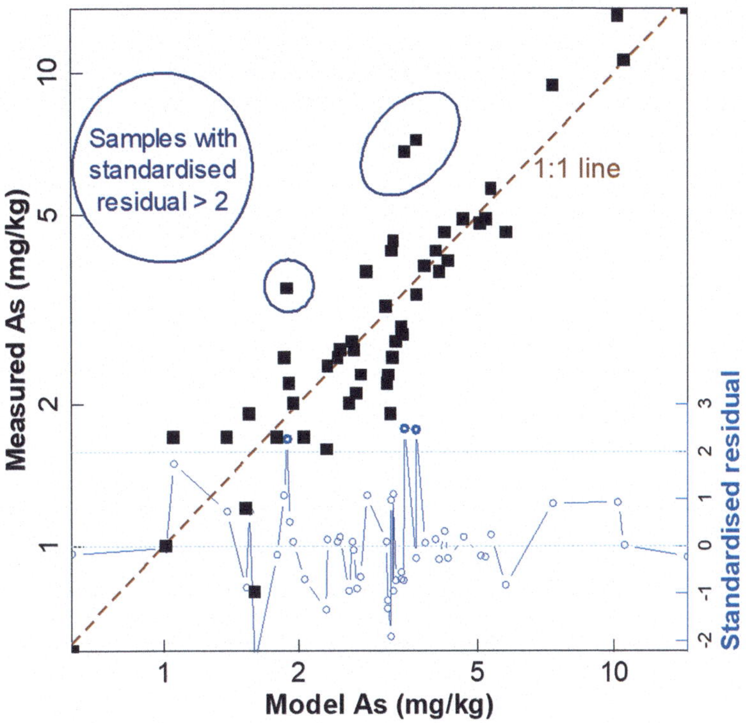

Fig. 6.9 Example of using analysis of residuals from a regression model (see Chap. 3) to identify unusual observations in a dataset with a variable background (graphic by Andrew W. Rate). Note that the unusual samples do not have the highest arsenic (As) concentration

background concentrations of the contaminant of interest (Table 6.2). Other indices recognise that soil contamination is very often multivariate and is essentially sample/background ratios for several contaminants which are averaged in some way (e.g. using arithmetic mean or geometric mean values; see Table 6.2).

More complex composite indices such as the Nemerow integrated pollution index (Yang et al. 2011) have also been used to assess pollution in urban soils.

A complement, or an alternative, to calculation of contamination indices is the use of multivariate numerical methods to analyse multielement soil data. The most common methods include principal component analysis (PCA – see Chap. 3) and one of the various types of cluster analysis. The main use of these multivariate statistical methods appears to be to identify groupings of contaminants, which are then used to infer contaminant sources (usually in combination with other data or statistical analysis – a detailed explanation is given by Reimann et al. 2008). More recently the positive matrix factorisation (PMF) technique, more commonly used for characterising atmospheric particulates, has been applied to urban soils for identification of the source of inorganic chemical contaminants (Li et al. 2016;

Table 6.2 Examples of contamination indices used in urban soil studies

Index name	Calculation and parameters		Reference
Geoaccumulation index, I_{geo}	$I_{geo} = \log_2(C_n / 1.5B_n)$	C_n = measured concentration of element, B_n = background concentration	Wei and Yang (2010)
Pollution index, PI	$PI = C_i / S_i$	C_i = measured concentration of element, S_i = background concentration	Sun et al. (2010)
Integrated pollution index, IPI	$IPI = \left(\overset{n}{\underset{i=1}{?}} (C_i / S_i) \right) / n$	$\sum$ means the *sum* of terms 1 to n, C_i = measured concentration of ith element, S_i = background concentration of ith element, n = number of elements	Sun et al. (2010)
Pollution load index, PLI	$PLI = \left(\overset{n}{\underset{i=1}{?}} (C_i / S_i) \right)^{1/n}$	$\prod$ means the *product* of terms 1 to n, C_i = measured concentration of ith element, S_i = background concentration of ith element, n = number of elements	Madrid et al. (2002)
Enrichment factor, EF	$EF = [C_n / C_{ref}] \div [B_n / B_{ref}]$	C_n = concentration of nth element in soil, C_{ref} = concentration of reference element in soil, B_n = background concentration of nth element, B_{ref} = background concentration of reference element	Yongming et al. (2006)
Nemerow index, P	$P = \sqrt{\frac{\left(\frac{C_i}{S_i}\right)_{max}^2 + \left(\frac{C_i}{S_i}\right)_{avg}^2}{2}}$	C_i = measured concentration of ith element, S_i = background concentration of ith element, subscript 'max' refers to the maximum value, and subscript 'avg' refers to the mean value	Gong et al. (2008)

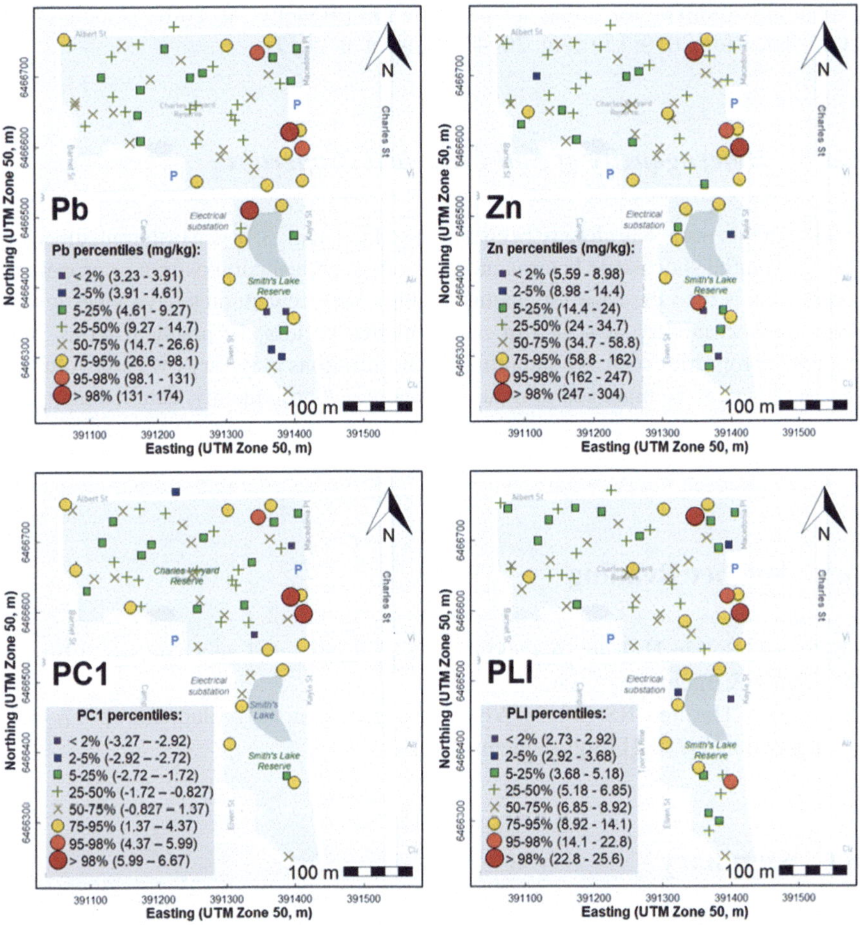

Fig. 6.10 A comparison of maps of single element concentrations for Pb and Zn with indices derived from calculation of principal component (PC1) and the pollution load index (PLI) from Table 6.2 (graphic by Andrew W. Rate). Calculation of principal components was done on centred-log-ratio transformed concentrations, and PC1 (plotted) has its greatest variable weightings from Pb, Cu, Zn, and Cd. Calculation of PLI was based on concentrations of As, Ba, Cr, Cu, Mn, Pb, and Zn using background concentrations from Rate (2018)

Rastegari Mehr et al. 2017). Figure 6.10 shows a comparison of using single-element concentrations, a multivariate index, and a contamination index to map pollution in an urban parkland. While the spatial patterns based on each type of data are similar, Fig. 6.10 emphasises the importance of different analyses of data; for example, both the indices plotted miss the high Pb concentration near an electrical substation.

The specialised discipline of risk analysis also indexes the health risk index, hazard quotient, and hazard index. These parameters are based on a likely proportion of toxic dose for humans, based on risk analysis of possible pathways for human ingestion (examples for urban soils include De Miguel et al. 2007; Nabulo et al. 2010; Thornton 2010; Liu et al. 2014; Abbasi et al. 2017).

6.4.8 *Heterogeneity of Soil Chemical Properties*

Although this issue has already been addressed in Chap. 3, it is worth remembering that substantial short-scale heterogeneity in soil chemical composition or properties exists (this is also the case for 'natural' soils). Very few studies have investigated urban soil variability at scales of a few metres or less. In practical terms, soil chemical properties such as contaminant concentrations may vary over soil volumes smaller than can be managed or treated efficiently. Consequently, detailed mapping of soil chemical properties is unlikely to result in changes in the management or classification of urban soil.

6.5 Further Reading

DeVivo B, Belkin H, Lima A (eds) (2017) Environmental geochemistry. Elsevier Science & Technology, San Diego

Lyons WB, Harmon RS (2012) Why urban geochemistry? Elements 8:417–422. https://doi.org/10.2113/gselements.8.6.417

6.6 Summary

- The main inorganic contaminants that we consider in urban soils are nutrients (mostly nitrogen and phosphorus), trace elements (metals and metalloids), asbestos, radionuclides, plastics, manufactured nanoparticles, and cyanide.
- The sources of contaminants in urban soils are very numerous and differ from site to site. Common sources include historical agriculture and horticulture, including manures, fertilisers, and pesticides; solid wastes and wastewater; mining and related activities; various industries; fossil fuels; road traffic; building construction; and weathering.
- Acid sulphate soils are a specific example of soil 'contamination' (the actual added substance is oxygen!) which can result in export of contaminants into receiving environments.
- The chemical behaviour of inorganic contaminants is governed by reactions with fine-grained solid phases: adsorption, ion exchange, and (co)precipitation. In turn, these reactions are controlled by system variables such as pH, redox potential, contaminant concentrations, a wide range of other soil properties, and time.

- Urban soil contaminants in high concentrations cause adverse effects on living organisms. Nutrients are rarely toxic but alter ecological balances. Trace elements, asbestos, radionuclides, and cyanide can be extremely toxic to a wide range of organisms and may also have other deleterious effects such as carcinogenicity. The adverse effects of plastics and manufactured nanoparticles are poorly defined at present.
- The total concentration of an inorganic contaminant in soil is often poorly related to bioavailability, since total concentrations include forms of elements which are inaccessible to any organisms. Various forms of partial analysis of soil contaminants may be more useful to assessing the risk to biota including humans.
- To distinguish between natural 'geogenic' concentrations of inorganic contaminants and the amounts added by humans, various comparisons or numerical analyses are possible. These include determining background concentrations, comparison with regulatory guidelines, calculation of contamination indices, or use of statistical (e.g. regression) models.

6.7 Questions

6.7.1 Checking Your Understanding

1. Identify the main differences and similarities between sources for nutrients, trace elements (metals and metalloids), asbestos, and cyanide in urban soils.
2. Explain how acid sulphate soils form in urban environments, and suggest some ways in which their formation could be avoided.
3. Why does the mobility of cationic metals in soils approximately increase in the order Fe^{3+} < Pb^{2+}, Hg^{2+} < Al^{3+}, Cu^{2+}, Cr^{3+} < Co^{2+}, Mn^{2+}, Ni^{2+}, Zn^{2+}, Fe^{2+} < Cd^{2+} Ca^{2+} Sr^{2+} < Na^+, K^+, NH_4^+ (i.e. Fe^{3+} least mobile; Na^+, K^+, NH_4^+ most mobile)? Explain in terms of chemical reactions.
4. Explain the differences between total and partial analyses for inorganic contaminants in soils and the situations in which we would use each type of analysis.
5. Describe the various methods for distinguishing natural 'geogenic' concentrations of inorganic contaminants in soils and the amounts of inorganic contaminants added to soils by humans. Is any of these approaches the 'best' one to use?

6.7.2 Thinking About the Topics More Deeply

6. Describe a sequence of events for a soil which begins in aerobic conditions and ends with long-term submergence of the soil.

7. Radionuclides (such as some isotopes of caesium or thorium) occur naturally in some soils in significant concentrations. Should we be concerned about this? Why (or why not)?
8. Can we manipulate chemical reactions in soils to make inorganic contaminants less bioavailable or mobile? How might this work?
9. If the bioavailability of some inorganic contaminants in urban soils decreases with increasing time, could one management strategy simply be to leave urban land unused or seal the surface? Explain why this might (or might not) be successful and/or practical.

6.7.3 A Question 'Out of Left Field'

10. Shouldn't we just leave all the plastic waste from urban environments in landfills? Discuss the issues involved!

References

Abanda PA, Hannigan RE (2007) Mineral control of minor, trace and rare earth elements during black shale weathering at near-neutral pH. Dev Environ Sci 5:273–301

Abbasi S, Keshavarzi B, Moore F, Delshab H, Soltani N, Sorooshian A (2017) Investigation of microrubbers, microplastics and heavy metals in street dust: a study in Bushehr city, Iran. Environ Earth Sci 76. https://doi.org/10.1007/s12665-017-7137-0

Abrahams PW (2002) Soils: their implications to human health. Sci Total Environ 291:1–32

Acosta JA, Gabarrón M, Faz A, Martínez-Martínez S, Zornoza R, Arocena JM (2015) Influence of population density on the concentration and speciation of metals in the soil and street dust from urban areas. Chemosphere 134:328–337. https://doi.org/10.1016/j.chemosphere.2015.04.038

Adachi K, Tainosho Y (2004) Characterization of heavy metal particles embedded in tire dust. Environ Int 30:1009–1017

Adamo P, Dudka S, Wilson MJ, McHardy WJ (1996) Chemical and mineralogical forms of cu and Ni in contaminated soils from the Sudbury mining and smelting region, Canada. Environ Pollut 91:11–19. https://doi.org/10.1016/0269-7491(95)00035-P

Adriano DC (2001) Trace elements in terrestrial environments: biogeochemistry, bioavailability and risks of metals. Springer, New York, 867 pp

Ahern CR, McElnea AE, Sullivan LA (2004) Acid sulfate soils laboratory methods guidelines. Queensland Department of Natural Resources, Mines and Energy, Indooroopilly, Australia

Alloway BJ (1995) The origins of heavy metals in soils. In: Alloway BJ (ed) Heavy metals in soils. Blackie Academic and Professional, Glasgow, pp 38–57

Alloway BJ (2004) Contamination of soils in domestic gardens and allotments: a brief overview. Land Contam Reclam 12:179–187

Alloway BJ (ed) (2012) Heavy metals in soils: trace metals and metalloids in soils and their bioavailability, Environmental Pollution, vol 22, 3rd edn. Springer, Dordrecht

Andersson KG, Roed J (1994) The behaviour of Chernobyl ^{137}Cs, ^{134}Cs and ^{106}Ru in undisturbed soil: implications for external radiation. J Environ Radioact 22:183–196. https://doi.org/10.1016/0265-931X(94)90080-9

Anikwe MAN, Nwobodo KCA (2002) Long term effect of municipal waste disposal on soil properties and productivity of sites used for urban agriculture in Abakaliki, Nigeria. Bioresour Technol 83:241–250. https://doi.org/10.1016/S0960-8524(01)00154-7

Appleyard S, Wong S, Willis-Jones B, Angeloni J, Watkins R (2004) Groundwater acidification caused by urban development in Perth, Western Australia: source, distribution, and implications for management. Aust J Soil Res 42:579–585

Arogunjo AM, Höllriegl V, Giussani A, Leopold K, Gerstmann U, Veronese I, Oeh U (2009) Uranium and thorium in soils, mineral sands, water and food samples in a tin mining area in Nigeria with elevated activity. J Environ Radioact 100:232–240. https://doi.org/10.1016/j.jenvrad.2008.12.004

Aruoja V, Dubourguier HC, Kasemets K, Kahru A (2009) Toxicity of nanoparticles of CuO, ZnO and TiO_2 to microalgae *Pseudokirchneriella subcapitata*. Sci Total Environ 407:1461–1468

Backes CA, McLaren RG, Rate AW, Swift RS (1995) Kinetics of cadmium and cobalt desorption from iron and manganese oxides. Soil Sci Soc Am J 59:778–785

Baker DE, Senft JP (1995) Copper. In: Alloway BJ (ed) Heavy metals in soils. Blackie Academic and Professional, Glasgow, pp 179–205

Banerjee ADK (2003) Heavy metal levels and solid phase speciation in street dusts of Delhi, India. Environ Pollut 123:95–105

Barrow NJ (1998) Effects of time and temperature on the sorption of cadmium, zinc, cobalt or nickel by a soil. Aust J Soil Res 36:941–950

Batista AH, Melo VF, Rate AW, Gilkes RJ, Saunders M, Dodd A (2018) Scanning and transmission analytical electron microscopy (STEM-EDX) can identify structural forms of lead by mapping of clay crystals. Geoderma 310:191–200. https://doi.org/10.1016/j.geoderma.2017.09.026

Bennett EM, Carpenter SR, Clayton MK (2005) Soil phosphorus variability: scale-dependence in an urbanizing agricultural landscape. Landsc Ecol 20:389–400. https://doi.org/10.1007/s10980-004-3158-7

Bergkvist P, Jarvis N (2004) Modeling organic carbon dynamics and cadmium fate in long-term sludge-amended soil. J Environ Qual 33:181–191

Biasioli M, Fabietti G, Barberis R, Ajmone-Marsan F (2012) An appraisal of soil diffuse contamination in an industrial district in northern Italy. Chemosphere 88:1241–1249. https://doi.org/10.1016/j.chemosphere.2012.03.081

Bierwirth PN, Brodie RS (2005) Identifying acid sulfate soil hotspots from airborne gamma-radiometric data and GIS analysis. Australian government Bureau of Rural Sciences, Canberra

Bohn HL, McNeal BL, O'Connor GA, Myer RA (2002) Soil chemistry. John Wiley & Sons, Inc., New York, 321 pp

Bolan NS, Makino T, Kunhikrishnan A, Kim PJ, Ishikawa S, Murakami M et al (2013) Cadmium contamination and its risk management in rice ecosystems. Adv Agron 119:183–273

Boman A, Fröjdö S, Backlund K, Åström ME (2010) Impact of isostatic land uplift and artificial drainage on oxidation of brackish-water sediments rich in metastable iron sulfide. Geochim Cosmochim Acta 74:1268–1281. https://doi.org/10.1016/j.gca.2009.11.026

Browne MA, Crump P, Niven SJ, Teuten E, Tonkin A, Galloway T, Thompson R (2011) Accumulation of microplastic on shorelines woldwide: Sources and sinks. Environ Sci Technol 45:9175–9179. https://doi.org/10.1021/es201811s

Buatier MD, Sobanska S, Elsass F (2001) TEM-EDX investigation on Zn- and Pb-contaminated soils. Appl Geochem 16:1165–1177. https://doi.org/10.1016/S0883-2927(01)00015-4

Bundschuh M, Filser J, Lüderwald S, McKee MS, Metreveli G, Schaumann GE et al (2018) Nanoparticles in the environment: where do we come from, where do we go to? Environ Sci Eur 30:6–6. https://doi.org/10.1186/s12302-018-0132-6

Carpenter SR, Caraco NF, Correll DL, Howarth RW, Sharpley AN, Smith VH (1998) Nonpoint pollution of surface waters with phosphorus and nitrogen. Ecol Appl 8:559–568. https://doi.org/10.1890/1051-0761(1998)008[0559:nposww]2.0.co;2

Carr R, Zhang C, Moles N, Harder M (2008) Identification and mapping of heavy metal pollution in soils of a sports ground in Galway City, Ireland, using a portable XRF analyser and GIS. Environ Geochem Health 30:45–52. https://doi.org/10.1007/s10653-007-9106-0
Carrillo-González R, Šimůnek J, Sauvé S, Adriano DC (2006) Mechanisms and pathways of trace element mobility in soils. Adv Agron 91:111–178
Chang AC, Hyun HN, Page AL (1997) Cadmium uptake for Swiss Chard grown on composted sewage sludge treated field plots – plateau or time bomb. J Environ Qual 26:11–19
Chao TT (1984) Use of partial dissolution techniques in geochemical exploration. J Geochem Explor 20:101–135. https://doi.org/10.1016/0375-6742(84)90078-5
Chardon ES, Bosbach D, Bryan ND, Lyon IC, Marquardt C, Romer J et al (2008) Reactions of the feldspar surface with metal ions: Sorption of Pb(II), U(VI) and Np(V), and surface analytical studies of reaction with Pb(II) and U(VI). Geochim Cosmochim Acta 72:288–297. https://doi.org/10.1016/j.gca.2007.10.026
Chen TB, Wong JWC, Zhou HY, Wong MH (1997) Assessment of trace metal distribution and contamination in surface soils of Hong Kong. Environ Pollut 96:61–68
Clark MW, McConchie DM (2004) Development of acid sulfate soil in sub-aerially disposed dredge spoil at Fisherman Islands, Brisbane, Australia. Aust J Soil Res 42:553–567
Da Costa JP, Paço A, Santos PSM, Duarte AC, Rocha-Santos T (2019) Microplastics in soils: assessment, analytics and risks. Environ Chem 16:18–30. https://doi.org/10.1071/EN18150
Daṃbkowska-Naskręt H (2004) The mobility of heavy metals in urban soils used for food production in Poland. Land Contam Reclam 12:205–212. https://doi.org/10.2462/09670513.648
Davidson DA, Dercon G, Stewart M, Watson F (2006) The legacy of past urban waste disposal on local soils. J Archaeol Sci 33:778–783. https://doi.org/10.1016/j.jas.2005.10.017
Davies BE (1988) Lead in soils: its sources and typical concentration. In: Davies BE, Wixson BG (eds) Lead in soil: issues and guidelines, Environmental Geochemistry and Health Monograph Series. Science Reviews Ltd, Northwood, pp 65–72
Davis AP, Shokouhian M, Ni S (2001) Loading estimates of lead, copper, cadmium, and zinc in urban runoff from specific sources. Chemosphere 44:997–1009
De Miguel E, Llamas JF, Chacón E, Berg T, Larssen S, Røyset O, Vadset M (1997) Origin and patterns of distribution of trace elements in street dust: unleaded petrol and urban lead. Atmos Environ 31:2733–2740
De Miguel E, Iribarren I, Chacón E, Ordoñez A, Charlesworth S (2007) Risk-based evaluation of the exposure of children to trace elements in playgrounds in Madrid (Spain). Chemosphere 66:505–513. https://doi.org/10.1016/j.chemosphere.2006.05.065
Deevey ES, Rice DS, Rice PM, Vaughan HH, Brenner M, Flannery MS (1979) Mayan urbanism: impact on a tropical karst environment. Science 206:298–306. https://doi.org/10.1126/science.206.4416.298
Degryse F, Smolders E, Zhang H, Davison W (2009) Predicting availability of mineral elements to plants with the DGT technique: a review of experimental data and interpretation by modelling. Environ Chem 6:198–218
Department of Health (WA) (2009) Guidelines for the assessment, remediation and Management of Asbestos-Contaminated Sites in Western Australia. Government of Western Australia, Environmental Health Directorate, Perth. https://ww2.health.wa.gov.au/~/media/Files/Corporate/general%20documents/Asbestos/PDF/Guidelines-Asbestos-Contaminated%20Sites-May2009.pdf
Dermatas D, Cao X, Tsaneva V, Shen G, Grubb DG (2006) Fate and behavior of metal(loid) contaminants in an organic matter-rich shooting range soil: implications for remediation. Water Air Soil Pollut Focus 6:143–155. https://doi.org/10.1007/s11267-005-9003-4
Dereaik JGB (2002) The pollution of the marine environment by plastic debris: a review. Mar Pollut Bull 44:842–852
Di Toro DM, Allen HE, Bergman HL, Meyer JS, Paquin PR, Santore RC (2001) Biotic ligand model of the acute toxicity of metals. 1. Technical basis. Environ Toxicol Chem 20:2383–2396

Du X, Rate AW, Gee MAM (2012) Redistribution and mobilization of titanium, zirconium and thorium in an intensely weathered lateritic profile in Western Australia. Chem Geol 330–331:101–115. https://doi.org/10.1016/j.chemgeo.2012.08.030

Du E, de Vries W, Liu X, Fang J, Galloway JN, Jiang Y (2015) Spatial boundary of urban 'acid islands' in southern China. Sci Rep 5:12625–12625. https://doi.org/10.1038/srep12625

Easton ZM, Petrovic AM (2008a) Determining nitrogen loading rates based on land use in an urban watershed. In: Nett MT, Carroll MJ, Horgan BP, Petrovic AM (eds) The fate of nutrients and pesticides in the urban environment, ACS symposium series. American Chemical Society, pp 19–42. https://doi.org/10.1021/bk-2008-0997.ch002

Easton ZM, Petrovic AM (2008b) Determining phosphorus loading rates based on land use in an urban watershed. In: Nett MT, Carroll MJ, Horgan BP, Petrovic AM (eds) The fate of nutrients and pesticides in the urban environment, ACS symposium series. American Chemical Society, pp 43–62. https://doi.org/10.1021/bk-2008-0997.ch003

Egendorf SP, Cheng Z, Deeb M, Flores V, Paltseva A, Walsh D et al (2018) Constructed soils for mitigating lead (Pb) exposure and promoting urban community gardening: The New York City Clean Soil Bank pilot study. Landsc Urban Plan 175:184–194. https://doi.org/10.1016/j.landurbplan.2018.03.012

Endo S, Kimura S, Takatsuji T, Nanasawa K, Imanaka T, Shizuma K (2012) Measurement of soil contamination by radionuclides due to the Fukushima Dai-ichi Nuclear Power Plant accident and associated estimated cumulative external dose estimation. J Environ Radioact 111:18–27. https://doi.org/10.1016/j.jenvrad.2011.11.006

Essington ME, Mattigod SV (1991) Trace-element solid-phase associations in sewage-sludge and sludge-amended soil. Soil Sci Soc Am J 55:350–356

Ettler V, Šebek O, Grygar T, Klementová M, Bezdička P, Slavíková H (2008) Controls on metal leaching from secondary Pb smelter air-pollution-control residues. Environ Sci Technol 42:7878–7884. https://doi.org/10.1021/es801246c

Evans CV, Fanning DS, Short JR (2000) Chapter 2: Human-influenced soils. In: Brown RB, Huddleston JH, Anderson JL (eds) Managing soils in an urban environment, Agronomy Monograph. American Society of Agronomy, Crop Science Society of America, Soil Science Society of America, Madison, pp 33–67. https://doi.org/10.2134/agronmonogr39.c2

Fabietti G, Biasioli M, Barberis R, Ajmone-Marsan F (2010) Soil contamination by organic and inorganic pollutants at the regional scale: the case of Piedmont, Italy. J Soils Sediments 10:290–300. https://doi.org/10.1007/s11368-009-0114-9

Fanning DS (1990) Salinity problems in acid sulfate coastal soils. In: Lieth H, Al Masoom AA (eds), Towards the rational use of high salinity tolerant plants. Kluwer Academic Publishers, Dordrecht, Netherlands, pp 491–500

Fanning DS, Rabenhorst MC (1990) Micromorphology of acid sulfate soils in Baltimore Harbor dredged materials. In: Douglas LA (ed) Developments in soil science. Elsevier, pp 279–288. https://doi.org/10.1016/S0166-2481(08)70339-8

Fernández-Caliani JC (2012) Risk-based assessment of multimetallic soil pollution in the industrialized peri-urban area of Huelva, Spain. Environ Geochem Health 34:123–139. https://doi.org/10.1007/s10653-011-9396-0

Franklin NM, Rogers NJ, Apte SC, Batley GE, Gadd GE, Casey PS (2007) Comparative toxicity of nanoparticulate ZnO, bulk ZnO, and $ZnCl_2$ to a freshwater microalga (*Pseudokirchneriella subcapitata*): the importance of particle solubility. Environ Sci Technol 41:8484–8490

Gallardo-Lara F, Nogales R (1987) Effect of the application of town refuse compost on the soil-plant system: a review. Biol Wastes 19:35–62. https://doi.org/10.1016/0269-7483(87)90035-8

Gambrell RP (1994) Trace and toxic metals in wetlands – a review. J Environ Qual 23:883–891. https://doi.org/10.2134/jeq1994.00472425002300050005x

Geboy NJ, Engle MA (2011) Quality assurance and quality control of geochemical data: a primer for the research scientist, Open-File Report 2011–1187. U.S. Geological Survey, Reston

Golab AN, Indraratna B (2009) Occurrence and consequences of acid sulphate soils and methods of site remediation. Geomech Geoeng 4:201–208. https://doi.org/10.1080/17486020902923387

Golyeva A, Khokhlova O, Lebedeva M, Shcherbakov N, Shuteleva I (2018) Micromorphological and chemical features of soils as evidence of Bronze Age ancient anthropogenic impact (Late Bronze Age Muradymovo Settlement, Ural region, Russia). Geosciences (Basel) 8:313. https://doi.org/10.3390/geosciences8090313

Gong Q, Deng J, Xiang Y, Wang Q, Yang L (2008) Calculating pollution indices by heavy metals in ecological geochemistry assessment and a case study in parks of Beijing. J China Univ Geosci 19:230–241. https://doi.org/10.1016/S1002-0705(08)60042-4

Gonnelli C, Renella G (2012) Chromium and nickel. In: Alloway BJ (ed) Heavy metals in soils: trace metals and metalloids in soils and their bioavailability, 3rd edn. Springer, Dordrecht, pp 313–333

Green DS, Boots B, Da Silva Carvalho J, Starkey T (2019) Cigarette butts have adverse effects on initial growth of perennial ryegrass (gramineae: Lolium perenne L.) and white clover (leguminosae: Trifolium repens L.). Ecotoxicol Environ Saf 182:109418. https://doi.org/10.1016/j.ecoenv.2019.109418

Gregory MR (2009) Environmental implications of plastic debris in marine settings- entanglement, ingestion, smothering, hangers-on, hitch-hiking and alien invasions. Philos Trans R Soc B Biol Sci 364:2013–2025. https://doi.org/10.1098/rstb.2008.0265

Gregory MM, Leslie TW, Drinkwater LE (2016) Agroecological and social characteristics of New York city community gardens: contributions to urban food security, ecosystem services, and environmental education. Urban Ecosyst 19:763–794. https://doi.org/10.1007/s11252-015-0505-1

Griffioen J (2001) Potassium adsorption ratios as an indicator for the fate of agricultural potassium in groundwater. J Hydrol 254:244–254. https://doi.org/10.1016/S0022-1694(01)00503-0

Groffman PM, Boulware NJ, Zipperer WC, Pouyat RV, Band LE, Colosimo MF (2002) Soil nitrogen cycle processes in urban riparian zones. Environ Sci Technol 36:4547–4552. https://doi.org/10.1021/es020649z

Grybos M, Davranche M, Gruau G, Petitjean P (2007) Is trace metal release in wetland soils controlled by organic matter mobility or Fe-oxyhydroxides reduction? J Colloid Interface Sci 314:490–501. https://doi.org/10.1016/j.jcis.2007.04.062

Hall GEM, Vaive JE, Beer R, Hoashi M (1996) Selective leaches revisited, with emphasis on the amorphous Fe oxyhydroxide phase extraction. J Geochem Explor 56:59–78

Hamon RE, McLaughlin MJ, Gilkes RJ, Rate AW, Zarcinas B, Robertson A et al (2004) Geochemical indices allow estimation of heavy metal background concentrations in soils. Glob Biogeochem Cycles 18. https://doi.org/10.1029/2003GB002063

Harrison RM, Laxen DPH, Wilson SJ (1981) Chemical associations of lead, cadmium, copper, and zinc in street dusts and roadside soils. Environ Sci Technol 15:1378–1383. https://doi.org/10.1021/es00093a013

Haynes RJ (1986) Mineral nitrogen in the plant-soil system. Academic, Orlando

He ZL, Yang XE, Stoffella PJ (2005) Trace elements in agroecosystems and impacts on the environment. J Trace Elem Med Biol 19:125–140

He D, Luo Y, Lu S, Liu M, Song Y, Lei L (2018) Microplastics in soils: analytical methods, pollution characteristics and ecological risks. TrAC – Trends Anal Chem 109:163–172. https://doi.org/10.1016/j.trac.2018.10.006

Hinsinger P (2001) Bioavailability of soil inorganic P in the rhizosphere as affected by root-induced chemical changes: a review. Plant Soil 237:173–195. https://doi.org/10.1023/a:1013351617532

Horton AA, Walton A, Spurgeon DJ, Lahive E, Svendsen C (2017) Microplastics in freshwater and terrestrial environments: evaluating the current understanding to identify the knowledge gaps and future research priorities. Sci Total Environ 586:127–141. https://doi.org/10.1016/j.scitotenv.2017.01.190

Huerta Lwanga E, Gertsen H, Gooren H, Peters P, Salánki T, van der Ploeg M et al (2017) Incorporation of microplastics from litter into burrows of Lumbricus terrestris. Environ Pollut 220:523–531. https://doi.org/10.1016/j.envpol.2016.09.096

Iglesias Jiménez E, Perez Garcia V (1989) Evaluation of city refuse compost maturity: a review. Biol Wastes 27:115–142. https://doi.org/10.1016/0269-7483(89)90039-6

Imperato M, Adamo P, Naimo D, Arienzo M, Stanzione D, Violante P (2003) Spatial distribution of heavy metals in urban soils of Naples city (Italy). Environ Pollut 124:247–256

Independent Working Group for PMSEIC (2005) Nanotechnology – enabling technologies for Australian innovative industries. Prime Minister's Science, Engineering and Innovation Council (PMSEIC), Australian Government. https://www.chiefscientist.gov.au/wp-content/uploads/20060311-Enabling-technologies-for-Australian-innovative-industries.pdf

IPCC (2007) *IPCC Fourth Assessment Report: Climate Change 2007*, Intergovernmental Panel on Climate Change. Cambridge University Press, Cambridge

Ivask A, Bondarenko O, Jepihhina N, Kahru A (2010) Profiling of the reactive oxygen species-related ecotoxicity of CuO, ZnO, TiO_2, silver and fullerene nanoparticles using a set of recombinant luminescent *Escherichia coli* strains: differentiating the impact of particles and solubilised metals. Anal Bioanal Chem 398:701–716

Jambeck JR, Andrady A, Geyer R, Narayan R, Perryman M, Siegler T et al (2015) Plastic waste inputs from land into the ocean. Science 347:768–771. https://doi.org/10.1126/science.1260352

Johnson CC, Ander EL (2008) Urban geochemical mapping studies: how and why we do them. Environ Geochem Health 30:511. https://doi.org/10.1007/s10653-008-9189-2

Johnston BD, Scown TM, Moger J, Cumberland SA, Baalousha M, Linge K et al (2010) Bioavailability of nanoscale metal oxides TiO_2, CeO_2, and ZnO to fish. Environ Sci Technol 44:1144–1151

Kabata-Pendias A (2004) Soil–plant transfer of trace elements—an environmental issue. Geoderma 122:143–149. https://doi.org/10.1016/j.geoderma.2004.01.004

Kabata-Pendias A (2011) Trace elements in soils and plants. CRC Press, Boca Raton, 505 pp

Karathanasis AD, Hajek BF (1996) Elemental analysis by X-ray fluorescence spectroscopy. In: Sparks DL et al (eds) Methods of soil analysis. Part 3 – chemical methods, SSSA book series. Soil science Society of America, Madison, pp 161–223

Kelly J, Thornton I, Simpson PR (1996) Urban geochemistry: a study of the influence of anthropogenic activity on the heavy metal content of soils in traditionally industrial and non-industrial areas of Britain. Appl Geochem 11:363–370. https://doi.org/10.1016/0883-2927(95)00084-4

Kjeldsen P (1999) Behaviour of cyanides in soil and groundwater: a review. Water Air Soil Pollut 115:279–307. https://doi.org/10.1023/A:1005145324157

Kuo S (1996) Phosphorus. In: Sparks DL et al (eds) Methods of soil analysis. Part 3 – chemical methods, SSSA book series. Soil Science Society of America, Madison, pp 869–919

Laidlaw MAS, Alankarage DH, Reichman SM, Taylor MP, Ball AS (2018) Assessment of soil metal concentrations in residential and community vegetable gardens in Melbourne, Australia. Chemosphere 199:303–311. https://doi.org/10.1016/j.chemosphere.2018.02.044

Lee RJ, Strohmeier BR, Bunker KL, Van Orden DR (2008) Naturally occurring asbestos—a recurring public policy challenge. J Hazard Mater 153:1–21. https://doi.org/10.1016/j.jhazmat.2007.11.079

Li J, Pu L, Zhu M, Zhang J, Li P, Dai X et al (2014) Evolution of soil properties following reclamation in coastal areas: a review. Geoderma 226–227:130–139. https://doi.org/10.1016/j.geoderma.2014.02.003

Li J, Chen H, Teng Y, Dong Q (2016) Contamination characteristics and source apportionment of soil heavy metals in Lalin River basin. Nongye Gongcheng Xuebao/Transactions of the Chinese Society of Agricultural Engineering 32:226–233. https://doi.org/10.11975/j.issn.1002-6819.2016.19.031

Lindsay WL, Norvell WA (1978) Development of a DTPA soil test for zinc, iron, manganese and copper. Soil Sci Soc Am J 42:421–428

Lindsay WL, Vlek PLG, Chien SH (1989) Phosphate minerals. In: Dixon JB, Weed SB, Dinauer RC (eds) Minerals in soil environments, 2nd edn. Soil Science Society of America, Madison, pp 1089–1130

Liu E, Yan T, Birch G, Zhu Y (2014) Pollution and health risk of potentially toxic metals in urban road dust in Nanjing, a mega-city of China. Sci Total Environ 476-477:522–531

Ljung K, Maley F, Cook A, Weinstein P (2009) Acid sulfate soils and human health-a millennium ecosystem assessment. Environ Int 35:1234–1242
Ljung K, Maley F, Cook A (2010) Canal estate development in an acid sulfate soil-implications for human metal exposure. Landsc Urban Plan 97:123–131
Lu AX, Zhang SZ, Shan XQ (2005) Time effect on the fractionation of heavy metals in soils. Geoderma 125:225–234. https://doi.org/10.1016/j.geoderma.2004.08.002
Ma WK, Smith BA, Stephenson GL, Siciliano SD (2009) Development of a simulated earthworm gut for determining bioaccessible arsenic, copper, and zinc from soil. Environ Toxicol Chem 28:1439–1446
Maaß S, Daphi D, Lehmann A, Rillig MC (2017) Transport of microplastics by two collembolan species. Environ Pollut 225:456–459. https://doi.org/10.1016/j.envpol.2017.03.009
Madrid L, Díaz-Barrientos E, Madrid F (2002) Distribution of heavy metal contents of urban soils in parks of Seville. Chemosphere 49:1301–1308. https://doi.org/10.1016/S0045-6535(02)00530-1
Majone M, Papini MP, Rolle E (1998) Influence of metal speciation in landfill leachates on kaolinite sorption. Water Res 32:882–890
Mann SS, Rate AW, Gilkes AJ (2002) Cadmium accumulation in agricultural soils in Western Australia. Water Air Soil Pollut 141:281–297. https://doi.org/10.1023/A:1021300228019
Mao Q, Huang G, Buyantuev A, Wu J, Luo S, Ma K (2014) Spatial heterogeneity of urban soils: the case of the Beijing metropolitan region, China. Ecol Process 3. https://doi.org/10.1186/s13717-014-0023-8
Martin I, Merrington G, Stutt E (2017) Derivation and use of soil screening values for assessing ecological risks. ShARE id26, Environment Agency, Bristol
Matera V, Le Hécho I, Laboudigue A, Thomas P, Tellier S, Astruc M (2003) A methodological approach for the identification of arsenic bearing phases in polluted soils. Environ Pollut 126:51–64. https://doi.org/10.1016/S0269-7491(03)00146-5
McBride MB (1994) Environmental chemistry of soils. Oxford University Press, New York
McBride MB (1995) Toxic metal accumulation from agricultural use of sludge – are USEPA regulations protective? J Environ Qual 24:5–18. https://doi.org/10.2134/jeq1995.00472425002400010002x
McBride MB, Sauvé S, Hendershot W (1997) Solubility control of cu, Zn, cd and Pb in contaminated soils. Eur J Soil Sci 48:337–346
McGrath SP, Zhao FJ, Dunham SJ, Crosland AR, Coleman K (2000) Long-term changes in the extractability and bioavailability of zinc and cadmium after sludge application. J Environ Qual 29:875–883
McLaren RG, Cameron KC (1990) Soil science – an introduction to the properties and Management of New Zealand Soils. Oxford University Press, Auckland
McLaren RG, Swift RS, Williams JG (1981) The adsorption of copper by soil materials at low equilibrium solution concentrations. J Soil Sci 32:247–256
McLaren RG, Clucas LM, Taylor MD, Hendry T (2004) Leaching of macronutrients and metals from undisturbed soils treated with metal-spiked sewage sludge. 2. Leaching of metals. Aust J Soil Res 42:459–471
McLaughlin MJ, Hamon RE, McLaren RG, Speir TW, Rogers SL (2000) Review: a bioavailability-based rationale for controlling metal and metalloid contamination of agricultural land in Australia and New Zealand. Aust J Soil Res 38:1037–1086
McQueen KG (2009) Regolith geochemistry. In: Scott K, Pain C (eds) Regolith science. CSIRO Publishing, Melbourne, pp 73–104
Medeiros FHM, Yoshimura EM (2005) Influence of soil and buildings on outdoor gamma dose rates in São Paulo, Brazil. Health Phys 88:65–70. https://doi.org/10.1097/01.HP.0000142499.92778.76
Meeussen JCL, Keizer MG, Van Riemsdijk WH, De Haan FAM (1994) Solubility of cyanide in contaminated soils. J Environ Qual 23:785–792. https://doi.org/10.2134/jeq1994.00472425002300040024x

Mielke HW, Gonzales CR, Smith MK, Mielke PW (1999) The urban environment and children's health: soils as an integrator of lead, zinc, and cadmium in New Orleans, Louisiana, U.S.A. Environ Res 81:117–129. https://doi.org/10.1006/enrs.1999.3966

Mielke HW, Laidlaw MAS, Gonzales C (2010) Lead (Pb) legacy from vehicle traffic in eight California urbanized areas: continuing influence of lead dust on children's health. Sci Total Environ 408:3965–3975

Mival K, Pump W, Dixon G (2006) From wasteland to green parkland, the remediation of the former West Melbourne Gasworks, Victoria, Australia. Land Contam Reclam 14:194–199. https://doi.org/10.2462/09670513.766

Moormann FR (1963) Acid sulfate soils (cat-clays) of the tropics. Soil Sci 95:271–275. https://doi.org/10.1097/00010694-196304000-00009

Morgan B, Rate AW, Burton ED (2012) Trace element reactivity in FeS-rich eutrophic estuarine sediments: influence of formation environment and acid sulfate soil drainage. Sci Total Environ 438:463–476. https://doi.org/10.1016/j.scitotenv.2012.08.088

Mortimer M, Kasemets K, Kahru A (2010) Toxicity of ZnO and CuO nanoparticles to ciliated protozoa *Tetrahymena thermophila*. Toxicology 269:182–189

Muezzinoglu A (2003) A study of volatile organic sulfur emissions causing urban odors. Chemosphere 51:245–252. https://doi.org/10.1016/S0045-6535(02)00821-4

Mulvaney RL (1996) Nitrogen – inorganic forms. In: Sparks DL et al (eds) Methods of soil analysis. Part 3 – chemical methods, SSSA book series. Soil Science Society of America, Madison, pp 1123–1184

Nabulo G, Young SD, Black CR (2010) Assessing risk to human health from tropical leafy vegetables grown on contaminated urban soils. Sci Total Environ 408:5338–5351. https://doi.org/10.1016/j.scitotenv.2010.06.034

Nagajyoti PC, Lee KD, Sreekanth TVM (2010) Heavy metals, occurrence and toxicity for plants: a review. Environ Chem Lett 8:199–216

Naidu R, Channey R, McConnell S, Johnston N, Semple KT, McGrath S et al (2015) Towards bioavailability-based soil criteria: past, present and future perspectives. Environ Sci Pollut Res 22:8779–8785. https://doi.org/10.1007/s11356-013-1617-x

National Environment Protection Council (2013a) Schedule B5b: guideline on methodology to derive ecological investigation levels in contaminated soils, *National Environment Protection (assessment of site contamination) measure (amended)*. Commonwealth of Australia, Canberra

National Environment Protection Council (2013b) Schedule B (1): guideline on the investigation levels for soil and groundwater, *National Environment Protection (assessment of site contamination) measure (amended)*. Commonwealth of Australia, Canberra

National Nanotechnology Initiative (2019) Applications and benefits. U.S. Government, Office of Science and Technology Policy. https://www.nano.gov/you/nanotechnology-benefits. Accessed 20190816

Nichols DG, Beardsell DV (1981) Interactions of calcium, nitrogen and potassium with phosphorus on the symptoms of toxicity in grevillea cv. 'Poorinda Firebird'. Plant Soil 61:437–445. https://doi.org/10.1007/BF02182024

Nizzetto L, Bussi G, Futter MN, Butterfield D, Whitehead PG (2016) A theoretical assessment of microplastic transport in river catchments and their retention by soils and river sediments. Environ Sci Process Impacts 18:1050–1059. https://doi.org/10.1039/c6em00206d

Nolan AL, Lombi E, McLaughlin MJ (2003) Metal bioaccumulation and toxicity in soils – why bother with speciation? Aust J Chem 56:77–91

Nriagu JO (1988) A silent epidemic of environmental metal poisoning? Environ Pollut 50:139–161. https://doi.org/10.1016/0269-7491(88)90189-3

Nyenje PM, Foppen JW, Uhlenbrook S, Kulabako R, Muwanga A (2010) Eutrophication and nutrient release in urban areas of sub-Saharan Africa — a review. Sci Total Environ 408:447–455. https://doi.org/10.1016/j.scitotenv.2009.10.020

Oporto C, Smolders E, Degryse F, Verheyen L, Vandecasteele C (2009) DGT-measured fluxes explain the chloride-enhanced cadmium uptake by plants at low but not at high cd supply. Plant Soil 318:127–135

Pardue JH, DeLaune RD, Patrick WH (1992) Metal to aluminum correlation in Louisiana coastal wetlands: identification of elevated metal concentrations. J Environ Qual 21:539–545. https://doi.org/10.2134/jeq1992.00472425002100040003x

Parfitt RL, Hume LJ, Sparling GP (1989) Loss of availability of phosphate in New Zealand soils. J Soil Sci 40:371–382. https://doi.org/10.1111/j.1365-2389.1989.tb01281.x

Parker DR, Pedler JF (1997) Reevaluating the free-ion activity model of trace metal availability to higher plants. Plant Soil 196:223–228

Pauget B, Gimbert F, Coeurdassier M, Crini N, Pérès G, Faure O et al (2013) Ranking field site management priorities according to their metal transfer to snails. Ecol Indic 29:445–454

Paulette L, Man T, Weindorf DC, Person T (2015) Rapid assessment of soil and contaminant variability via portable x-ray fluorescence spectroscopy: Copşa Mică, Romania. Geoderma 243-244:130–140. https://doi.org/10.1016/j.geoderma.2014.12.025

Peters NE, Donohue R (2001) Nutrient transport to the Swan-Canning Estuary, Western Australia. Hydrol Process 15:2555–2577

Plumlee GS, Benzel WM, Hoefen TM, Hageman PL, Morman SA, Reilly TJ et al (2016) Environmental implications of the use of sulfidic back-bay sediments for dune reconstruction — lessons learned post hurricane Sandy. Mar Pollut Bull 107:459–471. https://doi.org/10.1016/j.marpolbul.2016.04.051

Pons LJ, Van Der Molen WH (1973) Soil genesis under dewatering regimes during 1000 years of polder development. Soil Sci 116:228–235. https://doi.org/10.1097/00010694-197309000-00011

Praetorius A, Gundlach-Graham A, Goldberg E, Fabienke W, Navratilova J, Gondikas A et al (2017) Single-particle multi-element fingerprinting (spMEF) using inductively-coupled plasma time-of-flight mass spectrometry (ICP-TOFMS) to identify engineered nanoparticles against the elevated natural background in soils. Environ Sci Nano 4:307–314. https://doi.org/10.1039/C6EN00455E

Qiu H, Vijver MG, He E, Peijnenburg WJGM (2013) Predicting copper toxicity to different earthworm species using a multicomponent Freundlich model. Environ Sci Technol 47:4796–4803

Rabenhorst MC (2016) International acid sulfate soils conference recap. CSA News Magazine 61:24–27. https://doi.org/10.2134/csa2016-61-9-8

Rastegari Mehr M, Keshavarzi B, Moore F, Sharifi R, Lahijanzadeh A, Kermani M (2017) Distribution, source identification and health risk assessment of soil heavy metals in urban areas of Isfahan province, Iran. J Afr Earth Sci 132:16–26. https://doi.org/10.1016/j.jafrearsci.2017.04.026

Rate AW (2018) Multielement geochemistry identifies the spatial pattern of soil and sediment contamination in an urban parkland, Western Australia. Sci Total Environ 627:1106–1120. https://doi.org/10.1016/j.scitotenv.2018.01.332

Ravishankara AR, Daniel JS, Portmann RW (2009) Nitrous oxide (N_2O): the dominant ozone-depleting substance emitted in the 21st century. Science 326:123–125. https://doi.org/10.1126/science.1176985

Rayment GE, Lyons DJ (2010) Soil chemical methods – Australasia, Australian soil and land survey handbooks series. CSIRO Publishing, Clayton

Reimann C, Garrett RG (2005) Geochemical background – concept and reality. Sci Total Environ 350:12–27

Reimann C, Filzmoser P, Garrett RG, Dutter R (2008) Statistical data analysis explained: applied environmental statistics with R. Wiley, Chichester, 343 pp

Ren HM, Wang JD, Zhang XL (2006) Assessment of soil lead exposure in children in Shenyang, China. Environ Pollut 144:327–335. https://doi.org/10.1016/j.envpol.2005.11.011

Rillig MC (2012) Microplastic in terrestrial ecosystems and the soil? Environ Sci Technol 46:6453–6454. https://doi.org/10.1021/es302011r

Rose A, Hawkes HE, Webb JS (1979) Geochemistry in mineral exploration. Academic, London

Salminen R, Batista MJ, Bidovec M, Demetriades A, De Vivo B, De Vos W et al (2005) Geochemical atlas of Europe. Part 1: background information, methodology and maps. Geological Survey of Finland, Espoo. http://www.gtk.fi/publ/foregsatlas/

Salmon SU, Rate AW, Rengel Z, Appleyard SJ, Prommer H, Hinz C (2014) Reactive transport controls on sandy acid sulfate soils and impacts on shallow groundwater quality. Water Resour Res 50:4924–4952. https://doi.org/10.1002/2013WR014404

Sangare SK, Compaore E, Buerkert A, Vanclooster M, Sedogo MP, Bielders CL (2012) Field-scale analysis of water and nutrient use efficiency for vegetable production in a West African urban agricultural system. Nutr Cycl Agroecosyst 92:207–224. https://doi.org/10.1007/s10705-012-9484-2

Sauvé S, Hendershot W, Allen HE (2000) Solid-solution partitioning of metals in contaminated soils: dependence on pH, total metal burden, and organic matter. Environ Sci Technol 34:1125–1131

Save S, Kovacik J, Demarly-Cresp F, Issenmann R, Poirier S, Sedlbauer S, Teyssonneyre Y (2020) Large-scale geochemical survey by pXRF spectrometry of archaeological settlements and features: new perspectives on the method. Archaeol Prospect 27:203–218. https://doi.org/10.1002/arp.1773

Schindelbeck RR, van Es HM, Abawi GS, Wolfe DW, Whitlow TL, Gugino BK et al (2008) Comprehensive assessment of soil quality for landscape and urban management. Landsc Urban Plan 88:73–80. https://doi.org/10.1016/j.landurbplan.2008.08.006

Schodlok MC, Whitbourn L, Huntington J, Mason P, Green A, Berman M et al (2016) HyLogger-3, a visible to shortwave and thermal infrared reflectance spectrometer system for drill core logging: functional description. Aust J Earth Sci 63:929–940. https://doi.org/10.1080/08120099.2016.1231133

Schulthess CP, Huang CP (1990) Adsorption of heavy metals by silicon and aluminium oxide surfaces on clay minerals. Soil Sci Soc Am J 54:679–688

Schwarz K, Pickett STA, Lathrop RG, Weathers KC, Pouyat RV, Cadenasso ML (2012) The effects of the urban built environment on the spatial distribution of lead in residential soils. Environ Pollut 163:32–39. https://doi.org/10.1016/j.envpol.2011.12.003

Semple KT, Doick KJ, Jones KC, Burauel P, Craven A, Harms H (2004) Defining bioavailability and bioaccessibility of contaminated soil and sediment is complicated. Environ Sci Technol 38:228A–231A

Sen TK, Khilar KC (2006) Review on subsurface colloids and colloid-associated contaminant transport in saturated porous media. Adv Colloid Interf Sci 119:71–96. https://doi.org/10.1016/j.cis.2005.09.001

Shuman LM (1985) Fractionation method for soil microelements. Soil Sci 140:11–22

Singh B, Gilkes RJ (1992) Properties and distribution of iron oxides and their association with minor elements in the soils of South-Western Australia. J Soil Sci 43:77–98. https://doi.org/10.1111/j.1365-2389.1992.tb00121.x

Singh SP, Tack FMG, Verloo MG (1996) Extractability and bioavailability of heavy metals in surface soils derived from dredged sediments. Chem Speciat Bioavailab 8:105–110

Smith BA, Greenberg B, Stephenson GL (2010) Comparison of biological and chemical measures of metal bioavailability in field soils: test of a novel simulated earthworm gut extraction. Chemosphere 81:755–766

Smolders E, Oorts K, van Sprang P, Schoeters I, Janssen CR, McGrath SP, McLaughlin MJ (2009) Toxicity of trace metals in soil as affected by soil type and aging after contamination: using calibrated bioavailability models to set ecological soil standards. Environ Toxicol Chem 28:1633–1642. https://doi.org/10.1897/08-592.1

Snowdon R, Birch GF (2004) The nature and distribution of copper, lead, and zinc in soils of a highly urbanised sub-catchment (Iron Cove) of Port Jackson, Sydney. Aust J Soil Res 42:329–338. https://doi.org/10.1071/SR03017

Sparks DL, Page AL, Helmke PA, Loeppert RH, Soltanpour PN, Tabatabai MA et al (eds) (1996) Methods of soil analysis. Part 3 – chemical methods, SSSA Book Series. Soil Science Society of America, Madison

Stec AA, Dickens K, Barnes JLJ, Bedford C (2019) Environmental contamination following the Grenfell tower fire. Chemosphere:576–586. https://doi.org/10.1016/j.chemosphere.2019.03.153

Sterrett SB, Chaney RL, Gifford CH, Mielke HW (1996) Influence of fertilizer and sewage sludge compost on yield and heavy metal accumulation by lettuce grown in urban soils. Environ Geochem Health 18:135–142

Stevenson FJ (1994) Humus chemistry: genesis, composition, reactions, 2nd edn. Wiley, New York

Stone AT, Morgan JJ (1987) Reductive dissolution of metal oxides. In: Stumm W (ed) Aquatic surface chemistry, Environmental science and technology series. Wiley, New York, pp 221–254

Strauch G, Möder M, Wennrich R, Osenbrück K, Gläser HR, Schladitz T et al (2008) Indicators for assessing anthropogenic impact on urban surface and groundwater. J Soils Sediments 8:23–33

Street JJ, Lindsay WL, Sabey BR (1977) Solubility and plant uptake of cadmium in soils amended with cadmium and sewage sludge. J Environ Qual 6:72–77

Sullivan TS, McBride MB, Thies JE (2013) Rhizosphere microbial community and Zn uptake by willow (*Salix purpurea* L.) depend on soil sulfur concentrations in metalliferous peat soils. Appl Soil Ecol 67:53–60

Sun Y, Zhou Q, Xie X, Liu R (2010) Spatial, sources and risk assessment of heavy metal contamination of urban soils in typical regions of Shenyang, China. J Hazard Mater 174:455–462. https://doi.org/10.1016/j.jhazmat.2009.09.074

Szolnoki Z, Farsang A, Puskás I (2013) Cumulative impacts of human activities on urban garden soils: origin and accumulation of metals. Environ Pollut 177:106–115. https://doi.org/10.1016/j.envpol.2013.02.007

Taylor PD, Ramsey MH, Potts PJ (2004) Balancing measurement uncertainty against financial benefits: comparison of in situ and ex situ analysis of contaminated land. Environ Sci Technol 38:6824–6831. https://doi.org/10.1021/es049739p

Tessier A, Campbell PGC, Bisson M (1979) Sequential extraction procedure for the speciation of particulate trace metals. Anal Chem 51:844–851

Thakali S, Allen HE, Di Toro DM, Ponizovsky AA, Rooney CP, Zhao FJ, McGrath SP (2006) A terrestrial biotic ligand model. 1. Development and application to cu and Ni toxicities to barley root elongation in soils. Environ Sci Technol 40:7085–7093

Thangarajan R, Bolan NS, Tian G, Naidu R, Kunhikrishnan A (2013) Role of organic amendment application on greenhouse gas emission from soil. Sci Total Environ 465:72–96. https://doi.org/10.1016/j.scitotenv.2013.01.031

Thebo AL, Drechsel P, Lambin EF (2014) Global assessment of urban and peri-urban agriculture: irrigated and rainfed croplands. Environ Res Lett:9.:, 9pp. https://doi.org/10.1088/1748-9326/9/11/114002

Theng BKG, Yuan GD (2008) Nanoparticles in the soil environment. Elements 4:395–399

Thornton I (2010) Research in applied environmental geochemistry, with particular reference to geochemistry and health. Geochem Explor Environ Anal 10:317–329

U.S. EPA (2002) Guidance for comparing background and chemical concentrations in soil for CERCLA sites. EPA 540-R-01-003 / OSWER 9285.7-41, United States Environmental Protection Agency, Office of Emergency and Remedial Response, Washington, DC. https://www.epa.gov/sites/production/files/2015-11/documents/background.pdf

Unrine JM, Tsyusko OV, Hunyadi SE, Judy JD, Bertsch PM (2010) Effects of particle size on chemical speciation and bioavailability of copper to earthworms (*Eisenia fetida*) exposed to copper nanoparticles. J Environ Qual 39:1942–1953

USEPA (2008) Framework for investigating Asbestos-contaminated superfund sites. OSWER DIRECTIVE #9200.0-68, United States Environmental Protection Agency, Washington, DC. https://semspub.epa.gov/work/HQ/175329.pdf

Vogeler I, Cichota R, Snow VO, Dutton T, Daly B (2011) Pedotransfer functions for estimating ammonium adsorption in soils. Soil Sci Soc Am J 75:324–331. https://doi.org/10.2136/sssaj2010.0192

Wang L, Cheng Y, Lamb D, Naidu R (2020) The application of rapid handheld FTIR petroleum hydrocarbon-contaminant measurement with transport models for site assessment: a case study. Geoderma 361:114017. https://doi.org/10.1016/j.geoderma.2019.114017

Weber J, Karczewska A, Drozd J, Licznar M, Licznar S, Jamroz E, Kocowicz A (2007) Agricultural and ecological aspects of a sandy soil as affected by the application of municipal solid waste composts. Soil Biol Biochem 39:1294–1302. https://doi.org/10.1016/j.soilbio.2006.12.005

Wei B, Yang L (2010) A review of heavy metal contaminations in urban soils, urban road dusts and agricultural soils from China. Microchem J 94:99–107. https://doi.org/10.1016/j.microc.2009.09.014

Wenzel WW (2012) Arsenic. In: Alloway BJ (ed) Heavy metals in soils: trace metals and metalloids in soils and their bioavailability, 3rd edn. Springer, Dordrecht, pp 241–282

Whitten MG, Ritchie GSP (1991) Calcium chloride extractable cadmium as an estimate of cadmium uptake by subterranean clover. Aust J Soil Res 29:215–221

Wild A (ed) (1988) Russell's soil conditions and plant growth. Longman Scientific and Technical, Harlow, 990 pp

Wolff C, Gilhaus K, Hölzel N, Schneider S (2017) Status and restoration potential of heathlands and sand grasslands in the southwest of Luxembourg. Tuexenia 37:179–200. https://doi.org/10.14471/2017.37.007

Wolt JD (1994) Soil solution chemistry: applications to environmental science and agriculture. Wiley, New York, 345 pp

Wong SC, Li XD, Zhang G, Qi SH, Min YS (2002) Heavy metals in agricultural soils of the Pearl River Delta, South China. Environ Pollut 119:33–44. https://doi.org/10.1016/s0269-7491(01)00325-6

Xia K, Bleam W, Helmke PA (1997) Studies of the nature of Cu^{2+} and Pb^{2+} binding sites in soil humic substances using X-ray absorption spectroscopy. Geochim Cosmochim Acta 61:2211–2221

Xu N (2018) Geochemistry and mineralogy of trace and rare-earth elements In acidified estuarine dredge spoil and associated sulfidic drain sediments, South Yunderup, Western Australia. Ph.D. thesis, The University of Western Australia, Perth, Western Australia, 208 pp

Yang Z, Lu W, Long Y, Bao X, Yang Q (2011) Assessment of heavy metals contamination in urban topsoil from Changchun City, China. J Geochem Explor 108:27–38. https://doi.org/10.1016/j.gexplo.2010.09.006

Yarlagadda PS, Matsumoto MR, Vanbenschoten JE, Kathuria A (1995) Characteristics of heavy metals in contaminated soils. J Environ Eng (United States) 121:276–286. https://doi.org/10.1061/(ASCE)0733-9372(1995)121:4(276)

Yaylalı-Abanuz G (2011) Heavy metal contamination of surface soil around Gebze industrial area, Turkey. Microchem J 99:82–92. https://doi.org/10.1016/j.microc.2011.04.004

Yongming H, Peixuan D, Junji C, Posmentier ES (2006) Multivariate analysis of heavy metal contamination in urban dusts of Xi'an, Central China. Sci Total Environ 355:176–186. https://doi.org/10.1016/j.scitotenv.2005.02.026

Young SD (2012) Chemistry of heavy metals and metalloids in soils. In: Alloway BJ (ed) Heavy metals in soils: trace metals and metalloids in soils and their bioavailability, 3rd edn. Springer, Dordrecht, pp 51–95

Zhang H, Zhao FJ, Sun B, Davison W, McGrath SP (2001) A new method to measure effective soil solution concentration predicts copper availability to plants. Environ Sci Technol 35:2602–2607

Zhou Y, Liu X, Wang J (2019) Characterization of microplastics and the association of heavy metals with microplastics in suburban soil of Central China. Sci Total Environ 694. https://doi.org/10.1016/j.scitotenv.2019.133798

Chapter 7
Organic Contaminants in Urban Soils

Andrew W. Rate

Abstract Organic pollutants are predominantly synthetic compounds made or extracted by humans, and it is therefore not surprising that they are common contaminants in urban soils. This chapter provides details of the immense range of potential organic pollutants in urban soils, including their types, physical properties, chemical structures, and sources. The behaviour of non-polar, polar, and ionic organic compounds in soils is explained in terms of their key chemical reactions, including abiotic and biological degradation, and transport phenomena are also addressed, before a brief discussion of toxicity. Sampling, extraction, and chemical analysis methods are covered briefly, together with background concentrations, regulatory contamination thresholds, and contamination indices. A comprehensive case study is focused on polycyclic aromatic hydrocarbons (PAH) in the highly urbanised Pearl River Delta region of China. This case study uses the PAH family of compounds to illustrate many of the concepts needed to understand the behaviour of persistent organic pollutants in urban soils, concluding with their fluxes and a conceptual PAH cycle.

Keywords Organic contaminants · Urban soils · Hydrocarbons · Persistent organic pollutants · PAH · Pearl River Delta

---

A. W. Rate (✉)
School of Agriculture and Environment, University of Western Australia, Crawley, WA, Australia
e-mail: andrew.rate@uwa.edu.au

A. W. Rate (ed.), *Urban Soils*, Progress in Soil Science,
https://doi.org/10.1007/978-3-030-87316-5_7

What you could learn from this chapter:

- The types of organic contaminants that are of concern are in urban soils and their sources.
- The processes controlling immobilisation, mobility, and bioavailability of organic contaminants.
- What the fluxes of organic contaminants are between air, water, sediment, and soil and the factors controlling these, including a case study of the Pearl River Delta, Guangdong, PR China.
- Being aware of chemical analysis methods for organic contaminants in soils.

7.1 Sources of Organic Contaminants in Urban Soils

An enormous diversity of organic compounds has been added to soils in urban environments. Very few of these compounds are naturally occurring in soils, and therefore even low concentrations represent contamination. The much less common natural occurrence of organic pollutants is an important point of difference from inorganic pollutants such as metals. There is some natural occurrence of compounds in the PAH and dioxin categories from phenomena such as wildfires, but even for completely anthropogenic compounds, there are still 'background' concentrations in soils in minimally disturbed natural environments (Biasioli et al. 2012). Background concentrations of organic contaminants represent widespread dispersion of anthropogenic compounds by global atmospheric circulation by the late twentieth century (Wania and MacKay 1996). For example, Fuoco et al. (2009) describe the occurrence of persistent organic pollutants in Antarctica, derived from both remote sources and local use of fuels and other organic chemicals. Some important categories of organic pollutants and their main sources are listed on Table 7.1.

Table 7.1 Categories and sources of organic contaminants which may be found in urban soils

Category of organic contaminant	Important sources	Reference(s)
Petroleum hydrocarbons (subdivided into fractions based on number of C atoms)	Fuel/oil leaks; drainage and stormwater from impermeable surfaces; some pesticides	Kostecki et al. (2005)
Volatile organic chlorinated compounds (e.g. trichloroethylene (TCE), vinyl chloride)	Emissions from industrial use as solvents/cleaners, precursors for other products (e.g. plastics)	National Pollutant Inventory (2020)
Monocyclic aromatic hydrocarbons (MAH), e.g. BTEX (benzene, toluene, ethylbenzene, xylene)	Fuel/oil leaks; improper industrial solvent disposal; vehicle and industrial emissions	Kostecki et al. (2005)
Polycyclic aromatic hydrocarbons (PAH), e.g. naphthalene, anthracene, benzo(a)pyrene, etc.	Emissions from incomplete combustion of coal, petroleum hydrocarbons, wood, etc.; road bitumen	Johnson (2005)
Phenols and **phthalates**	plastics; industrial products and wastes; landfills; oil refineries; coal conversion plants; spills	National Pollutant Inventory (2020)
Pesticides (e.g. organochlorine pesticides (OCPs), phenoxyacetic acid herbicides, triazines)	Pesticide usage and residues; leaks and spills	Bromilow (2005)
Polychlorinated biphenyls (PCB)	Industrial and consumer products and wastes; electrical transformer leakage; e-waste	Johnson (2005)
Dioxins and dibenzofurans (PCDD/F)	Unintentionally produced from combustion (e.g. of waste) and some industrial processes	Weber et al. (2008)
Flame retardants (e.g. PBDEs, PFAS)	Flame retardant products; industrial and consumer products and wastes; e-waste; waste incineration	Barceló (2012)
Pharmaceuticals, cosmetics, household products, etc. (e.g. antibiotics, steroids, oils, detergents)	Biosolids; septic tanks; wastewater and wastewater treatment	Barceló (2012) and Rodríguez-Eugenio et al. (2018)
POPs – persistent organic pollutants	Category includes PAHs, OCPs, PCBs, PCDD/F, PBDEs, PFAS	See Sect. 7.5

7.1.1 Types of Organic Compounds

McBride (1994) identifies some important properties of organic compounds which affect their behaviour, such as retention or mobility, in soils. These include the type and chemical reactivity of organic functional groups, the size and shape of organic molecules, and polarity/polarisability or charge of organic molecules and ions. The susceptibility of an organic compound to oxidation or reduction is also important. These molecular properties also affect physical properties such as volatility (e.g. vapour pressure or boiling point), density, or solubility in water or other solvents which, in turn, affect their behaviour in soils and similar environments. One of the more important aspects of the behaviour of organic compounds in soils is whether or not they are readily degraded by soil (micro)organisms, a distinction which contributes to separating the *persistent organic pollutants* (POPs) from other organic chemicals.

7.1.1.1 Categories of Organic Contaminants Based on General Chemical Properties

Polarity and charge (ionisation) of organic molecules are some of the more important properties affecting their behaviour (McBride 1994). We also distinguish organic molecules based on whether they consist of linear or branched chains of carbon atoms (aliphatic) or contain ring structures with carbon-carbon double bonding that involves delocalisation of electrons (aromatic); some examples of chemical structures are shown in Table 7.2 together with information on ionisability and polarity and potential sources. In both aliphatic and aromatic compounds, carbon atoms are bonded either to other carbon atoms or to hydrogen atoms. Much differentiation between organic compounds also depends on the presence in the molecules of specific arrangements of atoms, or *functional groups*. These functional groups often define a class of organic compounds and may include atoms other than carbon and hydrogen: oxygen, nitrogen, sulphur, phosphorus, or halogens such as fluorine, chlorine, or bromine.

The chemical structure of organic pollutants is an important determinant of their toxicity and persistence in environments. Of the 12 'persistent organic pollutant' (POP) compounds originally listed by the United Nations' Stockholm Convention, all contain both chlorine atoms and rings of carbon atoms. In some cases, the compounds were designed to be toxic (the original 12 POPs included 8 pesticides), but their effects on non-target species, including humans, were not initially considered. Other chemicals such as the PCBs, hexachlorobenzene, and the dioxins and dibenzofurans (see Tables 7.1 and 7.2) were either intended for other uses or, in the case of the PCDD/F compounds, were unintentional by-products of other anthropogenic processes. In humans and other mammals, exposure to the Stockholm Convention POPs can result in adverse effects to multiple physiological systems including endocrine, immune, digestive, and reproductive systems. Many of the compounds are also known or suspected to increase the incidence of cancers and have been shown to produce birth defects.

Much of the risk associated with persistent organic pollutants relates to their tendency to bioaccumulate and biomagnify – that is, they are not excreted efficiently from the body, so they accumulate and also tend to increase in concentration towards higher levels in food chains which increases the risk to predators and humans.

Persistent organic pollutants (POPs) are anthropogenic organic chemicals which persist for many years once released into the environment. In addition, POPs tend to *bioaccumulate* and are toxic to humans and other organisms (Rodríguez-Eugenio et al. 2018). The POPs (Table 7.3) are generally highly regulated, based on the

Table 7.2 Categories of organic compounds showing structure diagrams as 'ball and stick' models or conventional structure drawings

Category	Ionisation or polarity	Structure examples		Examples of pollutants or polluting activities
Aliphatic hydrocarbons	Non-ionisable, non-polar	Hexane $CH_3(CH_2)_4CH_3$ 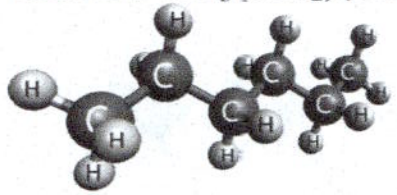	2-Methylpropane $CH_3CH(CH_3)CH_3$ 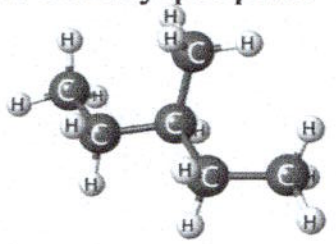	Petroleum hydrocarbons (NAPLs). Fuel providers and refiners.
Aromatic hydrocarbons These compounds contain one or more 6-carbon rings with alternating double bonds, drawn 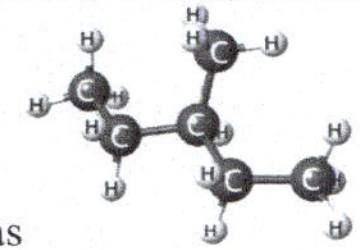 as	Non-ionisable, non-polar	1,2-Xylene (o-xylene) 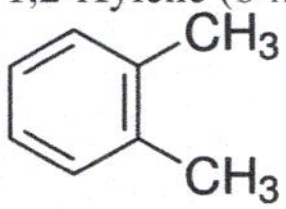	Phenanthrene (one of the PAHs) 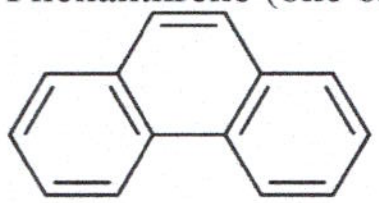	Compounds from the BTEX and PAH groups Use of petrochemicals, combustion processes
Chlorinated hydrocarbons	Non-ionisable, slightly polar	Trichloroethane $Cl_3C–CH_3$ 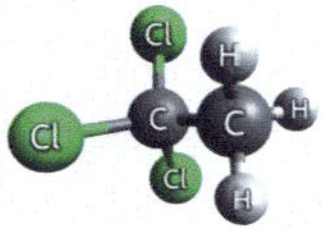	1,2,4-Trichlorobenzene 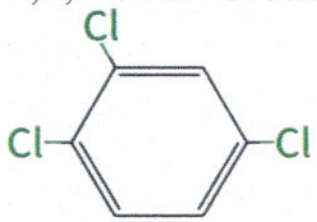	Solvents, paints, adhesives and industries using or producing them
PCBs	Non-ionisable, slightly or non-polar	2,4,4′-Trichlorobiphenyl (PCB-28) 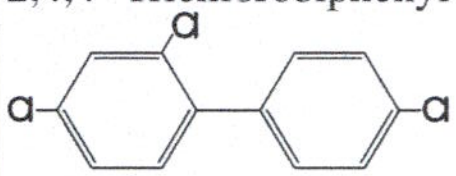		Transformer oils

(continued)

Table 7.2 (continued)

Category	Ionisation or polarity	Structure examples		Examples of pollutants or polluting activities
Organochlorine pesticides	Non-ionisable, polar	p,p′-DDT 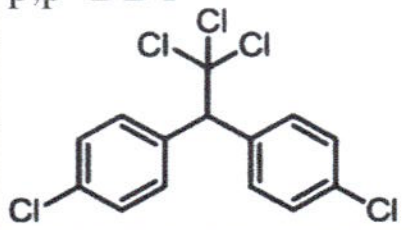	Heptachlor 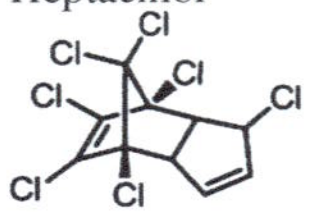	Current and legacy pesticide usage
Other pesticides	Ionisable,polar	2,4-Dichlorophenoxyacetic acid 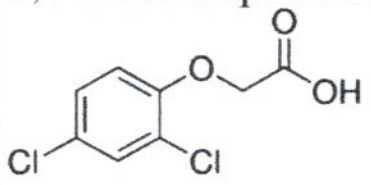	Atrazine 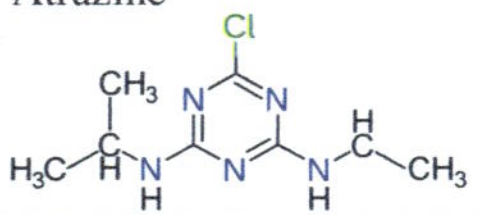	Phenoxyacetic acid herbicides, triazines used for weed control
Phenols, phthalates, phthalate esters	Ionisable (non-ester phthalates, some phenols) or non-ionisable, polar	Bisphenol-A 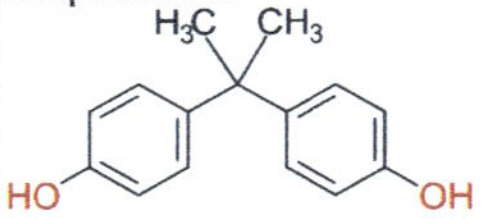	Diethyl phthalate 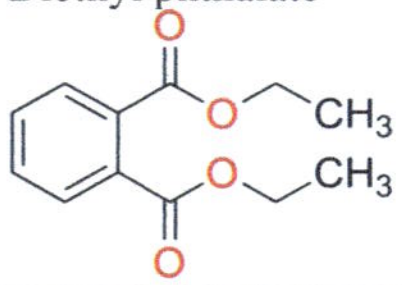	Plasticisers, cosmetic products; common in urban wastewaters
Fire retardants	Ionisable (PFAS) or non-ionisable (PBDE), polar	Perfluorooctanesulfonic acid 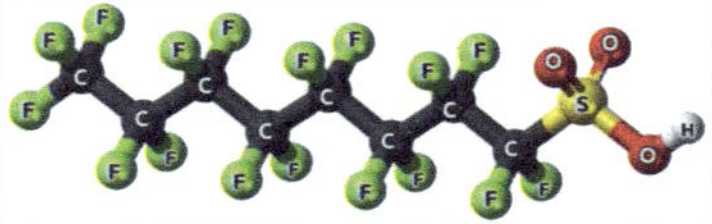	Pentabromodiphenyl ether 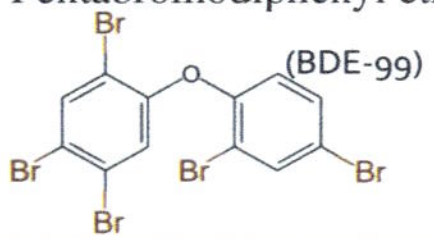	PFOA, PFAS, PBDE Airports, military bases, fire services

Category	Ionisation or polarity	Structure examples		Examples of pollutants or polluting activities
Dioxins, dibenzofurans(PCDDs, PCDFs) (also equivalent polybrominated compounds)	Non-ionisable, polar or non-polar	2,3,7,8-Tetrachlorodibenzo-*p*-dioxin 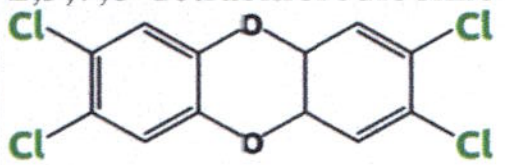	2,3,7,8-Tetrachlorodibenzofuran 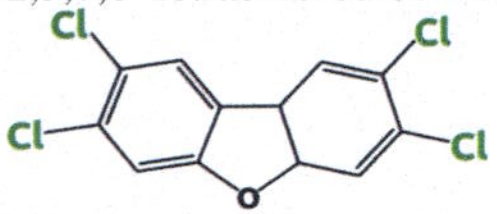	Combustion by-products, impurities
Pharmaceuticals	Ionisable or non-ionisable, polar	Oestradiol 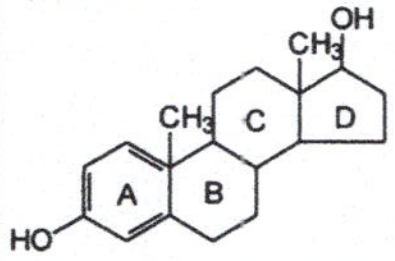	Penicillin-type antibiotic 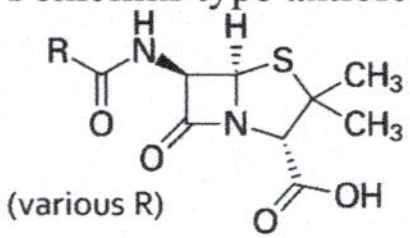	Oral contraceptives, hospital effluents, urban wastewaters

Table 7.3 List of the 28 persistent organic pollutants (POPs) recognised by the Stockholm Convention

Category of POP	Recognised persistent organic pollutants
Pesticides	Aldrin; chlordane; DDT; dieldrin; endrin; heptachlor; hexachlorobenzene; mirex; toxaphene; lindane; pentachlorophenol and its salts and esters; technical endosulfan and its related isomers
Industrial chemicals	Hexachlorobenzene; polychlorinated biphenyls (PCBs); decabromodiphenyl ether (commercial mixture = c-decaBDE); alpha hexachlorocyclohexane; beta hexachlorocyclohexane; chlordecone; dicofol; pentachlorobenzene; hexabromobiphenyl; hexabromocyclododecane; hexabromodiphenyl ether and heptabromodiphenyl ether (commercial octabromodiphenyl ether); perfluorooctanesulfonic acid, its salts, and perfluorooctane sulfonyl fluoride; perfluorooctanoic acid (PFOA), its salts, and PFOA-related compounds; short-chain chlorinated paraffins (SCCPs); tetrabromodiphenyl ether and pentabromodiphenyl ether (commercial pentabromodiphenyl ether)
By-products (where not included above)	Polychlorinated dibenzo-p-dioxins and polychlorinated dibenzofurans (PCDD/PCDF); hexachlorobutadiene; polychlorinated naphthalenes

revised Stockholm Convention of the United Nations Environment Programme (Secretariat of the Stockholm Convention 2018).

7.2 Controls on Organic Contaminant Behaviour in Urban Soils

7.2.1 *Non-polar, Non-ionic Compounds*

The organic compounds which are both non-ionic and non-polar include the unsubstituted hydrocarbons (i.e. those composed of only C and H) and some halogenated hydrocarbons having molecules with high symmetry such as carbon tetrachloride, tetrachloroethylene, or 1,3,5-trichlorobenzene. Some molecules with low polarity (e.g. 1,2-dichlorobenzene, DCB) behave similarly and are also classified as non-polar in some contexts (Kile et al. 1995). A good operational measure of whether an organic compound behaves as polar or not is the octanol-water partition coefficient, K_{ow}, (Eq. 7.1):

$$K_{ow} = \frac{C_o}{C_w} \tag{7.1}$$

where C_o is the concentration of compound in n-octanol and C_w is the concentration of compound in n-octanol, both in contact and at equilibrium.

A very non-polar compound will be much more soluble in the low-polarity liquid n-octanol compared with water and have a large K_{ow} (or, commonly, $\log_{10}(K_{ow})$) value compared with a polar compound.

Very non-polar organic compounds, if present in soils in large enough concentrations (e.g. from a leak or spill), can be present as a separate liquid phase, commonly called a non-aqueous-phase liquid or NAPL. The NAPL group is subdivided on the basis of density relative to water into the LNAPL (light non-aqueous-phase liquid) and DNAPL (dense non-aqueous-phase liquid) subcategories (Fig. 7.1). The separation of LNAPLs and DNAPLs on the basis of density relative to water is important, since their density is one of the primary factors affecting how separate NAPL phases will behave in a soil-groundwater system. Light NAPLs, such as petrol/gasoline, will 'float' on top of the groundwater; conversely, dense NAPLs will sink through the groundwater until reaching a confining layer such as a clay lens or impermeable bedrock. This tendency to float or sink also affects how contamination with LNAPLs or DNAPLs is treated, as we discuss in Chap. 11.

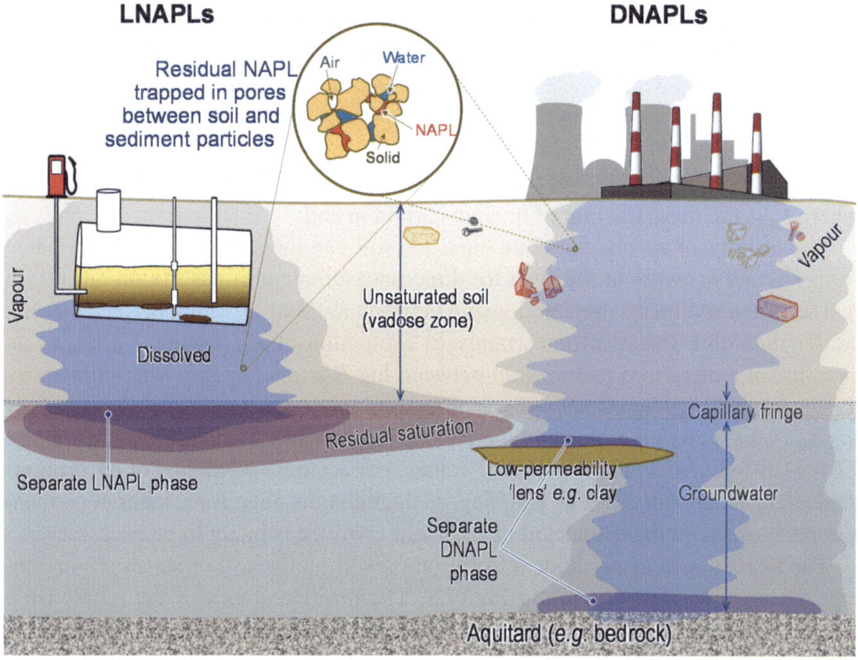

Fig. 7.1 Schematic of the behaviour of light non-aqueous-phase liquids (LNAPL) and dense non-aqueous-phase liquids (DNAPL) in soils, sediment, and groundwater below release zones. (Graphic by Andrew W. Rate)

7.2.1.1 Adsorption of Non-polar Organics

The K_{ow} value is also extremely useful in explaining the retention of organic chemicals by soils or sediments. Natural organic matter in soils and sediments behaves as an organic 'solvent' into which non-polar organic compounds preferentially partition into, by a related mechanism to their partitioning into n-octanol. The contents of soil organic matter and the associated *black carbon* are therefore the main factor affecting adsorption of non-polar organic compounds in soils. The partition coefficient for soil organic matter is related to K_{ow} by a linear log-log relationship, and so retention can be predicted if the organic carbon content of soil or sediment is known. Retention of organic contaminants in soils is commonly described by K_D, the adsorption (or partition) coefficient:

$$K_D = \frac{C_s}{C_w} \tag{7.2}$$

where C_s is the concentration of compound adsorbed to soil and C_w is the concentration of compound in water at equilibrium with soil.

The organic carbon normalised adsorption coefficient, K_{OC}, is defined by (Eq. 7.3):

$$K_{oc} = \frac{K_D}{f_{oc}} \tag{7.3}$$

where f_{oc} is the mass fraction of organic carbon in soil.

The affinity of an organic compound for soil can then, in theory, be estimated from K_{ow} and f_{oc}, without the need for direct measurement of adsorption on soil. In real soils and sediments, the relationship between K_{OC} and K_{ow} is not perfectly linear, conferring some statistical uncertainty to estimation of K_D. Figure 7.2 shows an example of a measured relationship between log K_{OC} and log K_{ow}, which illustrates this uncertainty (Li and Ran 2012). In this example, prediction of log K_{OC} from any log K_{ow} value may fall within a range 1.2–1.33 units wide, corresponding to a 16- to 22-fold difference in possible log K_{OC} values. For accurate prediction of adsorption-dependent behaviour such as leaching or degradation, therefore, laboratory measurement of K_D on the actual soil or sediment involved is likely to be necessary.

The dependence of pesticide adsorption on soil on the soil organic carbon content is shown by the example in Fig. 7.3. The positive slopes of the relationship between ($\log_{10}$)-K_D and soil organic carbon suggest that, for a 1% increase in soil organic carbon content, the strength of adsorption as measured by K_D increases by about fivefold. More recently, it has been recognised that organic matter in the form of [micro]plastic contamination can also act as an adsorbent for organic contaminants (Rodríguez-Eugenio et al. 2018). If the organic matter itself is mobile (as is the case for some naturally occurring humic macromolecules), adsorption may actually increase transport of organic pollutants (Enfield et al. 1989).

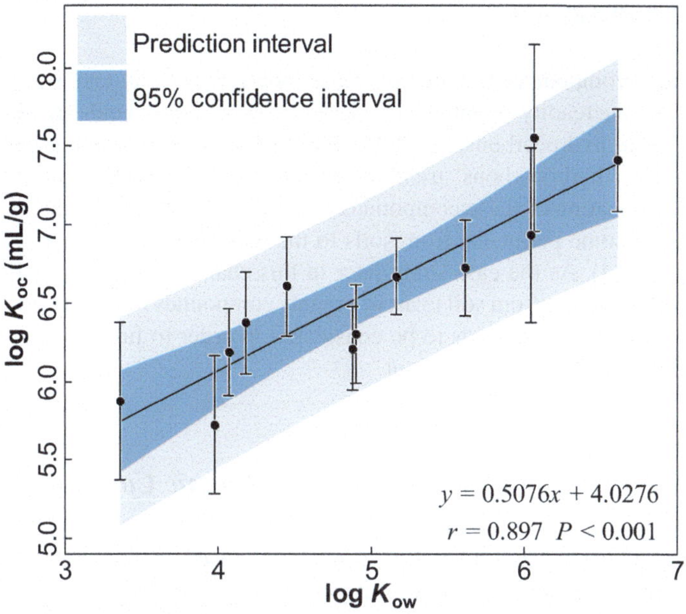

Fig. 7.2 Relationship between $\log_{10}K_{ow}$ and $\log_{10}K_{oc}$ for PAH adsorption on sediment from the Pearl River, Guangdong, China, showing the prediction and confidence intervals for the linear regression model. (Redrawn from Li and Ran (2012); used under the terms of a CC-BY-3.0 license)

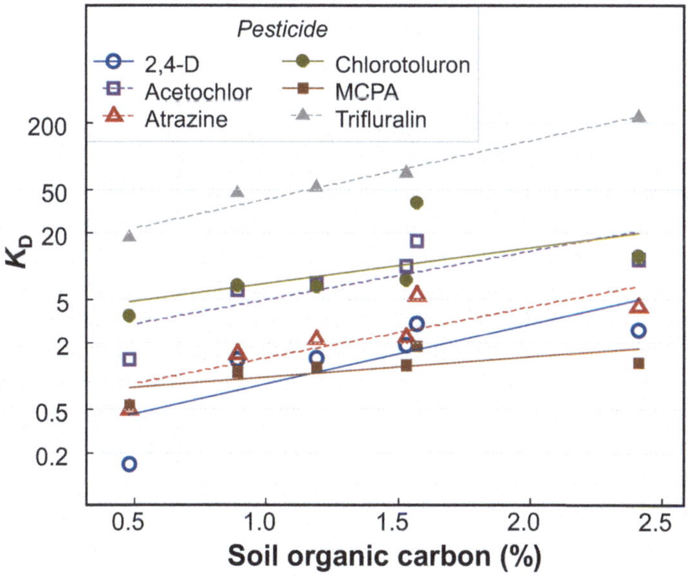

Fig. 7.3 Dependence of adsorption coefficient K_D on soil organic carbon content for various organic pollutants (pesticides) on several soils. (Drawn by Andrew W. Rate from data in Hiller et al. (2008))

7.2.1.2 Volatilisation

Many organic compounds, particularly those with low polarity and low molecular weight, evaporate readily at ambient temperatures. Such compounds are often called volatile organic hydrocarbons, or VOCs, and include common pollutants such as light petroleum hydrocarbons, trichloroethylene, and the BTEX group of compounds. Volatilisation of these compounds is known to be an important mechanism for transfer of some pollutants from soils to the atmosphere (Petersen et al. 1996; Martí et al. 2014). As the case study later in this chapter (Sect. 7.5) will discuss, however, volatilisation from soil to air of organic compounds other than VOCs, such as PAHs, does occur and needs to be considered in order to fully understand their behaviour and fate in the environment.

7.2.2 *Behaviour of Polar or Ionisable Organic Compounds*

Compounds which have high polarity, or readily ionisable compounds, have much greater solubility in water (such as soil pore water) than do the non-polar compounds discussed above. High polarity compounds tend to be those having short (1–4 atom) carbon chains, with functional groups in asymmetric positions containing atoms such as oxygen, nitrogen, or sulphur with different electronegativity than carbon, such as simple alcohols or ketones. Readily ionisable compounds include carboxylic acids, amides, and amines, and the tendency to ionise is related to acid-base reactions with water as a solvent.

Volatilisation of polar organic solvents tends to be less than for non-polar organic solvents, since intermolecular attraction of polar molecules lowers their vapour pressure. Polar compounds do volatilise, however, such as low molecular weight alcohols (e.g. methanol, ethanol) and ketones (e.g. acetone).

Ionisation. Organic molecules containing carboxylic, sulfonic, or phosphoric acid functional groups are able to ionise in water, by acid dissociation reactions, to form carboxylate or sulfonate anions, leading to high water solubility, since water can easily solvate most ionic chemical species. Similarly, organic molecules containing amino functional groups are basic and can accept hydrogen ions from water to form cationic species. Ionisable pollutants include many pesticides (e.g. anionic 2,4-D or metsulfuron-methyl; cationic triazine herbicides) (McBride 1994; Kah and Brown 2006). In addition, some organic compounds (e.g. glyphosate, the active ingredient of the widely used herbicide Roundup®) can form 'zwitterions', having combinations of functional groups which can allow molecules to have both positive and negative charge.

7.2.3 *Adsorption of Ionisable and Polar Organics*

Organic compounds can accumulate as ions or molecules at the surface of finely particulate substances in soils, such as secondary minerals and organic matter, by similar mechanisms to those described for inorganic substances in Chaps. 4 and 6.

7.2.3.1 Cation Exchange

Organic compounds which are ionisable to form organic cations tend to be adsorbed onto the surfaces of negatively charged soil solids such as clays and organic matter by cation exchange (McBride 1994). Other mechanisms such as hydrophobic interactions (i.e. non-polar partitioning as described in Sect. 7.2.1), hydrogen bonding, and charge transfer can also result in adsorption of cationic pesticides (Kah and Brown 2006). Adsorption of organic cations such as the triazine herbicides on soils generally increases with increasing soil organic carbon content, since much of the CEC of many soils is conferred by negative charge on soil organic matter and since adsorption by non-polar partitioning also increases with soil organic carbon content (Fig. 7.3). The effect of soil pH on organic cation adsorption is complex; generally, organic cations have decreased adsorption as pH increases, which is counter-intuitive if we only consider increasing pH as increasing the pH-dependent negative charge on soil colloids. The decrease in adsorption of organic cations with increasing pH relates to the ionisation of the organic molecules themselves; at low pH the molecules are cationic since their weakly basic amino or pyridyl functional groups accept a hydrogen ion to become positively charged. As the pH increases beyond the pK_a value ($-\log_{10}$ acid dissociation constant) of the organic cation, the molecule loses its charge and can no longer interact electrostatically with a negatively charged soil surface. Adsorption of organic cations on clay minerals such as smectites is probably by physical processes rather than chemical bonding. The ability of clays such as smectites with expanding interlayers to adsorb organic cations is restricted by the molecular size of the cations, such that larger cations (e.g. quaternary ammonium compounds) are excluded from interlayer spaces (McBride 1994).

7.2.3.2 Adsorption of Anionic Organics

Molecules with weakly acidic functional groups (carboxylic acids, $-NHSO_2-$ acids, sulfonic acids) become progressively more dissociated as pH increases, producing organic anions and hydrogen ions. The acid dissociation constants of the undissociated acids are usually large enough, however, that the molecules are almost completely dissociated within normal soil pH ranges. Adsorption occurs by ion exchange if positively charged soil colloids such as ferric oxides exist, by *ligand exchange* on clays or oxide minerals, or by other mechanisms such as cation bridging (Kah and Brown 2006). Organic anion adsorption generally decreases with increasing pH,

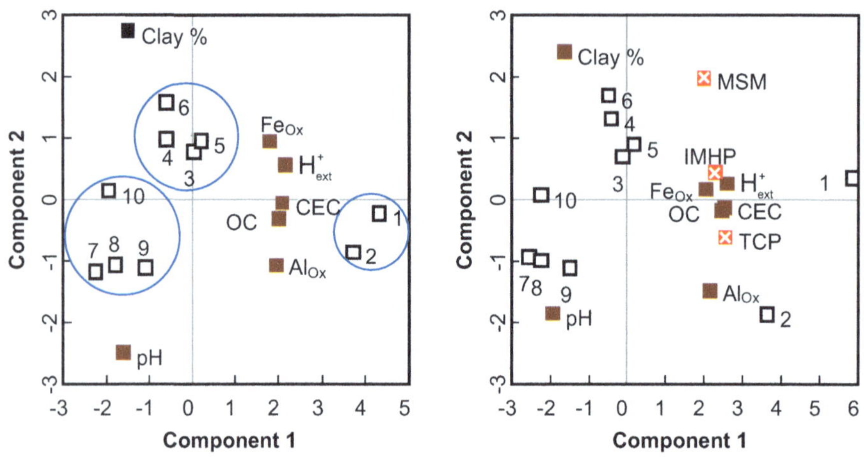

Fig. 7.4 Association, based on component weightings from principal component analysis, of adsorption constants for anionic organic contaminants (☒) and soil properties (■), with observation scores for individual soils 1–10 (□). *MSM* metsulfuron-methyl, *IMHP* 2-isopropyl-4-methyl-6--hydroxypyrimidine (a metabolite of diazinon), *TCP* 3,5,6-trichloro-2-pyridinol (a metabolite of chlorpyrifos). The clustering of adsorption constants with organic carbon (OC), amorphous iron oxides (Fe_{Ox}), and CEC in principal component space suggests that these soil properties exert the most control on adsorption. (Redrawn from Báez et al. (2015); used with permission from Springer)

due to the increasing negative charge on soil colloids as pH increases. Adsorption of anionic organic contaminants increases with increasing content of soil organic carbon, reactive ferric oxides, or cation exchange capacity (Fig. 7.4).

7.2.4 *Chemical Degradation (Abiotic) of Organic Compounds*

Organic contaminants degrade in soils, with the main cause of degradation being metabolism by soil microorganisms, a topic that will be addressed in sections of Chaps. 8 and 11. The biochemical degradation processes do not always convert organic contaminants directly to CO_2, and intermediary degradation products, or *metabolites*, are produced which may have similar toxicity to the original compound. For example, the organochlorine insecticide DDT (now banned in many jurisdictions) is metabolised to several products including *o,p′*-DDE, *p,p′*-DDE, and *o,p′*-DDD. The persistence and toxicity of DDE, the predominant DDT metabolite, is greater than for DDT itself; DDD has also been used as a pesticide. Conversely, the metabolites of 2,3,7,8-TCDD, the most toxic of the dioxin (PCDD/PCDF) group of compounds, are considerably less toxic than the parent 2,3,7,8-TCDD.

Many organic compounds are oxidisable or reducible, and chemical oxidation or reduction is pathway for abiotic degradation of organic pollutants in soils (Mulligan and Yong 2004). The principal oxidising agent, or electron acceptor, in soils is oxygen; other substances, such as Fe^{III} or Mn^{IV} oxides, can also react as oxidising agents, and some clays can catalyse oxidation by oxygen by forming reactive oxygen species (Yong and Mulligan 2003). In the presence of aluminosilicate clays, for example, reactive oxygen species can degrade phenols by oxidation to quinones. Barrett and McBride (2005) showed that glyphosate (N-(phosphonomethyl)glycine) could also be degraded by an oxidation reaction where a manganese oxide was the electron acceptor. Reductive processes are perhaps better known in soils, especially reductive dechlorination of chlorinated hydrocarbons such as PCE (perchloroethylene = tetrachloroethene), TCE (trichloroethene), and carbon tetrachloride. Reduction of chlorinated organic contaminants can occur if other substances (e.g. polycarboxylic acids such as oxalate or citrate, which can be released by plant roots, or natural organic matter) reduce Fe^{3+} to Fe^{2+} at a ferric oxide surface; the surface Fe^{2+} ions can then facilitate the reductive degradation (Li et al. 2008a). Zerovalent (metallic) iron (ZVI) is a powerful reducing agent which can also reductively dechlorinate organic compounds, and the use of ZVI in remediation is discussed in Chap. 11. Other naturally occurring substances such as Fe^{2+} or pyrite can also reduce organic compounds; their presence in soil would normally require anoxic conditions.

Many organic molecules, including persistent and other pollutants, can be degraded by ultraviolet light derived from solar radiation (e.g. Zhong et al. 1994; John and Shaike 2015). The efficiency of photodegradation is lower in soil than in air or water (Kromer et al. 2004), and the importance of photodegradation for organic pollutant removal in urban soils is essentially unknown, but not considered to be significant relative to other removal mechanisms.

7.2.5 *Biological Degradation*

Some soil microorganisms (including species of bacteria and fungi) can use contaminant compounds as a metabolic energy source and growth substrate, decomposing the contaminants in the process. We call this biological decomposition process *biodegradation*, and the end products are carbon dioxide (or intermediary carbon compounds, or metabolites). We discuss the biodegradation process in more detail in Chap. 8, including explanations of the factors affecting bioavailability of different organic compounds.

From a perspective of chemical kinetics, the biodegradation rates of organic contaminants in soils are commonly first-order processes, meaning that the degradation rate is dependent only on the concentration of the contaminant compound, and the *half-life* is constant. First-order behaviour was confirmed for a range of pesticides and soils by Jury et al. (1987), whose analysis showed that the reaction rates observed were dependent on the half-life for a specific compound (i.e. related to the pollutant) and the K_{OC} value (Eq. 7.3) for adsorption (i.e. related to soil properties). Other soil properties are also known to affect the rates (and therefore amounts) or biodegradation that occurs. In many cases, the ability of microorganisms to degrade pollutants is limited by the external supply of water and/or nutrients such as nitrogen and phosphorus (Shahsavari et al. 2017). In many pollutant environments, organic contaminants coexist with other contaminants such as metals. The presence any one of several metals (e.g. Cd, Cr, Cu, Hg, Pb, or Zn) at sufficient concentrations can inhibit biodegradation of compounds such as pesticides, PAHs, or petroleum hydrocarbons (Sandrin and Maier 2003).

As with inorganic contaminants (Chap. 6), the bioavailability of organic pollutants is affected by their residence time in soil, and this is known as an 'aging effect'. Biodegradation is slower and occurs to a lesser extent, for contaminants which have been in contact with soil for longer, compared with very recently added contaminants (Hatzinger and Alexander 1995). In processes such as *composting*, where microbial stabilisation of organic materials is the primary objective, the bioavailability of organic pollutants is also known to decrease (Chen et al. 2015b).

7.2.6 Transport of Organic Contaminants

Organic compounds can be transferred within soils or between soils and other environmental compartments (air, surface water, groundwater, biota) by a number of processes (Fig. 7.5). The existence of widespread diffuse pollution of Earth's environments with organic contaminants is strong evidence that transport of organic contaminants is a significant issue (Fuoco et al. 2009; Rodríguez-Eugenio et al. 2018; Sect. 7.1). Some transport phenomena specifically related to polycyclic aromatic hydrocarbons (PAH) are described in Sect. 7.5 later in this chapter, with Sect. 7.5.7 explicitly covering fluxes of PAH compounds to and from soils.

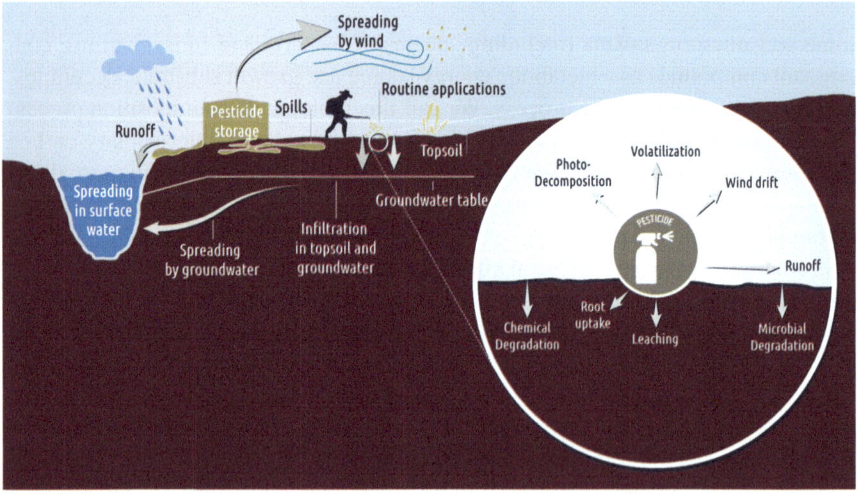

Fig. 7.5 Transport mechanisms for pesticides within and between environmental compartments. (Adapted from Rodríguez-Eugenio et al. 2018; reproduced with permission from UNEP)

Fluxes from direct human use or production of organic compounds to soils (also called 'primary emissions'; Wang et al. 2016) include:

- Accidental loss, leakage, or spillage (e.g. of fuel hydrocarbons or during manufacture or use of most classes of organic compound – see Table 7.1).
- Deliberate application to soils (e.g. pre-emergent herbicides).
- Incidental addition to soils from other deliberate activities (e.g. spraying of pesticides onto crops, use of fire retardants).
- Unintentional generation during combustion (e.g. formation of PAHs or dioxins from coal or biomass burning).
- Entry of organic pollutants into wastewater streams which may eventually be deliberately or accidentally applied to soil (e.g. detergents or antibiotics in sewage sludge).

Atmosphere-soil fluxes include additions to soils from wet and dry deposition and emissions from soils by volatilisation. The additions to soils are from the primary emission processes described above. Secondary emissions from soil to atmosphere occur for volatile organic pollutants, depending on several factors. Higher temperatures favour volatilisation (Rodríguez-Eugenio et al. 2018), and greater soil water contents suppress volatilisation by reducing air-filled pore space and by Henry's Law partitioning of organic compounds into soil water (Grathwohl 2011). Emissions from soil by volatilisation are also limited by adsorption of organic compounds (see Sect. 7.5.7) or if biodegradation is rapid (Grathwohl 2011).

Soil-surface water fluxes are mainly from run-off, although wind erosion and subsequent deposition of atmospheric particulates containing organic pollutants are also possible. The partitioning of organic pollutants into organic matter may increase the transport of these pollutants in the dissolved form by association with dissolved organic matter (Enfield et al. 1989). Transport from soil to water can also occur when pollutants are adsorbed to particulate organic matter, by soil erosion or loss of soil during urbanisation (Wang et al. 2016).

Soil-groundwater fluxes occur via leaching, that is, vertical transport through soil by percolating water. As we would expect, leaching is greater for organic compounds with greater water solubility (Mulligan et al. 2001) and depends on sufficient precipitation or other water input to soil to allow transport of water through the soil matrix or preferential flow through macropores. Reactions between organic contaminants and soil sold phases suppress leaching to groundwater, as do biodegradation (Grathwohl 2011). The partitioning of hydrophobic organic compounds into dissolved organic matter macromolecules may increase the transport of pollutants in soil water, even if the pollutants have low water solubility (Enfield et al. 1989).

7.3 Effects of Organic Soil Contaminants on Ecosystem Services

The toxicity of organic pollutants to a range of organisms is well established (Rodríguez-Eugenio et al. 2018), and we address some of the effects and issues involved in Chap. 8, particularly for soil microorganisms. Organic pollutants such as petroleum hydrocarbons, PAHs, PCBs, organochlorine or organophosphate pesticides, and PCDDs/PCDFs have also been shown to be toxic to organisms other than microorganisms in soils or sediments, including plants, insects, earthworms, crustaceans, and burrowing vertebrates. In addition, many of the persistent organic pollutants (POPs) are known to undergo biomagnification as pollutants taken up by organisms lower in food webs are consumed by predators (Rodríguez-Eugenio et al. 2018). The toxicity of POPs varies widely with type and species of organism. While some plants show adverse effects such as reduced growth and photosynthetic activity (Ahammed et al. 2012), others are more resistant to toxic effects and may be able to take up organic pollutants to an extent that they can be used for phytoremediation (Wyrwicka et al. 2014). Similarly, while some microbial processes such as respiration or nitrification may be affected by organic pollutants such as PAHs or pesticides (Sverdrup et al. 2007; John and Shaike 2015), some microorganisms are able to degrade organic pollutants, a phenomenon which may also be used for remediation of contaminated soils (Chap. 11).

The toxicity of many organic pollutants to humans is well-known; for example, benzo[*a*]pyrene is a recognised human carcinogen. In contrast, however, benzo[*a*]pyrene in soil does not always show adverse effects on plants or invertebrate animals (Sverdrup et al. 2007). Vertebrates other than humans have negative responses to organic pollutants; for example, Gonzalez-Mille et al. (2013) showed DNA damage could occur to a toad (amphibian) species exposed to urban soil contaminated with POPs (mainly organochlorine pesticides such as hexachlorocyclohexane and DDT and also PCBs).

7.4 Measurements and Data

7.4.1 Total and Partial Analyses

7.4.1.1 Soil Sampling for Organic Compounds

The sampling methods for hydrocarbons differ depending on whether the target organic compound(s) are volatile or semi/non-volatile. For volatile compounds such as C5–C10 alkanes, BTEX, and 2–3-ring PAHs (see Sect. 7.5.2), sampling techniques must avoid the loss of target compounds by evaporation. Such methods include zero-headspace sampling in specialised containers, or solvent extraction (Sadler and Connell 2003). Less rigorous sampling procedures are required for semi- or non-volatile compounds.

7.4.1.2 Extraction of Organic Compounds from Soil

Materials sampled for volatile organic compounds by zero-headspace or solvent extraction methods need no further extraction prior to analysis. Non-volatile or semi-volatile compounds, however, may be strongly adsorbed to soil materials (Sect. 7.2.1) and require extraction with a solvent able to remove target compounds from adsorbed forms. These include the commonly used Soxhlet extraction using solvents such as acetone/hexane or methylene chloride/acetone (US EPA 1996a) or extraction in specialised high-pressure vessels with supercritical fluids such as liquid CO_2 (e.g. US EPA 1996b). The supercritical fluid extraction methods may also be applied to volatile organic compounds. Adaptations of these methods are applicable to several other categories of organic contaminants, such as organochlorine pesticides, phenols, PCBs, and so on.

It is worthwhile noting that solvent extraction procedures do not extract all of the organic contaminants present in soil or sediment samples. The extractability, for example, of PAHs, is controlled by organic carbon and black carbon content (Ma et al. 2010). He et al. (2008) found that between 34% and 57% of PAHs in soils and sediments was non-extractable in dichloromethane in a Soxhlet procedure, with between 235 and 1425 mg/kg of total PAHs not extracted in some samples.

7.4.1.3 Analysis of Organic Compounds in Extracts

Since soils usually contain a range of potential organic contaminant compounds, analytical methods usually involve a chromatography technique which separates compounds according to their polarity or other chemical property. The most common of these is gas chromatography (US EPA 2018), which may be coupled with a mass-spectrometer detector to increase the method's ability to identify individual compounds unambiguously. This textbook does not have space to cover all the details of chemical analytical techniques, so readers are referred to a general analytical chemistry text such as the excellent open-access book by Harvey (2016).

7.4.2 Background Concentrations

Many potentially polluting organic compounds have natural sources, and so background concentrations in urban soils must be considered. Natural phenomena such as volcanism, forest fires, crude oil seeps, decomposition of organic residues, or plant emissions can emit organic compounds such as hydrocarbons including those in the BTEX and PAH categories, phenols, dioxins, and others. In contrast, numerous organic compounds such as many chlorinated hydrocarbons including PCBs, PFAS, and most pesticides (i.e. most of the compounds identified as POPs in Table 7.3) have no known natural sources. Even for compounds which do not occur naturally, however, measurable background concentrations may occur in urban soils due to the widespread use and emissions of these substances.

Any assessment of organic contaminants should therefore account for unavoidable background concentrations of potential pollutants caused by diffuse source emissions from urban areas. This type of background concentration is called the 'anthropic background' by Biasioli et al. (2012), who used the 90th percentile (excluding outliers) of multiple observations to calculate background concentrations of benzo[*a*]pyrene (31.5 μg/kg), PCBs (8.9 μg/kg), and PCDDs+PCDFs (3.5 μg/kg). Most other studies which report background concentrations for organic pollutants in urban or peri-urban areas rely either on measurements at non-urban sites or concentrations cited in other scientific literature.

7.4.3 Regulatory Contamination Thresholds

Contamination thresholds for organic contaminants soil are based on an analysis of risk, including the desired level of protection, exposure pathways, and the types of organisms which may potentially be affected (e.g. National Environment Protection Council 2013a). More detail on risk assessment processes is presented in Chap. 9. Some examples of regulatory contamination threshold concentrations in soil, for organic contaminants, appear in Table 7.4, for compounds from the BTEX, PAH, and PCDD/PCDF categories.

Table 7.4 Examples of human health-based regulatory concentrations of organic pollutants from different jurisdictions

	Country/region			
Parameter	Australia	UK	New Zealand	Canada
Guideline type	Trigger values	Guidelines	Soil contaminant standards	Remediation criteria
Applicable land use	Residential	Residential	Residential	
Reference	1	2	3	4, 5
	Regulatory concentration in soil (mg/kg)			
Pollutants				
Benzene	0.5–0.8[a]	0.33		0.05
Benzo[*a*]pyrene	3	3.6[b]	10	20
Total PCDD/PCDF		8	0.27	0.00001

References: 1. National Environment Protection Council (2013b); 2. CL:AIRE (2009-); 3. Ministry for the Environment (Manatū Mō Te Taiao) (2012); 4. Canadian Council of Ministers of the Environment (1991); 5. Canadian Council of Ministers of the Environment (2010)

[a]Depends on soil depth, soil texture, and residential density

[b]Urban background value

7.4.4 Contamination Indices

The concept of contamination indices, as discussed in Chap. 6 for trace elements (metals and metalloids), is seldom used explicitly, probably because the practice of summing concentrations of pollutants in a particular category is so common. As the case study below (see Sect. 7.5.2) observes, it is normal for polycyclic aromatic hydrocarbon (PAH) concentrations to be presented as the sum of several (typically 9–28) critical compounds (even though there are hundreds of PAH compounds). The same practice is used for other classes of compounds; for example, Biasioli et al. (2012) summed the concentrations of polychlorinated biphenyls (PCBs) and also polychlorinated dibenzo-dioxins and furans (PCDDs/PCDFs) rather than presenting the concentrations of individual compounds. Similarly, summed concentrations of phthalic acid esters (Mo et al. 2009) and organochlorine pesticides (You 2015) have been reported instead of separate compounds. Effectively, a sum of concentrations of a class of contaminants (e.g. $\sum$PAH) is analogous to a simple additive contamination index, as presented in Chap. 6.

7.5 Case Study: Polycyclic Aromatic Hydrocarbons (PAHs) in the Pearl River Delta

The focus on the Pearl River Delta region of China in this textbook is quite deliberate – we think that it is representative of many urban environments on Earth. As well as being one of the largest conurbations worldwide, it is also one of the fastest growing in terms of population, and like many large urban areas, it has a legacy of soil contamination with a range of substances including metals and persistent organic pollutants. Similarly, we have chosen to focus on polycyclic aromatic hydrocarbons (PAHs) as representative organic pollutants. This does not mean that other organic pollutants are less important or relevant, but PAHs have been studied now for a long time, and we think that many aspects of their behaviour in the Pearl River Delta are more widely applicable to other pollutants or other regions.

7.5.1 The Pearl River Delta

The Pearl River Delta (PRD) in Guangdong Province, People's Republic of China, is a large (54,156 km^2) and industrialised area which contains a series of interlinked megacities, including Guangzhou, Shenzhen, and Hong Kong. Table 7.5 shows the sizes of these urban areas in the PRD region in terms of their land areas and human populations. The area, extending from Yangjiang in the west to Shanwei in the east, to approximately Qingyuan and Heyuan in the north, and bounded in the south by the South China Sea (Nan Hai) (Fig. 7.6), is still rapidly urbanising, and an

Table 7.5 Population and projected population of large urban centres in the Pearl River Delta, China

City	Land area (km^2)	Population (millions)	Projected 2030 (millions)
Dongguan	2469.4	8.179	8.279
Foshan	3763	7.197	8.35
Guangzhou	7690.75	12.530	16.024
Hong Kong	1078.5	7.070	7.987
Huizhou	9187.73	4.451	–
Jiangmen	9183.55	4.451	1.956
Macao	30.3	0.540	–
Shenzhen	1944.36	10.136	14.537
Zhaoqing	14,822.17	3.805	–
Zhongshan	1783.9	3.121	3.302
Zhuhai	2202.27	1.563	2.12
TOTAL	54,155.93	63.043	72.7

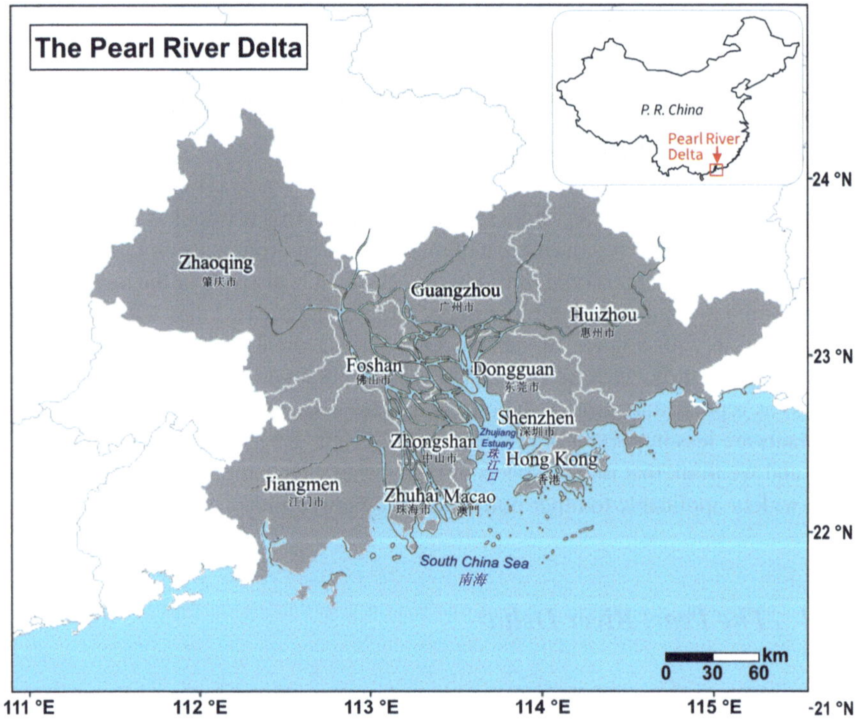

Fig. 7.6 Location map of the Pearl River (珠江口, Zhujiang) Delta, showing the major waterways and urban centres. (Modified by Andrew W. Rate from Liang et al. (2019), used under the terms of a CC-BY-4.0 license)

estimated 120 million people live in the greater Pearl River Delta area (UN-Habitat 2016). The PRD region is known to be heavily polluted with a wide range of contaminants and other anthropogenic substances, including nutrients, trace metals, a range of organic pollutants including PAHs and many other POPs, as well as emerging contaminants such as pharmaceuticals, perfluoroalkyl compounds, and microplastics (Wang and Rainbow 2020).

7.5.2 Polycyclic Aromatic Hydrocarbons

The polycyclic aromatic hydrocarbons (PAHs) are a group of persistent organic pollutant compounds with chemical structures based on chemically bonded benzene (6-carbon aromatic) ring structures sharing at least two carbon atoms (see the example of phenanthrene in Table 7.2). They range in molecular size from the smallest, naphthalene, with two rings, up to molecules with six to nine rings such as benzo[*ghi*]perylene (see Table 7.6) and numerous other compounds which are

Table 7.6 List of the 16 US EPA priority polycyclic aromatic hydrocarbons (PAHs) with abbreviations and selected physico-chemical properties

PAH compound	Abbreviations[a]	Number of rings	Molar mass (g/mol)	Boiling point (°C)[b]	Log K_{OW}[c]
Naphthalene	NAP, NA, NaP	2	128	281	3.35
Acenaphthene	ACE, AC, ANA	3	154	278	3.92
Acenaphthylene	ACY, CAN, ANY	3	152	265	3.94
Fluorene	FLU, Flo, FL	3	166	295	4.18
Phenanthrene	PHE, Ph, PHEN	3	178	339	4.46
Anthracene	ANT, An	3	178	340	4.45
Fluoranthene	FLA, Flu, FLT, FA	4	202	375	5.16
Pyrene	PYR, Py	4	202	360	4.88
Benzo[*a*]anthracene[d]	BaA, B[a]A	4	228	435	5.76
Chrysene[d]	CHR, Ch, CRY	4	228	448	5.84
Benzo[*b*] fluoranthene[d]	BbF, B[b]F	5	252	481	5.78
Benzo[*k*] fluoranthene[d]	BkF, B[k]F	5	252	481	6.11
Benzo[*a*]pyrene[d]	BaP, B[a]P	5	252	495	6.13
Benzo[*ghi*]perylene	BghiP, B[ghi]P	6	276	NA (>500)	6.63
Indeno[1,2,3-*cd*] pyrene[d]	IcdP, IP	6	276	536	6.70
Dibenzo[*a,h*] anthracene[d]	DahA, D[ah]A, DB(ah)A	6	278	524	6.75

[a]Abbreviations *in* the literature are not always case-sensitive; e.g. Pyr is equivalent to PYR; initial letters always capitalised
[b]Boiling points are from Joa et al. (2009)
[c]K_{OW} is octanol-water partition coefficient; values are from Achten and Andersson (2015b)
[d]Listed as probable human carcinogens (USEPA 2020)

generally less common as environmental pollutants (Achten and Andersson 2015a). In some of the compounds, 5-membered rings are also present (e.g. acenaphthene, fluorene, indeno[1,2,3-*cd*]pyrene, etc.).

The concentrations of PAHs in soils may be reported individually, and it is very common for a sum of PAH concentrations to be reported, such as the 16 PAHs prioritised by the US EPA (2020; see Table 7.6 and Fig. 7.7). The abbreviation Σ_{16}PAHs is commonly used to represent the combined concentrations of the 16 US EPA priority PAHs (or Σ_{15}PAHs since naphthalene is sometimes omitted), or a different number subscript may be used to show how many individual PAH compounds were analysed for (e.g. Σ_{28}PAHs). Seven of the 16 US EPA priority PAHs are known to be carcinogenic: benzo[*a*]anthracene, chrysene, benzo[*b*]fluoranthene, benzo[*k*]fluoranthene, benzo[*a*]pyrene, indeno[1,2,3-*cd*]pyrene, and dibenzo[*a,h*]anthracene (abbreviated variously as Σ_7PAHs, Σ_7CarPAHs, ΣcarcPAHs, etc.).

The PAHs occur naturally in many soils in small concentrations due to their presence in geogenic hydrocarbon deposits such as oil shales or crude oil and emission by natural wildfires or volcanic activity (National Pollutant Inventory 2020). In environments affected by urbanisation, however, PAHs are released in greater amounts into various environmental compartments due to their presence in fossil fuels and solvents and their release during combustion of coal, other fossil fuels, and biomass for energy production (Chen et al. 2005). Another source of PAH pollution is waste disposal, especially disposal or burning of e-waste (Leung et al. 2006; Wang et al. 2017).

Lower molecular weight PAHs may volatilise significantly or be transported in the atmosphere by sorption to airborne particulates. Contamination of soils, surface water, and groundwater with PAHs is well known in many urban environments worldwide (Menzie et al. 1992; Wang and Rainbow 2020), and exchange between environmental compartments does occur, including uptake by plants from soils.

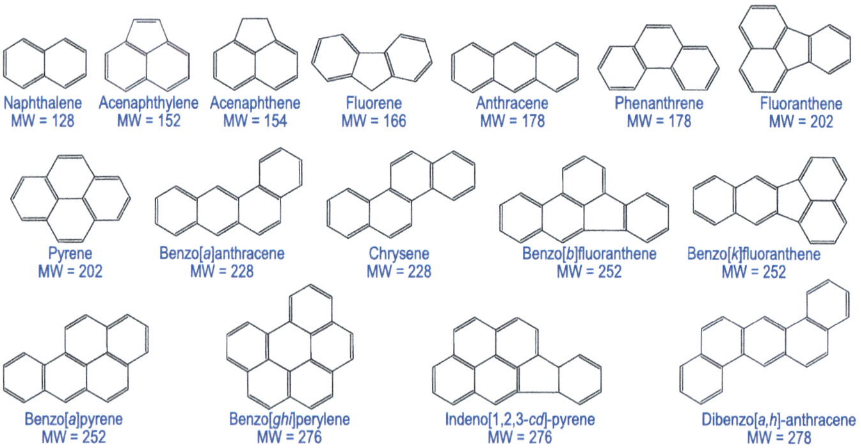

Fig. 7.7 Schematic chemical structures for the 16 US EPA priority PAHs, in increasing order of molecular weight

7.5.3 PAH Concentrations in Soils of the Pearl River Delta

The concentrations of PAHs in soils are generally reported in units of μg/kg (or equivalently ng/g); both units correspond to parts per billion to reflect the low concentrations of PAHs in uncontaminated environments. In a range of studies across the PRD, in urban and peri-urban soils used for vegetable growing, parklands, urban forest, and even e-waste recycling, $\sum_{16}$PAH concentrations ranged from undetectable (Li et al. 2007) to >12,000 μg/kg (Wei et al. 2014 and Fig. 7.8). The mean concentrations in urban surface soils, if presented, were generally in the range 150–320 μg/kg. The focus on surface soils is reasonable since the risk of transfers by wind erosion, plant uptake, or direct soil contact is greatest in the soil surface. Some of the difference between findings of different studies may be caused by the differences in sampling depth for 'surface' soils; while most studies sampled 0–20 cm depth, a few sampled 0–10 cm (e.g. Leung et al. 2006; Ni et al. 2011), and one study sampled soil from 0 to 5 cm (Zhang et al. 2016).

The concentration ranges measured in PRD soils are often lower than other urbanised areas worldwide (Ma et al. 2009; Cao et al. 2010), perhaps reflecting the shorter history of intense urbanisation. Chen et al. (2005) found that background concentrations of $\sum_{16}$PAHs in vegetable-growing soils in Guangzhou were between 1 and 10 μg/kg, lower than background concentrations between 20 and 50 μg/kg in the Netherlands. Menzie et al. (1992) reviewed 15 studies of PAHs in urban soils and found a median concentration of Σ_7CarPAHs (carcinogens) of 1100 μg/kg with

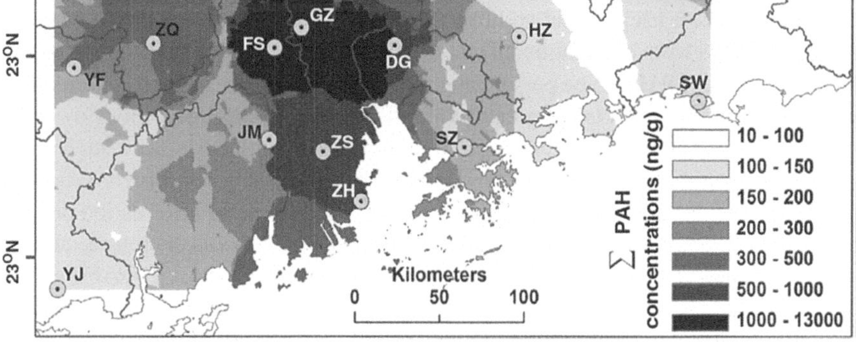

Fig. 7.8 Map of the distribution of $\sum_{28}$PAHs in soils of the Pearl River Delta. (From Wei et al. (2014) showing PAH pollution centred on Foshan, Guangzhou, and Dongguan. Reproduced with permission from Elsevier)

a range of 60–5800 µg/kg, suggesting that the Pearl River Delta soils are genuinely less contaminated with PAHs than elsewhere (especially since Σ_7CarPAHs omits 9 of the US EPA priority PAHs and so underestimates Σ_{16}PAHs). Nevertheless, Ke et al. (2017) calculated that the PAH concentrations in urban parks in Guangzhou represented a ‘potentially high’ cancer risk for people using the parks, based on established exposure pathways and likely dosage for adults and children.

PAHs in street dust. Street dusts can contribute PAHs to air, water, and soils in urban areas and have been suggested to do so in the Pearl River Delta region (Mai et al. 2003; Liang et al. 2008; Wu et al. 2014). These dusts are not PAH sources in themselves but are a transitional reservoir of PAHs and other contaminants derived from traffic (e.g. liquid fossil fuel combustion, direct petroleum emissions) and other sources. The concentrations of PAHs are commonly greater in street dusts than in soils; Menzie et al. (1992) gave a range of measured concentration of Σ_7CarPAHs in street dusts between 8000 and 336,000 µg/kg and suggested that very high ΣPAHs measured in the literature for urban soils probably include a contribution from street dusts. Apart from emissions or incomplete combustion of vehicle fuels and oils, PAHS may also be derived from asphalt-based road materials (Abdel-Shafy and Mansour 2016).

7.5.4 Fractionation of Individual PAHs

Fractionation refers to changes in relative amounts of related substances, such as individual PAHs as a result of environmental processes. The concepts of fractionation are very well developed in inorganic geochemistry, with considerable focus on fractionation of isotopes of various elements, the rare-earth elements, and many others. The concentrations of different PAH compounds in urban soils will differ, since different PAHs will be released into receiving environments in various amounts depending on the nature of the source and physical and chemical properties of the PAHs themselves. For example, dominance of PAHs with two or three rings may suggest background additions (Li et al. 2007; Xiao et al. 2014) or more recent additions from a contaminant source (Cai et al. 2007). The tendency of lower molecular weight PAHs to be present at greater concentrations, when contamination is recent, probably reflects the greater volatilisation and water solubility (lower K_d values) of these compounds relative to the larger PAH molecules. Low molecular weight PAHs also have a greater tendency to be chemically reduced or oxidised (Abdel-Shafy and Mansour (2016), and see Sect. 7.5.7 for more information on factors affecting PAH fluxes). The smaller PAH molecules would therefore tend to persist for shorter times in soils, before being lost relative to heavier PAHs as the contamination ages. Biodegradation of PAHs is also more rapid for compounds having lower molecular weight; for example, Zhu et al. (2019) found the degradation rate of phenanthrene (3 rings, MW = 178 g/mole) to be much greater than for benzo[a]pyrene (5 rings, MW = 252 g/mole). The same factors lead to greater concentrations of 2–3-ring PAHs in water and in plant tissue following uptake from PAH-contaminated soil. You (2015) found that 2–3-ring PAHS were dominant in water and that in vegetable tissue 2–4-ring PAHs were most common. In relation to fractionation of PAHs by plants, Zhang et al. (2008) found that PAH uptake by vegetables from (peri)urban soils was greater for PAHs with lower K_{OC} values, with 4-ring PAHs being most common in soils but 2–3-ring PAHs taken up by vegetables in greatest amounts. The relative amounts of PAH compounds also reflect the source of contamination. At an e-waste recycling/disposal facility in Guiyu, Leung et al. (2006) found mainly 2–3-ring PAHs associated with a printer roller dump, but where wastes had been burned, 3–5-ring PAHs were the most common. In general, combustion yields PAHs with greater molecular weights than other sources (Chen et al. 2005). Most commonly, however, additional information is necessary to identify the probable source of PAH pollution, and the next section discusses how we can use PAH data to obtain information about their sources.

7.5.5 Sources of PAH Pollution in the Pearl River Delta

The sources of PAH pollution in soil can be assessed to some extent by analysing the relative concentrations of PAHs of different ring number or molecular weight. More commonly, however, additional information from the molar ratios of PAH isomers is used, where isomers are one or more compounds which share the same chemical formula (and therefore the same molecular weight) but have a different chemical structure (see Table 7.6 for molecular weights and Fig. 7.7 for chemical structures of PAHs). The sets of isomers commonly used are anthracene-phenanthrene (ANT-PHE, MW 178), benzo[*a*]anthracene-chrysene (BaA-CHR, MW 228), benzo[*b*]fluoranthene-benzo[*k*]fluoranthene-benzo[*a*]pyrene (BbF-BkF-BaP, MW 252), and benzo[*ghi*]perylene-indeno[1,2,3-*cd*]pyrene (BghiP-IcdP, MW 278). For example, Cao et al. (2010) interpreted the ratios listed in Table 7.7 to infer a combustion source for PAH contamination of soil in Shenzhen. We should note that isomeric ratios of PAHs should always be interpreted with caution, since even

Table 7.7 PAH isomeric ratios and their interpretation thresholds for source identification in soils of the Pearl River Delta (Cai et al. 2007; Li et al. 2007; Cao et al. 2010)

	Threshold and interpretation		
Ratio	Low	Medium	High
ANT/(ANT + PHE)	< 0.1 petroleum	–	> 0.1 combustion
FLU/(FLU + PYR)	< 0.4 petroleum	0.4–0.5 liquid fuel combustion	> 0.5 coal, wood, grass combustion
BaA/(BaA + CHR)	< 0.20 petroleum	0.20–0.35 petroleum or combustion	> 0.35 combustion
IcdP/(IcdP + BghiP)	< 0.20 petroleum	0.2–0.5 liquid fuel combustion	> 0.5 coal, wood, grass combustion

PAHs which are members of the same isomer set can have different mobilities, transformation rates, or degradation rates in various environmental compartments (Zhang et al. 2005).

In most studies, two or more PAH isomeric ratios are plotted in bivariate 'cross plots', so that more than one ratio informs the identification of likely PAH source(s). Figure 7.9 presents an example of the plots from Ma et al. (2008), who used the PAH isomeric ratios to deduce that the PAHs in the Dongguan, Guangzhou, and Foshan areas of the PRD had multiple sources. These sources were mainly combustion of solid fuels (coal, plant matter), with fewer samples suggesting that PAHs were derived from liquid fuel combustion or direct pollution by petroleum (Fig. 7.9). This interpretation was supported by relative concentrations of phenanthrene (3-ring) and other 4- or 5-ring PAHs accounting for, on average, 68% of PAHs by mass, also suggesting combustion sources.

In a variation on the analysis of PAH isomer ratios, source information can also be obtained from the concentrations of substituted PAHs, in which a hydrogen bonded to the PAH carbon framework is replaced by another atom or functional group (such as a methyl group, $-CH_3$, or a hydroxyl group –OH). The ratio of methyl-phenanthrene/phenanthrene, for example, was used to infer a mainly petroleum-based origin for PAHs from vegetable-growing soils in urban Guangzhou (Chen et al. 2005). In a somewhat different context, Gao et al. (2019) concluded that the spatially different ratios between individual PAHs and their hydroxyl-substituted counterparts represented different degrees of microbial transformation depending on land use.

Multivariate statistical analyses (see Chap. 3) have also been used to identify the potential sources of PAHs in contaminated urban soils of the PRD. For example, Ke et al. (2017) used principal component analysis (PCA) in combination with multiple regression to assess the contributions of traffic emissions and coal combustion to PAH contamination in urban parklands in Guangzhou. Ma et al. (2009) also analysed data for PAHs in soils from Huizhou using principal component analysis, in

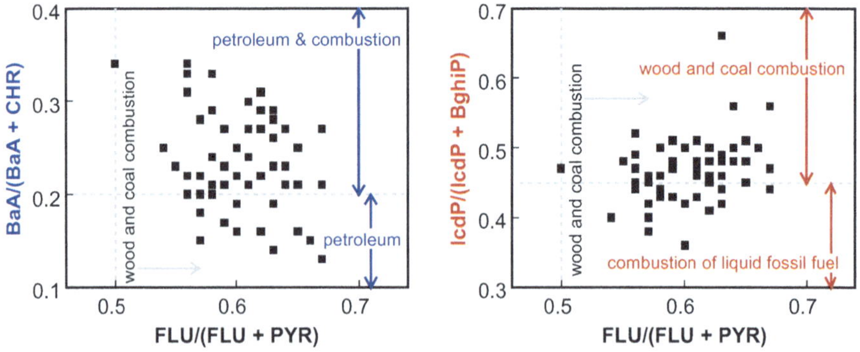

Fig. 7.9 Graphical analysis of PAH isomeric ratios using 'cross plots', implying a dominance of combustion sources for PAHs in urban soils in Dongguan and Foshan. (From Ma et al. 2008; used with permission from Springer)

combination with hierarchical cluster analysis (another multivariate statistical method). The cluster analysis broadly grouped PAHs by ring number, which Ma et al. (2009) attributed to the trends in parameters affecting environmental behaviour of PAHs having different molecular weights (e.g. see Table 7.6 and Fig. 7.13).

7.5.6 Controls on PAH Concentrations and Fractionation

Soil organic carbon and 'black carbon'. As we discussed in Sect. 7.2, soil organic carbon content strongly affects the adsorption of non-polar (or low-polarity) organic compounds such as PAHs. In many examples of PAH contamination of urban soils in the Pearl River Delta, a positive relationship between PAH concentration and soil organic carbon content is observed (Li et al. 2008b; Ma et al. 2008; Zhu et al. 2019). Similarly, there can also be a positive relationship between PAH concentration and *black carbon* content of soil, where black carbon is composed of partially combusted solid materials such as soot or finely divided charcoal (Ma et al. 2008). Figure 7.10 shows examples of relationships of $\sum_{16}$PAHs concentration with both total soil organic carbon and black carbon, in soils of the Pearl River Delta, which suggest that PAHs in these soils are adsorbed to one or both forms of organic carbon. (It is worth noting here that not all methods of measuring soil organic carbon

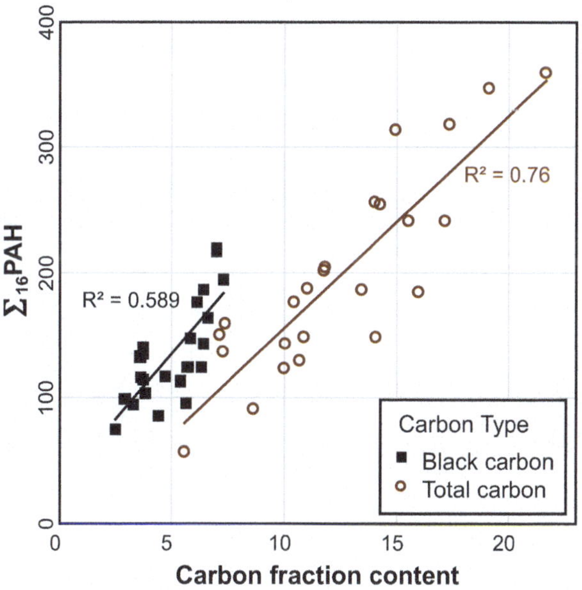

Fig. 7.10 Relationships between $\sum_{16}$PAHs concentration and both total soil organic carbon and black carbon, in urban and peri-urban soils in Guangzhou, Dongguan, and Foshan (Pearl River Delta, Guangdong, China). (Redrawn from Ma et al. (2008); used with permission from Springer)

measure total carbon. For example, the widely used methods employing wet oxidation of soil organic matter with acid dichromate solution are known to exclude black carbon, so do not measure a true 'total' organic carbon content (TOC).)

In some urban environments, PAH concentrations are *not* (or poorly) correlated with either soil organic carbon or black carbon. The lack of a strong correlation between $\sum_{16}$PAHs concentration and soil organic/black carbon is usually attributed to recent additions of PAHs or proximity to ongoing PAH sources, so that PAHs are not in equilibrium with the soil environment (Cai et al. 2007; Ke et al. 2017).

The microbial community composition of urban soil, although not strictly an effect of soil organic matter (which is usually defined as the non-living organic material), may also affect the concentrations of fractionation of PAHs. For example, Zhu et al. (2019) found shifts in bacterial community structure when exposed to phenanthrene and benzo[*a*]pyrene, suggesting that some groups of soil bacteria may be able to degrade PAHs more effectively than others.

Urban spatial and land use effects. The concentrations of PAHs in urban soils should reflect spatial factors such as distance from point sources or from the urban centre, since PAH concentrations into soils should be greater where industrial or traffic emissions are greater. For similar reasons, land use should also have an effect on PAH concentrations in urban soils. Unless there have been significant changes in land use, we would expect lower PAH concentrations in less urbanised environments, such as residential suburbs or peri-urban agriculture, than in industrial areas or in the urban centre.

In the Pearl River Delta, PAH concentrations in urban and peri-urban soils seem to follow these expected trends. For example, Liu et al. (2011), Xiao et al. (2014), and You (2015) all found that PAH concentrations decreased with increasing distance from urban centres across several cities in the PRD (Fig. 7.11). The greatest $\sum$PAH concentrations observed in Shenzhen (Zhang et al. 2014) in traffic-affected

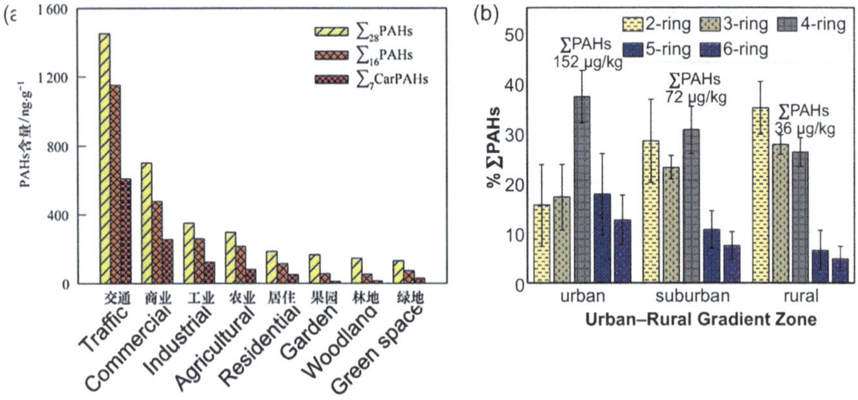

Fig. 7.11 (**a**) Decreasing trends in $\sum$PAH *concentrations* with decreasing intensity of urban land use, redrawn and translated from Zhang et al. (2014); (**b**) trends in the *proportions* of $\sum$PAHs by ring number, on an urban-rural gradient. (Redrawn from Xiao et al. (2014) and used under the terms of a CC-BY-3.0 license (http://creativecommons.org/licenses/by/3.0/))

soils are consistent with the greater expected concentrations of PAHs in street dusts (Menzie et al. 1992). Low-intensity land uses such as gardens, urban woodland, and urban green space tend to have the lowest PAH concentrations (Fig. 7.11a).

When individual PAHs, or smaller categories based on ring number, are considered, the trends do not necessarily match those of the $\sum$PAHs. As the data from Xiao et al. (2014) show (Fig. 7.11b), the proportions of 2-ring and 3-ring PAHs (e.g. [2-ring PAHs] / [$\sum_{16}$PAHs]) are both greater in soil in rural land than in urban land, tending to increase from urban, to suburban, to rural. In contrast, the proportions of 4-, 5-, and 6-ring PAHs follow the same trend as $\sum_{16}$PAHs in soil and decrease from urban, to suburban, to rural land. The combination of low $\sum_{16}$PAHs and greater proportions of 2- and 3-ring PAHs in rural soils may represent small ongoing background additions from other environmental compartments such as wet or dry deposition from air.

Uptake of PAHs from soil by plants. The concentrations of PAHs in plants are generally lower than in the soils they are grown in. Song et al. (2013) determined mean $\sum_{16}$PAHs of 318 µg/kg in soil, which was approximately 1.4 times greater than the 234 µg/kg in vegetables. Similarly, Zhang et al. (2008) measured approximately 1.3-fold greater concentrations of $\sum_{16}$PAHs in soils than in vegetables. In contrast, You (2015) found greater $\sum_{16}$PAH concentrations in vegetable matter (Fig. 7.12), with mean $\sum_{16}$PAH concentrations across 5–6 sites of 232 µg/kg in soils and 641 µg/kg in vegetables.

In addition, plant uptake is not the same for all PAH compounds, resulting in fractionation during the transfer between soil and plant. For example, Song et al. (2013) showed that vegetable tissues were depleted in 5- and 6-ring PAHs relative to soils (Fig. 7.12). The differences in uptake for different PAHs are consistent with their water solubilities. The PAH compounds with greater molecular weights (i.e. greater ring numbers) have lower water solubilities and larger K_{OW} values, meaning

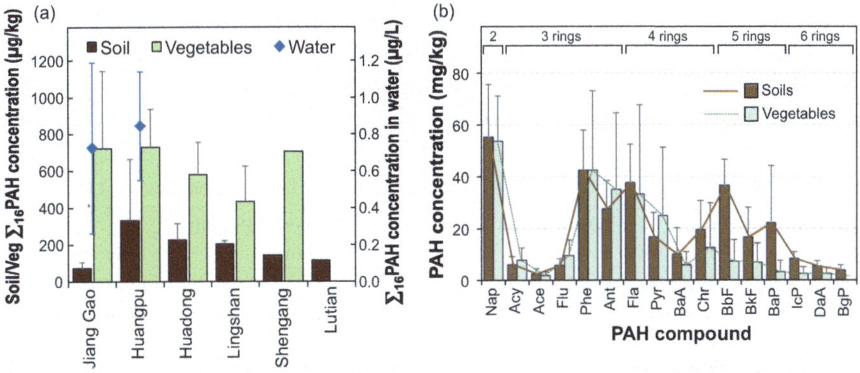

Fig. 7.12 (**a**) $\sum_{16}$PAH concentrations in soils, vegetables, and water (drawn by Andrew W. Rate from data in You (2015) and (**b**) mean concentrations of individual 16 US EPA priority PAHs in soils and vegetables from the Pearl River Delta. (Drawn by Andrew W. Rate from data in Song et al. 2013).

that they are more strongly associated with soil solids (mainly soil organic carbon), a situation which restricts plant uptake.

Within plant biomass, PAHs partition differently between above-ground (shoots) and below-ground (roots) tissues. Transfer within plants is in water-based media (the phloem and xylem), so, being non-polar compounds, PAHs would be expected to have limited mobility in plants. Individual PAHs with greater water solubility (lower K_{OW}) are more likely to move from soil to roots to shoots. For a range of plant species grown at e-waste recycling sites in Longtang, Guangdong, Wang et al. (2017) measured mean $\sum_{16}$PAH concentrations in shoots of 196 μg/kg compared with 284 μg/kg in roots. In addition, there was some evidence of differential accumulation of individual PAHs in root and shoot tissues of plants.

7.5.7 *Fluxes of PAHs to and from Soils in the Pearl River Delta*

Emissions of PAHs from industry and other sources are mainly into the atmosphere from combustion processes, so that atmospheric deposition represents the dominant input flux into environmental compartments such as soils (Abdel-Shafy and Mansour 2016). The soil budget for PAHs in the Pearl River Delta region is not known completely, but the amounts are large; Ma et al. (2008) calculated values for Guangzhou of 920,000 kg, for Dongguan of 143,000 kg, and for Foshan of 229,000 kg. In Shenzhen, Ni et al. (2011) calculated the soil $\sum_{15}$PAH inventory to be 152,000 kg. With such large reservoirs of PAHs in urban and peri-urban soils of the PRD, the transfers to and from the soil environment are therefore very significant.

Atmosphere to soil fluxes of PAHs. The inputs of PAHs across the entire Pearl River Delta region are substantial (Table 7.8); for example, Li et al. (2010) sampled sites across the PRD including urban and rural locations and calculated that particle-phase deposition of $\sum_{15}$PAHs was 2950 kg/y (excluding Hong Kong), with $\sum_{15}$PAHs deposition of 86 kg/y in Hong Kong (both fluxes are possibly underestimates, given that only particle-phase deposition was measured). Including Hong Kong in the regional estimate, this represents 0.054 kg/km²/y over the 54,156 km² of the PRD,

Table 7.8 Atmosphere to soil fluxes of PAHs in the Pearl River Delta region, Southern China

Location in PRD	Land use	Deposition type	PAH analysis	Air to soil PAH flux (kg/km²/y)	References
Whole region	Multiple	Particulate	$\sum_{15}$PAHs[a]	0.054	Li et al. (2010)
Shenzhen	Multiple	Total	$\sum_{15}$PAHs[a]	0.56	Ni et al. (2011)
Guangzhou	Urban forest		$\sum_{9}$PAHs[b]	0.206	Chen et al. (2015a)

[a]The 16 EPA priority PAHs excluding naphthalene

[b]Phenanthrene, anthracene, fluoranthene, pyrene, benzo[*a*]anthracene, chrysene, benzo[*b*]fluoranthene, benzo[*k*]fluoranthene, and benzo[*a*]pyrene

which includes both urban and non-urban land. In Shenzhen, Ni et al. (2011) calculated that total atmospheric fluxes (wet and dry deposition) accounted for inputs of 1156 kg/y over the 2050 km^2 area of Shenzhen (i.e. 0.56 $kg/km^2/y$). The differences are likely to be due to more comprehensive measurement in Shenzhen, which included 'rain washing', wet deposition, and dry deposition; the rain washing flux was dominant, being about 98% of total atmosphere to soil transfer (Ni et al. 2011). In addition, the proportion of non-urban land in Shenzhen may be greater than in the PRD region as a whole. PAH fluxes to soil via precipitation were also important in a peri-urban forest in Guangzhou (Chen et al. 2015a); importantly, interception of precipitation by tree canopies decreased PAH inputs to soil by nearly 25% relative to the original precipitation.

Soil to atmosphere fluxes of PAHs. Despite the large atmospheric deposition of PAHs to soil, Ni et al. (2011) found that diffusion of PAHs from soil into air was also an important pathway, accounting for over 10,000 kg/y of $\sum_{15}$PAHs over the 2050 km^2 area of Shenzhen (i.e. approximately 4.9 $kg/km^2/y$). As Fig. 7.13 shows, the net flux may be positive or negative for individual PAHs (Wei et al. 2014). Lower molecular weight PAH molecules such as acenaphthene or fluorene are more volatile and have less affinity for soil organic matter so have positive soil to air fluxes. In contrast, high molecular weight PAHs such as benzo[*a*]pyrene or indeno[1,2,3-*cd*]pyrene have net air-soil fluxes (Fig. 7.13).

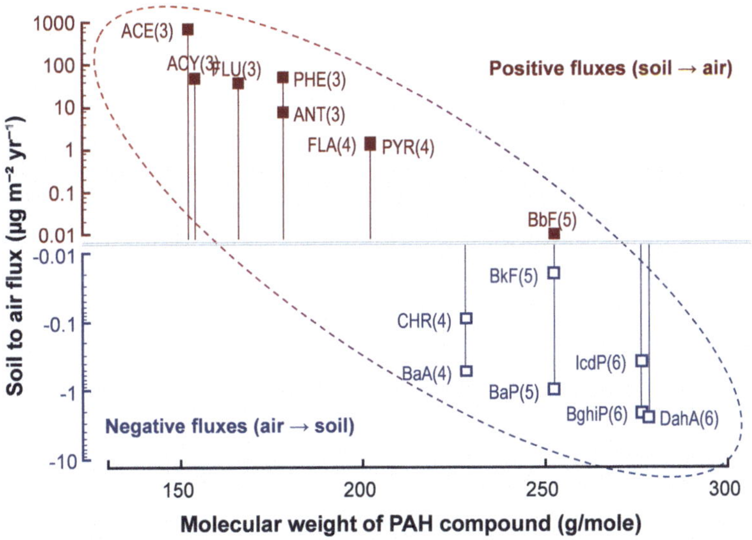

Fig. 7.13 Relationship of soil-atmosphere flux for priority pollutant PAHs to their molecular mass (redrawn by Andrew W. Rate using data for central Pearl River Delta soils from Wei et al. 2014). Abbreviations for individual PAHs are from Table 7.6 and are followed by the number of rings in parentheses; the trend ellipse is just a visual guide. A similar plot is obtained for soil-air flux vs. $\log_{10}K_{OW}$ (octanol-water partition coefficient). Larger PAH molecules are less volatile and have greater affinity for soil organic matter so have small or negative soil to air fluxes

Other soil fluxes of PAHs: run-off, leaching, and erosion. In their Guangzhou peri-urban forest study, Chen et al. (2015a) calculated a loss of $\sum_9$PAHs in run-off of 0.034 kg/km^2/y. In the same forest environment, leaching of $\sum_9$PAHs through soil transferred 0.065 kg/km^2/y to a depth of 30 cm and 0.008 kg/km^2/y to 60 cm. The decrease in PAHs during leaching through soil was attributed to a combination of processes including biodegradation and adsorption (and presumably also volatilisation from surface soil, although the authors did not consider this). A component of the run-off flux was soil erosion, which also removed PAHs from the soil environment. Run-off was separated into water and particulate (i.e. erosion) components by Ni et al. (2011), who estimated a water (dissolved) run-off flux of $\sum_{15}$PAHs of 0.94 kg/km^2/y and soil erosion (particulate run-off) flux of 0.32 kg/km^2/y.

7.5.8 Other Issues for PAHs in the Pearl River Delta

The PAHs released by combustion, petroleum, and other sources into urban soils may react to give new compounds which are not measured in routine PAH analyses. For example, small but significant concentrations of several hydroxylated PAHs were measured in 15 cities across the Pearl River Delta region by Gao et al. (2019). Ni and Zeng (2012) found non-negligible concentrations of a range of chlorinated and brominated PAHs in urban soils in Shenzhen. For hydroxylated, chlorinated, and brominated PAHs, the absence of correlations between concentrations of substituted PAHs and their probable parent compounds inferred that they were formed by reactions in the soil environment. The toxicity profiles of these compounds are not currently well understood, but both studies recommended that the potential risks to human health be considered.

The occurrence and behaviour of polycyclic aromatic hydrocarbons in China's Pearl River Delta is a globally relevant case study. Based on the material covered in this chapter, Fig. 7.14 presents a generalised conceptual model for cycling of organic pollutants in urban environments.

7.6 Summary

- A wide range of organic compounds are potential contaminants in urban soils and sediments, with these compounds having widely varying structures and properties. Soils are one of the largest reservoirs of organic pollutants.
- Organic compounds can be highly toxic to a wide range of organisms, including humans.

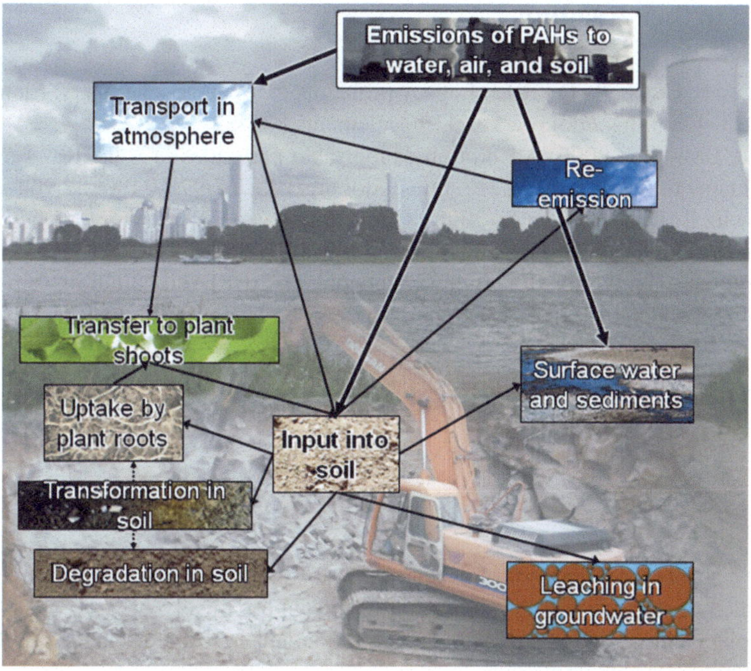

Fig. 7.14 Fate of persistent organic pollutants such as polycyclic aromatic hydrocarbons (PAHs) in urban soil-plant-atmosphere systems, based on studies in the Pearl River Delta region, China

- The behaviour of organic contaminants in soils is affected by the properties of the individual compounds, or categories of compounds, involved, including: polarity, volatility, density, ionisation, substitution with atoms other than C and H, and molecular weight.
- The natural organic carbon content of soils (including black carbon) is very important for the retention of organic contaminants by adsorption and cation exchange.
- Organic contaminants can degrade chemically in soils (as opposed to biologically mediated degradation), but the most effective and common form of degradation is mediated by soil microorganisms such as bacteria and fungi.
- Using the behaviour of polycyclic aromatic hydrocarbons in soils of the large urban agglomeration of the Pearl River delta as a case study, it is apparent that:
  - Categories of closely related organic contaminants such as PAHs can have systematic differences in properties which lead to fractionation in soils. This fractionation can provide information about the sources of contamination.
  - Transfers of organic contaminants such as PAHs between urban soils and the atmosphere can be particularly important, as can transfers between soils and other environmental compartments.

7.7 Further Reading

Albanese S, Cicchella D (2012) Legacy problems in urban geochemistry. Elements 8:423–428. https://doi.org/10.2113/gselements.8.6.423

Kah M, Brown C (2006) Adsorption of ionisable pesticides in soils. Rev Environ Contam Toxicol 188:149–217. https://doi.org/10.1007/978-0-387-32,964-2_5

Berkowitz B, Dror I, Yaron B (2008). Contaminant geochemistry: interactions and transport in the subsurface environment. Springer, Berlin

Rodríguez-Eugenio N, McLaughlin M, Pennock D (2018) Soil pollution – a hidden reality. Food and Agriculture Organization of the United Nations, Rome

7.8 Questions

7.8.1 Checking Your Understanding

1. Give as many examples as you can of non- or weakly polar organic pollutants.
2. Which organic pollutants could be derived from use of solvents?
3. Explain the acronyms LNAPL and DNAPL and how they relate to soil and groundwater contamination.
4. What are the properties of persistent organic pollutants that distinguish them from other organic compounds?
5. Why are K_{OW} values of organic compounds useful – what do they represent, and what can they predict?
6. What are the two main steps in analysis of organic compounds in soils, and what is involved in each step?

7.8.2 Thinking About the Topics More Deeply

7. Explain how fractionation of polycyclic aromatic hydrocarbons occurs and what information that fractionation can provide.
8. Why is biodegradation of organic pollutants usually more significant for attenuation than abiotic degradation?
9. The intensity of soil pollution with organic contaminants is often related to the intensity of land use, especially along a rural-urban gradient, with land in industrial or urban-core zones being the most contaminated. Can you think of any scenarios where this may *NOT* be the case?

7.8.3 *A Question 'Out of Left Field'*

10. Different countries have different limits for acceptable concentrations of organic contaminants in soils (even for the same organic compound). Do you think that there is a scientific reason for this or do other factors contribute to the diversity in contaminant limits?

References

Abdel-Shafy HI, Mansour MSM (2016) A review on polycyclic aromatic hydrocarbons: source, environmental impact, effect on human health and remediation. Egypt J Pet 25:107–123. https://doi.org/10.1016/j.ejpe.2015.03.011

Achten C, Andersson JT (2015a) Overview of polycyclic aromatic compounds (PAC). Polycycl Aromat Compd 35:177–186. https://doi.org/10.1080/10406638.2014.994071

Achten C, Andersson JT (2015b) Overview of polycyclic aromatic compounds (PAC) [supplementary material]. Polycycl Aromat Compd 35:177–186. https://doi.org/10.1080/10406638.2014.994071

Ahammed GJ, Wang MM, Zhou YH, Xia XJ, Mao WH, Shi K, Yu JQ (2012) The growth, photosynthesis and antioxidant defense responses of five vegetable crops to phenanthrene stress. Ecotoxicol Environ Saf 80:132–139. https://doi.org/10.1016/j.ecoenv.2012.02.015

Báez ME, Espinoza J, Silva R, Fuentes E (2015) Sorption-desorption behavior of pesticides and their degradation products in volcanic and nonvolcanic soils: interpretation of interactions through two-way principal component analysis. Environ Sci Pollut Res 22:8576–8585. https://doi.org/10.1007/s11356-014-4036-8

Barceló D (ed) (2012) Emerging organic contaminants and human health, The handbook of environmental chemistry, vol 20. Springer, Heidelberg, 466 pp

Barrett KA, McBride MB (2005) Oxidative degradation of glyphosate and aminomethylphosphonate by manganese oxide. Environ Sci Technol 39:9223–9228. https://doi.org/10.1021/es051342d

Biasioli M, Fabietti G, Barberis R, Ajmone-Marsan F (2012) An appraisal of soil diffuse contamination in an industrial district in northern Italy. Chemosphere 88:1241–1249. https://doi.org/10.1016/j.chemosphere.2012.03.081

Bromilow RH (2005) Pesticides. In: Hillel D (ed) Encyclopedia of Soils in the environment. Elsevier, Oxford, pp 188–195. https://doi.org/10.1016/B0-12-348530-4/00110-7

Cai QY, Mo CH, Li YH, Zeng QY, Katsoyiannis A, Wu QT, Férard JF (2007) Occurrence and assessment of polycyclic aromatic hydrocarbons in soils from vegetable fields of the Pearl River Delta, South China. Chemosphere 68:159–168. https://doi.org/10.1016/j.chemosphere.2006.12.015

Canadian Council of Ministers of the Environment (1991) Interim Canadian environmental quality criteria for contaminated sites. Canadian Council of Ministers of the Environment, Winnipeg

Canadian Council of Ministers of the Environment (2010) Canadian soil quality guidelines for the protection of environmental and human health – polycyclic aromatic hydrocarbons. Canadian Council of Ministers of the Environment, Winnipeg

Cao SP, Ni HG, Qin PH, Zeng H (2010) Occurrence and human non-dietary exposure of polycyclic aromatic hydrocarbons in soils from Shenzhen, China. J Environ Monit 12:1445–1450. https://doi.org/10.1039/c000549e

Chen L, Ran Y, Xing B, Mai B, He J, Wei X et al (2005) Contents and sources of polycyclic aromatic hydrocarbons and organochlorine pesticides in vegetable soils of Guangzhou, China. Chemosphere 60:879–890. https://doi.org/10.1016/j.chemosphere.2005.01.011

Chen B, Pei N, Huang J, Liu S, Zhang N, Xiao Y, Pan Y (2015a) Removal of polycyclic aromatic hydrocarbons from precipitation in an urban forest of Guangzhou, South China. Bull Environ Contam Toxicol 95:240–245. https://doi.org/10.1007/s00128-015-1567-8
Chen M, Xu P, Zeng G, Yang C, Huang D, Zhang J (2015b) Bioremediation of soils contaminated with polycyclic aromatic hydrocarbons, petroleum, pesticides, chlorophenols and heavy metals by composting: applications, microbes and future research needs. Biotechnol Adv 33:745–755. https://doi.org/10.1016/j.biotechadv.2015.05.003
CL:AIRE (2009) Soil guideline values – soil guideline value reports. Contaminated Land: Applications in Real Environments, Haddenham. https://www.claire.co.uk/information-centre/water-and-land-library-wall/44-risk-assessment/178-soil-guideline-values. Accessed 27 Aug 2020
Enfield CG, Bengtsson G, Lindqvist R (1989) Influence of macromolecules on chemical transport. Environ Sci Technol 23:1278–1286
Fuoco R, Capodaglio G, Muscatello B, Radaelli M (2009) Persistent organic pollutants (POPs) in the Antarctic environment: a review of findings. Scientific Committee on Antarctic Research (SCAR), Cambridge
Gao CJ, Xia LL, Wu CC, Shen HM, Guo Y (2019) Hydroxylated polycyclic aromatic hydrocarbons in surface soil in an emerging urban conurbation in South China. Sci Total Environ 692:1250–1256. https://doi.org/10.1016/j.scitotenv.2019.07.334
Gonzalez-Mille DJ, Espinosa-Reyes G, Rivero-Pérez NE, Trejo-Acevedo A, Nava-Montes AD, Ilizaliturri-Hernández CA (2013) Persistent organochlorine pollutants (POPs) and DNA damage in Giant Toads (*Rhinella marina*) from an industrial area at Coatzacoalcos, Mexico. Water Air Soil Pollut 224:1781. https://doi.org/10.1007/s11270-013-1781-0
Grathwohl P (2011) Fate and transport of organic compounds in (to) the subsurface environment. In: Xing B, Senesi N, Huang PM (eds) Biophysico-chemical processes of anthropogenic organic compounds in environmental systems, Wiley online books. Wiley, Hoboken, pp 215–231. https://doi.org/10.1002/9780470944479.ch8
Harvey DT (2016) Analytical chemistry 2.1 (Open access; CC-BY-VC-SA 4.0). http://dpuadweb.depauw.edu/harvey_web/eTextProject/version_2.1.html. Accessed 18 June 2021
Hatzinger PB, Alexander M (1995) Effect of aging of chemicals in soil on their biodegradability and extractability. Environ Sci Technol 29:537–545. https://doi.org/10.1021/es00002a033
He L, Song J, Peng P (2008) Characterization of extractable and non-extractable polycyclic aromatic hydrocarbons in soils and sediments from the Pearl River Delta, China. Environ Pollut 156:769–774. https://doi.org/10.1016/j.envpol.2008.06.003
Hiller E, Krascsenits Z, Čerňanský S (2008) Sorption of acetochlor, atrazine, 2,4-D, chlorotoluron, MCPA, and trifluralin in six soils From Slovakia. Bull Environ Contam Toxicol 80:412–416. https://doi.org/10.1007/s00128-008-9430-9
Joa K, Jefimova J, Irha N, Teinemaa E, Lintelmann J, Kirso U (2009) Determination of polycyclic aromatic hydrocarbons (PAHs) in oil shale processing wastes: current practice and new trends. Oil Shale 26. https://doi.org/10.3176/oil.2009.1.07
John EM, Shaike JM (2015) Chlorpyrifos: pollution and remediation. Environ Chem Lett 13:269–291. https://doi.org/10.1007/s10311-015-0513-7
Johnson D (2005) Pollutants | persistent organic (POPs). In: Hillel D (ed) Encyclopedia of soils in the environment. Elsevier, Oxford, pp 264–271. https://doi.org/10.1016/B0-12-3485 30-4/00560-9
Jury WA, Focht DD, Farmer WJ (1987) Evaluation of pesticides groundwater pollution potential from standard indices of soil-chemical adsorption and biodegradation. J Environ Qual 16:422–428
Kah M, Brown C (2006) Adsorption of ionisable pesticides in soils. Rev Environ Contam Toxicol 188:149–217. https://doi.org/10.1007/978-0-387-32964-2_5
Ke CL, Gu YG, Liu Q (2017) Polycyclic aromatic hydrocarbons (PAHs) in exposed-lawn soils from 28 urban parks in the megacity Guangzhou: occurrence, sources, and human health implications. Arch Environ Contam Toxicol 72:496–504. https://doi.org/10.1007/s00244-017-0397-6

Kile DE, Chiou CT, Zhou H, Li H, Xu O (1995) Partition of nonpolar organic pollutants from water to soil and sediment organic matters. Environ Sci Technol 29:1401–1406. https://doi.org/10.1021/es00005a037

Kostecki P, Morrison R, Dragun J (2005) Hydrocarbons. In: Hillel D (ed) Encyclopedia of soils in the environment. Elsevier, Oxford, pp 217–226. https://doi.org/10.1016/B0-12-348530-4/00098-9

Kromer T, Ophoff H, Stork A, Führ F (2004) Photodegradation and volatility of pesticides: chamber experiments. Environ Sci Pollut Res 11:107–120. https://doi.org/10.1007/BF02979710

Leung A, Cai ZW, Wong MH (2006) Environmental contamination from electronic waste recycling at Guiyu, southeast China. J Mater Cycles Waste Manage 8:21–33. https://doi.org/10.1007/s10163-005-0141-6

Li H, Ran Y (2012) Distribution and bioconcentration of polycyclic aromatic hydrocarbons in surface water and fishes. Sci World J 2012:Article 632910. https://doi.org/10.1100/2012/632910

Li Y, Li F, Zhang T, Yang G, Chen J, Wan H (2007) Pollution assessment, distribution and sources of PAHs in agricultural soils of Pearl River Delta – the biggest manufacturing Base in China. J Environ Sci Health A Toxic Hazard Subst Environ Eng 42:1979–1987. https://doi.org/10.1080/10934520701628890

Li FB, Wang XG, Li YT, Liu CS, Zeng F, Zhang LJ et al (2008a) Enhancement of the reductive transformation of pentachlorophenol by polycarboxylic acids at the iron oxide-water interface. J Colloid Interface Sci 321:332–341. https://doi.org/10.1016/j.jcis.2008.02.033

Li YT, Li FB, Chen JJ, Yang GY, Wan HF, Zhang TB et al (2008b) The concentrations, distribution and sources of PAHs in agricultural soils and vegetables from Shunde, Guangdong, China. Environ Monit Assess 139:61–76. https://doi.org/10.1007/s10661-007-9816-x

Li J, Liu X, Zhang G, Li XD (2010) Particle deposition fluxes of BDE-209, PAHs, DDTs and chlordane in the Pearl River Delta, South China. Sci Total Environ 408:3664–3670. https://doi.org/10.1016/j.scitotenv.2010.04.048

Liang Y, Fung PK, Tse MF, Hong HC, Wong MH (2008) Sources and seasonal variation of PAHs in the sediments of drinking water reservoirs in Hong Kong and the Dongjiang River (China). Environ Monit Assess 146:41–50. https://doi.org/10.1007/s10661-007-0058-8

Liang Y, Zhou Z, Li X (2019) Dynamic of regional planning and sustainable development in the Pearl River Delta, China. Sustainability 11:6074. https://doi.org/10.3390/su11216074

Liu G, Yu L, Li J, Liu X, Zhang G (2011) PAHs in soils and estimated air-soil exchange in the Pearl River Delta, South China. Environ Monit Assess 173:861–870. https://doi.org/10.1007/s10661-010-1429-0

Ma X, Ran Y, Gong J, Zou M (2008) Concentrations and inventories of polycyclic aromatic hydrocarbons and organochlorine pesticides in watershed soils in the Pearl River Delta, China. Environ Monit Assess 145:453–464. https://doi.org/10.1007/s10661-007-0054-z

Ma J, Zhou Y, Wan H (2009) Study on polycyclic aromatic hydrocarbons (PAHs) contents and sources in the surface soil of Huizhou City, South China, based on multivariate statistics analysis. Chin J Geochem 28:335–339. https://doi.org/10.1007/s11631-009-0335-z

Ma X, Ran Y, Gong J, Chen D (2010) Sequential accelerated solvent extraction of polycyclic aromatic hydrocarbons with different solvents: performance and implication. J Environ Qual 39:2072–2079. https://doi.org/10.2134/jeq2010.0163

Mai B, Qi S, Zeng EY, Yang Q, Zhang G, Fu J et al (2003) Distribution of polycyclic aromatic hydrocarbons in the coastal region off Macao, China: assessment of input sources and transport pathways using compositional analysis. Environ Sci Technol 37:4855–4863. https://doi.org/10.1021/es034514k

Martí V, Jubany I, Pérez C, Rubio X, De Pablo J, Giménez J (2014) Human health risk assessment of a landfill based on volatile organic compounds emission, immission and soil gas concentration measurements. Appl Geochem 49:218–224. https://doi.org/10.1016/j.apgeochem.2014.06.018

McBride MB (1994) Environmental chemistry of soils. Oxford University Press, New York

Menzie CA, Potocki BB, Santodonato J (1992) Exposure to carcinogenic PAHs in the environment. Environ Sci Technol 26:1278–1284. https://doi.org/10.1021/es00031a002

Ministry for the Environment (Manatū Mō Te Taiao) (2012) Users' guide: national environmental standard for assessing and managing contaminants in soil to protect human health. New Zealand Government, Wellington

Mo CH, Cai QY, Tang SR, Zeng QY, Wu QT (2009) Polycyclic aromatic hydrocarbons and phthalic acid esters in vegetables from nine farms of the pearl river delta, South China. Arch Environ Contam Toxicol 56:181–189. https://doi.org/10.1007/s00244-008-9177-7

Mulligan CN, Yong RN (2004) Natural attenuation of contaminated soils. Environ Int 30:587–601. https://doi.org/10.1016/j.envint.2003.11.001

Mulligan CN, Yong RN, Gibbs BF (2001) Surfactant-enhanced remediation of contaminated soil: a review. Eng Geol 60:371–380. https://doi.org/10.1016/S0013-7952(00)00117-4

National Environment Protection Council (2013a) Schedule B5b: guideline on methodology to derive ecological investigation levels in contaminated soils, national environment protection (assessment of site contamination) measure (amended). Commonwealth of Australia, Canberra

National Environment Protection Council (2013b) Schedule B (1): guideline on the investigation levels for soil and groundwater, national environment protection (assessment of site contamination) measure (amended). Commonwealth of Australia, Canberra

National Pollutant Inventory (2020) Substances. Department of Agriculture, Water and the Environment, Canberra. http://www.npi.gov.au/substances. Accessed 17 June 2020

Ni HG, Zeng EY (2012) Environmental and human exposure to soil chlorinated and brominated polycyclic aromatic hydrocarbons in an urbanized region. Environ Toxicol Chem 31:1494–1500. https://doi.org/10.1002/etc.1854

Ni HG, Qin PH, Cao SP, Zeng H (2011) Fate estimation of polycyclic aromatic hydrocarbons in soils in a rapid urbanization region, Shenzhen of China. J Environ Monit 13:313–318. https://doi.org/10.1039/c0em00470g

Petersen LW, El-Farhan YH, Moldrup P, Rolston DE, Yamaguchi T (1996) Transient diffusion, adsorption, and emission of volatile organic vapors in soils with fluctuating low water contents. J Environ Qual 25:1054–1063. https://doi.org/10.2134/jeq1996.00472425002500050018x

Rodríguez-Eugenio N, McLaughlin M, Pennock D (2018) Soil pollution – a hidden reality. Food and Agriculture Organization of the United Nations, Rome. http://www.fao.org/3/I9183EN/i9183en.pdf

Sadler R, Connell D (2003) Analytical methods for the determination of total petroleum hydrocarbons in soil. In: Langley A, Gilbey M, Kennedy B (eds) Proceedings of the fifth national workshop on the assessment of site contamination. National Environment Protection Council Service Corporation, Adelaide. http://www.nepc.gov.au/system/files/resources/27b740ab-880f-d7f4-d10e-bab0fa964e48/files/asc-wkshoppaper-09-tphs-sadler-petrol-hydro-soil-200301.pdf

Sandrin TR, Maier RM (2003) Impact of metals on the biodegradation of organic pollutants. Environ Health Perspect 111:1093–1101. https://doi.org/10.1289/ehp.5840

Secretariat of the Stockholm Convention (2018) Stockholm convention on persistent organic pollutants: texts and annexes; revised in 2017. United Nations Environment Programme, Châtelaine

Shahsavari E, Poi G, Aburto-Medina A, Haleyur N, Ball AS (2017) Bioremediation approaches for petroleum hydrocarbon-contaminated environments. In: Anjum NA, Gill SS, Tuteja N (eds) Enhancing cleanup of environmental pollutants: volume 1: biological approaches. Springer, Cham

Song Y, Hao R, Peng S, Wan H (2013) Polycyclic aromatic hydrocarbons in different soils and vegetables from the Pearl River Delta, South China. Environ Eng Manag J 12:1891–1895

Sverdrup LE, Hagen SB, Krogh PH, van Gestel CAM (2007) Benzo(a)pyrene shows low toxicity to three species of terrestrial plants, two soil invertebrates, and soil-nitrifying bacteria. Ecotoxicol Environ Saf 66:362–368. https://doi.org/10.1016/j.ecoenv.2006.01.007

U.S. EPA (1996a) Method 3540C: soxhlet extraction, Test methods for evaluating solid waste, physical/chemical methods, EPA publication SW-846. United States Environmental Protection Agency, Washington, DC, pp 1–8. https://www.epa.gov/sites/production/files/2015-12/documents/3540c.pdf

U.S. EPA (1996b) Method 3560: supercritical fluid extraction of total recoverable petroleum hydrocarbons, Test methods for evaluating solid waste, physical/chemical methods, EPA publication SW-846. United States Environmental Protection Agency, Washington, DC, pp 1–8. https://www.epa.gov/sites/production/files/2015-12/documents/3540c.pdf

U.S. EPA (2018) Method 8000D: determinative chromatographic analysis, Test methods for evaluating solid waste, physical/chemical methods, EPA publication SW-846 Update VI. United States Environmental Protection Agency, Washington, DC. https://www.epa.gov/sites/production/files/2015-12/documents/8000d.pdf

U.S. EPA (2020) Integrated risk information system (IRIS). United States Environmental Protection Agency, Washington. https://www.epa.gov/iris. Accessed 22 July 2020

UN-Habitat (2016) World cities report 2016: urbanization and development – emerging futures. United Nations Human Settlements Programme, Nairobi. https://unhabitat.org/sites/default/files/download-manager-files/WCR-2016-WEB.pdf

Wang W-X, Rainbow PS (2020) Environmental pollution of the Pearl River Estuary, China, status and impact of contaminants in a rapidly developing region. Springer, Berlin/Heidelberg, 125 pp. https://doi.org/10.1007/978-3-662-61834-9

Wang X, Sun D, Yao T (2016) Climate change and global cycling of persistent organic pollutants: a critical review. Sci China Earth Sci 59:1899–1911. https://doi.org/10.1007/s11430-016-5073-0

Wang Y, He J, Wang S, Luo C, Yin H, Zhang G (2017) Characterisation and risk assessment of polycyclic aromatic hydrocarbons (PAHs) in soils and plants around e-waste dismantling sites in southern China. Environ Sci Pollut Res 24:22173–22182. https://doi.org/10.1007/s11356-017-9830-7

Wania F, MacKay D (1996) Tracking the distribution of persistent organic pollutants. Environ Sci Technol 30:390A–396A. https://doi.org/10.1021/es962399q

Weber R, Gaus C, Tysklind M, Johnston P, Forter M, Hollert H et al (2008) Dioxin- and POP-contaminated sites – contemporary and future relevance and challenges: overview on background, aims and scope of the series. Environ Sci Pollut Res 15:363–393. https://doi.org/10.1007/s11356-008-0024-1

Wei YL, Bao LJ, Wu CC, He ZC, Zeng EY (2014) Association of soil polycyclic aromatic hydrocarbon levels and anthropogenic impacts in a rapidly urbanizing region: spatial distribution, soil-air exchange and ecological risk. Science of the Total Environment, 473-474:676–684. https://doi.org/10.1016/j.scitotenv.2013.12.106

Wu SP, Yang BY, Wang XH, Yuan CS, Hong HS (2014) Polycyclic aromatic hydrocarbons in the atmosphere of two subtropical cities in Southeast China: seasonal variation and gas/particle partitioning. Aerosol Air Qual Res 14:1232–1246. https://doi.org/10.4209/aaqr.2013.01.0015

Wyrwicka A, Steffani S, Urbaniak M (2014) The effect of PCB-contaminated sewage sludge and sediment on metabolism of cucumber plants (*Cucumis sativus* L.). Ecohydrol Hydrobiol 14:75–82. https://doi.org/10.1016/j.ecohyd.2014.01.003

Xiao Y, Tong F, Kuang Y, Chen B (2014) Distribution and source apportionment of polycyclic aromatic hydrocarbons (PAHs) in forest soils from urban to rural areas in the Pearl River Delta of southern China. Int J Environ Res Public Health 11:2642–2656. https://doi.org/10.3390/ijerph110302642

Yong RN, Mulligan CN (2003) Natural attenuation of contaminants in soils. CRC Press, Baton Rouge

You Y (2015) Concentrations and distributions of organic pollutants in typical vegetable fields in Guangzhou, China. Jilin Daxue Xuebao (Diqiu Kexue Ban)/Journal of Jilin University (Earth Science Edition), 45:1205–1216. https://doi.org/10.13278/j.cnki.jjuese.201504207

Zhang XL, Tao S, Liu WX, Yang Y, Zuo Q, Liu SZ (2005) Source diagnostics of polycyclic aromatic hydrocarbons based on species ratios: a multimedia approach. Environ Sci Technol 39:9109–9114. https://doi.org/10.1021/es0513741

Zhang T, Wan H, Yang G, Gao Y, Luo W (2008) Distribution of polycyclic aromatic hydrocarbons in agricultural soil and vegetables of Foshan city in the Pearl River Delta. Huanjing Kexue Xuebao/Acta Sci Circumst 28:2375–2384

Zhang D, Cao SP, Sun JL, Zeng H (2014) Occurrence and spatial differentiation of polycyclic aromatic hydrocarbons in surface soils from Shenzhen, China. Huanjing Kexue/Enviro Sci 35:711–718

Zhang D, Wang J, Zeng H (2016) Soil polycyclic aromatic hydrocarbons across urban density zones in Shenzhen, China: occurrences, source apportionments, and spatial risk assessment. Pedosphere 26:676–686. https://doi.org/10.1016/S1002-0160(15)60076-5

Zhong Y, Overcash MR, McPeters AL (1994) Photodegradation of TCDD in soil containing organic solvents. Environ Geochem Health 16:235–240. https://doi.org/10.1007/BF01747921

Zhu Q, Wu Y, Zeng J, Wang X, Zhang T, Lin X (2019) Influence of bacterial community composition and soil factors on the fate of phenanthrene and benzo[a]pyrene in three contrasting farmland soils. Environ Pollut 247:229–237. https://doi.org/10.1016/j.envpol.2018.12.079

Chapter 8
Soil Biological Processes in Urban Soils

Deirdre Bridget Gleeson

Abstract Globally soils harbour a significant diversity and abundance of organisms, ranging in size from nanometre- and micrometre-sized bacteria and archaea, to larger soil fauna such as mites, springtails, ants and earthworms. Together these organisms are responsible for a significant proportion of the ecosystem services provided by soils. Urban soils are a growing proportion of the global soil atlas, and thus it is critical that we understand key dynamics that sustain urban ecosystems. Urban soils can be highly variable in terms of the degree to which they have been influenced by anthropogenic activities; each of these differing urban soil types has its own unique challenges and thus differing impacts on the types of organisms present and the functions they perform. In this chapter we discuss different types of urban soil environments and provide an overview of the diversity of organisms present with a focus on how features of urban soils impact on soil biological diversity. We include an overview of the different methods available to measure soil organisms as well as an insight into which soil organisms may benefit human health and well-being in the long term.

Keywords Ecosystem services · Urban soils · Soil habitat · Habitat structure · Biodiversity · Urban green infrastructure · Measurement · Metagenomics · Bioinformatics · Multivariate · Ordination

---

D. B. Gleeson (✉)
UWA School of Agriculture and Environment, The University of Western Australia, Perth, WA, Australia
e-mail: deirdre.gleeson@uwa.edu.au

A. W. Rate (ed.), *Urban Soils*, Progress in Soil Science,
https://doi.org/10.1007/978-3-030-87316-5_8

What you could learn from this chapter:

- The general features of urban soils which affect the populations and diversity of soil organisms.
- The different types of urban soil environment and what their characteristics are.
- An overview of the diverse range of organisms present in soils, with definitions and examples of each group.
- The advantages and disadvantages of various methods for measuring soil biological populations and diversity, including advanced molecular techniques, and how to analyse soil biological data.
- How the diverse community of soil organisms interacts with different urban environments.
- New ways in which soil organisms may benefit human health and well-being.

8.1 Urban Soils

Urbanisation will have a significant effect on the biological properties and functioning of soils in urban environments. As discussed in Chap. 1, currently approximately 55% of humans reside in urban areas (Schmidt et al. 2019) with this figure projected to move closer to 70% by the year 2050 (United Nations 2018). Thus, urban land cover globally is projected to grow substantially indicating that the role of cities, and urban spaces more generally, in contributing to soil ecosystem services will increase (Seto et al. 2012; Norton et al. 2019).

Soils contain a huge diversity and abundance of organisms, and, with urban soils now encompassing more of the soil environment than ever before, it is critical that we understand the key dynamics at work in sustaining urban ecosystems (Mummey et al. 2006; Maron et al. 2011). Urban soils are highly heterogeneous with soil conditions varying from relatively undisturbed soils, for example, older parks and remnant vegetation patches, to completely degraded soils, for example, those contaminated with pollutants. We also saw in Chap. 3 how this soil variability can occur over small spatial scales, even within patches of a particular land use type. Such soils are significantly impacted by local environments, local land uses and local anthropogenic conditions (Fig. 8.1). In terms of their ability to support soil biological functions, urban soils can be characterised by (1) sealing via concrete, asphalt or bitumen, (2) significant compaction and potential to become *anoxic*, (3) contamination by a wide range of pollutants and (4) a potentially high degree of mixing and/or movement (De Kimpe and Morel 2000; Lehmann and Stahr 2007). The conceptual diagram in Fig. 8.1 depicts a typical urban scene and forms the basis for our descriptors of soil biology in an urban context. Compaction and sealing (e.g. from roads and bitumen) both decrease infiltration of water and air, as well as organic matter decomposition and nutrient cycling, while contamination by pollutants can have a wide diversity of effects across the soil biosphere (McKinney 2002; Sikorski et al. 2013). Many soils in urban environments are deficient in nutrients such as nitrogen (N) and potassium (K) (Guilland et al. 2018). As a result the topsoil, the area of most biological activity, is usually the most impacted. As a consequence of soil physical and chemical constraints, there may be a decrease in habitat

Fig. 8.1 Conceptual diagram highlighting the diversity of urban environments, the habitats they provide, organisms present in those habitats and their functions in relation to ecosystem service provision. Illustration by Ooid Scientific

availability for soil organisms resulting in a decrease in overall soil function and soil quality (Didden et al. 1996; Amosse et al. 2016; Guilland et al. 2018). In addition, modified resource availability, human disturbance and competition between exotic and native species can influence the types of organisms present in urban environments and the ecosystem processes they mediate (McKinney 2006).

8.2 Ecosystem Services Provided by Urban Soils

As outlined in Chap. 4, soils provide critical ecosystem services that underpin human society (Fig. 8.2). These include, but are not limited to, regulation of water infiltration, provision of a physical matrix and nutrients for the growth of plants and soil biota, biogeochemical and nutrient cycling (e.g. *nitrification* and *denitrification*, carbon sequestration and moderation of greenhouse gas emissions) as well as

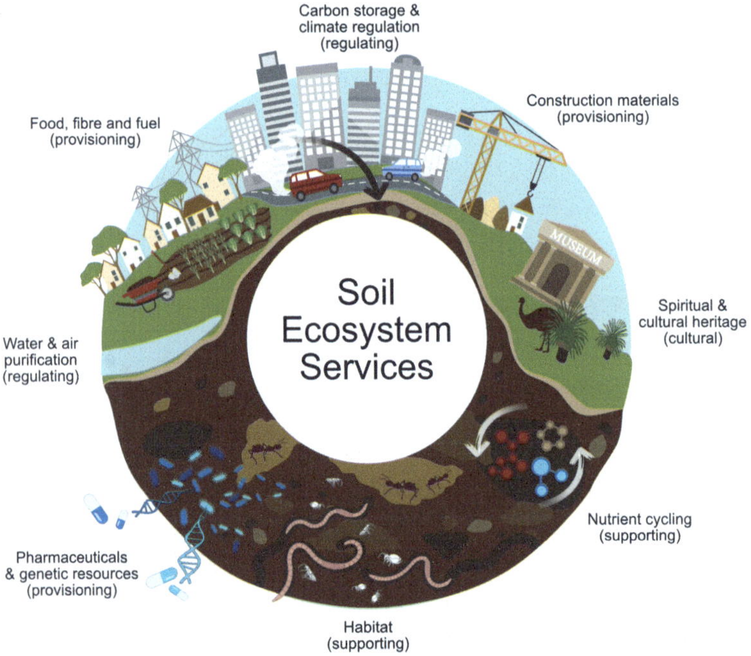

Fig. 8.2 Schematic diagram of soil ecosystem services. Ecosystem services are classically divided into four categories [MEA 2005: https://www.millenniumassessment.org/en/index.html]: (1) support services, e.g. supporting development of plant communities or construction; (2) provisioning services, e.g. food, fibre, etc.; (3) regulating services, e.g. regulation of air quality and climate; (4) cultural and social services, e.g. heritage and educational services. Soil organisms are critical for the functioning of soil ecosystems, and in fact the key to maintaining soil ecosystems and their functions is not simply the contribution of individual organisms, but rather the combined actions and interactions of the enormous diversity of organisms found in soil. Illustration by Ooid Scientific

waste decomposition and pollutant decontamination (Devigne et al. 2016; Rodríguez-Eugenio et al. 2018; Fig. 8.2). The delivery of these ecosystem services depends on the biodiversity contained within the soil, and even small or localised modifications to the soil environment can affect the provision of ecosystem services. Although we understand quite well the links between biodiversity and ecosystem services in natural and agricultural environments, these are less well defined for urban soils which now make up 2.6% of the globe, with 0.65% in built-up areas and 0.45% impervious surfaces (Liu et al. 2014). Thus knowledge that we have gained about rural and agricultural systems cannot be simply transferred to, and conclusions drawn about, urban environments. We should also acknowledge that any disruption of the soil ecosystem resulting in a degradation of these ecosystem services could have significant adverse effects on human well-being (European Commission 2006). The following material (Sect. 8.3) therefore addresses the important types of urban soil habitat and their characteristics.

8.3 Types of Urban Soil Environments and the Habitat they Provide

In urban environments soils are commonly formed during the urbanisation process and can be referred to as *anthropogenic* soils. That is, urban soils are modified by human activities (Chap. 2) and are fundamentally different, in terms of providing a biological habitat, from natural soils (Guilland et al. 2018). Such soils can be characterised by the combination of a wide variety of substrates, for example, nutrient rich materials, sand and/or clay, brick debris and compacted rubble. Gradients in such substrates can result in heterogeneous soil characteristics, for example, moisture, penetrability, pH or mineral content. Current and former industrial activities in urban environments include an array of potential contaminant sources such as fossil fuel combustion, industrial wastes and discharges of various chemicals. Urban soils are thus often enriched with trace metals and persistent organic pollutants (POPs: see Table 7.2) which can find their way into the food chain and *bioaccumulate* within distinct *trophic levels* (Wei and Yang 2010; Luo et al. 2012). Urban soil environments are a patchwork of different sized habitats that are subject to varying degrees of environmental stress including, but not limited to, heat, pollutants, disturbances and habitat fragmentation that can alter the resident organisms and their functions. Additionally, since many soils in urban environments can be highly compacted, and/or deficient in major nutrients, they are often not able to support biodiverse communities. To restore soil fertility and suitable physical properties, urban soils may therefore require amendment, for example, the addition of fertile topsoil or composts. Such amendments enhance available soil nutrients and soil structure which can lead to increased colonisation by soil organisms (Guilland et al. 2018). In addition, the amendments themselves may contain a diverse soil biological community which may prove to be beneficial to the already resident community (e.g. Meng et al. 2019).

In the urban context parkland soils are the most studied, but parklands do not always constitute the largest component of green space that houses soil organisms (Thompson and Kao-Kniffin 2019). Soils in built-up areas that are contained within or below impermeable surface cover, such as street medians and tree pits, can often constitute an understudied component of urban soil biodiversity (Reese et al. 2016). These spaces are often not considered to be relevant as they are unlikely to provide habitat to many vertebrates which are the focus of most urban biodiversity research. However, such habitats can support a variety of soil invertebrates and microorganisms that contribute to ecosystem functioning, and thus they may be more important in that context. There are six main types of soil habitats generally found in urban environments. Each of these habitats provides a specific environment for soil fauna (invertebrates and microorganisms) with unique characteristics that influence the types of organism present, their diversity and abundance. The six habitat types we will consider are grasslands (including parks, gardens and sports grounds); urban gardens (including community and collective gardens); urban wastelands and unoccupied land; road verges, street medians and tree pits; covered soils; and urban green infrastructure (including green roofs, bioswales and other engineered habitats). See Fig. 8.3 for exemplar images of each of these different habitat types.

8.3.1 Grasslands (Parks, Gardens and Sports Grounds)

Urban grassland ecosystems are generally comprised of perennial grasses grown and managed for aesthetic, recreational or functional purposes. These grasslands range from extensively managed golf courses to infrequently maintained institutional or residential grassed areas (Thompson and Kao-Kniffin 2019). Such spaces can provide functional and biodiverse green areas, depending on their individual management, that are essential for creating liveable cities. They can be heterogeneous in structure, depending on the type of vegetation present, and can provide favourable habitats for some invertebrates and soil microorganisms. Despite urban parks generally being dominated by a small number of grass species, these parks can be responsible for a significant proportion of urban plant diversity. It has been reported that per unit area lawns harbour a *species richness* (i.e. number of plant species) that is equivalent to some semi-natural grasslands (Bertoncini et al. 2012). However, some of these park grasses provide limited vegetation structure which can lead to a decrease in the *diversity* of many invertebrate species relative to the more structurally complex natural grasslands (Jerrentrup et al. 2015). This lower biodiversity results from direct effects of decreased habitat complexity, in addition to trampling by humans, fertiliser use, mowing and seed set (e.g., Humbert et al. 2010). The biodiversity of urban grassland habitats may be increased by introducing areas of 'meadow' vegetation, that is, areas of infrequently mown grasslands

Fig. 8.3 Above-ground views of the six main types of urban soil habitat: (1 = grassland; 2 = garden; 3 = wasteland; 4 = road verge/tree pit, etc.; 5 = sealed; 6 = green infrastructure). (Photo credits (all CC-BY licences): 1 Ivan Rohovchenko on unsplash; 2 Linda on flickr; 3 Matt Brown on flickr; 4 Tom Zimmerman on unsplash; 5 John Peterson on flickr; 6 Chuttersnap on unsplash)

containing a mix of flowering natives which would replace conventional park grass (Norton et al. 2019).

8.3.2 Urban Gardens (Residential, Community and Collective Gardens)

Gardens in urban areas form a significant proportion (up to 36%) of urban green space, mainly in the form of private residential gardens (Farahani et al. 2018). Gardening contributes to improved individual and community well-being, can be a significant source of food production and may introduce biodiversity which is absent from other urban habitats (Orsini et al. 2013; Taylor and Lovell 2015). Despite the positive aspects of urban gardens, however, there may be constraints on their suitability as habitats for soil organisms. Many urban garden soils are known to be contaminated with a range of substances including metals and persistent organic pollutants from historical activities (Alloway 2004; Laidlaw et al. 2018). Contamination of urban gardens may also be ongoing, for example, if airborne pathways from contaminant sources exist (Clark et al. 2008). To overcome contamination issues for food production, imported and/or manufactured soil materials may be added and gardening performed in raised beds. Amendment with organic materials such as composts and manures is very common, and may result in increased nutrient contents and favourable soil physical properties, as well as avoiding legacy contamination. In some cases, nutrient content in urban garden soils may be excessive (Gregory et al. 2016). The frequent cultivation and amendments applied to urban garden soil habitats create a particular physical structure which affects the range of habitats available for soil organisms. For example, mulching affects soil physical properties such as temperature and water content, with consequent effects on the abundance and activity of soil organisms (Byrne 2007). Soil temperature and water content are also expected to be modified in raised garden beds, with the increased surface area of soil exposed to air temperatures causing daytime warming with consequent greater loss of soil water. Biodiversity of soil organisms may be decreased in urban gardens by the presence of large numbers of invasive species, such as weeds or introduced fauna. This in turn increases the abundances of invasive invertebrates (e.g. garden snails). For example, Uno et al. (2010) found species richness of ants to be lower in urban gardens than in urban forests, and suggested that this may have been related to competition by an invasive ant species.

8.3.3 Urban Wastelands and Unoccupied Land

Vacant land or wasteland represents a substantial and growing habitat globally that is potentially of great importance in terms of biodiversity, as they can contain a rich flora and fauna including rare or cryptic species (Bonthoux et al. 2014). Wasteland areas are usually dynamic across both space and time. That is, the location of the patches of wasteland harbouring biodiversity can change depending on the time scales of construction and demolition, and successional changes can take place within vegetated patches of long-term wastelands. Some wastelands have a greater species diversity than other urban green spaces, for example, lawns, managed parklands and some urban forests (e.g. Robinson and Lundholm 2012). One reason for this could be the minimal or absent management of urban wastelands which encourages an enhanced plant diversity (bearing in mind though that some vacant plots may be mown semi-regularly for fire management purposes, for example). Vacant land is often ignored in conservation planning, possibly because many waste or vacant land areas in cities are also brownfield sites which may be contaminated (Thornton et al. 2008) and many are subject to complex land ownership issues. In addition to the acknowledged vegetative biodiversity, several invertebrate studies across Europe have reported that urban wastelands, if uncontaminated, can behave a little like so-called natural habitat analogues and thus could be valuable reservoirs of biodiversity (Eversham et al. 1996; Harrison and Davies 2002; Kattwinkel et al. 2011). A number of studies have shown vacant lands or former industrial sites to harbour rare and vulnerable species (Eyre et al. 2003; Gardiner et al. 2013). For example, in the United Kingdom Eyre et al. (2003) reported 182 sightings of 46 rare beetle species across urban wasteland sites. Usually these beetle species have only been sighted in natural habitats, for example, grasslands, lake and river edges, and heathlands.

8.3.4 Road Verges, Street Medians and Tree Pits

Road verges, street medians and tree pits, often located next to paved, concreted or covered surfaces, are receptors for runoff and contaminants, and yet they can harbour significant soil faunal and microbial diversity (Kim and Yoo 2021; Guilland et al. 2018). Soils in these roadside areas usually suffer from compaction, erosion and potentially dumping of waste. Roadside soils are also usually confined in terms of individual area, meaning they are not often considered as valuable contributors to urban soil biodiversity due to their inherently degraded quality (Ow and Ghosh 2017). The likelihood of contaminants being present in urban roadside soils has meant that research has primarily focussed on metals such as lead, cadmium and nickel, with less research focussing on biological properties (e.g. organic matter, microbial communities or even nutrients) of these soils. Classical lawns which are found alongside streets and roads are usually carefully structured and maintained by regular mowing, fertilisation and watering. In contrast, other roadside verges are left to wild and maintained only by irregular mowing and provide a quite different habitat for soil organisms. Ornamental flower cover or native vegetation is also used, particularly within roundabouts, or along highways and street verges – this type of plant portfolio will also influence the types of soil dwelling organisms. This is particularly the case for soil microorganisms, which will be influenced by fertilisation, soil amendments, plant root type and root exudate profile (Guilland et al. 2018).

8.3.5 Covered Soils

Soil sealing may involve total impermeability, as caused by concrete or asphalt, or soil may be sealed with a semi-pervious surface such as paving slabs, which allow partial penetration of water and air. Soil sealing has a significant impact on the functioning of soil, causing an irreversible loss of its biological functions and loss of biodiversity through landscape fragmentation. Most previous studies have focussed on effects of sealed areas on water movement, gas diffusion and vertebrate biodiversity (e.g. Morgenroth and Buchan 2009; Grella et al. 2018), while the effects of soil sealing on nutrient cycling and biological components, such as soil microbial diversity, have seldom been investigated (Bhaduri et al. 2001; Peffy and Nawaz 2008; Zhao and Xia 2012). The sealing of soils prevents the exchange of gases, water and nutrients between the soil and the atmosphere, which results in a negative effect on their physicochemical properties (Zhao and Xia 2012). This, in turn, would be expected to have a negative influence on invertebrate and microbial activities. Piotrowska-Dlugosz and Charzynski (2015) showed that artificial sealing in urban areas significantly altered the soil environment, resulting in less favourable biological and physicochemical properties where soils under impervious surfaces had lower water content, less soil carbon and nitrogen, and lower microbial biomass than adjacent unsealed soils. Recent innovations in using semi-permeable materials for roads, pavements, driveways and carparks to increase water infiltration can be expected to have a positive impact on soil biodiversity and functioning, but to date this technology has only been applied on a small scale.

8.3.6 Urban Green Infrastructure (Green Roofs, Bioswales and Other Engineered Habitats)

The growth of cities naturally results in a loss of green spaces; this may be overcome via the use of *green infrastructure* (GI), defined as engineered green habitats (Francis and Lorimer 2011; Gill et al. 2020). Green infrastructure can include green roofs, *bioswales* (storm water runoff areas) and any habitat engineered to contain soil, improve stormwater infiltration and restore urban biodiversity. GI sites are generally designed to provide semi-natural environments that could help increase the diversity of vegetation and fauna as well as that of soil dwelling invertebrates and microorganisms. GI installations are usually filled with engineered *Technosols* that have an optimum texture, nutrient content and pH (Joyner et al. 2019). Compared with other urban soils, GI sites are usually planted with specific vegetation, are protected from pedestrians and traffic, have prescribed resource and water fluxes associated with stormwater intake and specified monitoring and maintenance, depending on the type of GI installed (NYC Department of Environmental Protection 2017). Bioswales and green roofs, as examples of typical GI installations, have been the subject of a number of studies of biogeochemical cycling (e.g. Berndtsson 2010), although there has been a lack of studies on their potential as habitats for biodiversity (see Schrader and Böning 2006 as an example). Green roofs in particular are usually protected from the usual types of contaminants observed in urban soils due to their raised locations (Braaker et al. 2014). However, excluding green roofs, it has also been reported that other GI infrastructure soils have unexpectedly low concentrations of metal contaminants or organic contaminants such as petroleum derived polycyclic aromatic hydrocarbons (PAHs) despite their proximity to heavily used roads (Deeb et al. 2020).

8.4 Organisms Present in Urban Soils

Although there has been a significant accumulation of knowledge around soil biodiversity, research has focussed more on agricultural and natural systems with little work focussing on urban environments (De Kimpe and Morel 2000). This is despite urban soil habitats being among the most rapidly expanding biomes on Earth. We do, however, know that there is an enormous diversity of organisms, from animals and plants to microorganisms, present in urban environments (Fig. 8.4). Most biodiversity studies focus on above-ground organisms (e.g. Koh and Sodhi 2004; Cannon et al. 2005); however, the majority of terrestrial biodiversity is buried in the soil (Bardgett and van der Putten 2014). Thus, although urban soils support a variety of animals and plants, for the purposes of this text we will focus only on soil dwelling fauna and the urban soil microbiome which includes bacteria, archaea, fungi and protists as well as non-living viruses that do not possess all of the traits of living organisms.

Soil organisms are generally divided by size into three categories (Fig. 8.5):

1. *Microfauna* including nematodes and microorganisms (bacteria, archaea and fungi) that are only visible with a microscope (Fig. 8.5a, b, c).
2. *Mesofauna*, for example, mites and springtails, that are visible with a magnifying glass (Fig. 8.5d–h).
3. *Macrofauna* such as ants and termites, which are visible with the naked eye (Fig. 8.5i-o).

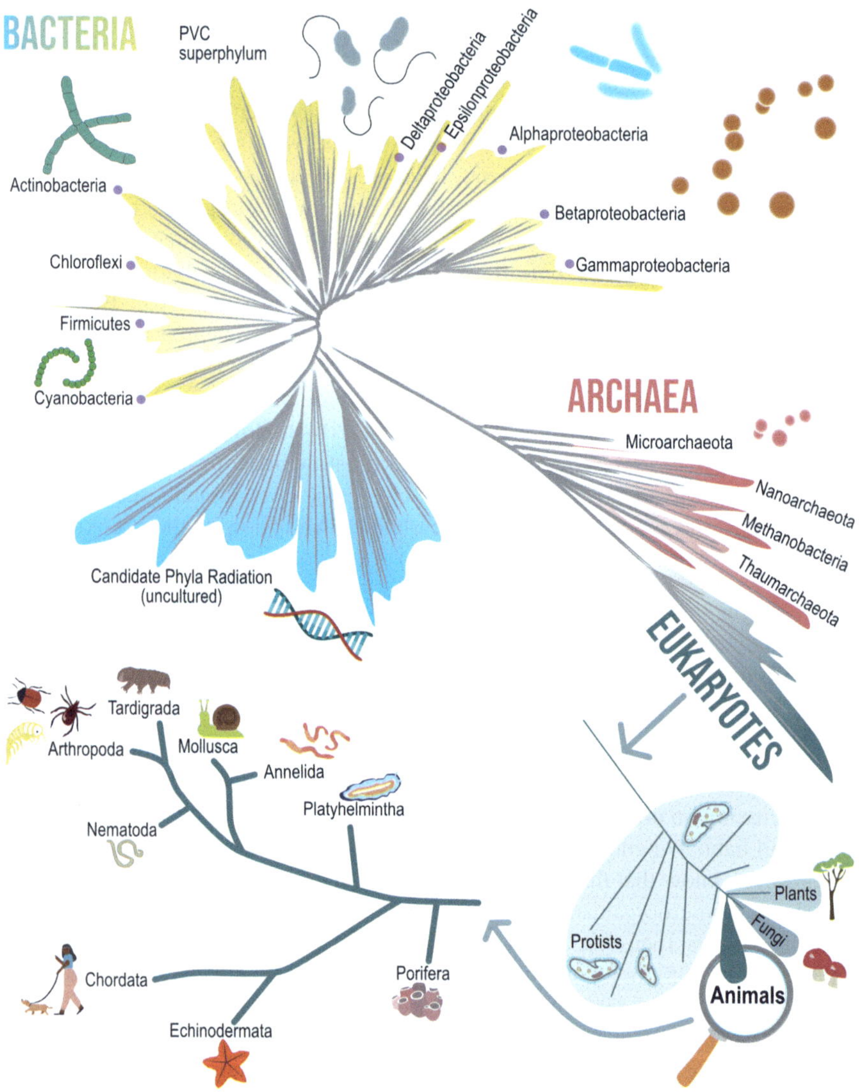

Fig. 8.4 This figure shows the three main domains of the Tree of Life: Bacteria, Archaea and Eukaryotes. There are seven main taxonomic ranks: kingdom (now regularly replaced by domain), phylum, class, order, family, genus and species. When an organism is given a species name, it is assigned to a genus, and the genus name is part of the species name, for example, *Homo sapiens* (humans): Homo is the genus name and sapiens indicates the species. Here the main structure of the Tree of Life has been adapted from Hug et al. (2016) and has been constructed using the total diversity represented by sequenced genomes. This includes the 92 named Bacterial phyla, 26 named Archaeal phyla and all 5 of the Eukaryotic supergroups. The additional detail of the Eukaryotic branch is not phylogenetically coherent and has been provided by way of reference only to highlight some of the soil organisms that are covered in this text. Illustration by Ooid Scientific

Fig. 8.5 Examples of the range of organisms present in soils: (**a**) mycorrhizal fungi; (**b**) decomposer fungi; (**c**) bacteria; (**d**) nematode; (**e**) tardigrade; (**f**) collembola; (**g**) mite; (**h**) enchytraeids (potworms); (**i**) millipede; (**j**) centipede; (**k**) earthworm; (**l**) ant; (**m**) woodlouse; (**n**) flatworm; (**o**) wombat. Image credits: (**a**) Armen Phelps on Wikimedia Commons (CC-BY-4.0); (**b**) Marlon Winger (USDA); (**c**) CDC/PHIL on Wikimedia Commons; (**d**) K-State Research and Extension on flickr; (**e**) Philippe Garcelon on flickr; (**f**) Erop Kamelev on unsplash; (**g**) Jean&Fred on flickr; (**h**) Tamas Salanki (© used with permission); (**i**) Rashid Grey on unsplash; (**j**) Schizoform on flickr; (**k**) Katja Schulz on flickr; (**l**) Mikhail Vasilyev on unsplash; (**m**) Katja Schulz on flickr; (**n**) Jean&Fred on flickr; (**o**) Steven Penton on flickr. (All flickr licenses CC-BY-2.0; otherwise if no license is shown, images are in the public domain)

Please refer to the images displayed in Fig. 8.5 when reading the material presented in Sects. 8.4.1 and 8.4.2.

8.4.1 Soil Invertebrates

Invertebrate communities in urban soils include a wide variety of micro- and macro-invertebrates, for example, ants, earthworms, arthropods and nematodes. These organisms are generally between 25 μm and 4 cm in size (see Fig. 8.6) and play a significant role in the decomposition and redistribution of soil organic matter and cycling of nutrients as well as regulating the soil microbiome (bacteria, archaea, fungi and viruses) and interacting with plant roots (Table 8.1).

Earthworms and Potworms are soil dwelling worms with segmented bodies that both belong to the phylum Annelida (Domain: Eukaryote). Both are widespread in urban soils, and adverse effects on their populations and diversity are used extensively as indicators of soil contamination. Potworms (enchytraeids) are generally

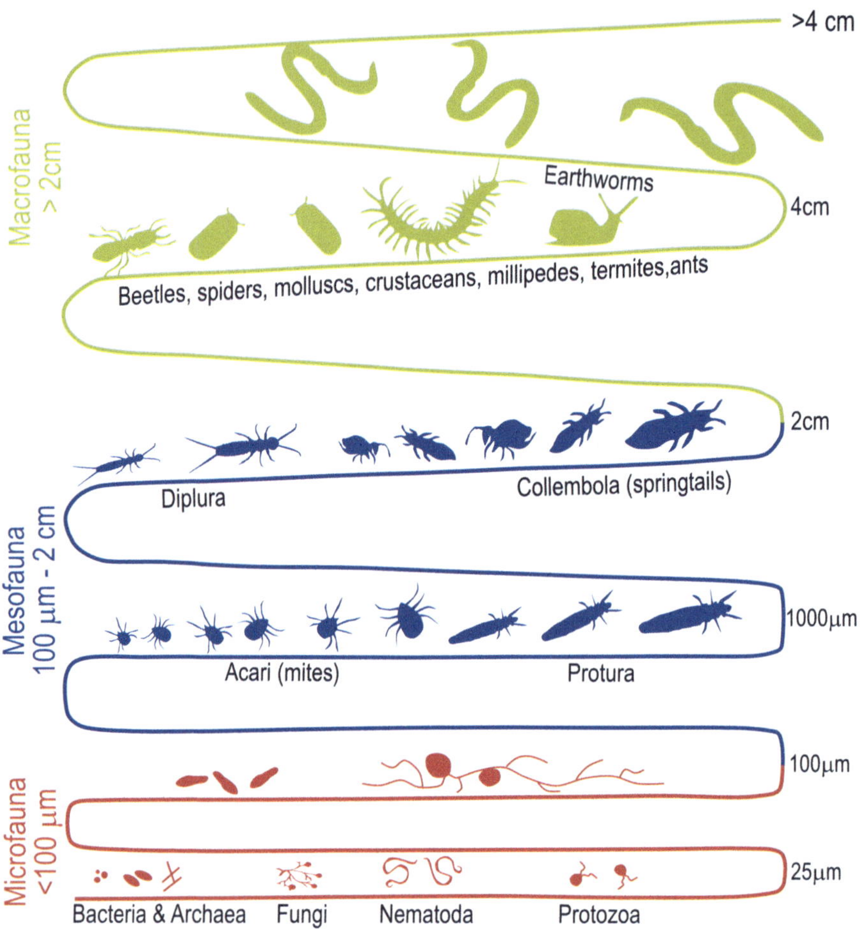

Fig. 8.6 Classification of soil organisms according to size (adapted from Decaens 2010). Illustration by Ooid Scientific

smaller than earthworms and are less useful as *bioindicator* species as they have been reported to be more tolerant of environmental conditions (Didden et al. 1996). Potworms are similar to earthworms in their mode of action in soil: they modify their environment through burrowing and feeding activities, and they contribute significantly to litter fragmentation and organic matter deposition, as well as aerating the soil. They make a substantial contribution to soil structure and aggregation and, in addition to earthworms, are often referred to as soil engineers (Amosse et al. 2016). Globally there are approximately 3500 described earthworm species, although most have a restricted distribution (Csuzdi 2012). One of the largest earthworms in the world is the Giant Gippsland earthworm (*Megascolides australis*) which is found in Victoria, Australia, and grows up to 2–3 m long. Due to their

Table 8.1 Selected factors affecting invertebrates in urban soils; invertebrate response is to an increase in the factor (table adapted from Bray and Wickings 2019)

Factor	Invertebrate	Metric	Response	Urban environment	Reference
Soil pH	Macroinvertebrates	Species totals (Lumbricidae, Chilopoda, Diplopoda, isopoda, Formicidae)	Positive	Horticultural borders, mown grass lawns	Smith et al. (2006)
Soil temperature	Springtails	Total density and diversity	Negative	Green roofs	Rumble and Gange (2013)
Soil moisture	Springtails	Total density and diversity	Positive	Green roofs	Rumble and Gange (2013)
Soil bulk density (indicative of compaction)	Earthworms	Total density (average number per unit area)	Negative	Residential garden turfgrass, urban park turfgrass	Smetak et al. (2007)
Plant input chemistry (carbon/nitrogen ratio)	Springtails	Total abundance	Negative	Landfill, city garden	Vauramo and Setala (2010)
	Springtails	Total abundance	Positive	Landfill, urban park	Vauramo and Setala (2011)
	Nematodes	Abundance parasitic nematodes	Negative	Landfill, city garden	Vauramo and Setala (2010)
		Abundance fungal feeders	None	Landfill, city garden	Vauramo and Setala (2010)
Management intensity	Nematodes	Total, free-living, plant parasites, number of genera	Negative	Community garden	Grewal et al. (2010)
	Nematodes	Total, free-living, plant parasites, number of genera	None	Vacant land	Grewal et al. (2010)

(continued)

Table 8.1 (continued)

Factor	Invertebrate	Metric	Response	Urban environment	Reference
Addition of organic matter	Earthworms	Total number	Positive	Lawn, mulched (bark or gravel)	Byrne et al. (2008)
	Earthworms	Total density	Positive	Residential garden turfgrass, urban park turfgrass	Smetak et al. (2007)
	Macroinvertebrates	Species totals (Lumbricidae, Chilopoda, Diplopoda, isopoda, Formicidae)	Positive	Horticultural borders, mown grass lawns	Smith et al. (2006)
Metal contamination	Earthworms	Total abundance	Negative	Residential lawns	Pouyat et al. (2015)
	Earthworms	Species density (*Aporrectodea caliginosa*)	Negative	Polluted (cadmium, zinc) site	Nahmani and Lavelle (2002)
Pesticides	Springtails	Total abundance	Negative	Golf turfgrass	Gan and Wickings (2017)
	Mites	Total oribatid abundance	Negative	Golf turfgrass	Gan and Wickings (2017)
	Springtails	Total abundance	Negative	Residential lawn	Peck (2009)

relatively long life spans (taking 5 years to reach sexual maturity), these worms are now listed as Endangered on the IUCN list.

Arthropods have an exoskeleton, segmented bodies and usually paired jointed appendages. They belong in the phylum Arthropoda (Domain: Eukaryotes) and include, but are not limited to, spiders; millipedes; ticks; mites; insects including ants, termites, and beetles; tardigrades; isopods; and springtails. Arthropods can be divided according to their function in soil, for example, decomposers such as millipedes consume dead plant matter, predators such as mites prey on a variety of small soil organisms including nematodes, while others feed on soil fungi, for example, springtails (Gobat et al. 2004). Soil-dwelling ants and termites, which often replace earthworms in hotter drier habitats, disturb soils while making tunnels and nests and again are sometimes referred to as *soil engineers*. This soil movement, or *bioturbation*, has a major effect on horizonation, structure and distribution of organic matter as well as increasing aeration and water infiltration in soil (Gobat et al. 2004; Amosse et al. 2016).

Nematodes are multicellular organisms with smooth unsegmented bodies. They belong in the phylum Nematoda (Domain: Eukaryotes) and consist of both parasitic and free living species. Nematodes are a key component of soil food webs. They are involved in nutrient cycling as well as regulation of the microbial biomass, and their structure and composition in soil are often used as indicators of environmental disturbance (Gobat et al. 2004; Amosse et al. 2016).

Protozoa are single-celled eukaryotes that feed on organic matter found in soil such as organic debris, tissues, other protozoa or microorganisms including bacteria and archaea and less often fungi. They can be either free-living or parasitic, are grouped together and informally classified into three groups based on their shape: (1) Flagellates are the smallest of the protozoa and use a few whip-like flagella to move; (2) Amoebae are the next largest of the protozoa and move by means of a temporary foot or "pseudopod"; (3) Ciliates are the largest of the protozoa and move by means of hair-like cilia and can consume the other two types of protozoa as well as microorganisms.

8.4.2 Soil Microbiome

The urban soil microbiome is a diverse assemblage of microorganisms that includes bacteria, archaea, fungi and viruses that together are responsible for a vast number of ecosystem services, including both global and local biogeochemical cycling, decomposition and mediation of greenhouse gas emissions (Falkowski et al. 2008; Bray and Wickings 2019).

Bacteria are single-celled, microscopic organisms that inhabit virtually all environments including soils, oceans, in and on animals, plants and humans. There are anywhere from 100 million to one billion bacteria in just a teaspoon (5–10 g) of moist, fertile soil (Churchman and Landa 2014). They are classified as prokaryotes, as they are single celled, have no nucleus and contain DNA that floats freely as a mass called the nucleoid, or as circular pieces of DNA known as plasmids. *Ribosomes*, small spherical organelles where proteins are made, are present in all bacterial cells. Most bacteria also have an outer cell wall and an inner cell membrane, and some bacteria have *flagella* or short pili that allow them to move around and attach to surfaces. It is virtually impossible to estimate the number of different bacterial species on the planet, but it is generally accepted that there are likely over 60,000 different species of bacteria, most which have yet to be even named. The advent of advanced molecular biology technologies has meant that bacterial phylogeny is now a rapidly moving field, and currently there are 89 formally recognised bacterial phyla within the global Silva database (https://www.arb-silva.de/) including recognisable soil associated phyla such as Actinobacteria, Proteobacteria, Acidobacteria and Cyanobacteria as well as less well-recognised phyla including Dadabacteria, Woesebacteria and Torobacteria (Hug et al. 2016). It is estimated that as many as 72% of widely recognised bacterial phyla have no cultured representative and these are referred to as "candidate phyla" (e.g. see Candidate Phylum Radiation – Fig. 8.4) (Hug et al. 2016).

Archaea are single-celled organisms with no nucleus or organelles but also containing a nucleoid and ribosomes. It is the analysis of their ribosomes that identified them as a separate domain of organisms from the bacteria (Woese et al. 1990). Originally they were thought to dominate in extreme environments such as hot springs and hypersaline salt lakes. New genetic analysis techniques, however, have identified these organisms across all environments on earth, although they are not as numerically dominant as the bacteria. Archaea have different cell membrane biochemistry which separates them from bacteria, and they also possess a range of metabolic pathways that are more closely related to those of eukaryotes, in particular those involved in *transcription* (conversion of DNA to messenger RNA) and *translation* (conversion of messenger RNA to proteins). Again the identification and classification of archaea is a rapidly moving field, and estimates of the number of phyla range from 18 to 23, of which only 8 have cultured representatives (Castelle and Banfield 2018).

Fungi form part of the eukaryotic domain and are distinct from plants and bacteria in that they have chitin (a polysaccharide, or long chain polymeric carbohydrate) in their cell walls. There are three main functional groups: (1) multicellular filamentous moulds; (2) macroscopic filamentous fungi that form large fruiting bodies (e.g. mushrooms); and (3) single-celled microscopic yeasts. Classically, fungi were divided based on morphology (e.g. spore colour or other microscopic features). As for bacteria and archaea, however, advances in molecular genetics and DNA analysis of fungi have resulted in new classifications that sometimes challenge historical groupings. There are now at least 7 recognised phyla and numerous subphyla including the Basidiomycota (mushrooms) and, for example, the well-known soil fungal phyla Glomeromycota (arbuscular mycorrhiza). Fungi are found in just about any habitat but most live on the land, mainly in soil and plant material. They perform an essential role in the decomposition of organic matter and play a significant role in nutrient cycling in the environment, although some fungi are parasites of plants or animals that cause disease.

Viruses are infectious agents that replicate only when they are inside the living cells of an organism. They infect all life forms, from animals to plants, and including microorganisms (bacteria, archaea and fungi). They have a wide diversity of shapes and sizes, but are generally much smaller than bacteria and require scanning electron microscopes to visualise them. Viruses consist of *nucleic acid* surrounded by a protein coat called a capsid. They do not have their own metabolic mechanisms and require a host cell to survive. Once infected, a host cell is forced to rapidly produce thousands of identical copies of the original viral genetic material. Viruses in the urban environment can exist in soil organisms, and, given close proximity of these organisms to human life, it is important to understand implications for impacts of viruses in urban areas.

8.5 Measuring Soil Organisms

There is a general lack of clarity on defined methodologies to measure soil organisms, including invertebrates and microorganisms, and there are a variety of methodologies in general use. The issue of methodology is important, as there have been a number of efforts to review and establish a set of biological indicators of relevance to assist policy-makers in land management (Ritz et al. 2009; Pulleman et al. 2012; Faber et al. 2013). Monitoring of these biological indicators requires a consistent methodology to be used. The consensus, however, remains that major efforts are still required to standardise operational procedures and to validate them for different types of land use. This applies equally across land management types, but is particularly relevant in the urban environment where there has been less effort applied to date. Some methods, for example, in situ trapping methods used to sample arthropods, are easily transportable and widely used. However the number of individuals trapped, and variety of species trapped, depends on the sampling effort (i.e. number of traps used, the length of the trapping period as well as the expertise of the taxonomist in identifying the *taxa* present).

8.5.1 Measurement of Soil Fauna

To assess the types of soil fauna present in an urban soil sample, visual identification, often using optical microscopy, is generally required (see Table 8.2). Although the types of methodologies involved are well established (even if not thoroughly benchmarked), they are usually time-consuming and require specialist skills in taxonomy and identification that have become more scarce in recent years (Ritz and Trudgill 1999; Lear et al. 2018). Using traditional identification methods, cryptic and/or rare species can sometimes be misidentified. In addition, whole taxa may be underrepresented or overlooked due to factors such as their small size; nocturnal habits; occurrence in areas which are difficult to access; or difficulty in obtaining representative samples of individual organisms against which to make comparisons (Lear et al. 2018). Additional factors that should be considered include the size of the sample needed, and the method of extraction that will give an unbiased count of the organisms present. More recent molecular methods involving the use of *environmental DNA* (*eDNA*) have been developed to overcome these issues; however, these methods come with their own caveats and biases.

Deoxyribonucleic acid (DNA) is found in all organisms and is the genetic code formed by the association of nucleotides (adenine, guanine, cytosine and thymine) that constitutes the genome containing all the genes that code for proteins. The analysis of DNA extracted from any environment can be used to identify the organisms present in an environmental sample as well as their physiological and metabolic potential. There are a number of caveats that need to be considered; for example, when we extract DNA from an environment, we are capturing both living

Table 8.2 Standard methodologies described for surveying urban soil fauna and the soil microbiome

Invertebrate or microorganism	Metric and title	Methodology (ISO listed)[a]	Reference
Earthworm (natural community, e.g. *Lumbricidae*, *Glossocolecidae*)	Combination of hand-sorting and formalin extraction	1. Dig-out and hand-sorting of the soil within an area of 50 cm × 50 cm × 20 cm depth	Rombke et al. (2006)
	Abundance, biomass, species composition	2. Application of 5–10 L (several times) of a 0.5% aqueous formalin solution into the dug-out hole, waiting 30 min until the worms appear at the soil surface	Smetak et al. (2007)
		3. Collection and fixation in ethanol (70%) for 1–2 days, followed by 1–2 weeks in 4% formalin, then final storage in 70% ethanol	
Potworms Natural enchytraeid field community	Extraction of animals from soil samples using behavioural methods	1. Collect soil samples with a soil-corer in the field	Rombke et al. (2006)
	Abundance, species composition, dominance spectrum, biomass	2. Extract worms using a simple wet-extraction method	Amosse et al. (Amosse et al. 2016)
		3. Perform microscopical species determination	
Microarthropods Natural community of springtails (*Collembola*) and different groups of mites (*Acarina*), e.g. oribatids or gamasids	Dry extraction of animals from soil samples using behavioural methods	1. Collect soil samples with a split-corer (diameter about 6 cm) in the field	Rombke et al. (2006)
	Abundance, biomass, species composition	2. Extract arthropods in a modified Macfadyen apparatus for several days	Rumble and Gange (2013)
		3. Fix in ethanol (70%)	
Nematodes Natural nematodes field community	Wet extraction of animals from soil samples using floatation, sieving and behavioural methods	1. Collect soil samples with a small soil-corer in the field	Oostenbrink (1960)
	Abundance, species composition, functional groups, ecological indexes (maturity index)	2. Create composite samples	Rombke et al. (2006)
		3. Extract worms in a simple wet-extraction method based on the Oostenbrink method (1960)	Vauramo and Setala (2011)
		4. Perform microscopical species identification	

(continued)

Table 8.2 (continued)

Invertebrate or microorganism	Metric and title	Methodology (ISO listed)[a]	Reference
Bacteria, Archaea and Fungi	DNA extraction and analysis	1. Collect soil samples with a small soil-corer in the field	Petric et al. (2011)
	Relative abundance, community structure, species composition	2. Create composition samples and remove 250 mg for extraction	Martin-Laurent et al. (2001) (modified in ISO-11063)
		3. DNA should be extracted using (i) microbial cell lysis by chemical (SDS) and physical (bead beating) action,	
		(ii) deproteination and (iii) alcohol precipitation and washing of the nucleic acids extracted (see ISO-11063)	
		4. Sequencing of DNA (see Sect. 8.5.2 for details)	

and dead organisms. This is due to the relatively long half-life of DNA which can remain in soil long after the organism has died. Although we might consider that, in assessing the DNA present in an environment, the majority of this DNA would most likely come from the most active organisms, this is not always the case. That is, a greater abundance of an organism's DNA does not correspond to increased activity of that organism (Blazewicz et al. 2013). In addition, a large proportion of the DNA extracted from environmental samples has been shown to correspond to extracellular DNA that is actively excreted by organisms or passively released after cell death (although in some applications we actively harness this phenomenon – see text on eDNA). Care should always be taken to understand that detection of DNA in a soil sample does not imply that the organisms from which this DNA originates are living, or that genes associated with the DNA are functional. DNA measures can also not be used to identify the genes being expressed in an environment at any particular moment in time, and thus cannot be used to determine the activity of the organisms or their functions. In terms of the application of DNA analyses to assess biodiversity, DNA is useful for macro- and meso-soil fauna, as well as the soil microbiome.

Ribonucleic acid (RNA) is also present in all organisms and represents the transfer of the genetic code (DNA) into a message that is used to create proteins in a process known as transcription. RNA is formed by a succession of nucleotides similar to DNA, with the exception that thymine is replaced by uracil. The uracil nucleotide is much less stable than thymine, and thus RNA is not as stable in the environment as DNA, but has the advantage of being useful in establishing a more direct link between the presence of an organism and its activity. RNA analysis allows us to characterise the taxonomic diversity and composition of the active or viable organisms in a sample, which can provide information on the organisms involved in the cycling of nutrients or chemicals, or which organisms are involved in the degradation of pollutants. Although RNA analysis techniques have grown in popularity, they are generally applied only to the soil microbiome, rather than soil macro- or meso-fauna. The functional attributes of the soil microbiome are not always obvious, and thus require an understanding of the RNA and even protein and metabolite production (see Sect. 8.5.2 on the soil microbiome).

Environmental DNA or eDNA in the urban soil context can be defined as the complex mixture of DNA extracted from that urban soil whereby the DNA is representative of the organisms present including macro (e.g. earthworms), meso (e.g. springtails) and micro (e.g. bacteria) organisms. The methodology which eDNA analyses are based on has its roots in the analysis of the soil microbiome, where this type of approach has been commonplace for the past 40 years or so. More recently eDNA and related tools have been used more to enhance biodiversity studies of larger soil organisms including invertebrates. The advantages of methods such as eDNA include the possibility for samples to be collected by non-specialists. Sampling of environmental materials for eDNA is less invasive, i.e. generally organisms themselves do not need to be collected, rather the environment in which they inhabit (e.g. soil or water). Many soil organisms excrete or shed large quantities of eDNA which can be used to detect their presence; thus DNA extracted from urban soils can be used to provide information on the occurrence, distribution and diversity of organisms across multiple branches of the tree of life (see Fig. 8.4). The detection of organisms from eDNA has the added benefit that it may allow an insight into transient or rare organisms that may be missed in more traditional faunal collection surveys, as well as help detect problematic invasive species. Once collected, eDNA has the advantage that it can be stored long term, thus providing a library of samples that can be returned to in the future and re-investigated as new techniques become available. However, there are still a number of issues that need to be overcome in this area, for example, a better understanding of how much DNA individual organisms can shed across time, the longevity of DNA in the environment and the impact of environmental conditions, for example, pH and temperature, on the persistence of eDNA. Despite these issues it is likely that in the future eDNA may become the tool of choice to overcome time and personnel pressures in undertaking large monitoring programs across geographical or taxonomic scales.

8.5.2 Measuring the Soil Microbiome

The soil microbiome can be measured using a number of different methods, including culture-dependent techniques based on growing microorganisms and culture-independent techniques based on nucleic acid or other cellular component extractions. Much of the work on developing methods for microbiome characterisation has been performed by private organisations, so this section will mention several proprietary products. The mentions of these products do not constitute an endorsement, but the proprietary methods are presented as common and convenient solutions for analysis of the soil microbiome. Culture-dependent techniques involve isolating microorganisms from an environment and subsequent growing and identifying individual *colony-forming units (CFUs)* that can be counted or fully characterised (Hill et al. 2000). Culture-dependent methods have the advantage that organisms of interest can be mass-produced once isolated and identified and their functions and *metabolites* studied in depth, or they may produce useful medical or pharmaceutical products that can be exploited for human use (see Sect. 8.7 on novel uses). Culturing is relatively inexpensive and easily performed, although it is labour intensive, and generally less than 5% of the soil microbiome can be cultured (Torsvik et al. 1998). There are a number of other traditional measures of the soil microbiome that are based in culturing without explicit isolation of individual pure cultures. These include measurements of biological activity, for example, basal and substrate-induced respiration (carbon dioxide respired by soil organisms either naturally, or induced, e.g. via glucose addition) and enzyme activity, for example, measurement of denitrification enzyme activity (DEA). Others include the measurement of the size of the microbial community via chloroform-fumigation measurement of microbial biomass carbon or nitrogen, or the use of fatty acid biomarkers (Bending et al. 2004). Additional culture-based methods include community level physiological profiling (CLPP) techniques, such as the proprietary products MicroResp™ or Biolog™. In CLPP methods the response of the microbiome to specific substrates is measured creating an overview of the potential functional diversity of the microbiome, i.e. the diversity of functions it can perform (Garland and Mills 1991; Campbell et al. 2003). Results from these types of assays can provide direct information on the ability of the soil microbiome to metabolise specific compounds of interest, for example, contaminants present in urban soils. All culture-dependent methods, however, are biased toward organisms that are fast growing and thus do not represent the microbiome as a whole. The soil microbiome may be altered during the culturing process, and the conditions under which physiological profiling is undertaken (e.g. pH, temperature, etc.) may have a profound impact on the response observed and thus results obtained may no longer represent the community that was present in the original environment.

Culture-independent based methods are performed by extracting DNA or RNA from the soil environment and utilising quantitative polymerase chain reaction (qPCR), barcoded sequencing (for specific domains, phyla or functions) or shot gun sequencing techniques (*metagenomics* or *metatranscriptomics*) to provide information on diversity, abundances and functions within the soil microbiome. The extraction of nucleic acids from soil samples generally involves some form of in situ cell lysis: either chemical with a detergent, or physical with bead beating or a combination of both, followed by separation and purification of the nucleic acids from the rest of the sample constituents. This section will focus on DNA extraction (see section on metatranscriptomics for RNA), in which proprietary DNA extraction kits are often used to extract DNA from soil samples (e.g. the DNeasy PowerSoil kit by Qiagen), giving users methodological consistency across samples and projects. For example, the earth microbiome project (www.earthmicrobiome.org) has specified kits and protocols to use so that different groups can apply the same techniques. This consistency of techniques allows more detailed integration of data outputs, and may provide a good starting point for any methodology being developed in the urban soil space. Once DNA has been extracted, there are a number of possible options for analysis via either direct sequencing or polymerase chain reaction (PCR) targeted sequencing. The PCR method is used to amplify a target piece of DNA to potentially generate thousands of millions of copies, and is used in combination with primers (short pieces of DNA) that target specific regions with the extracted DNA. For example, primers can be designed that target different *domains* (Bacteria, Archaea, Eukaryotes), different *phyla* within those domains (e.g. Proteobacteria within the domain Bacteria) or even to genera or species. PCR can also be used to target functional attributes of the soil microbiome, for example, nitrogen or phosphorus cycling capacity, or the capacity to break down pollutants via production of specific *enzymes*. The use of PCR is widespread and provides opportunity to investigate, in detail, many elements of the same community. The use of quantitative PCR (qPCR) allows us the ability to quantify the abundance of specific populations or genes in an environmental sample. This can therefore reflect, for example, the abundance of bacteria if primers that amplify all bacteria in a given environment are used. PCR is additionally used as a mechanism to generate *amplicons* that can be sequenced generating nucleotide combinations, and the sequencing data can then be compared with databases that allow us to determine the identity of the species or functions present. In generating amplicon sequencing data, a number of *bioinformatics* routines need to be employed to assign taxonomy to the sequences obtained. There are a number of freely available software packages that can be used, for example, within the R coding environment the DADA2 pipeline (https://benjjneb.github.io/dada2/index.html) or Mothur pipeline (https://mothur.org/) can be used, or within the python coding environment the QIIME pipeline (http://qiime.org/) can be used, all free of charge, to assign taxonomy and generate datasets that can then be further interrogated using multivariate statistical tools. Care needs to be taken in the interpretation of data derived from PCR methods, as biases can be introduced during the amplification process that may affect the outcomes. Generally speaking, these biases are understood and data can still be interpreted meaningfully.

Options for DNA analysis which do not involve PCR are also available, in that the DNA extracted from the environment can be sequenced directly (*metagenomics*) without the need for PCR to amplify or target a region of interest. The costs involved in metagenome sequencing have been decreasing to an extent that they are now affordable to many working in this space. Metagenomics allows us a thorough understanding of the potential taxa and functions present in a typical urban soil community, and there are a number of *sequencing* platforms operating in this space. For example, second generation sequencing capabilities such as Illumina Hi Seq or Nova seq use a sequencing-by-synthesis approach to generate short read sequences (approximately 100–500 base pairs of sequence at a time). Other options include third generation sequencing via single-molecule real-time (SMRT) by Pacific Biosciences, or nanopore sequencing developed by Oxford Nanopore, both of which offer longer sequence read lengths (see Nkongolo and Narendrula-Kotha 2020 for a full review of advances in this area). Once DNA sequences have been generated, *bioinformatic* analysis needs to be applied – this can be via annotation using applications such as the MG-RAST server (www.mg-rast.org) which can automatically *annotate* phylogeny and functional attributes of submitted sequences. Alternatively, analysis can proceed via generation of a metagenome assembled genome (MAG) in which sequences are assembled and *binned* into *contigs*, a set of overlapping DNA segments that together represent a consensus region of DNA, allowing us to recover the genome sequences of abundant organisms which can then be further interrogated (Orellana et al. 2019; Schmidt et al. 2019).

Metatranscriptomics involves the extraction of RNA from an environmental sample and the use of RNASeq (RNA sequencing) technology to determine the sequence of expressed **transcripts** within a microbiome, under a specific set of environmental parameters, which provide a closer look at active members of the community at that particular point in time (Shakya et al. 2019). As with DNA, proprietary RNA extraction kits are often used to extract RNA from soil samples (e.g. the RNeasy PowerSoil kit by Qiagen) with additional precautions being required to ensure quality and quantity of RNA yield. For example, all materials should be RNase free to ensure that this enzyme does not degrade any RNA that is isolated from a sample. RNA has widely been used in the past to quantify gene expression via qPCR, but the advent of cheaper sequencing technologies has led to significant interest in the use of RNA to provide an assessment not only of the active members of a community but more broadly of the active functions being expressed. In particular metatranscriptomics has been applied across a variety of environments (including soils and water) to discover novel functions and to track gene expression and to identify novel microbial interactions (Bikel et al. 2015; Bashiardes et al. 2016; Moniruzzaman et al. 2017). However useful this methodology is, though, it is not without its drawbacks. RNA has a short half-life, and its extraction from soil in quantity and quality sufficient for metatranscriptomic analysis is challenging. Rigorous experimental design is critical to ensure activity can be sufficiently attributed, and samples require destructive analysis. In addition, due to the complexity of the soil microbiome (i.e. high diversity and abundance of particular species), the wide range of transcript expressions and technology-specific limitations, metatranscriptomics is not always able to capture the entire metatranscriptome.

Metaproteomics is an option to characterise the urban soil microbiome via the analysis of proteins that are produced via translation of the messenger RNA into a chain of *amino acids* and thus a *protein* (see Fig. 8.7). In contrast to DNA and RNA, most proteins have an intrinsic metabolic function, and thus, in a complex microbial environment such as an urban soil, proteins can be linked to specific organisms. However, methods for the extraction of proteins from soil lag behind those for nucleic acids. Protein extraction is challenging because of the presence of both organic (e.g. complex carbohydrates, lipids, phenolic compounds (e.g. lignin) and

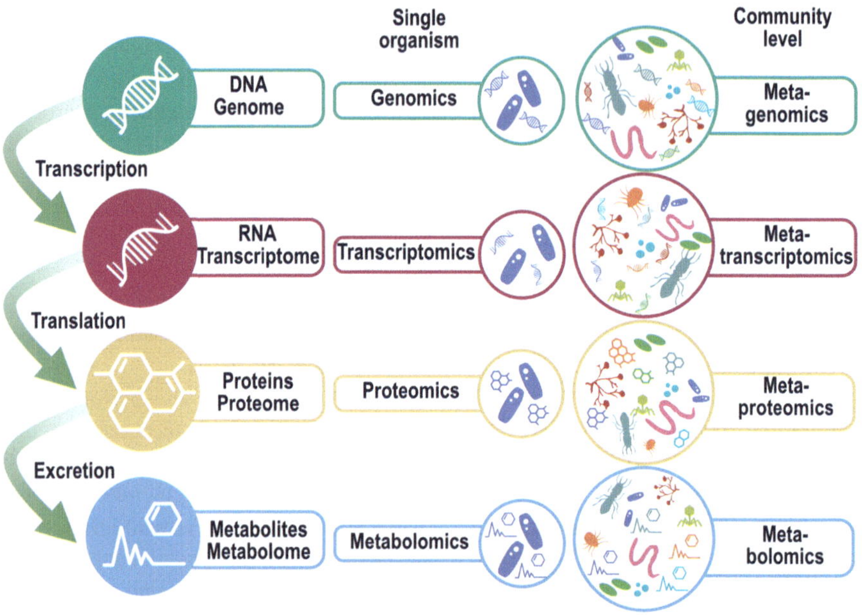

Fig. 8.7 This figure highlights links between nucleic acid, protein and metabolite methodologies for measuring the urban soil microbiome. Illustration by Ooid Scientific

humic substances) and inorganic (e.g. silt and clay minerals) compounds that can affect extraction efficiency. While co-extraction of humic substances has been an issue in DNA and RNA extraction, it has largely been overcome though ongoing optimisation work. Co-extraction of humic substances is an additional problem for metaproteomics, as humic substances interfere with the separation of peptides, and protein identification and quantification (Adamczyk et al. 2008; Bastida et al. 2009). Despite these issues, however, the identification of microbial proteins within a specific habitat, coupled with an analysis of their phylogenetic origin, could provide new insights into the role played by microorganisms in the urban environment (Schulze et al. 2005; Wang et al. 2011).

Metabolomics is the analysis of the many naturally occurring low molecular weight (< 1000 Daltons) metabolites, or molecules, such as sugars, organic acids, amino acids, lipids, etc. that are produced as part of an organism's normal cellular metabolism for maintenance and growth (Jones et al. 2013). Recent advances in spectroscopy have made it possible to both identify and quantify the relative abundance of metabolites present in biological samples (Patti et al. 2012). This type of approach is considered to be similar in terms of data complexity and cost to genomics and proteomics approaches; however it allows for more rapid sample processing and has the capacity to identify molecules involved in metabolic pathways including intermediate metabolites and thus potentially can improve our understanding of biological processes in soil that involve soil organisms (Withers et al. 2020).

8.5.3 Multivariate Analysis of Soil Organisms

You may be familiar with univariate statistics that are typically used to assess variability and to test for significant differences in univariate measures across experimental treatments, for example, pH, temperature or even abundances of specific organisms, e.g. total abundance of earthworms. Some of the statistical methods we apply in these situations are described in Chap. 3. In ecological datasets multivariate statistics are used when the data collected are complex and information across a variety of different organisms has been collected. That is, multivariate statistics are used to assess high-dimensional data typically collected in ecological studies where multiple species data are collected – this applies equally to urban soils as well as other environment types. One example of this type of data collection is of DNA sequencing data across the bacterial or archaeal domains where the methodologies result in the collection of relative abundances of a large list of organisms, for example, a typical soil microbial community analysis might yield data on the relative abundance of 400 different bacterial families. Analysing such data across multiple samples requires significant computational power, particularly when it comes to testing hypotheses involving not only multiple species data but also multiple response variables (i.e. a variety of environmental metadata is also collected such as soil pH, temperature, moisture, nutrients, contaminants, etc.).

A range of measures are used in the analysis of multivariate data. Alpha diversity is generally described as the 'within sample' diversity and is a measure of both the *richness* (i.e. number of species present) and *evenness* (i.e. variation in the abundance of each species present) of species in a single sample. Alpha diversity can be simply viewed as the number of species present in a single sample and can easily be applied to species datasets that are constructed using traditional approaches. Alpha diversity measures such as the Shannon diversity index can be equally applied here to calculate the within sample species diversity (Shannon 1948). Although alpha diversity can be relatively easily determined for species that are assessed by traditional methods, it becomes more difficult to determine when we use DNA sequencing approaches as often we cannot be sure we have identified every single species present. In this case we need to calculate values known as estimated alpha diversity; one such example is the Chao1 estimator (Chao 1984). Both the traditional Shannon diversity type index and the estimator Chao1 index give equal weight to the presence of each species and do not take into account phylogenetic placement, i.e. this means that the same alpha diversity is yielded from a community consisting of 10 species from a single genus as it would one composed of 10 different phyla. Other alpha diversity measures such as Faith's phylogenetic diversity take into account the sum of the branches of the phylogenetic tree (refer to Fig. 8.4) and thus provide a more accurate measure of the within-sample alpha diversity (Faith 1992). Although the calculation of a diversity index results in a univariate value representing diversity, it should not be forgotten that this value was calculated based on the collection of a multivariate dataset. Since the alpha diversity index is a univariate measure of diversity, it can be analysed using standard univariate statistical methodologies just as you would pH or temperature (see Chap. 3).

Beta diversity is described as the 'between sample' diversity, i.e. the extent of change of community composition or diversity across an environmental gradient or set of experimental treatments. In other words beta-diversity can be described as the degree to which two samples differ in terms of the species present and their abundances, i.e. richness and evenness. A variety of beta diversity indices have been developed including distance-based metrics such as Euclidean distance and dissimilarity-based metrics such as Bray Curtis (Kuczynski et al. 2010). It is not the purpose of this text to provide background on all the distance and similarity/dissimilarity measures available – please review Anderson (2001), Clarke and Warwick (2001) and Anderson and Willis (2003) for a more detailed overview of the area. Euclidean distance is widely used in the multivariate analysis of environmental data; however it requires very large effect sizes for statistical significance and doesn't perform well with datasets containing many zeros, which inevitably species datasets will contain and thus it is not recommended or widely used for the analysis of species beta diversity. Bray Curtis, a dissimilarity measure used to quantify the species compositional dissimilarity between samples based on species counts and abundances, is more widely used as it has a high degree of tolerance for zero-rich datasets. However, again Bray Curtis does not take into account phylogenetic diversity – in order that we assess sample's differences based on the phylogeny of their communities, indices such as Unifrac should be applied (Lozupone and Knight 2005). In general a combination of non-phylogenetic and phylogenetic approaches can be useful in interrogating complex community datasets in order that we gain a deeper understanding of the similarities and differences contained within these communities.

To visualise the similarities or differences obtained using measures of similarity, whether they be *phylogeny* based like Unifrac or non-phylogeny based like Bray Curtis, *ordination* approaches are usually applied. One of the oldest approaches to visualise multivariate data is principal component analysis or PCA (Ringner 2008). In this type of approach variables are treated as axes in a Euclidean multidimensional space, and the first principal component is by definition placed on the direction representing the largest variation of the data. The second component placed in a direction orthogonal to this that explains the largest remaining variation and so on (Chap. 3 has some additional explanations of principal component ordination). Usually the majority of variation is explained within the first few components, with the percentage of variation explained by each axis indicating whether they are the dominant drivers present or not. However, as mentioned previously Euclidean distance is rarely used for species data, and the related PCoA or Principal Coordinate Analysis method is used, in combination with Bray Curtis, Unifrac or other similarity or dissimilarity measures suitable for species data containing zero-rich datasets. PCoA makes use of actual measures of similarity between samples; however, other methods such as multidimensional scaling (MDS) make use of ranked similarities, i.e. samples that are most similar to one another are given the rank of 1, the next most similar set of two samples is ranked second and so on until all samples have been ranked against one another. This technique can be useful to decrease the complexity of the data and can also be considered a mechanism to de-emphasise the absolute measure of similarity chosen to represent beta diversity. In addition to visualisation of beta diversity a priori, there are also routines available that allow samples to be 'grouped' according to the experimental treatment applied (Canonical Analysis of Principal Components: CAP) or according to environmental measurements collected (Distance based redundancy analysis: DISTLM).

In addition to visualisation of data, it is important to be able to test hypotheses within a multivariate environment in a similar way to the approach taken with univariate data. For example, it might be useful to question whether an a priori grouping of samples (e.g. across environmental gradients or due to experiment treatments imposed) corresponds to the presence of statistically different biological or microbial communities. You may be familiar with the analysis of variance approach (ANOVA) for univariate data, or due to lack of normality (usually the case for biological data) the application of the Kruskal-Wallis rank sum test. For multiple comparisons, permutational ANOVA where permutations are used to assess significance, called PERMANOVA, is usually applied. A review of these methods as they are applied to microbial datasets can be found in Hugerth and Andersson (2017). All of the analysis methodologies mentioned above can be implemented using free-to-use R packages such as 'vegan' (Oksanen et al. 2020), or the for-purchase Primer-E package (https://www.primer-e.com/).

8.6 Relationships Between Urban Habitats and Resident Organisms

There are numerous studies focussing on assessment of soil organisms against a rural-urban divide – generally the argument is made that urban environments harbour less biodiversity than their rural counterparts. However, this assessment may not always be true as it will depend on the urban habitat being considered, the scale of the habitat (μm^2 to cm^2 to km^2) as well as the specific organisms in question (Szlavecz et al. 2006; Guilland et al. 2018). Given the great global variety of urban environments (See Sect. 8.3), in this text we will focus on the relationship between urban environments and their resident organisms and vice versa, rather than the differences between urban habitats and other habitat types. Studies of urban soils have also generally focussed on disturbed anthropogenic *technosols* (i.e. soils constructed by humans; see Chap. 2), and thus such soils have been considered as being of low biodiversity (Deeb et al. 2020). However although there is no doubt that soil organisms are vulnerable to many human induced changes in urban environments, for example, habitat disturbance, impervious surfaces, changes to land use and presence of pollutants such as metals, petro-chemicals and pesticides, there remains a vast unexplored biodiversity within urban environments (McIntyre 2000; Pavao-Zuckerman 2008; Jones and Leather 2012).

8.6.1 Habitat Structure

Habitat structure can be described as (i) physical aspects of the landscape including buildings, tree canopy and grass cover which mediate habitat characteristics such as soil temperature; (ii) soil structure, that is, the arrangement of the physical components of the soil; and (iii) vegetation, that is, the diversity of trees, shrubs and grasses in terms of both species and morphology or growth habit.

Physical aspects of the landscape can result in alteration of macroclimate conditions, for example, changes in temperature (Chap. 5). Alteration of macroclimate conditions by urbanisation is defined as the 'urban heat island' effect where urbanised environments have a greater air temperature than the surrounding land. Therefore shaded areas within these environments, for example, those shaded by trees and shrubs, usually have a lower ambient temperature (Geiger et al. 2009). In fact belowground 'heat islands' have also been described in the context of direct soil cover, where, for example, soils beneath pavement and gravel are warmer than those underneath organic materials such as mulches, lawns and unmowed fields, due to decreased heat exchange between the soil and the atmosphere (Byrne 2007). This effect is temporal, with maximum habitat temperature differences at the hottest part of the day and the effect lessening overnight as aboveground temperatures cool. Organisms that inhabit these environments are directly impacted by this temporal variability. Generally speaking an increase in temperature will result in an increase in an organism's biological activity. Most organisms have an ideal temperature range under which they optimally operate, for example, most soil organisms are active between 0 °C and 50 °C (although individual species will have optima that sit more tightly within this wide range). The temperature optimum for individual species relates to the temperature at which biological functions perform best, and outside of these optimum ranges cellular processes are restricted both at the upper and lower bounds of the optimum range. Temperature both directly and indirectly affects soil organisms. For example, a direct effect of temperature might influence physiological activity of an organism, e.g. enzyme activity, while indirect effects include those impacting local soil physico-chemical properties, e.g. nutrient diffusion and solubility and local evaporation rates. For example, increasing temperatures can result in an increased microbial activity, thus resulting in a shift in the microbial community in favour of organisms adapted to higher temperatures and faster growth rates (Zogg et al. 1997). In addition, there is a maximum temperature after which microbial activity will decrease again as the upper range of an organism's physiological activity is reached.

Habitat fragmentation and the theory of island biogeography can also explain some of the physical impacts on soil organisms in an urban context. Essentially island biogeography states that larger islands have more species than smaller more isolated islands. This can be applied to habitat fragments that exist as an 'island' within an urban setting, for example, it has been shown that arthropod communities were affected by the patch sizes of urban habitats with greater species richness observed in larger patches. However the effect was species-specific as some species persisted in smaller patches but not larger ones and vice versa (Gibb and Hochuli 2002). Species size can also play a role, for example, the body size of spiders influenced which species survived in urban patches of forest of differing size, i.e. species with larger body size were not present in smaller forest patches (Miyashita et al. 1998). It has also been suggested that the time since fragmentation was apparent is also an important physical factor, for example, Bolger et al. (2000) reported that arthropod abundance and diversity negatively correlated with fragment age.

Soil structure, incorporating elements of solid soil particles, plant roots and void spaces (Chap. 4), is an important driver of many ecological processes in urban environments. It can regulate community structure of both macro and micro fauna by providing resources such as nutrients, moisture and niches for colonisation, as well as mediating interactions, for example, *competition* and *predation*, between organisms (Byrne et al. 2008). Soil is structurally complex and consists of a matrix of solid mineral and organic components that are organised into *aggregates*. That is, soil structure is part of habitat structure, but they are not the same thing! This aggregation creates a wide variety of microenvironments that offer distinct niches for soil organisms to reside in (O'Donnell et al. 2007). The soil mineral component influences micro-niche development and thus diversity of organisms in soil, and in particular the microbial component (Gleeson et al. 2005; Gleeson et al. 2006; Carson et al. 2007; Carson et al. 2009). The spatial arrangement of the solid soil matrix ensures the creation of a complex network of pores that vary in their *connectivity* and which contributes to the formation of diverse niches for colonisation (Carson et al. 2010). This pore structure can influence interactions between the various types of organisms found in soil, for example, predation of bacteria by protozoa can be influenced by soil pore size distribution. This soil pore geometry can be adversely affected, particularly in urban soils where compaction and sealing are an issue (see Fig. 8.8).

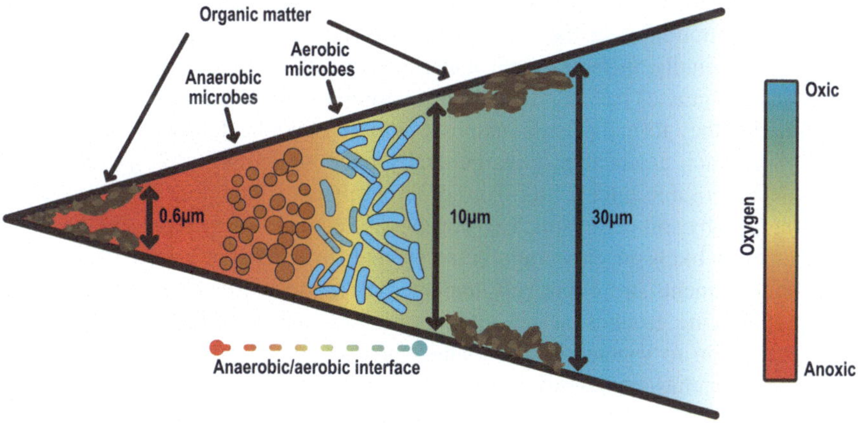

Fig. 8.8 Impact of soil pore structure on the interaction of soil microorganisms, organic matter accumulation and oxygen diffusion (adapted from Strong et al. 1998). Illustration by Ooid Scientific

Soil organisms themselves also contribute to soil structure in a meaningful way. Soil *ecosystem engineers* change the physical state of soil such as earthworms that burrow and tunnel through soil mixing soil layers and enhancing soil structure. Earthworms produce casts that assist in attaching soil particles together to form aggregates. The activities of earthworms leaving their casts on the soil surface, and isopods transporting their faecal pellets to the surface, contribute to the rebuilding of topsoil (Jones et al. 2006). It is widely understood that organic matter in soil contributes to soil structure by decreasing bulk density and increasing water holding capacity. What is less well understood is the contribution to soil structure of the decomposition of organic matter by soil organisms. We do know that the process of microbial decomposition of organic matter contributes positively to soil structure and improvement of aggregation. During breakdown of organic matter, soil microorganisms secrete various carbohydrates and polysaccharides that enhance aggregation in soil and thus contribute to the formation of soil structure. These *extracellular polymeric substances* are present only in small quantities in soil (less than 0.1%) but are highly effective at binding soil particles (Costa et al. 2018). Aggregates help improve the fertility of soils, and thus the presence of microorganisms, for example, strains of *Pseudomonas*, *Bacillus* and *Paenibacillus*, that contribute to soil aggregation are important. Fungi contribute to soil structure via the production of hyphae that assist in anchoring soil particles. Polysaccharides on the surface of these hyphae also assist in binding soil particles together thus contributing to soil structure, with the greatest benefit observed in sandy soils (Tisdall et al. 2012).

Vegetation present in a habitat can have a variety of impacts on resident organisms. For example, plants can have different rooting architectures, *root exudate* composition, or leaf/shoot nutrient contents. First, the rooting architecture of plants (density and depth distribution of roots) can moderate moisture availability in different soil layers as plants take up water from soil. Next, the composition of plant root exudates (i.e. the concentrations of low molecular weight organic molecules released by plant roots, such as organic acids) provides differing substrates for microbial communities in the rhizosphere (i.e. the soil zone immediately beside plant roots). Finally, different plant materials can have different nutrient profiles. Plants such as grasses have a relatively low carbon to nitrogen ratio in their tissues, meaning that they form easily decomposable litter. In contrast, tissues from sclerophyll plants such as eucalypt trees generally have high C:N ratios, yielding less easily decomposable litter. Litter from different plants is therefore decomposed at different rates by soil organisms. Heterogeneous vegetation may represent the environment most likely to encourage belowground diversity, but a large amount of urban environments are effectively managed as monocultures (e.g. lawns, urban parks and playing grounds) which may not promote increased belowground diversity. In addition to vegetation, the management of, for example, lawn clippings, leaves and mulches, can significantly impact nutrient availability to belowground soil organisms. These mulches can also assist in increasing soil water content via the addition of mulches to the soil surface which can decrease evaporation rates (Byrne 2007). The diversity of vegetation in urban environments can also impact soil moisture patterns as they can impact local evapotranspiration rates (Eviner 2004). An additional issue is the removal of native vegetation and replacement with exotic or new combinations of plant species that often focus on non-natives. Designed urban environments have not been extensively examined in terms of the influence of plant species richness and identity on soil organisms – it may transpire that these human designed environments have unique impacts on soil biodiversity and ecosystem functioning (Wolfe and Klironomos 2005).

8.6.2 *Soil Characteristics (pH, Nutrients, Moisture)*

As previously described for temperature, organisms have a pH range within which they are optimally active. Some organisms can tolerate extremes of pH, but most organisms have a pH optimum of around 7.0, with some microorganisms (bacteria and archaea in particular) tolerating pH values as low as 1.0 or as high as 12.0. Larger soil fauna, for example, earthworms, are less tolerant of extremes in pH and are not very active in, for example, forest soils where the pH can be lower than 5.0. The activity of earthworms in urban forests may be replaced by other organisms (e.g. enchytraeids, or potworms). In soil the solubility of elements is affected by pH, for example, at low (acidic) pH, aluminium-bearing minerals become more soluble, and dissolved aluminium ions are therefore more bioavailable, resulting in increased toxicity to soil organisms. Conversely some essential elements can become unavailable at high pH (alkaline), e.g. phosphorus and manganese. The urban soil microbiome is similar to its rural counterpart, in that abiotic factors such as pH are important drivers of microbial diversity (Gleeson et al. 2016; Schmidt et al. 2017; O'Brien et al. 2019). In general, although not always, urban soils have a pH greater than 7 which has been attributed to the use of filling materials contaminated with building waste (e.g. concrete and cement) which results in the release of carbonate-rich solutions into the soil thus increasing the pH (Yang and Zhang 2015). Although soil pH has long been recognised as a driver of the soil microbiome in terms of the species present and their functions, differing effects have been reported across the bacteria, archaea and fungi. The highest bacterial diversity has been reported in soils around pH 7.0 (neutral pH) with diversity being significantly lower in acidic soils (Fierer and Jackson 2006), although there are numerous conflicting reports. For example, Roesch et al. (Roesch et al. 2007) suggested that forest soils (that generally have a pH less than 7.0) have higher microbial diversity than in three different agricultural soils that were also investigated. In the urban context there have been relatively few studies assessing the composition of the microbiome with some noteworthy exceptions. For example, in New York City (USA), research in Central Park has indicated that soil microbial composition is related to microsite differences, particularly soil edaphic characteristics like pH (Ramirez et al. 2014; Gill et al. 2020).

Water in urban soils is modified by limited infiltration and by increased run-off into storm drainage infrastructure (Chap. 5). Infiltration in urban soils is limited both because of soil sealing and the compaction of soils that have not been sealed. Thus, runoff is increased by the existence of sealing, either via pavement, roads, and other surfaces that drain into storm sewers. Therefore it can generally be said that urban soils are drier than non-urban soils (Pickett et al. 2011). As always there are exceptions to this, specifically in the arid and semi-arid areas where irrigation is used meaning that soil water content can be higher in urban soils (Brazel et al. 2000). Soil water contents in urban areas can be further decreased due to local higher air temperatures (heat island effect), which generate higher evaporation and transpiration rates. Related to poor infiltration is the potential for excess water to run-off onto compacted soils (rather than into stormwater infrastructure) resulting in potential anoxic soil conditions (Pickett et al. 2011). Thus, soil water content likely has a significant impact on soil organisms in urban environments – and this impact will not be uniform. Soil water content significantly affects soil organisms in a number of ways. Considering soil structure, the degree to which soil pores are filled with water affects the movement and predation of microorganisms in soil (see Fig. 8.9). Although it has been suggested that fungi have greater resistance to drought than bacteria and archaea, some bacteria do form *endospores* which allow these organisms to survive until more suitable conditions arise (for instance, there is a rainfall event). Water availability is also known to directly affect the *osmotic status* of microbial cells. Extended lack of water can affect soil microorganisms and induce osmotic stress, although some soil microorganisms, for example, *Actinomyces*, can produce chemicals, in this case the amino acid proline, which protects the osmotic balance of cells and allows survival under drought conditions (Griffiths et al. 2003; Drenovsky et al. 2004).

Plants require nutrients to be in a usable form and optimal concentration to foster strong growth. This is particularly important when attempting to restore and re-vegetate urban environments. However, concentrations of nutrients in urban soils can be highly varied across both city centre and suburban locations. Urban soils in city centre locations are relatively enriched in both nitrogen and phosphorus, for example, high levels of atmospheric deposition of both nitrate and ammonium have been reported in the urban forests of New York City (Lovett et al. 2000). This is likely due to atmospheric nitrous oxides from fossil fuel combustion that can react with dust from urban construction work. Other sources of urban pollution include leaky sewers, leaching from landfills, storm water runoff and industrial spills. Fertiliser use on lawns and city parks is likely to lead to higher nutrient availabilities in suburban locations and parklands (Paul and Meyer 2001). Soil nutrient status strongly influences microbial communities, and in natural ecosystems soil eutrophication has been linked to decreases in the number and abundance of microbial species. It has been reported that there is a greater abundance of bacteria in the genus *Nitrospira* in residential areas, likely related to private lawns receiving greater nitrogen fertilizer than other soils thus promoting the growth of these nitrifying microorganisms. In urban green spaces *Verrucomicrobia*, which are generally slow-growing and oligotrophic bacteria, were more abundant in nitrogen poor urban soils (Wang

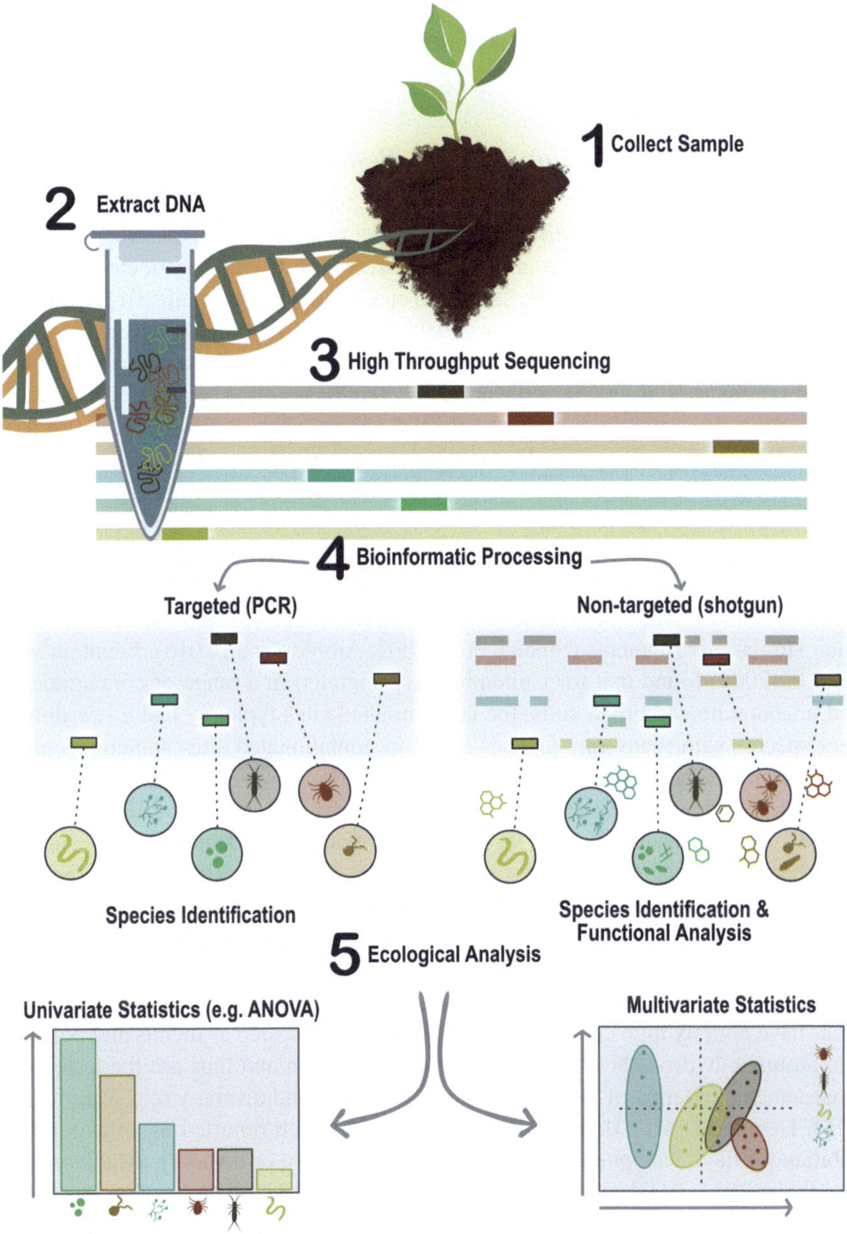

Fig. 8.9 Flowchart illustrating steps involved in sampling of urban soils through to analysis of DNA extracted from that soil. This process can be applied to environmental DNA (eDNA) extracted to identify soil fauna or soil microorganisms. Illustration by Ooid Scientific

et al. 2018). The issue of nutrients in soil is also linked with pH as mentioned above – with elevated soil pH influencing the form and availability of most nutrients.

8.6.3 Soil Contamination (Metals, Pesticides, PAHs)

Soil contaminants can have differing impacts on resident organisms; for example, most invertebrates decrease in abundance in response to both pesticides and soil metal contamination (Peck 2009; Santorufo et al. 2012; Pouyat et al. 2015; Gan and Wickings 2017; Bray and Wickings 2019). There are reports, however, of metal contamination increasing the abundance of some soil macroinvertebrates such as isopods (Pouyat et al. 2015). Szlavecz et al. (2006) reported a greater abundance and biomass of earthworms in more urbanised sites, whereas Smetak et al. (2007) reported greater abundance and biomass in less urbanised sites. This lack of consistency in reporting of earthworm communities may be related to the type of urban environment that is sampled and the context of the city in terms of population and level of industrial activity. Data collected from different types of urban land use suggest that earthworm abundance and diversity are highest in urban gardens with a gradient running from gardens to urban forests to grass covered road verges and other similar environments (Smetak et al. 2007; Amosse et al. 2016). Fountain and Hopkin (2004) found that for Collembola (springtails) in a range of contaminated and uncontaminated urban soils, the contaminated sites typically had a few dominant species with many rare species. Soils in contaminated sites sometimes contained more species than in uncontaminated sites. Pavao-Zuckerman and Coleman (2007) found that, while the functions performed by soil nematodes differed with varying degrees of urbanisation, the taxonomic diversity of nematodes was similar for urbanised and non-urbanised soils. In contrast, Uno et al. (2010) found that the diversity of ant species was greater in forested areas than in urban environments; however, the lower urban ant diversity may have been related to the colonisation of urban soils by an introduced ant species.

Grass verges located alongside freeways and motorways and in highly trafficked areas have notably high concentrations of contaminants such as metals and polycyclic aromatic hydrocarbons, as well as soil compaction and thus are the least rich environments in terms of macroarthropod abundance and diversity (e.g. Varet et al. 2014; Devigne et al. 2016). Green roofs have also been reported as unfavourable habitats for the development of macroarthropod density or diversity (MacIvor and Lundholm 2011; Braaker et al. 2014). The very low levels of colonisation observed in green roofs may be due to the limited plant diversity on these surfaces and the fact that they are usually constructed from manufactured soils. Microarthropods (including Collembla and Acari) are often considered as reliable bioindicators of urbanisation, for example, pollution and contamination (Joimel et al. 2017) as they are abundant, relatively easy to sample and respond quickly to soil disturbance, and so have been selected as one of the main bio-indicators to be used in soil surveys (McIntyre 2000).

One of the most important pathways for *attenuation* (i.e. removal) of organic contaminants in soils is *biodegradation*. The biodegradation process occurs where soil microorganisms use contaminant compounds as a metabolic energy source and growth substrate. Alternatively, pollutant removal can occur in a process of *co-metabolism*, in which organic contaminant compounds are degraded while microorganisms are metabolising another (presumably less toxic) substrate (Alexander 1981). Bacteria, archaea and fungi are involved in biodegradation of various contaminants, with fungi often being more effective at degrading persistent organic pollutants (Harms et al. 2011). For biodegradation to occur, the pollutant molecules must be in a chemical form, or physical location, that allows interaction with microorganisms. In other words, the organic pollutant compounds must be both *bioavailable*, in a form that allows biochemical processes to occur, and *bioaccessible*, in a location which allows microorganisms to come into contact with pollutant molecules (Semple et al. 2004). If *rhizosphere* microorganisms, existing in association with plant roots are involved, then bioaccessibility also applies to the ability of roots or associated fungal hyphae to contact or closely approach the pollutant molecules. Numerous chemical forms of organic contaminants are not bioavailable, for example, when contaminant molecules are physically or chemically bound to clay minerals or organic matter in soils. Biodegradation also occurs at different rates and to different extents depending on the types of organic contaminant involved (McBride 1994). For example, among the pollutants in the petroleum hydrocarbon category, simple unsubstituted compounds such as alkanes and light aromatics (e.g. BTEX) are degraded more rapidly by microorganisms than more complex compounds (Shahsavari et al. 2017). Compounds with more complex or larger molecules such as PAHs, PCBs or PCDDs (see Table 7.2) are more slowly degraded, and their biological attenuation in soils may be mediated mainly by fungi (Harms et al. 2011). Slow or even negligible degradation is a defining property for the group of persistent organic pollutants (POPs; see Table 7.3).

Some of the other issues around biodegradation have already been discussed in Chap. 6. You will recall that biodegradation is characterised by first-order kinetics, and that biodegradation rates depend on both the identity of the pollutant (as implied above), and soil properties such as temperature, soil water content, soil pH and nutrient availability. These are the factors that affect the activity of microorganisms in general. Since other toxic substances such a metals can affect biological activity, high concentrations of metals also decrease the rates and amounts of biodegradation that can occur. Finally, we should remember that in general bioavailability in soil decreases with increasing time since contaminant addition.

8.6.4 Urban Soils and Human Health

Most soil organisms are beneficial and present no risk to human health. However, in some instances soil can serve as a reservoir for human disease vectors, pathogens and parasites. These organisms represent a poorly studied group, and in fact the

World Health Organization does not even recognise soil-borne human diseases as a distinct group (with the exception of soil-transmitted helminths), as they do with food-borne diseases or zoonoses. Urbanisation can cause disturbances in the soil environment that may favour an increase in pathogenic microorganisms. For example, as already noted, urban areas usually have higher temperatures due to the urban heat island effect. These higher temperatures have the potential to stimulate the growth of some pathogenic species (Charron et al. 2004). Some soil-borne pathogens such as *Pseudomonas* and *Enterobacter* are opportunistic, and, although they can infect and cause disease in humans, their main functions are as decomposers in the soil, antagonists against plant root pathogens and as plant growth promoters (Wall et al. 2015). In urban areas water can be contaminated with pathogens via a number of pathways ranging from pets, storm water run-off, wastewater effluent and urban agriculture. Water containing pathogens can often find its way onto the soil thus providing an additional source of potential soil pathogens (e.g. Walters et al. 2011). Other urban pathogens are obligate parasites that require a host in order to complete their life cycle. For example, many studies in urbanized landscapes have reported locally high tick abundance, the vector for lyme disease and tick typhus – both caused by different tick-borne bacteria (Diuk-Wasser et al. 2020). In general it can be observed that degraded, low biodiversity soils that perhaps may be typical of urban environments may favour more opportunistic and potentially pathogenic microorganisms. One option to remedy this is to ensure that future urban developments incorporate novel technologies and planning such as Water Sensitive Urban Design (Chap. 5) (including increasing infiltration of impervious surfaces, reducing fertiliser use on green lawns and planting native trees). Sustainable management practices such as these promote a biodiverse soil community that ensures a fertile and healthy soil (Liddicoat et al. 2019; Mills et al. 2019).

8.7 Novel Uses/Metabolisms or Urban Soil Organisms

All soils, including urban soils, provide a diverse range of habitats for soil microorganisms in particular. It is this diversity of habitat that likely leads to the development of the adaption strategies that microorganisms require to survive and reproduce. Thus microorganisms produce a range of microbial products that enhance their survival. For example, *Bacillus thuringiensis* is a soil bacterium that produces insecticidal crystal proteins (Bt toxin) that are toxic. The genes that encode the Bt toxin have now been engineered into many crop plants for use as a bio-insecticide, for example, Bt-corn. Use of these genetically engineered crop plants decreases the need for additional chemical insecticides to be applied in the field. However, we should not forget that there may be unintended consequences such as an impact of the Bt crop on root exudates which may in turn influence rhizosphere microorganisms or horizontal gene transfer (HGT) whereby the Bt gene could be transferred to other organisms (see Turrini et al. 2015 for a review of the area). The most well-known antimicrobials produced by soil organisms are antibiotics. Soil bacteria can produce tetracyclines, streptomycins and β-lactams while fungi produce penicillin and cephalosporin antibiotics – all of which are widely used in the medical field. However, blanket widespread use of antibiotics has given rise to a new issue – that of antibiotic resistance. Pathogenic microorganisms have developed a series of defence mechanisms over time that protect them from the action of antibiotics. Therefore new effective antibiotics should be a focus for those working in this area which includes better understanding of microbial natural products, their derivatives and synthetic analogues that are widely used in medicine (Rutledge and Challis 2015). Microorganisms produce both generic and specialised metabolites with the majority of specialised metabolites resulting from metabolic pathways encoded by a suite of genes at the same chromosomal locus generally referred to as biosynthetic gene clusters (BGCs). These BGCs generally are difficult to express in laboratory culturing experiments, as they often require specific environmental triggers for their expression (Rutledge and Challis 2015). Despite this challenge, microbial metabolites or synthetic analogues have been approved as drugs to treat emerging health threats, such as daptomycin (to treat life-threatening infections caused by methicillin-resistant *Staphylococcus aureus* (MRSA)), fidaxomicin (to combat gastrointestinal infection by *Clostridium difficile*) and carfilzomib (a chemotherapy drug for the treatment of relapsed or refractory multiple myeloma) (Rutledge and Challis 2015). In more recent years a wealth of metagenomic data have been generated, allowing us to identify novel biosynthetic gene clusters that do not have the same restrictions as culturing based experiments, and thus allow us to select promising BGCs for further research and exploitation (Loureiro et al. 2018). The types of microbial products listed above have generally been produced using microorganisms and their DNA isolated from extreme environments, for example, hot springs, deep ocean and tundra. However, there has been a recent effort to investigate urban environments for potentially useful microbial products. For example, Charlop-Powers et al. (2016) reported that it is likely that urban environments, even small urban parks, could be significant reservoirs of natural bacterial product biosynthetic diversity. They report that many gene cluster families (that were first found in samples in extreme environments around the world) are predicted to be present in the collective urban soil microbiome. Soils also harbour an enormous diversity of viruses with a metagenomic and/or metatranscriptomic approach needed to understand viral community ecology because there is no universal marker gene for viruses as there is for bacteria (e.g. the 16S rRNA ribosomal subunit gene present in all bacteria). In marine systems, where viral ecology has been ongoing for the past 20 years or so, ecologists have reported that viruses can affect global ocean food webs, carbon cycling as well as climate. This type of research in the soil environment is only starting to take off – it is likely that viruses play a significant and as yet undiscovered role in our soils (Emerson 2019).

8.8 Additional Reading

Bardgett RD, van der Putten WH (2014) Belowground biodiversity and ecosystem functioning. Nature 515:505–511

Byrne LB (2007) Habitat structure: a fundamental concept and framework for urban soil ecology. Urban Ecosyst 10:255–274

Guilland C, Maron PA, Damas O, Ranjard L (2018) Biodiversity of urban soils for sustainable cities. Environ Chem Lett 16:1267–1282

8.9 Review and Study Questions

8.9.1 Checking Your Understanding

1. Soil biological functioning is affected by which four characteristics of urban soils? Explain why each characteristic is important.
2. List as many ecosystem services as you can think of which are provided directly by soil-dwelling organisms.
3. Give an example of the six main types of soil habitat found in urban environments.
4. Discuss the advantages and disadvantages of culture-based vs. culture-independent methods for measuring microbiological properties of soils.
5. How does soil pore structure affect soil microbial diversity and functioning?
6. Identify the possible mechanisms for biodegradation of urban soil contaminants, and describe the factors affecting the effectiveness of biodegradation.

8.9.2 Thinking About the Issues

7. What are the "–omics" methods used in soil biology, what information does each one provide, and what are the limitations of each technique?
8. Describe the important components of soil habitat structure, and how these components might be expressed in urban environments.
9. Discuss some of the ways in which urban soils can adversely AND beneficially affect human health.

8.9.3 Contemplating Urban Soil Biology More Creatively

10. It's often stated that a handful of soil can contain many millions of organisms. How do we know this? What measurements and data have been used?

References

Adamczyk B, Kitunen V, Smolander A (2008) Protein precipitation by tannins in soil organic horizon and vegetation in relation to tree species. Biol Fertil Soils 45:55–64. https://doi.org/10.1007/s00374-008-0308-0

Alexander M (1981) Biodegradation of chemicals of environmental concern. Science 211:132–138. https://doi.org/10.1126/science.7444456

Alloway BJ (2004) Contamination of soils in domestic gardens and allotments: a brief overview. Land Contam Reclam 12:179–187

Amosse J, Dozsa-Farkas K, Boros G, Rochat G, Sandoz G, Fournier B et al (2016) Patterns of earthworm, enchytraeid and nematode diversity and community structure in urban soils of different ages. Eur J Soil Biol 73:46–58. https://doi.org/10.1016/j.ejsobi.2016.01.004

Anderson MJ (2001) Permutation tests for univariate or multivariate analysis of variance and regression. Can J Fish Aquat Sci 58:626–639

Anderson MJ, Willis TJ (2003) Canonical analysis of principal coordinates: a useful method of constrained ordination for ecology. Ecology 84:511–525

Bardgett RD, van der Putten WH (2014) Belowground biodiversity and ecosystem functioning. Nature 515:505–511. https://doi.org/10.1038/nature13855

Bashiardes S, Zilberman-Schapira G, Elinav E (2016) Use of metatranscriptomics in microbiome research. Bioinform Biol Insights 10:19–25. https://doi.org/10.4137/bbi.s34610

Bastida F, Moreno JL, Nicolas C, Hernandez T, Garcia C (2009) Soil metaproteomics: a review of an emerging environmental science. Significance, methodology and perspectives. Eur J Soil Sci 60:845–859. https://doi.org/10.1111/j.1365-2389.2009.01184.x

Bending GD, Turner MK, Rayns F, Marx M-C, Wood M (2004) Microbial and biochemical soil quality indicators and their potential for differentiating areas under contrasting agricultural management regimes. Soil Biol Biochem 36:1785–1792. https://doi.org/10.1016/j.soilbio.2004.04.035

Berndtsson JC (2010) Green roof performance towards management of runoff water quantity and quality: a review. Ecol Eng 36:351–360. https://doi.org/10.1016/j.ecoleng.2009.12.014

Bertoncini AP, Machon N, Pavoine S, Muratet A (2012) Local gardening practices shape urban lawn floristic communities. Landsc Urban Plan 105:53–61. https://doi.org/10.1016/j.landurbplan.2011.11.017

Bhaduri B, Minner M, Tatalovich S, Harbor J (2001) Long-term hydrologic impact of urbanization: a tale of two models. J Water Resour Plan Manage-Asce 127:13–19. https://doi.org/10.1061/(asce)0733-9496(2001)127:1(13)

Bikel S, Valdez-Lara A, Cornejo-Granados F, Rico K, Canizales-Quinteros S, Soberon X et al (2015) Combining metagenomics, metatranscriptomics and viromics to explore novel microbial interactions: towards a systems-level understanding of human microbiome. Comput Struct Biotechnol J 13:390–401. https://doi.org/10.1016/j.csbj.2015.06.001

Blazewicz SJ, Barnard RL, Daly RA, Firestone MK (2013) Evaluating rRNA as an indicator of microbial activity in environmental communities: limitations and uses. ISME J 7:2061–2068. https://doi.org/10.1038/ismej.2013.102

Bolger DT, Suarez AV, Crooks KR, Morrison SA, Case TJ (2000) Arthropods in urban habitat fragments in southern California: area, age, and edge effects. Ecol Appl 10:1230–1248. https://doi.org/10.1890/1051-0761(2000)010[1230:aiuhfi]2.0.co;2

Bonthoux S, Brun M, Di Pietro F, Greulich S, Bouche-Pillon S (2014) How can wastelands promote biodiversity in cities? A review. Landsc Urban Plan 132:79–88. https://doi.org/10.1016/j.landurbplan.2014.08.010

Braaker S, Ghazoul J, Obrist MK, Moretti M (2014) Habitat connectivity shapes urban arthropod communities: the key role of green roofs. Ecology 95:1010–1021. https://doi.org/10.1890/13-0705.1

Bray N, Wickings K (2019) The roles of invertebrates in the urban soil microbiome. Front Ecol Evol 7:9. https://doi.org/10.3389/fevo.2019.00359

Brazel A, Selover N, Vose R, Heisler G (2000) The tale of two climates - Baltimore and Phoenix urban LTER sites. Clim Res 15:123–135. https://doi.org/10.3354/cr015123

Byrne LB (2007) Habitat structure: a fundamental concept and framework for urban soil ecology. Urban Ecosyst 10:255–274. https://doi.org/10.1007/s11252-007-0027-6

Byrne LB, Bruns MA, Kim KC (2008) Ecosystem properties of urban land covers at the aboveground-belowground interface. Ecosystems 11:1065–1077. https://doi.org/10.1007/s10021-008-9179-3

Campbell CD, Chapman SJ, Cameron CM, Davidson MS, Potts JM (2003) A rapid microtiter plate method to measure carbon dioxide evolved from carbon substrate amendments so as to determine the physiological profiles of soil microbial communities by using whole soil. Appl Environ Microbiol 69:3593–3599. https://doi.org/10.1128/aem.69.6.3593-3599.2003

Cannon AR, Chamberlain DE, Toms MP, Hatchwell BJ, Gaston KJ (2005) Trends in the use of private gardens by wild birds in Great Britain 1995-2002. J Appl Ecol 42:659–671. https://doi.org/10.1111/j.1365-2664.2005.01050.x

Carson JK, Rooney D, Gleeson DB, Clipson N (2007) Altering the mineral composition of soil causes a shift in microbial community structure. FEMS Microbiol Ecol 61:414–423. https://doi.org/10.1111/j.1574-6941.2007.00361.x

Carson JK, Campbell L, Rooney D, Clipson N, Gleeson DB (2009) Minerals in soil select distinct bacterial communities in their microhabitats. FEMS Microbiol Ecol 67:381–388. https://doi.org/10.1111/j.1574-6941.2008.00645.x

Carson JK, Gonzalez-Quinones V, Murphy DV, Hinz C, Shaw JA, Gleeson DB (2010) Low pore connectivity increases bacterial diversity in soil. Appl Environ Microbiol 76:3936–3942. https://doi.org/10.1128/aem.03085-09

Castelle CJ, Banfield JF (2018) Major new microbial groups expand diversity and alter our understanding of the tree of life. Cell 172:1181–1197. https://doi.org/10.1016/j.cell.2018.02.016

Chao A (1984) Nonparametric-estimation of the number of classes in a population. Scand J Stat 11:265–270

Charlop-Powers Z, Pregitzer CC, Lemetre C, Ternei MA, Maniko J, Hover BM et al (2016) Urban park soil microbiomes are a rich reservoir of natural product biosynthetic diversity. Proc Nat Acad Sci U S A 113:14811–14816. https://doi.org/10.1073/pnas.1615581113

Charron DF, Thomas MK, Waltner-Toews D, Aramini JJ, Edge T, Kent RA et al (2004) Vulnerability of waterborne diseases to climate change in Canada: a review. J Toxicol Environ Health-Part A Curr Issues 67:1667–1677. https://doi.org/10.1080/15287390490492313

Churchman GJ, Landa ER (2014) The soil underfoot: infinite possibilities for a finite resource. CRC Press

Clark HF, Hausladen DM, Brabander DJ (2008) Urban gardens: lead exposure, recontamination mechanisms, and implications for remediation design. Environ Res 107:312–319. https://doi.org/10.1016/j.envres.2008.03.003

Clarke KR, Warwick RM (2001) A further biodiversity index applicable to species lists: variation in taxonomic distinctness. Mar Ecol Prog Ser 216:265–278. https://doi.org/10.3354/meps216265

Costa OYA, Raaijmakers JM, Kuramae EE (2018) Microbial extracellular polymeric substances: ecological function and impact on soil aggregation. Front Microbiol 9. https://doi.org/10.3389/fmicb.2018.01636

Csuzdi C (2012) Earthworm species, a searchable database, 43

De Kimpe CR, Morel JL (2000) Urban soil management: a growing concern. Soil Sci 165:31–40. https://doi.org/10.1097/00010694-200001000-00005

Decaens T (2010) Macroecological patterns in soil communities. Glob Ecol Biogeogr 19:287–302. https://doi.org/10.1111/j.1466-8238.2009.00517.x

Deeb M, Groffman PM, Blouin M, Egendorf SP, Vergnes A, Vasenev V et al (2020) Using constructed soils for green infrastructure - challenges and limitations. Soil 6:413–434. https://doi.org/10.5194/soil-6-413-2020

Devigne C, Mouchon P, Vanhee B (2016) Impact of soil compaction on soil biodiversity - does it matter in urban context? Urban Ecosyst 19:1163–1178. https://doi.org/10.1007/s11252-016-0547-z

Didden W, Fründ HC, Graefe U (1996) Enchytraeids. Fauna in soil ecosystems. Recycling processes, nutrient fluxes, and agricultural production:135–172

Diuk-Wasser MA, VanAcker MC, Fernandez MP (2020) Impact of land use changes and habitat fragmentation on the eco-epidemiology of tick-borne diseases. J Med Entomol. https://doi.org/10.1093/jme/tjaa209

Drenovsky RE, Vo D, Graham KJ, Scow KM (2004) Soil water content and organic carbon availability are major determinants of soil microbial community composition. Microb Ecol 48:424–430

Emerson JB (2019) Soil viruses: a new hope. mSystems 4:e00120-19. https://doi.org/10.1128/mSystems.00120-19

European Commission (2006) Directive of the European Parliament and of the council establishing a framework for the protection of soil and amending directive 2004/35/EC. European Union, Brussels. https://eur-lex.europa.eu/LexUriServ/LexUriServ.do?uri=COM:2006:0232:FIN:EN:PDF

Eversham BC, Roy DB, Telfer MG (1996) Urban, industrial and other manmade sites as analogues of natural habitats for Carabidae. Ann Zool Fenn 33:149–156

Eviner VT (2004) Plant traits that influence ecosystem processes vary independently among species. Ecology 85:2215–2229

Eyre MD, Luff ML, Woodward JC (2003) Beetles (Coleoptera) on brownfield sites in England: an important conservation resource? J Insect Conserv 7:223–231. https://doi.org/10.1023/B:JICO.0000021020.66549.1e

Faber JH, Creamer RE, Mulder C, Rombke J, Rutgers M, Sousa JP et al (2013) The practicalities and pitfalls of establishing a policy-relevant and cost-effective soil biological monitoring scheme. Integr Environ Assess Manag 9:276–284. https://doi.org/10.1002/ieam.1398

Faith DP (1992) Conservation evaluation and phylogenetic diversity. Biol Conserv 61:1–10. https://doi.org/10.1016/0006-3207(92)91201-3

Falkowski PG, Fenchel T, Delong EF (2008) The microbial engines that drive Earth's biogeochemical cycles. Science 320:1034–1039. https://doi.org/10.1126/science.1153213

Farahani LM, Maller C, Phelan K (2018) Private gardens as urban greenspaces: can they compensate for poor greenspace access in lower socioeconomic Neighbourhoods? Landscape Online 59:1–18. https://doi.org/10.3097/LO.201859

Fierer N, Jackson RB (2006) The diversity and biogeography of soil bacterial communities. Proc Nat Acad Sci U S A 103:626–631

Fountain MT, Hopkin SP (2004) Biodiversity of collembola in urban soils and the use of Folsomia candida to assess soil 'quality'. Ecotoxicology 13:555–572. https://doi.org/10.1023/B:ECTX.0000037192.70167.00

Francis RA, Lorimer J (2011) Urban reconciliation ecology: the potential of living roofs and walls. J Environ Manag 92:1429–1437. https://doi.org/10.1016/j.jenvman.2011.01.012

Gan HJ, Wickings K (2017) Soil ecological responses to pest management in golf turf vary with management intensity, pesticide identity, and application program. Agric Ecosyst Environ 246:66–77. https://doi.org/10.1016/j.agee.2017.05.014

Gardiner MM, Burkman CE, Prajzner SP (2013) The value of urban vacant land to support arthropod biodiversity and ecosystem services. Environ Entomol 42:1123–1136. https://doi.org/10.1603/en12275

Garland JL, Mills AL (1991) Classification and characterization of heterotrophic microbial communities on the basis of patterns of community-level sole-carbon-source utilization. Appl Environ Microbiol 57:2351–2359. https://doi.org/10.1128/aem.57.8.2351-2359.1991

Geiger R, Aron RH, Todhunter P (2009) The climate near the ground. Rowman & Littlefield

Gibb H, Hochuli DF (2002) Habitat fragmentation in an urban environment: large and small fragments support different arthropod assemblages. Biol Conserv 106:91–100. https://doi.org/10.1016/s0006-3207(01)00232-4

Gill AS, Purnell K, Palmer MI, Stein J, McGuire KL (2020) Microbial composition and functional diversity differ across urban green infrastructure types. Front Microbiol 11:13. https://doi.org/10.3389/fmicb.2020.00912

Gleeson DB, Clipson N, Melville K, Gadd GM, McDermott FP (2005) Characterization of fungal community structure on a weathered pegmatitic granite. Microb Ecol 50:360–368

Gleeson DB, Kennedy NM, Clipson N, Melville K, Gadd GM, McDermott FP (2006) Characterization of bacterial community structure on a weathered pegmatitic granite. Microb Ecol 51:526–534

Gleeson D, Mathes F, Farrell M, Leopold M (2016) Environmental drivers of soil microbial community structure and function at the Avon River Critical Zone Observatory. Sci Total Environ 571:1407–1418. https://doi.org/10.1016/j.scitotenv.2016.05.185

Gobat JM, Aragno M, Matthey W (2004) The living soil, fundamentals of soil science and soil biology. Science Publishers, Enfield

Gregory MM, Leslie TW, Drinkwater LE (2016) Agroecological and social characteristics of New York city community gardens: contributions to urban food security, ecosystem services, and environmental education. Urban Ecosyst 19:763–794. https://doi.org/10.1007/s11252-015-0505-1

Grella C, Renshaw A, Wright IA (2018) Invasive weeds in urban riparian zones: the influence of catchment imperviousness and soil chemistry across an urbanization gradient. Urban Ecosyst 21:505–517. https://doi.org/10.1007/s11252-018-0736-z

Grewal SS, Cheng Z, Masih S, Wolboldt M, Huda N, Knight A, Grewal PS (2010) Are urban vacant lots suitable for food production? An assessment of soil nematode food webs and nutrient pools in community gardens and vacant lots. J Nematol 42:244–244

Griffiths RI, Whiteley AS, O'Donnell AG, Bailey MJ (2003) Physiological and community responses of established grassland bacterial populations to water stress. Appl Environ Microbiol 69:6961–6968

Guilland C, Maron PA, Damas O, Ranjard L (2018) Biodiversity of urban soils for sustainable cities. Environ Chem Lett 16:1267–1282. https://doi.org/10.1007/s10311-018-0751-6

Harms H, Schlosser D, Wick LY (2011) Untapped potential: exploiting fungi in bioremediation of hazardous chemicals. Nat Rev Microbiol 9:177–192. https://doi.org/10.1038/nrmicro2519

Harrison C, Davies G (2002) Conserving biodiversity that matters: practitioners' perspectives on brownfield development and urban nature conservation in London. J Environ Manag 65:95–108. https://doi.org/10.1006/jema.2002.0539

Hill GT, Mitkowski NA, Aldrich-Wolfe L, Emele LR, Jurkonie DD, Ficke A et al (2000) Methods for assessing the composition and diversity of soil microbial communities. Appl Soil Ecol 15:25–36. https://doi.org/10.1016/s0929-1393(00)00069-x

Hug LA, Baker BJ, Anantharaman K, Brown CT, Probst AJ, Castelle CJ et al (2016) A new view of the tree of life. Nat Microbiol 1:6. https://doi.org/10.1038/nmicrobiol.2016.48

Hugerth LW, Andersson AF (2017) Analysing microbial community composition through amplicon sequencing: from sampling to hypothesis testing. Front Microbiol 8:22. https://doi.org/10.3389/fmicb.2017.01561

Humbert JY, Ghazoul J, Richner N, Walter T (2010) Hay harvesting causes high orthopteran mortality. Agric Ecosyst Environ 139:522–527. https://doi.org/10.1016/j.agee.2010.09.012

Jerrentrup JS, Seither M, Petersen U, Isselstein J (2015) Little grazer species effect on the vegetation in a rotational grazing system. Agric Ecosyst Environ 202:243–250. https://doi.org/10.1016/j.agee.2015.01.007

Joimel S, Schwartz C, Hedde M, Kiyota S, Krogh PH, Nahmani J et al (2017) Urban and industrial land uses have a higher soil biological quality than expected from physicochemical quality. Sci Total Environ 584:614–621. https://doi.org/10.1016/j.scitotenv.2017.01.086

Jones EL, Leather SR (2012) Invertebrates in urban areas: a review. Eur J Entomol 109:463–478 https://doi.org/10.14411/eje.2012.060

Jones CG, Gutierrez JL, Groffman PM, Shachak M (2006) Linking ecosystem engineers to soil processes: a framework using the Jenny state factor equation. Eur J Soil Biol 42:S39–S53. https://doi.org/10.1016/j.ejsobi.2006.07.017

Jones OAH, Maguire ML, Griffin JL, Dias DA, Spurgeon DJ, Svendsen C (2013) Metabolomics and its use in ecology. Austral Ecol 38:713–720. https://doi.org/10.1111/aec.12019

Joyner JL, Kerwin J, Deeb M, Lozefski G, Prithiviraj B, Paltseva A et al (2019) Green infrastructure design influences communities of urban soil bacteria. Front Microbiol 10:982–982. https://doi.org/10.3389/fmicb.2019.00982

Kattwinkel M, Biedermann R, Kleyer M (2011) Temporary conservation for urban biodiversity. Biol Conserv 144:2335–2343. https://doi.org/10.1016/j.biocon.2011.06.012

Kim YJ, Yoo G (2021) Suggested key variables for assessment of soil quality in urban roadside tree systems. J Soils Sediments 11. https://doi.org/10.1007/s11368-020-02827-5

Koh LP, Sodhi NS (2004) Importance of reserves, fragments, and parks for butterfly conservation in a tropical urban landscape. Ecol Appl 14:1695–1708. https://doi.org/10.1890/03-5269

Kuczynski J, Liu ZZ, Lozupone C, McDonald D, Fierer N, Knight R (2010) Microbial community resemblance methods differ in their ability to detect biologically relevant patterns. Nat Methods 7:813–U67. https://doi.org/10.1038/nmeth.1499

Laidlaw MAS, Alankarage DH, Reichman SM, Taylor MP, Ball AS (2018) Assessment of soil metal concentrations in residential and community vegetable gardens in Melbourne, Australia. Chemosphere 199:303–311. https://doi.org/10.1016/j.chemosphere.2018.02.044

Lear G, Dickie I, Banks J, Boyer S, Buckley HL, Buckley TR et al (2018) Methods for the extraction, storage, amplification and sequencing of DNA from environmental samples. New Zealand J Ecol 42:10–+. https://doi.org/10.20417/nzjecol.42.9

Lehmann A, Stahr K (2007) Nature and significance of anthropogenic urban soils. J Soils Sediments 7:247–260. https://doi.org/10.1065/jss2007.06.235

Liddicoat C, Weinstein P, Bissett A, Genie NJC, Mills JG, Waycott M, Breed MF (2019) Can bacterial indicators of a grassy woodland restoration inform ecosystem assessment and microbiota-mediated human health? Environ Int 129:105–117. https://doi.org/10.1016/j.envint.2019.05.011

Liu Z, He C, Zhou Y, Wu J (2014) How much of the world's land has been urbanized, really? A hierarchical framework for avoiding confusion. Landsc Ecol 29. https://doi.org/10.1007/s10980-014-0034-y

Loureiro C, Medema MH, van der Oost J, Sipkema D (2018) Exploration and exploitation of the environment for novel specialized metabolites. Curr Opin Biotechnol 50:206–213. https://doi.org/10.1016/j.copbio.2018.01.017

Lovett GM, Traynor MM, Pouyat RV, Carreiro MM, Zhu WX, Baxter JW (2000) Atmospheric deposition to oak forests along an urban-rural gradient. Environ Sci Technol 34:4294–4300. https://doi.org/10.1021/es001077q

Lozupone C, Knight R (2005) UniFrac: a new phylogenetic method for comparing microbial communities. Appl Environ Microbiol 71:8228–8235. https://doi.org/10.1128/aem.71.12.8228-8235.2005

Luo XS, Yu S, Zhu YG, Li XD (2012) Trace metal contamination in urban soils of China. Sci Total Environ 421:17–30. https://doi.org/10.1016/j.scitotenv.2011.04.020

MacIvor JS, Lundholm J (2011) Insect species composition and diversity on intensive green roofs and adjacent level-ground habitats. Urban Ecosyst 14:225–241. https://doi.org/10.1007/s11252-010-0149-0

Maron PA, Mougel C, Ranjard L (2011) Soil microbial diversity: methodological strategy, spatial overview and functional interest. C R Biol 334:403–411. https://doi.org/10.1016/j.crvi.2010.12.003

Martin-Laurent F, Philippot L, Hallet S, Chaussod R, Germon JC, Soulas G, Catroux G (2001) DNA extraction from soils: old bias for new microbial diversity analysis methods. Appl Environ Microbiol 67:2354–2359. https://doi.org/10.1128/aem.67.5.2354-2359.2001

McBride MB (1994) Environmental chemistry of soils. Oxford University Press, New York

McIntyre NE (2000) Ecology of urban arthropods: a review and a call to action. Ann Entomol Soc Am 93:825–835. https://doi.org/10.1603/0013-8746(2000)093[0825:eouaar]2.0.co;2

McKinney ML (2002) Urbanization, biodiversity, and conservation. Bioscience 52:883–890. https://doi.org/10.1641/0006-3568(2002)052[0883:ubac]2.0.co;2

McKinney ML (2006) Urbanization as a major cause of biotic homogenization. Biol Conserv 127:247–260. https://doi.org/10.1016/j.biocon.2005.09.005

Meng Q, Yang W, Men M, Bello A, Xu X, Xu B et al (2019) Microbial community succession and response to environmental variables during cow manure and corn straw composting. Front Microbiol 10. https://doi.org/10.3389/fmicb.2019.00529

Mills JG, Brookes JD, Gellie NJC, Liddicoat C, Lowe AJ, Sydnor HR et al (2019) Relating urban biodiversity to human health with the 'Holobiont' concept. Front Microbiol 10:8. https://doi.org/10.3389/fmicb.2019.00550

Miyashita T, Shinkai A, Chida T (1998) The effects of forest fragmentation on web spider communities in urban areas. Biol Conserv 86 (3):357–364

Moniruzzaman M, Wurch LL, Alexander H, Dyhrman ST, Gobler CJ, Wilhelm SW (2017) Virus-host relationships of marine single-celled eukaryotes resolved from metatranscriptomics. Nat Commun 8:16054. https://doi.org/10.1038/ncomms16054

Morgenroth J, Buchan GD (2009) Soil moisture and aeration beneath pervious and impervious pavements. Arboriculture and Urban Forestry 35:135–141

Mummey D, Holben W, Six J, Stahl P (2006) Spatial stratification of soil bacterial populations in aggregates of diverse soils. Microb Ecol 51:404–411

Nahmani J, Lavelle P (2002) Effects of heavy metal pollution on soil macrofauna in a grassland of northern France. Eur J Soil Biol 38:297–300. https://doi.org/10.1016/s1164-5563(02)01169-x

Nkongolo KK, Narendrula-Kotha R (2020) Advances in monitoring soil microbial community dynamic and function. J Appl Genet 61:249–263. https://doi.org/10.1007/s13353-020-00549-5

Norton BA, Bending GD, Clark R, Corstanje R, Dunnett N, Evans KL et al (2019) Urban meadows as an alternative to short mown grassland: effects of composition and height on biodiversity. Ecol Appl 29:21. https://doi.org/10.1002/eap.1946

NYC Department of Environmental Protection (2017) NYC green infrastructure 2017 annual report. NYC DEP, New York City. https://www1.nyc.gov/assets/dep/downloads/pdf/water/stormwater/green-infrastructure/gi-annual-report-2017.pdf

O'Brien FJM, Almaraz M, Foster MA, Hill AF, Huber DP, King EK et al (2019) Soil salinity and pH drive soil bacterial community composition and diversity along a lateritic slope in the Avon River critical zone observatory, Western Australia. Front Microbiol 10:20. https://doi.org/10.3389/fmicb.2019.01486

O'Donnell AG, Young IM, Rushton SP, Shirley MD, Crawford JW (2007) Visualization, modelling and prediction in soil microbiology. Nat Rev Microbiol 5:689–699. https://doi.org/10.1038/nrmicro1714

Oksanen J, Blanchet FG, Friendly M, Kindt R, Legendre P., McGlinn D et al (2020) 'Vegan': community ecology package (R package version 2.5-7) https://cran.r-project.org/web/packages/vegan/index.html. Accessed 20210119

Oostenbrink M (1960) Estimating nematode populations by some selected methods. In: Sasser JN, Jenkins WR (eds) Nematology. The University of North Carolina Press, Chapel Hill, pp 85–102

Orellana LH, Ben Francis T, Kruger K, Teeling H, Muller MC, Fuchs BM et al (2019) Niche differentiation among annually recurrent coastal marine group II Euryarchaeota. ISME J 13:3024–3036. https://doi.org/10.1038/s41396-019-0491-z

Orsini F, Kahane R, Nono-Womdim R, Gianquinto G (2013) Urban agriculture in the developing world: a review. Agron Sustain Dev 33:695–720. https://doi.org/10.1007/s13593-013-0143-z

Ow LF, Ghosh S (2017) Urban tree growth and their dependency on infiltration rates in structural soil and structural cells. Urban For Urban Green 26:41–47. https://doi.org/10.1016/j.ufug.2017.06.005

Patti GJ, Yanes O, Siuzdak G (2012) Metabolomics: the apogee of the omics trilogy. Nat Rev Mol Cell Biol 13:263–269

Paul MJ, Meyer JL (2001) Streams in the urban landscape. Annu Rev Ecol Syst 32:333–365

Pavao-Zuckerman MA (2008) The nature of urban soils and their role in ecological restoration in cities. Restor Ecol 16:642–649. https://doi.org/10.1111/j.1526-100X.2008.00486.x

Pavao-Zuckerman MA, Coleman DC (2007) Urbanization alters the functional composition, but not taxonomic diversity, of the soil nematode community. Appl Soil Ecol 35:329–339. https://doi.org/10.1016/j.apsoil.2006.07.008

Peck DC (2009) Long-term effects of imidacloprid on the abundance of surface- and soil-active nontarget fauna in turf. Agric For Entomol 11:405–419. https://doi.org/10.1111/j.1461-9563.2009.00454.x
Peffy T, Nawaz R (2008) An investigation into the extent and impacts of hard surfacing of domestic gardens in an area of Leeds, United Kingdom. Landsc Urban Plan 86:1–13. https://doi.org/10.1016/j.landurbplan.2007.12.004
Petric I, Philippot L, Abbate C, Bispo A, Chesnot T, Hallin S et al (2011) Inter-laboratory evaluation of the ISO standard 11063 "soil quality - method to directly extract DNA from soil samples". J Microbiol Methods 84:454–460. https://doi.org/10.1016/j.mimet.2011.01.016
Pickett STA, Cadenasso ML, Grove JM, Boone CG, Groffman PM, Irwin E et al (2011) Urban ecological systems: scientific foundations and a decade of progress. J Environ Manag 92:331–362. https://doi.org/10.1016/j.jenvman.2010.08.022
Piotrowska-Dlugosz A, Charzynski P (2015) The impact of the soil sealing degree on microbial biomass, enzymatic activity, and physicochemical properties in the Ekranic Technosols of Torun (Poland). J Soils Sediments 15:47–59. https://doi.org/10.1007/s11368-014-0963-8
Pouyat RV, Szlavecz K, Yesilonis ID, Wong CP, Murawski L, Marra P et al (2015) Multi-scale assessment of metal contamination in residential soil and soil fauna: a case study in the Baltimore-Washington metropolitan region, USA. Landsc Urban Plan 142:7–17. https://doi.org/10.1016/j.landurbplan.2015.05.001
Pulleman M, Creamer R, Hamer U, Helder J, Pelosi C, Peres G, Rutgers M (2012) Soil biodiversity, biological indicators and soil ecosystem services-an overview of European approaches. Curr Opin Environ Sustain 4:529–538. https://doi.org/10.1016/j.cosust.2012.10.009
Ramirez KS, Leff JW, Barberan A, Bates ST, Betley J, Crowther TW et al (2014) Biogeographic patterns in below-ground diversity in New York City's Central Park are similar to those observed globally. Proc R Soc B Biol Sci 281:9. https://doi.org/10.1098/rspb.2014.1988
Reese AT, Savage A, Youngsteadt E, McGuire KL, Koling A, Watkins O et al (2016) Urban stress is associated with variation in microbial species composition—but not richness—in Manhattan. ISME J 10:751–760. https://doi.org/10.1038/ismej.2015.152
Ringner M (2008) What is principal component analysis? Nat Biotechnol 26:303–304. https://doi.org/10.1038/nbt0308-303
Ritz K, Trudgill DL (1999) Utility of nematode community analysis as an integrated measure of the functional state of soils: perspectives and challenges - discussion paper. Plant Soil 212:1–11. https://doi.org/10.1023/a:1004673027625
Ritz K, Black HIJ, Campbell CD, Harris JA, Wood C (2009) Selecting biological indicators for monitoring soils: a framework for balancing scientific and technical opinion to assist policy development. Ecol Indic 9:1212–1221. https://doi.org/10.1016/j.ecolind.2009.02.009
Robinson SL, Lundholm JT (2012) Ecosystem services provided by urban spontaneous vegetation. Urban Ecosyst 15:545–557. https://doi.org/10.1007/s11252-012-0225-8
Rodríguez-Eugenio N, McLaughlin M, Pennock D (2018) Soil pollution - a hidden reality. Food and Agriculture Organization of the United Nations, Rome. http://www.fao.org/3/I9183EN/i9183en.pdf
Roesch LF, Fulthorpe RR, Riva A, Casella G, Hadwin AKM, Kent AD et al (2007) Pyrosequencing enumerates and contrasts soil microbial diversity. ISME J 1:283–290. https://doi.org/10.1038/ismej.2007.53
Rombke J, Sousa JP, Schouten T, Riepert F (2006) Monitoring of soil organisms: a set of standardized field methods proposed by ISO. Eur J Soil Biol 42:S61–S64. https://doi.org/10.1016/j.ejsobi.2006.07.016
Rumble H, Gange AC (2013) Soil microarthropod community dynamics in extensive green roofs. Ecol Eng 57:197–204. https://doi.org/10.1016/j.ecoleng.2013.04.012
Rutledge PJ, Challis GL (2015) Discovery of microbial natural products by activation of silent biosynthetic gene clusters. Nat Rev Microbiol 13:509–523. https://doi.org/10.1038/nrmicro3496
Santorufo L, Van Gestel CAM, Rocco A, Maisto G (2012) Soil invertebrates as bioindicators of urban soil quality. Environ Pollut 161:57–63. https://doi.org/10.1016/j.envpol.2011.09.042

Schmidt DJE, Pouyat R, Szlavecz K, Setala H, Kotze DJ, Yesilonis I et al (2017) Urbanization erodes ectomycorrhizal fungal diversity and may cause microbial communities to converge. Nat Ecol Evol 1:9. https://doi.org/10.1038/s41559-017-0123

Schmidt DJE, Kotze DJ, Hornung E, Setala H, Yesilonis I, Szlavecz K et al (2019) Metagenomics reveals bacterial and archaeal adaptation to urban land-use: N catabolism, methanogenesis, and nutrient acquisition. Front Microbiol 10:17. https://doi.org/10.3389/fmicb.2019.02330

Schrader S, Böning M (2006) Soil formation on green roofs and its contribution to urban biodiversity with emphasis on collembolans. Pedobiologia 50:347–356. https://doi.org/10.1016/j.pedobi.2006.06.003

Schulze WX, Gleixner G, Kaiser K, Guggenberger G, Mann M, Schulze ED (2005) A proteomic fingerprint of dissolved organic carbon and of soil particles. Oecologia 142:335–343. https://doi.org/10.1007/s00442-004-1698-9

Semple KT, Doick KJ, Jones KC, Burauel P, Craven A, Harms H (2004) Defining bioavailability and bioaccessibility of contaminated soil and sediment is complicated. Environ Sci Technol 38:228A–231A

Seto KC, Guneralp B, Hutyra LR (2012) Global forecasts of urban expansion to 2030 and direct impacts on biodiversity and carbon pools. Proc Nat Acad Sci U S A 109:16083–16088. https://doi.org/10.1073/pnas.1211658109

Shahsavari E, Poi G, Aburto-Medina A, Haleyur N, Ball AS (2017) Bioremediation approaches for petroleum hydrocarbon-contaminated environments. In: Anjum NA, Gill SS, Tuteja N (eds) Enhancing cleanup of environmental pollutants: volume 1: biological approaches. Springer, Cham

Shakya M, Lo C-C, Chain PSG (2019) Advances and challenges in metatranscriptomic analysis. Front Genet 10. https://doi.org/10.3389/fgene.2019.00904

Shannon CE (1948) A mathematical theory of communication. Bell Syst Tech J 27:379–423. https://doi.org/10.1002/j.1538-7305.1948.tb01338.x

Sikorski P, Szumacher I, Sikorska D, Kozak M, Wierzba M (2013) Effects of visitor pressure on understory vegetation in Warsaw forested parks (Poland). Environ Monit Assess 185:5823–5836. https://doi.org/10.1007/s10661-012-2987-0

Smetak KM, Johnsn-Maynard JL, Lloyd JE (2007) Earthworm population density and diversity in different-aged urban systems. Appl Soil Ecol 37:161–168. https://doi.org/10.1016/j.apsoil.2007.06.004

Smith J, Chapman A, Eggleton P (2006) Baseline biodiversity surveys of the soil macrofauna of London's green spaces. Urban Ecosyst 9:337. https://doi.org/10.1007/s11252-006-0001-8

Strong DT, Sale PWG, Helyar KR (1998) The influence of the soil matrix on nitrogen mineralisation and nitrification. II The pore system as a framework for mapping the organisation of the soil matrix. Australian J Soil Res 36:855–872

Szlavecz K, Placella SA, Pouyat RV, Groffman PM, Csuzdi C, Yesilonis I (2006) Invasive earthworm species and nitrogen cycling in remnant forest patches. Appl Soil Ecol 32:54–62. https://doi.org/10.1016/j.apsoil.2005.01.006

Taylor JR, Lovell ST (2015) Urban home gardens in the global north: a mixed methods study of ethnic and migrant home gardens in Chicago, IL. Renew Agric Food Syst 30:22–32. https://doi.org/10.1017/S1742170514000180

Thompson GL, Kao-Kniffin J (2019) Urban grassland management implications for soil C and N dynamics: a microbial perspective. Front Ecol Evol 7:15. https://doi.org/10.3389/fevo.2019.00315

Thornton I, Farago ME, Thums CR, Parrish RR, McGill RAR, Breward N et al (2008) Urban geochemistry: research strategies to assist risk assessment and remediation of brownfield sites in urban areas. Environ Geochem Health 30:565–576. https://doi.org/10.1007/s10653-008-9182-9

Tisdall JM, Nelson SE, Wilkinson KG, Smith SE, McKenzie BM (2012) Stabilisation of soil against wind erosion by six saprotrophic fungi. Soil Biol Biochem 50:134–141. https://doi.org/10.1016/j.soilbio.2012.02.035

Torsvik V, Daae FL, Sandaa RA, Ovreas L (1998) Novel techniques for analysing microbial diversity in natural and perturbed environments. J Biotechnol 64:53–62. https://doi.org/10.1016/s0168-1656(98)00103-5

Turrini A, Sbrana C, Giovannetti M (2015) Belowground environmental effects of transgenic crops: a soil microbial perspective. Res Microbiol 166:121–131. https://doi.org/10.1016/j.resmic.2015.02.006

United Nations (2018) World urbanization prospects: the 2018 revision. Custom data acquired via https://population.un.org/

Uno S, Cotton J, Philpott SM (2010) Diversity, abundance, and species composition of ants in urban green spaces. Urban Ecosyst 13:425–441. https://doi.org/10.1007/s11252-010-0136-5

Varet M, Burel F, Petillon J (2014) Can urban consolidation limit local biodiversity erosion? Responses from carabid beetle and spider assemblages in Western France. Urban Ecosyst 17:123–137. https://doi.org/10.1007/s11252-013-0307-2

Vauramo S, Setala H (2010) Urban belowground food-web responses to plant community manipulation - impacts on nutrient dynamics. Landsc Urban Plan 97:1–10. https://doi.org/10.1016/j.landurbplan.2010.04.004

Vauramo S, Setala H (2011) Decomposition of labile and recalcitrant litter types under different plant communities in urban soils. Urban Ecosyst 14:59–70. https://doi.org/10.1007/s11252-010-0140-9

Wall DH, Nielsen UN, Six J (2015) Soil biodiversity and human health. Nature 528:69–76. https://doi.org/10.1038/nature15744

Walters SP, Thebo AL, Boehm AB (2011) Impact of urbanization and agriculture on the occurrence of bacterial pathogens and stx genes in coastal waterbodies of Central California. Water Res 45:1752–1762. https://doi.org/10.1016/j.watres.2010.11.032

Wang HB, Zhang ZX, Li H, He HB, Fang CX, Zhang AJ et al (2011) Characterization of metaproteomics in crop rhizospheric soil. J Proteome Res 10:932–940. https://doi.org/10.1021/pr100981r

Wang HT, Cheng MY, Dsouza M, Weisenhorn P, Zheng TL, Gilbert JA (2018) Soil bacterial diversity is associated with human population density in urban greenspaces. Environ Sci Technol 52:5115–5124. https://doi.org/10.1021/acs.est.7b0641

Wei BG, Yang LS (2010) A review of heavy metal contaminations in urban soils, urban road dusts and agricultural soils from China. Microchem J 94:99–107. https://doi.org/10.1016/j.microc.2009.09.014

Withers E, Hill PW, Chadwick DR, Jones DL (2020) Use of untargeted metabolomics for assessing soil quality and microbial function. Soil Biol Biochem 143:107758. https://doi.org/10.1016/j.soilbio.2020.107758

Woese CR, Kandler O, Wheelis ML (1990) Towards a natural system of organisms - proposal for the domains archaea, bacteria, and Eucarya. PRoc Nat Acad Sci U S A 87:4576–4579. https://doi.org/10.1073/pnas.87.12.4576

Wolfe BE, Klironomos JN (2005) Breaking new ground: soil communities and exotic plant invasion. Bioscience 55:477–487. https://doi.org/10.1641/0006-3568(2005)055[0477:bngsca]2.0.co;2

Yang J-L, Zhang G-L (2015) Formation, characteristics and eco-environmental implications of urban soils – a review. Soil Sci Plant Nutr 61:30–46. https://doi.org/10.1080/00380768.2015.1035622

Zhao X, Xia X (2012) Total nitrogen and total phosphorous in urban soils used for different purposes in Beijing, China. Procedia Environ Sci 13:95–104

Zogg GP, Zak DR, Ringelberg DB, MacDonald NW, Pregitzer KS, White DC (1997) Compositional and functional shifts in microbial communities due to soil warming. Soil Sci Soc Am J 61:475–481. https://doi.org/10.2136/sssaj1997.03615995006100020015x

Chapter 9
Urban Soil as a Source and Sink

Andrew W. Rate

Abstract Urban soils are well known to be receiving environments for material from other environmental compartments, particularly when human activity is involved. In this chapter, we start by discussing the transport of material to and from urban soils, and then consider urban soils as both sink and source for contaminants and other materials. Soils will act as a sink if the added materials persist in the soil environment. Urban soils can also be net contributors, or sources, of material to other environmental compartments, and the losses of material from soils are also strongly affected by human activities. In between these two extremes, soils (including those in urban environments) can act as a temporary sink, or transient storage, for substances which persist long enough to affect some soil functions, but are lost from the soil at a rate where significant accumulation does not occur.

To obtain a more complete understanding of the role of urban soils as a source and sink of materials, we will first discuss the most common transport mechanisms, and then move on to some discussion and examples of urban soils acting as either sinks for, or sources of, material to or from other environmental compartments. The possibility of contaminant transfers to and especially from soils means that the soils themselves, and the processes occurring in them, need to be considered when estimating the risk to humans or to receiving environments.

Keywords Contaminant transport · Urban soils · Contaminant sources · Conceptual site model · Risk analysis · Regulatory guidelines

---

A. W. Rate (✉)
School of Agriculture and Environment, University of Western Australia, Crawley, WA, Australia
e-mail: andrew.rate@uwa.edu.au

A. W. Rate (ed.), *Urban Soils*, Progress in Soil Science,
https://doi.org/10.1007/978-3-030-87316-5_9

What you could learn from this chapter:

- Various physical and biological mechanisms which transport materials in soils.
- What the external sources for materials or substances added to urban soils include.
- Different ways that urban soils can be a source of materials or substances to receiving environments.
- The basics of risk analysis as applied to soils in urban environments.

9.1 Transport Mechanisms Relating to Substances in Soils

9.1.1 Diffusion

Diffusion refers to the movement of substances down a concentration gradient of that substance, that is, from spaces with high concentration to spaces with low concentration. The process is driven solely by the concentration gradient (concentration ÷ distance), and is generally slow and acts over short distances such as very fine pores within soil microaggregates, or in water films surrounding soil particles (Fig. 9.1); see also Carrillo-González et al. (2006). There may be some confusion with differing usage of terminology, since the term "diffusion" is sometimes used to denote the gradual spreading of substances such as contaminants from a source to a wider spatial scale. In this textbook we will consistently use the small-scale, physicochemical definition above. Diffusion is only a significant transport process within fluid (air and water) phases in soils. For example, the transfer of low-volatility organic contaminants such as PAHs from soil to atmosphere is a process limited by diffusion in the vapour phase (Wei et al. 2014). In the aqueous (water) phase, diffusion rates may control transfer of trace element contaminants from urban river sediments to the water column (Garban et al. 1996), or limit the rate of transfer of

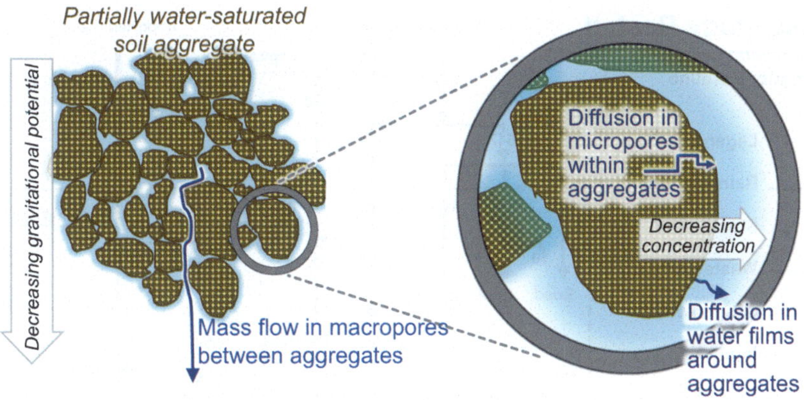

Fig. 9.1 Transport processes for gases, water, and solutes on the micro-scale within aggregates of particles in soils. Mass flow can also be driven by gradients in gas pressure, matric potential, etc. (graphic by Andrew W. Rate)

trace elements from solids to pore water in contaminated urban soils (Dubé et al. 2003).

9.1.2 *Advection (Mass Flow)*

Mass flow can potentially transfer contaminants in any phase: gases by aeolian (wind) processes, or flow in soil pores down a gravitational or pressure gradient; liquids by streamflow, or water flow within soils (such as leaching of dissolved and colloidal material); and solids by wind or water erosion. Advective transport depends on the factors controlling movement of the medium carrying the material (e.g. contaminant) of interest, such as atmospheric pressure gradients for wind, and gravitational potential difference and/or matric potential difference for vertical water flow in soils, stream flow, or water erosion. Anthropogenic disturbance such as vehicle traffic or excavation of soil can increase the amount of material transported by any of these mechanisms. Figure 9.2 shows a water balance diagram with examples of the various water fluxes in an urban environment and explicitly considers anthropogenic fluxes such as flow into urban soil from impervious surfaces or leakage of piped water.

It is well known that mass flows of nutrients such as N and P can occur in urban environments, with soils contributing by leaching (vertical water flow) or erosion to other environmental compartments, commonly surface water or groundwater (Carpenter et al. 1998). Metals and other trace elements can leach from soils into water bodies, usually in low concentrations (Parker et al. 1978), but acidification of soil (e.g. by acid sulphate soil oxidation) can greatly increase the concentrations and amounts of metals leached due to mineral dissolution and release of adsorbed ions

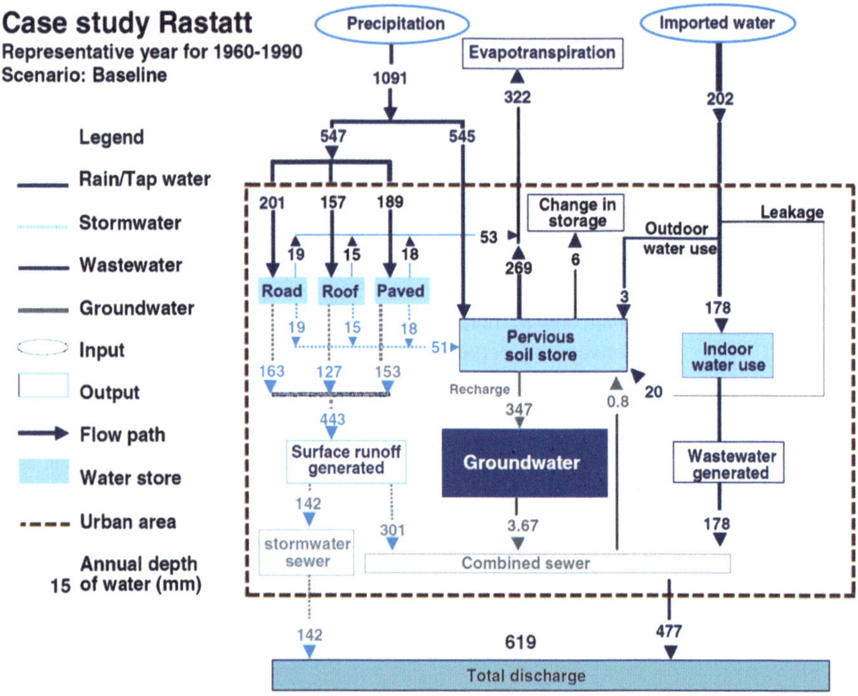

Fig. 9.2 Example of water fluxes, including into and out of storage in soils with no impervious surface cover, for an urban environment in Rastatt, Germany (Wolf et al. 2007; used with permission from Springer)

(Åström 2001). Some organic contaminants such as volatile hydrocarbons may be removed by soil-atmosphere fluxes or microbial degradation before leaching can occur, but less volatile or less degradable organic contaminants such as chlorinated hydrocarbons, pesticides, or pharmaceuticals may leach from soil into groundwater (Shepherd et al. 2006; Al-Rajab et al. 2008; Siemens et al. 2010; Slavens and Petrovic 2012). Mass flow can also result in transport of microorganisms in soils, such as the potentially pathogenic *Escherichia coli* and intestinal enterococci from faecal matter (Garcia-Armisen and Servais 2007).

9.1.3 Adhering

Soil and associated contaminants can adhere to a range of surfaces, such as human skin or clothing, harvested plants and food items, and vehicles. Urban soil adhering to human skin can transport metals (Siciliano et al. 2009) or organic contaminants (Siciliano et al. 2010), with adhering soil typically composed of finer particles than the bulk soil, making it more enriched in contaminants (Yamamoto et al. 2006). Soil adhering to vegetables is also known to transport metals (McBride et al. 2014) and plant pathogens (Ebrahimi et al. 2016). In some environments, soil adhered to vehicles is an important vector of the spores of parasitic fungi such as *Phytophthora* spp. which may be important pathogens of trees in urban environments (Lyons 2008; Barber et al. 2013).

9.1.4 Bioturbation

The mixing of soil by organisms (bioturbation), such as burrowing invertebrates or mammals, or soil inversion by tree fall, can transport material within or even beyond the soil environment. For example, Lévêque et al. (2019) showed that earthworm bioturbation could transport metals such as Pb, Cd, Cu, and Zn, changing the bioavailability of these elements in the process. Bioturbation was also an assumed process in a model of fate and transport of TCDD (a dioxin, which is a persistent chlorinated aromatic organic pollutant) in urban soils and dusts by Thibodeaux and Lipsky (1985).

9.2 Urban Soil as Source and Sink of Material

9.2.1 Point and Diffuse Sources of Soil Contamination

Cities contain soil contamination from both *diffuse sources*, where a single location cannot be identified as the origin of soil contaminants, and *point sources*, where a discrete location can be isolated as the pollution source (Fig. 9.3). In a sense, entire cities or urban areas can be considered to be diffuse sources of contamination, based on regional-scale soil surveys (e.g. Cicchella et al. 2015). We more usually, however, consider defined types of human activities as diffuse pollution sources (Fig. 9.3), such as road traffic, fuel combustion for domestic heating, poorly controlled waste disposal, weathering of metals and paints on buildings, and construction dusts (Norra et al. 2001; Manta et al. 2002; Fabietti et al. 2010; McIlwaine et al. 2017; Laidlaw et al. 2018). Diffuse sources may represent the combined effect of multiple point sources (DEC 2006).

9.2.2 Dusts and Other Airborne Particulates

Material from soils becomes entrained or resuspended in air, forming airborne dusts which have the potential to enter other environments. Allott et al. (1994) found mechanical transport of soil to be a significant source of dust to outdoor and indoor environments. While in some instances this dust can be transported long distances

Fig. 9.3 Schematic urban landscape diagram showing some examples of diffuse and point sources of contaminants (graphic by Andrew W. Rate)

from remote sources (Almeida et al. 2005), urban soil can also be a source of dust on a local scale (De Miguel et al. 1997). One of the sinks for dust derived from urban soil is house dust (Kelepertzis et al. 2016), and transfer of dust from soils to indoor living spaces is an important process, since the potential for exposure to humans is maximised. House dusts can also contain greater concentrations of trace elements than the local soils (Fergusson and Kim 1991). House dusts have been shown to represent a major pathway for human ingestion of potentially toxic contaminants such as lead (Lanphear et al. 1998) and some organic compounds (e.g. Jones-Otazo et al. 2005).

Dust from external sources can also enter soils; Fernández-Caliani (2012) describe dust from industrial sources contaminating soil in peri-urban Huelva, Spain. Similarly, Liu et al. (2014) concluded that dust from construction and demolition contributed potentially toxic metals to soils in Nanjing, China.

9.2.3 *Road Traffic*

Vehicles travelling on roads can contribute contaminants to soils directly, via volatile or particulate emissions from the vehicles themselves, or indirectly through accumulation and subsequent transport of road dusts (Fig. 9.4). The pollutants emitted include:

- ***Combustion products*** (including those from partial combustion), such as carbon dioxide, carbon monoxide, nitrogen oxides, and "black carbon" (sooty) particles.
- ***Evaporative losses*** from fuels and oils (mainly hydrocarbons including aliphatic alkanes and the hazardous BTEX (benzene, toluene, xylene, and related compounds) group).
- ***Abrasion losses*** from tyres, brakes, and other mechanical components, including the potentially toxic elements Al, As, Ba, Br, Cd, Cr, Cu, Fe, Mn, Pb, and Zn, poly-aromatic hydrocarbons (PAH), and also microrubbers and microplastics.

Combustion in vehicle engines, especially of diesel fuel, is a major source of black carbon in urban soils (Bucheli et al. 2004; Nehls and Shaw 2010). These black carbon particles may be associated with the toxic PAH family of persistent organic pollutants (see Chap. 7) and represent a significant human health issue on their own, being transported in air as very fine particles (Harrison et al. 1997; Castanho and Artaxo 2001).

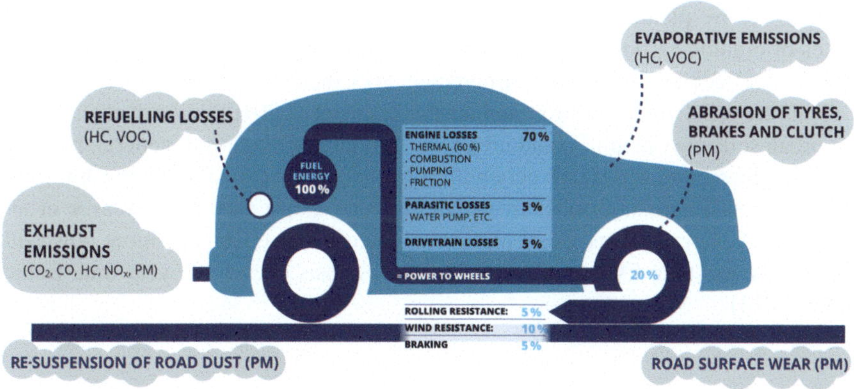

Fig. 9.4 Emission of pollutants by vehicles (redrawn from European Environment Agency 2016) (used under the terms of a CC-BY-2.5-DK licence). CO_2 = Carbon dioxide, HCs = hydrocarbons, VOC = volatile organic carbon (HCs and VOC include n-alkanes, benzene, toluene, xylene, etc.), CO = carbon monoxide, NO_X = nitrogen oxides, PM = particulate matter (e.g. Al, As, Ba, Br, Cd, Cr, Cu, Fe, Mn, Pb, Zn, polyaromatic hydrocarbons (PAH), microrubbers, microplastics)

9.2.4 *Atmospheric Deposition and Emissions from Urban Soils*

In Chap. 6 we introduced the idea that urban soils could generate fluxes of contaminants to the atmosphere. These fluxes are driven by biological processes such as respiration or denitrification, direct volatilisation of mainly organic contaminants, or from entrainment of fine particulate matter in air during wind erosion of soils. Volatilisation of organic contaminants was addressed in Chap. 7. Since biological respiration in soils produces carbon dioxide (predominantly) or methane, it is highly significant in the context of urban and global warming and anthropogenic climate change.

9.2.4.1 Carbon Fluxes to and from Urban Soils

Urban and peri-urban areas contribute more than 70% of global greenhouse gas emissions, which are dominated by fluxes of carbon dioxide. Of course, not all of these are related to urban soils; the largest urban fluxes of CO_2 or CH_4 to the atmosphere are from power generation and transport (Bellucci et al. 2012). Soils in urban environments do emit CO_2, however; for example, Byrne et al. (2008) and Livesley et al. (2010) (Fig. 9.5) both measured net CO_2 emissions from urban soils. The pathways for carbon uptake by soils are not, however, by atmosphere-to-soil fluxes of CO_2. Plants absorb CO_2 during photosynthesis and return carbon to soils via litter fall. Based on the heat and CO_2 islands in Phoenix, USA, Shen et al. (2008) concluded that increased CO_2 concentrations in air increased above-ground net primary productivity by plants (depending on water availability), which led to increased fluxes of organic carbon into soils. The increased temperatures in urban areas had less direct effect on soil organic carbon contents (Shen et al. 2008).

Soils are usually net sinks for methane (Livesley et al. 2010; Bellucci et al. 2012), but urban soils over landfills (and urban waste-related pathways in general, particularly for putrescible waste) are net emitters of methane, depending on the ability of the soil biota (methanotrophs) to metabolise or oxidise methane (Weaver et al. 2019). Methane emissions are of particular concern, since CH_4 is a more powerful greenhouse gas than CO_2, by 28-fold on a mass basis over a 100 year time frame (IPCC 2013).

9.2.4.2 Nitrogen Fluxes to and from Urban Soils

Urban soils contribute significant fluxes of nitrous oxide (N_2O) to air, similar in magnitude to fertilised agricultural soils which are the greatest anthropogenic source of N_2O to the atmosphere (Bellucci et al. 2012). Conversely, the oxides of nitrogen (including N_2O) in the atmosphere are precursors of nitrate, which is a source of nitrogen to soils through dry and wet deposition, and is one mechanism for enrichment of nitrogen in urban soils (Shen et al. 2008; Du et al. 2015).

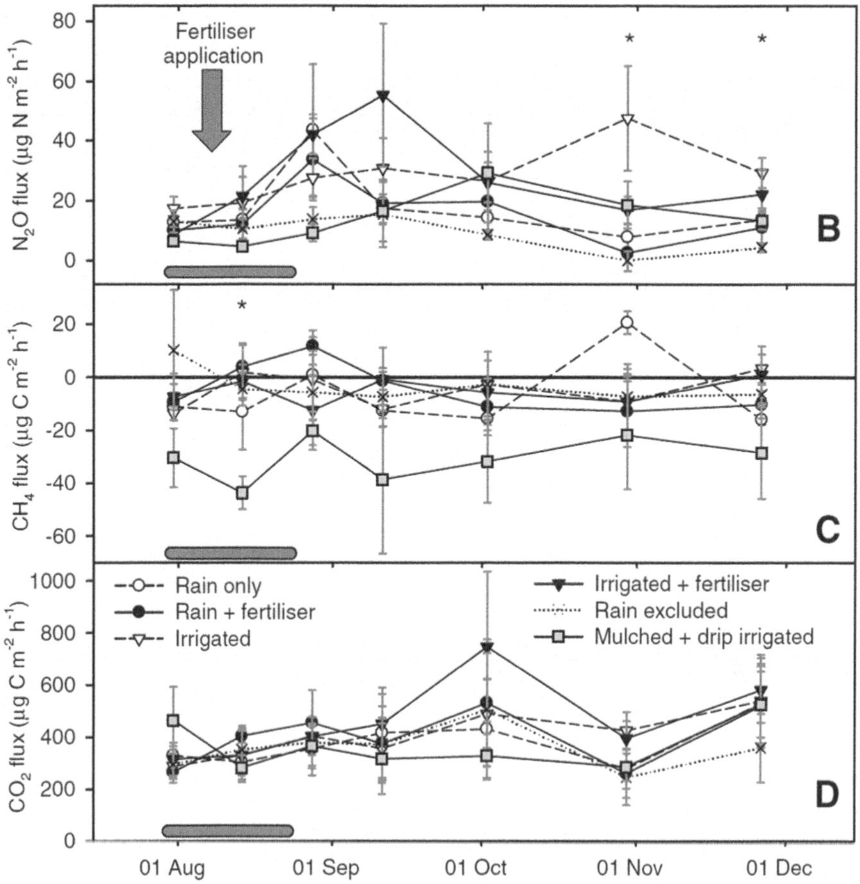

Fig. 9.5 Fluxes of greenhouse gases (**B** = N_2O, **C** = CH_4, **D** = CO_2) from soil in an urban park in Melbourne, Australia (from Livesley et al. 2010; used with permission from Springer)

9.2.4.3 Sulphur Fluxes to and from Urban Soils

The phenomenon of sulphur deposition to soils, commonly as SO_2, has been extensively studied in relation to pollution from combustion of sulphur-rich fossil fuels in urban and industrial contexts. The consequences for soils are usually greater rates of *acidification*, described comprehensively by van Breemen et al. (1984). While degradation of *natural* soils due to atmospheric contributions of SO_2 was the primary concern, sulphur deposition in *urban* environments and consequent soil acidification was described in a study by Du et al. (2015). Around cities in China, approximately 700,000 km^2 of land surface has soils which have acidified due to deposition of oxides of sulphur and nitrogen, with most acidification in a 70 km radius from urban centres (Du et al. 2015).

Although the ability of soils to act as a sink for SO_2 is well known, H_2S and SO_2 emissions from soils have also been measured, showing that soils can be a source of atmospheric sulphur compounds. Early research on a wide range of soils in the USA (Adams et al. 1981) showed that, along with H_2S, a number of other compounds (e.g. COS, CS_2) were involved in sulphur emissions from soils. These reduced sulphur compounds were considered to be precursors for atmospheric SO_2. Macdonald et al. (2004), however, measured direct emissions of SO_2 from oxidised acid sulphate soils. Although Macdonald et al.'s (2004) study was conducted outside an urban area, acid sulphate soils are known to occur in may urban environments (Fanning and Rabenhorst 1990; Appleyard et al. 2004; Clark and McConchie 2004). Emissions of H_2S from urban soils are a recognised issue wherever disposal of putrescible wastes occurs (Muezzinoglu 2003), but background H_2S emissions are also known from urban environments (Servant and Delapart 1982).

9.2.5 *Soil as a Source of Nutrients*

Concentrations of nutrient elements in soil in excess of the capacity of organisms to absorb them (for instance, uptake by plants or soil microorganisms) will often result in transfers out of the soil system, or losses. Nutrients can be lost from soils by leaching (mass transport with water), soil erosion, and for some elements by soil to atmosphere fluxes (such as volatilisation of gas-phase ammonia or hydrogen sulphide, or denitrification to produce nitrous oxide) (Carpenter et al. 1998; Groffman et al. 2002). Fig. 9.6 shows the contribution of various sources of nutrients, including soil erosion and urban areas, to a watershed in China (Du et al. 2014).

The effects of loss of nutrients (especially N and P) into waterways are commonly to cause excess concentrations of nutrients in water, or *eutrophication*. The

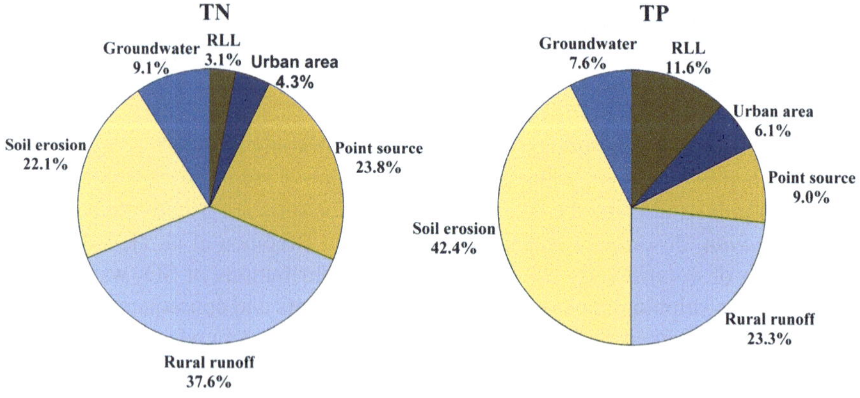

Fig. 9.6 Sources, including urban areas, of total nitrogen (TN) and total phosphorus (TP) in the semiarid watershed of a sub-basin of the Luan River in Hebei, China (recoloured from Du et al. (2014); used with permission from Springer). RLL is rural living and livestock

high concentrations of nutrients remove limitations to biomass production for aquatic photosynthetic organisms such as algae, causing phenomena such as algal blooms. In aquatic systems (such as streams, rivers, lakes, and estuaries), productivity is most commonly limited by the supply of phosphorus or nitrogen (Carpenter et al. 1998; Peters and Donohue 2001). Export of nutrients from soils, including urban areas, has also been implicated in the degradation of sensitive marine areas, such as the Great Barrier Reef in tropical Queensland, Australia (Tsatsaros et al. 2013). The specific practice of coastal soil reclamation may also result in nutrient export as soil conditions are substantially altered by drainage and other landscape modifications (Li et al. 2014). Eutrophication and algal blooms have their own adverse environmental effects, such as oxygen depletion, toxicity, and formation of sedimentary organic oozes, which are beyond the scope of this book.

Gas phase losses of nutrients from soils have different types of environmental consequences. Volatilisation of ammonia or hydrogen sulphide can cause toxicity to plants but, especially at higher concentrations, they are more commonly a nuisance odour to humans (Iglesias Jiménez and Perez Garcia 1989; Muezzinoglu 2003). Denitrification is more insidious; in theory, the end product of denitrification is the environmentally benign nitrogen gas (N_2). In reality, reaction intermediates such as oxides of nitrogen are released during denitrification in soils, in particular nitrous oxide (N_2O) which is a powerful greenhouse gas (on a mass basis, 265 times more effective at trapping solar radiation than CO_2 over a 100 year time frame (IPCC 2013)). Nitrous oxide is also an ozone-depleting substance, and is the most important ozone-depleting gas following the widespread international ban on use of halocarbons (the notorious "CFCs") in the late twentieth century (Ravishankara et al. 2009).

9.2.6 Soil as a Source and Sink for Potentially Toxic Substances

Potentially toxic elements such as some metals and metalloids, and potentially toxic compounds such as pesticides, hydrocarbons (including BTEX, PAH, PCB etc.), PFAS, and cyanide can in some cases be transferred from the soil to receiving environments by various pathways. Some examples of how urban soils can be a source of these contaminants are presented below.

9.2.6.1 Potentially Toxic Elements

The metals and metalloids usually have long residence times in soil, but their mobility can be enhanced by low pH, organic complexing agents, or colloid transport in coarse-textured soils. As discussed in Chap. 6, the intrinsic mobility of each element is different; for example, lead (Pb) and chromium (Cr) tend to be very immobile, while cadmium (Cd) and strontium (Sr) can be highly mobile. Kabata-Pendias (2011) shows that the same concepts apply to trace element bioavailability to plants, although the order of mobility and bioavailability may differ. Plant uptake of contaminant trace elements decreases in the approximate order Cd > Zn, Hg, Cu, Pb > As, Co > Mn, Ni > V, Cr > Fe, Zr, Sc, Ti, Ba. Plant uptake is obviously another pathway for contaminants to be transferred from soil into another environmental compartment, with the potential for further transfer due to plant harvest or natural biological cycling.

Dusts derived from wind erosion of soils also represent a pathway for transfer of trace elements out of soils into other environmental compartments, especially given that trace element concentrations are commonly greater in the finer grain-size fractions of soil (Harrison et al. 1997; Taylor et al. 2010), which are also the most easily transported in air. Metals and metalloids can leach into groundwater in low concentrations, as dissolved ions or complexes, or when associated with colloids or nanoparticles (Testa and Winegardner 2000; Appleyard et al. 2004; Grolimund and Borkovec 2005). The transport of metals in water is recognised, for example, in coastal soil reclamation, where practices such as drainage change redox and pH conditions (Li et al. 2014).

9.2.6.2 Hydrocarbons

Hydrocarbons such as fuels, solvents, and oils usually enter soil as a discrete phase, so the amount of contaminant can easily saturate the soil's sorption capacity (Kostecki et al. 2005). We have already discussed aspects of the environmental behaviour of hydrocarbons in Chap. 7. Liquid hydrocarbons with a density less than that of water (LNAPL, or light non-aqueous phase liquids) tend to travel vertically through soil and float at the upper surface of groundwater. In contrast, hydrocarbons with a density greater than that of water (DNAPL, or dense non-aqueous phase liquids) tend to travel vertically through soil and sink to confining layers below groundwater (Testa and Winegardner 2000). Many hydrocarbons have significant vapour pressures within ambient temperature and pressure ranges, so may volatilise from soils to air by simple evaporation (Kostecki et al. 2005). Hydrocarbons have low water solubility given their absence of significant ionisation and low polarity, but low concentrations may be present in groundwater (Riemann 1999; Testa and Winegardner 2000).

9.2.6.3 Persistent Organic Pollutants

The main groups of persistent organic pollutants (POPs – see Chap. 7) in soils are the polycyclic aromatic hydrocarbons (PAHs, such as naphthalene, anthracene, and chrysene), chlorobenzenes (CBs), polychlorinated biphenyls (PCBs), polychlorinated dibenzo-p-dioxins and furans (PCDD/Fs), with flame retardants such as polybrominated biphenyls (PBBs), polybrominated diphenyl ethers (PBDEs), and per- and poly-fluoroalkyl substances (PFAS). These compounds can be transferred from soil to other ecosystem compartments by being taken up by plants, leaching into groundwater in low concentrations, or being transported by wind or water erosion of soils (Johnson 2005; Rodríguez-Eugenio et al. 2018; U.S. EPA 2019). Other mechanisms for transfer of POPs are more specific to urban environments, such as fires and natural disasters which result in destruction of buildings and other structures (Plumlee et al. 2012). For example, Stec et al. (2019) describe the release of persistent organic pollutants such as dioxins, PAHs, fire retardants, and hydrocarbons during and following the tragically destructive Grenfell Tower fire in London, UK, in 2017.

9.2.6.4 Radionuclides

Radionuclides (i.e. radioactive isotopes of some elements including Cs, Ra, Th, U, Pu, and others) can be released into soils through nuclear accidents (Jacob et al. 1987; Endo et al. 2012) or weapons testing and the atmospheric fallout from such events. Soil-covered areas may act to concentrate radionuclides such as ^{137}Cs, as radioactive fallout is washed off impermeable surfaces by rainfall (Plumlee et al. 2012). The extraction and use of some mineral resources such as mineral sands (Arogunjo et al. 2009) or phosphate fertilisers (Al-Hamarneh and Awadallah 2009; Fernández-Caliani 2012) may also add radionuclides to soils. Urban soils will always contain background concentrations of radionuclides due to their presence in parent geological materials, including those used for construction purposes such as granites and brick-making clays (Arogunjo et al. 2009). Radioactive isotopes can be taken up by plants, and subsequently animal products destined for human consumption, which both represent a potential exposure pathway for humans (Rodríguez-Eugenio et al. 2018).

9.2.6.5 Mineral Contaminants Including Asbestos

The most hazardous transfer of asbestos involving urban soils is from soil to atmosphere, if asbestos is present both as fine, transportable particles, and is within surface layers of soil (Department of Health (WA) 2009). Disturbance of soil by construction or demolition activities can result in soil-atmosphere fluxes from deeper soil. Asbestos as atmospheric particulate material poses the risk of ingestion by inhalation by humans, with potentially severe health consequences. The incidence of other potential hazards, such as silicosis, from airborne soil particles is currently unknown. Most cases of silicosis involve prolonged exposure to airborne silicate mineral dusts, such as in agriculture or mining.

9.3 A First Look at Risk Analysis

9.3.1 The Conceptual Site Model Framework

This section will be framed around the concept of a *conceptual site model* used by environmental professionals to evaluate the risks of exposure to site contamination (including contaminated soils). We will not progress as far as formal Risk Assessment, but the widely used conceptual site model is a useful way to integrate many of the concepts covered so far.

A conceptual site model (CSM) is a text- or graphically based document that serves as a tool to allow people, involved in contaminated site assessment and remediation, to summarise, understand, and visualise the available information about a site. This information can be derived from a variety of sources such as desktop studies, field measurements, and numerical modelling. A CSM is an essential component of or complement to the Preliminary Site Investigation, described in Chap. 11. A CSM typically contains a source of contamination, receptor(s) which are affected by the contaminants, and pathways between source(s) and receptor(s), as follows (DEC 2006; NHBC 2008; U.S. EPA 2011).

In this respect, the Conceptual Site Model represents a story or narrative of a site, which has end points – the beginning and end that are common to all narratives. The beginning of the story of a contaminated site is the *source* (or sources) of contamination, and the end is the set of *receptors*, the object or objects at risk from the contaminants present. Between the source and receptors are the processes and fluxes, the crucial elements of the plot connecting the beginning and end. The elements of the CSM story are very often formalised into a list like the one below.

1. A ***source***, or sources, of contamination (e.g. landfill, leaking fuel tank or sewage pipe, chemical spill, etc.), including the contaminant(s) of concern.
2. A ***release mechanism*** (e.g. soil erosion, leaching to groundwater, volatilisation to the atmosphere).
3. ***Retention*** in the transport medium (or media) (e.g. adsorption to soil, filtration by fine soil pores, residence time in water or air).
4. An ***exposure point*** – a location where a receptor can come into contact with contaminated material (e.g. downwind of contaminated land which generates dust, or a groundwater-dependent wetland).
5. An ***exposure route***, also termed a "direct pathway" (e.g. plant root uptake, accumulation on fish gills, human inhalation, etc.)
6. ***Receptors*** – the human(s), or other biota, or ecosystem compartment ("receiving environment") which may be affected by or receive the contaminants derived from the source. The receptors may be existing (e.g. flora and fauna on-site), or likely to exist in the future (e.g. occupants of dwellings in a planned residential development). In some guidelines, structures such as buildings can also be receptors (NHBC 2008).

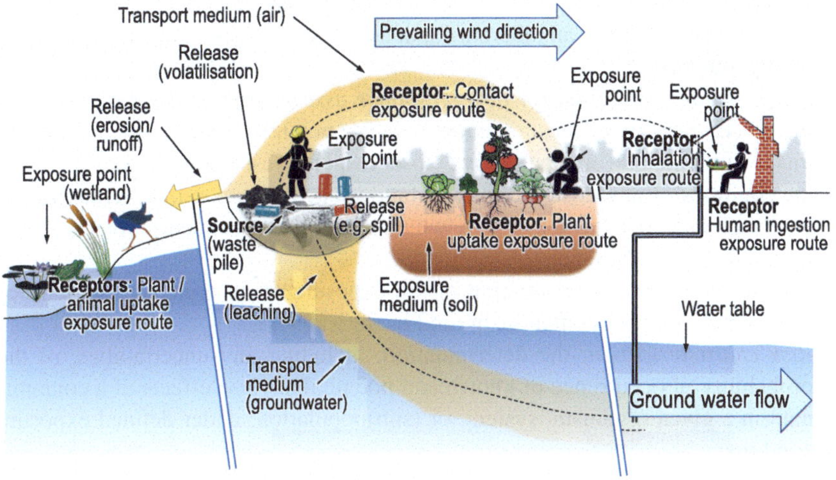

Fig. 9.7 Idealised and simplified initial conceptual site model for a contaminated site in diagrammatic form

Stages 2, 3, and 4 above may be aggregated into "indirect pathways" between source and receptors (e.g. NHBC 2008).

A conceptual site model is therefore an evolving summary, depending on the stage of investigation at a site. It is likely that a CSM will contain increasingly quantitative rather than conceptual or qualitative information as it evolves, and the purpose may change from characterisation through remediation to final land use (US EPA 2011). The diagram in Fig. 9.7 is an example of an early-stage conceptual site model, and diagrams and tables are commonly used to tell specific CSM stories. The initial CSM should identify and record any uncertainties, for example, in the source(s) or process(es) involved, which can be clarified as the CSM is refined. Some features of the initial CSM may be omitted after revision, or new sources, fluxes, and receptors added.

9.3.2 Risk Analysis: Dosages and Hazard Indices

A thorough analysis of risk often involves complex modelling of the **bioavailable** proportions of contaminants from different environmental compartments (US EPA 1996; DEC 2006). Since there is an emphasis on bioavailability, the **chemical species** of contaminant(s) present need to be considered. Risk assessment (of which hazard assessment is a subset) includes four main steps (from IPCS 2004, with wording made specific for contaminants):

7. *Hazard identification*: the identification of the type and nature of adverse effects that a contaminant has an inherent capacity to cause in an organism, system, or (sub)population.
8. *Hazard characterization* (or *dose–response assessment*): is the description of the inherent ability of a contaminant to cause potential adverse effects. Ideally this would include a quantitative dose–response assessment, including the uncertainties of any dose–response relationship. In more simple terms hazard can be characterised as a factor by which contaminant concentrations are multiplied (e.g. De Miguel et al. 2007).
9. *Exposure assessment*: evaluation of the exposure of an organism, system, or (sub)population to a contaminant (and its derivatives).
10. *Risk characterization*: the determination, including any uncertainties, of the probability of occurrence of known and potential adverse effects of a contaminant in a given organism, system, or (sub)population, under defined exposure conditions.

The exposure routes (as in a conceptual site model) are used to calculate dosages per day, for example, by ingestion, inhalation, skin (dermal) contact, and inhalation of vapour or atmospheric particulates. Calculation of dosage requires contaminant concentrations in each material, ingestion rates, frequency and duration of exposure to contaminated material, exposed skin area, and so on (De Miguel et al. 2007). Estimates of daily dosages can then be used to generate hazard quotients, by comparison with *acceptable daily intake* (also called "reference dose" or "tolerable daily intake") set by health authorities (where human exposure is involved) or environmental regulators (for other biota). In some assessment protocols the hazard quotients for different exposure pathways are combined into an overall hazard index. Proprietary software, such as the "Risk-Based Corrective Action" evaluation process, is available for formal risk assessment (e.g. Connor et al. 2001).

9.3.3 Soil-Based Guidelines: Threshold-Based Risk Assessment

A preliminary risk assessment which is less sophisticated, but much easier to implement, can be based on a simple comparison with health-based soil guideline concentrations (e.g. the health-based investigation levels from National Environment Protection Council 2013). Using this approach, some risk pathways are implied by different threshold concentrations for different land uses, since the risk of exposure is different for each land use, as shown by the example in Table 9.1. The data in Table 9.1 also reinforce the widely different toxicities of different contaminants; very toxic contaminants such as PCBs have low threshold concentrations, whereas contaminants which have lower toxicity such as Zn have greater threshold concentrations. Threshold-based risk analysis can also account for differences in bioavailability, or conceptual site model pathways such as retention or exposure routes, by

Table 9.1 Examples of health-based concentration thresholds from the Australian regulatory guidelines for selected contaminants in soil under different land uses (from National Environment Protection Council 2013). Used under the terms of a CC-BY-4.0 license

Chemical	Health-based investigation levels (mg/kg)			
	Residential (accessible soil)	Residential (low soil access)	Recreational (public open space)	Commercial/ industrial
Metals and inorganics				
Arsenic	100	500	300	3000
Cadmium	20	150	90	900
Chromium (VI)	100	500	300	3600
Copper	6000	30,000	17,000	240,000
Lead	300	1200	600	1500
Mercury (inorganic)	40	120	80	730
Methyl mercury	10	30	13	180
Nickel	400	1200	1200	6000
Zinc	7400	60,000	30,000	400,000
Cyanide (free)	250	300	240	1500
Polycyclic aromatic hydrocarbons (PAHs)				
Total PAHs	300	400	300	4000
Pesticides				
DDT + DDE + DDD	240	600	400	3600
Aldrin and dieldrin	6	10	10	45
2,4,5-T	600	900	800	5000
Atrazine	320	470	400	2500
Other organics				
PCBs	1	1	1	7
PBDE flame retardants (Br1-Br9)	1	2	2	10

making threshold values dependent on soil properties such as texture, pH, and cation exchange capacity (CEC) (see the example in Table 9.2 which also requires knowledge of the site-specific background concentration). Soil with low pH or low CEC is likely to have a greater proportion of total metal concentration bioavailable, so the concentration threshold needs to be lower (e.g. Table 9.2). Different bioavailability is also observed for aged and fresh contamination, with fresh contamination generally having greater contaminant bioavailability (Smolders et al. 2009).

Soil guideline concentrations may best be considered to be "trigger values", or concentrations above which action needs to be taken. Examples of the types of responses that are triggered include more detailed sampling and analysis leading to a more thorough characterisation of the extent of soil contamination at a site; remedial action to reduce risk from the urgent contamination (e.g. fine-grained asbestos

Table 9.2 Australian regulatory guidelines for added contaminant limits for aged zinc in soils of varying pH and cation exchange capacity (CEC) (National Environment Protection Council 2013). Used under the terms of a CC-BY-4.0 license

Zn added contaminant limits[a] (mg/kg) for urban residential/public open space							
		CEC ($cmol_c$/kg)					
	pH	**5**	**10**	**20**	**30**	**40**	**60**
	4.0	70	85	85	85	85	85
	4.5	100	120	120	120	120	120
	5.0	130	180	180	180	180	180
	5.5	180	270	270	270	270	270
	6.0	230	400	400	400	400	400
	6.5	230	400	590	590	590	590
	7.0	230	400	700	880	880	880
	7.5	230	400	700	960	1200	1300

[a]Concentration in soil after background concentration subtracted

in surface soil, or potentially flammable landfill gas). The most common response is for additional investigation of a site to be performed.

9.4 Further Reading

Carrillo-González, R., Šimůnek, J., Sauvé, S., Adriano, D.C., 2006. Mechanisms and pathways of trace element mobility in soils. *Advances In Agronomy*, 91: 111–178.

Rodríguez-Eugenio, N., McLaughlin, M., Pennock, D., 2018. *Soil Pollution - A Hidden Reality*, Food and Agriculture Organization of the United Nations Rome, Italy. (http://www.fao.org/3/I9183EN/i9183en.pdf).

Swartjes, F., 2015. Human health risk assessment related to contaminated land: state of the art. *Environmental Geochemistry and Health*, 37: 651–673, doi:https://doi.org/10.1007/s10653-015-9693-0.

9.5 Summary

Within any soil, transport of substances occurs by processes such as diffusion and mass flow (advection), which are driven by gradients in chemical concentration or potential energy, or mixing of soil by organisms (bioturbation).

Soil may also be transferred by adhering to surfaces, such those on humans, animals, plants, and vehicles (these pathways may be enhanced in urban environments).

Various external environments can contribute material to urban soils. We distinguish both diffuse sources, where a discrete source location cannot be identified, and point sources, where the external source of material can be attributed to a discrete location.

Important sources of material and substances to urban soils include dusts, vehicle emissions, atmospheric deposition other than dusts, and the various sources discussed in Chaps. 6, 7 and 8.

Soils may also contribute material to other environmental compartments, or receiving environments. The mechanisms include atmospheric emissions (respiration and volatilization), leaching in soil pore water, soil erosion by wind or water, plant uptake, and fire.

At a basic level, risk of hazards from substances in urban soils can be assessed using a Conceptual Site Model. The Conceptual Site Model framework includes identification of sources and receptors, plus the pathways connecting source(s) to receptor(s): release mechanism(s), retention, exposure point(s), and exposure route(s). Conceptual Site Models are site-specific and can be expressed in graphical or text form.

Both more sophisticated (e.g. dosage-based risk analysis) and more simple (comparison with guideline values) methods also exist for the assessment of risk from potential hazards associated with urban soils.

9.6 Questions

9.6.1 Checking your Understanding

1. Explain how (e.g. what causes diffusion to occur) and where (e.g. in which phases, and at which scale) transport of material by diffusion occurs in soils.
2. Identify the different types of mass flow that can occur in urban soils and explain why mass flow occurs in each instance.
3. Give three examples of diffuse sources, and three examples of point sources, listing the types of substances contributed by each source.
4. Draw a diagram showing the possible fate, through multiple environmental compartments, of nitrogen and phosphorus applied as fertilisers to urban soils.
5. Compare and contrast the ability of soil to be a sink and source for potentially toxic substances.
6. What are the end points and pathways for a Conceptual Site Model? Give examples, relevant to urban soils, of each stage in a conceptual site model.

9.6.2 *Thinking about the Topics more Deeply*

7. Can transport of soil material by adhering or bioturbation be considered as forms of diffusion and/or mass flow? If so, explain the chemical or potential gradients that are involved.
8. Given that urban soils are commonly found to emit carbon dioxide and methane, is it still possible that urban environments can function as carbon sinks? Explain your reasoning.
9. Discuss how a Conceptual Site Model could be used to guide sampling of an urban soil and its surrounding environment.

9.6.3 *Thinking Creatively about Urban Soils*

10. Which group(s) of people are at most risk of ingestion of soil-borne contaminants? Which group(s) of organisms will be affected most by soil contamination? Does the answer to either or both of these questions depend on the contaminant being considered?

References

Adams DF, Farwell SO, Pack MR, Robinson E (1981) Biogenic sulfur gas emissions from soils in eastern and southeastern United States. J Air Pollut Control Assoc 31:1083–1089. https://doi.org/10.1080/00022470.1981.10465330

Al-Hamarneh IF, Awadallah MI (2009) Soil radioactivity levels and radiation hazard assessment in the highlands of northern Jordan. Radiat Meas 44:102–110. https://doi.org/10.1016/j.radmeas.2008.11.005

Allott RW, Kelly M, Nicholas Hewitt C (1994) A model of environmental behaviour of contaminated dust and its application to determining dust fluxes and residence times. Atmos Environ 28:679–687. https://doi.org/10.1016/1352-2310(94)90044-2

Almeida SM, Pio CA, Freitas MC, Reis MA, Trancoso MA (2005) Source apportionment of fine and coarse particulate matter in a sub-urban area at the Western European coast. Atmos Environ 39:3127–3138. https://doi.org/10.1016/j.atmosenv.2005.01.048

Al-Rajab AJ, Amellal S, Schiavon M (2008) Sorption and leaching of 14C-glyphosate in agricultural soils. Agron Sustain Dev 28:419–428. https://doi.org/10.1051/agro:2008014

Appleyard S, Wong S, Willis-Jones B, Angeloni J, Watkins R (2004) Groundwater acidification caused by urban development in Perth, Western Australia: source, distribution, and implications for management. Aust J Soil Res 42:579–585

Arogunjo AM, Höllriegl V, Giussani A, Leopold K, Gerstmann U, Veronese I, Oeh U (2009) Uranium and thorium in soils, mineral sands, water and food samples in a tin mining area in Nigeria with elevated activity. J Environ Radioact 100:232–240. https://doi.org/10.1016/j.jenvrad.2008.12.004

Åström M (2001) Effect of widespread severely acidic soils on spatial features and abundance of trace elements in streams. J Geochem Explor 73:181–191. https://doi.org/10.1016/S0375-6742(01)00196-0

Barber PA, Paap T, Burgess TI, Dunstan W, Hardy GESJ (2013) A diverse range of Phytophthora species are associated with dying urban trees. Urban For Urban Green 12:569–575. https://doi.org/10.1016/j.ufug.2013.07.009

Bellucci F, Bogner JE, Sturchio NC (2012) Greenhouse gas emissions at the urban scale. Elements 8:445–450. https://doi.org/10.2113/gselements.8.6.445

Bucheli TD, Blum F, Desaules A, Gustafsson Ö (2004) Polycyclic aromatic hydrocarbons, black carbon, and molecular markers in soils of Switzerland. Chemosphere 56:1061–1076. https://doi.org/10.1016/j.chemosphere.2004.06.002

Byrne LB, Bruns MA, Kim KC (2008) Ecosystem properties of urban land covers at the aboveground–belowground interface. Ecosystems 11:1065–1077. https://doi.org/10.1007/s10021-008-9179-3

Carpenter SR, Caraco NF, Correll DL, Howarth RW, Sharpley AN, Smith VH (1998) Nonpoint pollution of surface waters with phosphorus and nitrogen. Ecol Appl 8:559–568. https://doi.org/10.1890/1051-0761(1998)008[0559:nposww]2.0.co;2

Carrillo-González R, Šimůnek J, Sauvé S, Adriano DC (2006) Mechanisms and pathways of trace element mobility in soils. Adv Agron 91:111–178

Castanho ADA, Artaxo P (2001) Wintertime and summertime São Paulo aerosol source apportionment study. Atmos Environ 35:4889–4902. https://doi.org/10.1016/S1352-2310(01)00357-0

Cicchella D, Giaccio L, Dinelli E, Albanese S, Lima A, Zuzolo D et al (2015) GEMAS: spatial distribution of chemical elements in agricultural and grazing land soil of Italy. J Geochem Explor 154:129–142. https://doi.org/10.1016/j.gexplo.2014.11.009

Clark MW, McConchie DM (2004) Development of acid sulfate soil in sub-aerially disposed dredge spoil at Fisherman Islands, Brisbane, Australia. Aust J Soil Res 42:553–567

Connor JA, Richard PE, Bowers L, Mchugh TE (2001) RBCA toolkit: comprehensive risk-based modelling system for soil and groundwater clean-up. In: Linders JBHJ (ed) Modelling of environmental chemical exposure and risk. Springer, Dordrecht, pp 165–181. https://doi.org/10.1007/978-94-010-0884-6_15

De Miguel E, Llamas JF, Chacón E, Berg T, Larssen S, Røyset O, Vadset M (1997) Origin and patterns of distribution of trace elements in street dust: unleaded petrol and urban lead. Atmos Environ 31:2733–2740

De Miguel E, Iribarren I, Chacón E, Ordoñez A, Charlesworth S (2007) Risk-based evaluation of the exposure of children to trace elements in playgrounds in Madrid (Spain). Chemosphere 66:505–513. https://doi.org/10.1016/j.chemosphere.2006.05.065

DEC (2006) The use of risk assessment in contaminated site assessment and management, Contaminated Sites Management Series, Department of Environment and Conservation, Government of Western Australia, Perth, Western Australia

Department of Health (WA) (2009) Guidelines for the assessment, remediation and Management of Asbestos-Contaminated Sites in Western Australia, Government of Western Australia, Environmental Health Directorate, Perth. https://ww2.health.wa.gov.au/~/media/Files/Corporate/general%20documents/Asbestos/PDF/Guidelines-Asbestos-Contaminated%20Sites-May2009.pdf

Du X, Li X, Zhang W, Wang H (2014) Variations in source apportionments of nutrient load among seasons and hydrological years in a semi-arid watershed: GWLF model results. Environ Sci Pollut Res 21:6506–6515. https://doi.org/10.1007/s11356-014-2519-2

Du E, de Vries W, Liu X, Fang J, Galloway JN, Jiang Y (2015) Spatial boundary of urban 'acid islands' in southern China. Sci Rep 5:12625–12625. https://doi.org/10.1038/srep12625

Dubé JS, Winiarski T, Galvez-Cloutier R (2003) Effect of the initial soil water saturation on the behaviour of a mixed LNAPL and heavy metal contaminated glaciofluvial deposit. Eur J Soil Sci 54:517–530. https://doi.org/10.1046/j.1365-2389.2003.00519.x

Ebrahimi N, Viaene N, Aerts J, Debode J, Moens M (2016) Agricultural waste amendments improve inundation treatment of soil contaminated with potato cyst nematodes, Globodera rostochiensis and G pallida. Eur J Plant Pathol 145:755–775. https://doi.org/10.1007/s10658-016-0864-3

Endo S, Kimura S, Takatsuji T, Nanasawa K, Imanaka T, Shizuma K (2012) Measurement of soil contamination by radionuclides due to the Fukushima Dai-ichi nuclear power plant accident and associated estimated cumulative external dose estimation. J Environ Radioact 111:18–27. https://doi.org/10.1016/j.jenvrad.2011.11.006

European Environment Agency (2016) Explaining road transport emissions: a non-technical guide, European Commission, Copenhagen, Denmark. https://www.eea.europa.eu/publications/explaining-road-transport-emissions/at_download/file

Fabietti G, Biasioli M, Barberis R, Ajmone-Marsan F (2010) Soil contamination by organic and inorganic pollutants at the regional scale: the case of Piedmont, Italy. J Soils Sediments 10:290–300. https://doi.org/10.1007/s11368-009-0114-9

Fanning DS, Rabenhorst MC (1990) Micromorphology of acid sulfate soils in Baltimore Harbor dredged materials. In: Douglas LA (ed) Developments in Soil Science. Elsevier, pp 279–288. https://doi.org/10.1016/S0166-2481(08)70339-8

Fergusson JE, Kim ND (1991) Trace elements in street and house dusts: sources and speciation. Sci Total Environ 100:125–150. https://doi.org/10.1016/0048-9697(91)90376-P

Fernández-Caliani JC (2012) Risk-based assessment of multimetallic soil pollution in the industrialized peri-urban area of Huelva, Spain. Environ Geochem Health 34:123–139. https://doi.org/10.1007/s10653-011-9396-0

Garban B, Ollivon D, Carru AM, Chesterikoff A (1996) Origin, retention and release of trace metals from sediments of the River Seine. Water Air Soil Pollut 87:363–381. https://doi.org/10.1007/BF00696848

Garcia-Armisen T, Servais P (2007) Respective contributions of point and non-point sources of E. coli and enterococci in a large urbanized watershed (the Seine river, France). J Environ Manag 82:512–518. https://doi.org/10.1016/j.jenvman.2006.01.011

Groffman PM, Boulware NJ, Zipperer WC, Pouyat RV, Band LE, Colosimo MF (2002) Soil nitrogen cycle processes in urban riparian zones. Environ Sci Technol 36:4547–4552. https://doi.org/10.1021/es020649z

Grolimund D, Borkovec M (2005) Colloid-facilitated transport of strongly sorbing contaminants in natural porous media: mathematical modeling and laboratory column experiments. Environ Sci Technol 39:6378–6386. https://doi.org/10.1021/es050207y

Harrison RM, Smith DJT, Pio CA, Castro LM (1997) Comparative receptor modelling study of airborne particulate pollutants in Birmingham (United Kingdom), Coimbra (Portugal) and Lahore (Pakistan). Atmos Environ 31:3309–3321. https://doi.org/10.1016/S1352-2310(97)00152-0

Iglesias Jiménez E, Perez Garcia V (1989) Evaluation of city refuse compost maturity: a review. Biol Wastes 27:115–142. https://doi.org/10.1016/0269-7483(89)90039-6

IPCC (2013) Climate change 2013: the physical science basis. Contribution of working group I to the fifth assessment report of the IPCC, Intergovernmental Panel on Climate Change, Cambridge University Press, Cambridge, UK

IPCS (2004) IPCS risk assessment terminology. Harmonization Project Document No. 1, Inter-Organization Programme for the Sound Management of Chemicals (IOMC) (A cooperative agreement among UNEP, ILO, FAO, WHO, UNIDO, UNITAR and OECD), Geneva, Switzerland. http://www.inchem.org/documents/harmproj/harmproj/harmproj1.pdf

Jacob P, Meckbach R, Muller HM (1987) Reduction of external exposure from deposited Chernobyl activity by run-off, weathering, street cleaning and migration in the soil. Radiat Prot Dosim 21:51–57

Johnson D (2005) Pollutants | persistent organic (POPs). In: Hillel D (ed) Encyclopedia of soils in the environment. Elsevier, Oxford, pp 264–271. https://doi.org/10.1016/B0-12-348530-4/00560-9

Jones-Otazo HA, Clarke JP, Diamond ML, Archbold JA, Ferguson G, Harner T et al (2005) Is house dust the missing exposure pathway for PBDEs? An analysis of the urban fate and human exposure to PBDEs. Environ Sci Technol 39:5121–5130. https://doi.org/10.1021/es048267b

Kabata-Pendias A (2011) Trace elements in soils and plants. CRC Press, Boca Raton, 505 pp

Kelepertzis E, Komárek M, Argyraki A, Šillerová H (2016) Metal(loid) distribution and Pb isotopic signatures in the urban environment of Athens, Greece. Environ Pollut 213:420–431. https://doi.org/10.1016/j.envpol.2016.02.049

Kostecki P, Morrison R, Dragun J (2005) Hydrocarbons. In: Hillel D (ed) Encyclopedia of soils in the environment. Elsevier, Oxford, pp 217–226. https://doi.org/10.1016/B0-12-348530-4/00098-9

Laidlaw MAS, Alankarage DH, Reichman SM, Taylor MP, Ball AS (2018) Assessment of soil metal concentrations in residential and community vegetable gardens in Melbourne, Australia. Chemosphere 199:303–311. https://doi.org/10.1016/j.chemosphere.2018.02.044

Lanphear BP, Matte TD, Rogers J, Clickner RP, Dietz B, Bornschein RL et al (1998) The contribution of lead-contaminated house dust and residential soil to children's blood lead levels: a pooled analysis of 12 epidemiologic studies. Environ Res 79:51–68. https://doi.org/10.1006/enrs.1998.3859

Lévêque T, Dumat C, Lagier L, Schreck E, Ruales J, Capowiez Y (2019) Influence of earthworm bioturbation on metals phytoavailability and human gastric bioaccessibility. Environ Sci Pollut Res 26:20052–20063. https://doi.org/10.1007/s11356-018-3010-2

Li J, Pu L, Zhu M, Zhang J, Li P, Dai X et al (2014) Evolution of soil properties following reclamation in coastal areas: a review. Geoderma 226-227:130–139. https://doi.org/10.1016/j.geoderma.2014.02.003

Liu E, Yan T, Birch G, Zhu Y (2014) Pollution and health risk of potentially toxic metals in urban road dust in Nanjing, a mega-city of China. Sci Total Environ 476-477:522–531

Livesley SJ, Dougherty BJ, Smith AJ, Navaud D, Wylie LJ, Arndt SK (2010) Soil-atmosphere exchange of carbon dioxide, methane and nitrous oxide in urban garden systems: impact of irrigation, fertiliser and mulch. Urban Ecosyst 13:273–293. https://doi.org/10.1007/s11252-009-0119-6

Lyons M (2008) A study into the risk of Phytophthora dieback in ten Peri-urban reserves within the Shire of Denmark, Shire of Denmark, Denmark, Western Australia. https://www.denmark.wa.gov.au/Profiles/denmark/Assets/ClientData/Documents/biosecurity/Risk_of_Dieback_Shire_Denmark_24_July_2008.pdf

Macdonald BCT, Denmead OT, White I, Melville MD (2004) Natural sulfur dioxide emissions from sulfuric soils. Atmos Environ 38:1473–1480. https://doi.org/10.1016/j.atmosenv.2003.12.005

Manta DS, Angelone M, Bellanca A, Neri R, Sprovieri M (2002) Heavy metals in urban soils: a case study from the city of Palermo (Sicily), Italy. Sci Total Environ 300:229–243. https://doi.org/10.1016/s0048-9697(02)00273-5

McBride MB, Shayler HA, Spliethoff HM, Mitchell RG, Marquez-Bravo LG, Ferenz GS et al (2014) Concentrations of lead, cadmium and barium in urban garden-grown vegetables: the impact of soil variables. Environ Pollut 194:254–261. https://doi.org/10.1016/j.envpol.2014.07.036

McIlwaine R, Doherty R, Cox SF, Cave M (2017) The relationship between historical development and potentially toxic element concentrations in urban soils. Environ Pollut 220:1036–1049. https://doi.org/10.1016/j.envpol.2016.11.040

Muezzinoglu A (2003) A study of volatile organic sulfur emissions causing urban odors. Chemosphere 51:245–252. https://doi.org/10.1016/S0045-6535(02)00821-4

National Environment Protection Council (2013) Schedule B (1): guideline on the investigation levels for soil and groundwater, National Environment Protection (Assessment of Site Contamination) Measure (Amended). Commonwealth of Australia, Canberra

Nehls T, Shaw R (2010) Black carbon in urban soils: relevance, analysis, distribution. Soil Survey Horizons 51:79–84. https://doi.org/10.2136/sh2010.3.0079

NHBC (2008) Guidance for the safe development of housing on land affected by contamination, National House-Building Council, Environment Agency, Chartered Institute of Environmental Health Amersham, UK

Norra S, Weber A, Kramar U, Stüben D (2001) Mapping of trace metals in urban soils: the example of Mühlburg/Karlsruhe, Germany. J Soils Sediments 1:77–97. https://doi.org/10.1065/iss2000.12.005

Parker GR, Mc Fee WW, Kelly JM (1978) Metal distribution in forested ecosystems in urban and rural Northwestern Indiana. J Environ Qual 7:337–342. https://doi.org/10.2134/jeq1978.00472425000700030008x

Peters NE, Donohue R (2001) Nutrient transport to the Swan-Canning Estuary, Western Australia. Hydrol Process 15:2555–2577

Plumlee GS, Morman SA, Cook A (2012) Environmental and medical geochemistry in urban disaster response and preparedness. Elements 8:451–457. https://doi.org/10.2113/gselements.8.6.451

Ravishankara AR, Daniel JS, Portmann RW (2009) Nitrous oxide (N_2O): the dominant ozone-depleting substance emitted in the 21st century. Science 326:123–125. https://doi.org/10.1126/science.1176985

Riemann U (1999) Impacts of urban growth on surface water and groundwater quality in the City of Dessau, Germany. In: Ellis B (ed) Impacts of urban growth on surface water and groundwater quality, IAHS-AISH Publication no. 259. International Association of Hydrological Sciences Press, Wallingford, pp 307–314

Rodríguez-Eugenio N, McLaughlin M, Pennock D (2018) Soil pollution - a hidden reality, food and agriculture Organization of the United Nations Rome, Italy. http://www.fao.org/3/I9183EN/i9183en.pdf

Servant J, Delapart M (1982) Daily variations of the H_2S content in atmospheric air at ground-level in France. Atmos Environ 16:1047–1052. https://doi.org/10.1016/0004-6981(82)90193-7

Shen W, Wu J, Grimm NB, Hope D (2008) Effects of urbanization-induced environmental changes on ecosystem functioning in the Phoenix metropolitan region, USA. Ecosystems 11:138–155. https://doi.org/10.1007/s10021-007-9085-0

Shepherd KA, Ellis PA, Rivett MO (2006) Integrated understanding of urban land, groundwater, baseflow and surface-water quality-the City of Birmingham, UK. Sci Total Environ 360:180–195. https://doi.org/10.1016/j.scitotenv.2005.08.052

Siciliano SD, James K, Zhang G, Schafer AN, Peak JD (2009) Adhesion and enrichment of metals on human hands from contaminated soil at an arctic urban brownfield. Environ Sci Technol 43:6385–6390. https://doi.org/10.1021/es901090w

Siciliano SD, Laird BD, Lemieux CL (2010) Polycyclic aromatic hydrocarbons are enriched but bioaccessibility reduced in brownfield soils adhered to human hands. Chemosphere 80:1101–1108. https://doi.org/10.1016/j.chemosphere.2010.04.061

Siemens J, Huschek G, Walshe G, Siebe C, Kasteel R, Wulf S et al (2010) Transport of pharmaceuticals in columns of a wastewater-irrigated Mexican clay soil. J Environ Qual 39:1201–1210. https://doi.org/10.2134/jeq2009.0105

Slavens MR, Petrovic AM (2012) Pesticide fate in sodded kentucky bluegrass lawns in response to irrigation. Acta Agric Scand Sect B 62:86–95. https://doi.org/10.1080/09064710.2012.685747

Smolders E, Oorts K, van Sprang P, Schoeters I, Janssen CR, McGrath SP, McLaughlin MJ (2009) Toxicity of trace metals in soil as affected by soil type and aging after contamination: using calibrated bioavailability models to set ecological soil standards. Environ Toxicol Chem 28:1633–1642. https://doi.org/10.1897/08-592.1

Stec AA, Dickens K, Barnes JLJ, Bedford C (2019) Environmental contamination following the Grenfell tower fire. Chemosphere:576–586. https://doi.org/10.1016/j.chemosphere.2019.03.153

Taylor MP, Mackay AK, Hudson-Edwards KA, Holz E (2010) Soil Cd, Cu, Pb and Zn contaminants around Mount Isa city, Queensland, Australia: potential sources and risks to human health. Appl Geochem 25:841–855

Testa SM, Winegardner DL (2000) Restoration of contaminated aquifers: petroleum hydrocarbons and organic compounds, 2nd edn. CRC Press, Boca Raton, 446 pp

Thibodeaux LJ, Lipsky D (1985) A fate and transport model for 2,3,7,8-tetrachlorodibenzo-p-dioxin in fly ash on soil and urban surfaces. Hazardous Waste Hazardous Mat 2:225–235. https://doi.org/10.1089/hwm.1985.2.225

Tsatsaros JH, Brodie JE, Bohnet IC, Valentine P (2013) Water quality degradation of coastal waterways in the wet tropics, Australia. Water Air Soil Pollut 224:1443. https://doi.org/10.1007/s11270-013-1443-2

U.S. EPA (1996) Soil screening guidance: User's guide (second edition) - attachment a: conceptual site model summary, Superfund, United States Environmental Protection Agency

U.S. EPA (2011) Environmental cleanup best management practices: effective use of the project life cycle conceptual site model. EPA 542-F-11-011, United States Environmental Protection Agency, Office of Superfund Remediation and Technology Innovation, Washington, DC, USA. https://www.epa.gov/sites/production/files/2015-04/documents/csm-life-cycle-fact-sheet-final.pdf

U.S. EPA (2019) *Chemicals and toxics topics*. United States Environmental Protection Agency, Arlington, VA, USA, https://www.epa.gov/environmental-topics/chemicals-and-toxics-topics (accessed 20190906)

van Breemen N, Driscoll CT, Mulder J (1984) Acidic deposition and internal proton sources in acidification of soils and waters. Nature 307:599–604

Weaver JE, Wang L, de los Reyes FL III, Barlaz MA (2019) Systems and methods for studying microbial processes and communities in landfills. In: Hurst CJ (ed) Understanding terrestrial microbial communities. Springer, Cham. https://doi.org/10.1007/978-3-030-10777-2_5

Wei YL, Bao LJ, Wu CC, He ZC, Zeng EY (2014) Association of soil polycyclic aromatic hydrocarbon levels and anthropogenic impacts in a rapidly urbanizing region: spatial distribution, soil-air exchange and ecological risk. Sci Total Environ 473-474:676–684. https://doi.org/10.1016/j.scitotenv.2013.12.106

Wolf L, Klinger J, Hoetzl H, Mohrlok U (2007) Quantifying mass fluxes from urban drainage systems to the urban soil-aquifer system. J Soils Sediments 7:85–95. https://doi.org/10.1065/jss2007.02.207

Yamamoto N, Takahashi Y, Yoshinaga J, Tanaka A, Shibata Y (2006) Size distributions of soil particles adhered to children's hands. Arch Environ Contam Toxicol 51:157–163. https://doi.org/10.1007/s00244-005-7012-y

Chapter 10
Urban Soil and Human Health

Andrew W. Rate

Abstract In cities, human health and well-being, socioeconomic status, food security, education, gender equity, employment, climate change, and biodiversity are interlinked, and one perhaps surprising common factor is urban soils. In this chapter we explore how a unifying framework for these interrelationships is presented by the United Nations' Sustainable Development Goals (SGDs). The analysis in this chapter suggests that soils have a role to play in the first fifteen of the seventeen Sustainable Development Goals. We suggest ways in which the knowledge and use of soils by urban inhabitants can help to address poverty, maintain a stable food supply, sustain physical, emotional, and social health, provide opportunities for education, promote gender equality and empowerment of women and girls, generate employment, maintain water quality, moderate climate change, and slow biodiversity and habitat loss. The chapter also addresses other soil-related effects on human health such as soil remediation and acid sulfate soils and has a particular focus on environmental justice issues related to urban soil contamination.

Keywords Human health · Urban soils · Sustainable Development Goals · Environmental justice

A. W. Rate (✉)
School of Agriculture and Environment, University of Western Australia, Crawley, WA, Australia
e-mail: andrew.rate@uwa.edu.au

A. W. Rate (ed.), *Urban Soils*, Progress in Soil Science,
https://doi.org/10.1007/978-3-030-87316-5_10

What you could learn from this chapter:

- How the relationship between urban soils and human health can be understood in the context of the United Nations' Sustainable Development Goals, especially the goals related to poverty, food security, human health and well-being, education, gender equality, water quality, employment and economic growth, climate change, and non-human ecosystems.
- How soil contamination and degradation, and soil remediation, can affect human health.
- What environmental justice means in the context of urban soils, and the types of environmental benefits and services that are inequitably distributed in cities.

10.1 Urban Soils and Sustainable Development

A highly relevant framework for considering the relationships between urban soils and human health are the seventeen Sustainable Development Goals (United Nations (2015) and Table 10.1). The United Nations Environment Program acknowledges that environmental issues underpin all seventeen Sustainable Development Goals (UNEP n.d.). An analysis of the role of urban soils in the context of the Sustainable Development Goals allows us to develop a holistic view of the multiple, interlinked components of human health and well-being (Table 10.1).

The remainder of this chapter will discuss the relationships between urban soils and human health by addressing some of the UN Sustainable Development Goals (SDGs): 1 No Poverty; 2 Zero Hunger; 3 Good Health and Well-Being; 4 Quality Education; 5 Gender Equality; 6 Clean Water and Sanitation; 8 Decent Work and Economic Growth; 13 Climate Action; and 15 Life on Land. This discussion is not intended to be a social or political agenda, but only to present what might be possible roles of urban soils in achieving international goals for sustainable

Table 10.1 How urban soils are related to the UN Sustainable Development Goals

Sustainable Development Goal		Opportunities related to urban soils	Threats related to urban soils
1	End poverty in all its forms everywhere	Soil as an income source, e.g. growing food, pottery	Soils tend to be more contaminated in less advantaged urban areas, so potential for perpetuation of poverty
2	End hunger, achieve food security and improved nutrition, and promote sustainable agriculture	Contributions of urban agriculture to food security (Siegner et al. 2018)	Soil continues to be lost to urbanization processes: Waste disposal, surface sealing
3	Ensure healthy lives and promote Well-being for all at all ages	Community and individual emotional/psychological/relational Well-being	Soil contamination and associated risks to human health (Chaps. 6, 7, 8 and 9)
4	Ensure inclusive and equitable quality education and promote lifelong learning opportunities for all	Focus on urban soils for environmental, health, and related education (Wortman and Lovell 2013; Kim et al. 2014; Gregory et al. 2016)	Without appropriate education, risks from soil contamination, etc. may be greater (Fett et al. 1992; Dietz et al. 2004; Lioy 2010), or adoption of sustainable practices may be less (Dhakal and Chevalier 2017)
5	Achieve gender equality and empower all women and girls	Women are commonly leaders of urban soil-related enterprises (Hovorka et al. 2009; Orsini et al. 2013; Wozniacka 2019)	Poverty related to urban soil loss may have more impact on women and girls; they may bear burden of soil-related work (Hovorka et al. 2009)
6	Ensure availability and sustainable management of water and sanitation for all	Permeable urban soils can act as physical/chemical/biological filters for groundwater	Transfers of contaminants can occur between urban soil and potable surface- or ground-water sources
7	Ensure access to affordable, reliable, sustainable, and modern energy for all	Urban soils used for insulation or heat exchange. Urban soils used to grow sustainable energy crops. Landfill gas extraction	–
8	Promote sustained, inclusive, and sustainable economic growth, full and productive employment, and decent work for all	Employment opportunities in urban agriculture, extension, environmental consultancy, urban soil remediation	Soil continues to be lost to urbanization
9	Build resilient infrastructure, promote inclusive and sustainable industrialization, and foster innovation	Disseminate soil knowledge to developers and engineers. Landfill mining	–

(continued)

Table 10.1 (continued)

Sustainable Development Goal		Opportunities related to urban soils	Threats related to urban soils
10	Reduce inequality within and among countries	Also through urban agriculture (Orsini et al. 2013) and restoration of degraded soil in low-socioeconomic areas	–
11	Make cities and human settlements inclusive, safe, resilient, and sustainable	Exposed urban soil can reduce urban heat island, especially when vegetated. Urban agriculture	–
12	Ensure sustainable consumption and production patterns	Urban agriculture and WSUD often emphasize practices such as recycling (Gathuru et al. 2009; Wortman and Lovell 2013)	Contamination or poor availability of urban soils may limit urban agriculture
13	Take urgent action to combat climate change and its impacts	Urban soils can provide a sink for carbon (with intentional management), and growth medium for urban forests to reduce urban heat islands	Urban soils (e.g. landfills, but also others) can be net GHG emitters
14	Conserve and sustainably use the oceans, seas, and marine resources for sustainable development	Urban soils can minimize downstream losses (with intentional management), e.g. rain gardens, constructed wetlands	Transfers of contaminants can occur between urban soil and potable surface- or ground-water sources
15	Protect, restore, and promote sustainable use of terrestrial ecosystems, sustainably manage forests, combat desertification, and halt and reverse land degradation and halt biodiversity loss	Urban soil rehabilitation, use of urban land for urban forests (Keesstra et al. 2018)	Soil continues to be lost to urbanization (sealing, compaction, contamination, etc.) causing loss of fertility, biodiversity, and soil itself. Poorly applied, UA may degrade soil
16	Promote peaceful and inclusive societies for sustainable development, provide access to justice for all and build effective, accountable, and inclusive institutions at all levels	–	–
17	Strengthen the means of implementation and revitalize the global Partnership for Sustainable Development	–	–

development, and acknowledges that some authors have considered different combinations of the Goals (e.g., Keesstra et al. 2018).

As we progress, it will become clear that practices that address one Sustainable Development Goal also address others (e.g. urban agriculture can address poverty, food security, climate change adaptation, and so on).

10.1.1 Poverty and its Relationships with Urban Soil

Ending poverty in all its forms everywhere (Sustainable Development Goal 1) is perhaps the most noble of the SDGs, and the Earth's urban soil resource, although significant, is just a part of any solution. The role of urban soils perhaps fits most neatly into two components of SDG Target 1.4, which addresses equal rights to economic and natural resources, and SDG Target 1.5 which aims to build resilience of poor people to extreme environmental, economic, and social risks (including those derived from climate extremes).

Soil management and use has the potential to provide an ongoing income source. For example, urban agriculture is a soil-based enterprise in which the primary objective is growing food, creating an economic advantage by savings in food expenditure, or by selling surplus produce (Hovorka et al. 2009). Other soil-based economic activities are possible; for example, Oladimeji et al. (2015) describe the use of clay collected from soil to make and sell pottery in a peri-urban area of Ilorin city in Nigeria.

In contrast, many studies in the discipline of *environmental justice* have shown that soils tend to be more contaminated in less economically or socially advantaged urban areas, such as those predominantly occupied by people who are poor or belong to a minority group (e.g. Mielke et al. 1999; Aelion et al. 2012; Zhuo et al. 2012; McClintock 2015). This uneven distribution of soil contamination limits the ability of poor people to use soil for food production or other enterprises (or may cause adverse health issues if the soil is used), and so polluted (or otherwise degraded) urban soils have the potential for perpetuation of poverty.

10.1.2 Food Security and Its Relationships with Urban Soil

Sustainable Development Goal 2 is to "End hunger, achieve food security and improved nutrition and promote sustainable agriculture," and so has clear links to soil use and management in urban environments. These links are most well-defined in SDG Target 2.3, which deals with increasing agricultural productivity of small-scale food producers. Just as it does for poverty (SDG 1), urban agriculture can make significant contributions to food security. For example, Siegner et al. (2018) concluded that urban agriculture could provide a number of benefits to communities in the USA, including food security, when allied with intentional policies to improve social justice and equity.

The rapid pace of urbanization in many parts of the world does create some threats to urban food security, in the form of limits to food access and safety, and effective distribution of food (Lal 2017). Urbanization will continue to remove soils from potential for food production. Inevitably, the pressures to develop more land for residential and commercial use will decrease the area of productive land by surface sealing. It is also likely that productive land will be compromised by ongoing soil degradation and contamination related to urban development, such as compaction and the need to dispose of ever-increasing amounts of wastes.

10.1.3 Physical Health and Well-being and Urban Soils

Sustainable Development Goal 3, which aims to "Ensure healthy lives and promote well-being…," has numerous connections to urban soils, most notably in SDG Target 3.9 which addresses health issues relating to pollution and contamination of soil and other environmental compartments. Some of the direct and indirect risks to human health were associated with soil contamination and have been addressed in Chaps. 6, 7, 8 and 9, and we present more details in this chapter. It is also important, however, to be aware of the positive effects on human health which relate to a connection to urban soils.

Nutrients Elements which are essential macronutrients are not generally considered to have direct human health effects when present as soil contaminants, although there may be deficiency symptoms in cases of insufficient supply by soils. However, in soils with high concentrations of nitrogen and phosphorus, for example, due to over-fertilization (Taylor and Lovell 2015) or organic waste disposal, leaching of nitrate and phosphate to groundwater and surface water can occur (Carpenter et al. 1998). Nitrate, in particular, can have adverse health effects if ingested at high concentrations, the most concerning of which is the potentially fatal blood disorder methemoglobinemia or "blue baby syndrome" in human infants (Croll and Hayes 1988).

Metals Contamination of soils with metals and metalloids may cause human health issues if a plausible pathway exists. Some metals or metalloids (e.g., As, Cd, Cr, Ni) can be carcinogens (Morgan 2012). Most metals or metalloids can also cause a wide range of chronic health effects if humans are exposed to contaminated soil (Morgan 2012; Pepper 2013). Historically, considerable concern has been raised about lead contamination in soils, since lead exposure in children can cause a range of neurological disorders, and children have a greater risk of direct or indirect soil ingestion than do adults (e.g., Aelion et al. 2009; Oliver and Gregory 2015; Li et al. 2018). Soil in public open space, including children's playgrounds, may be contaminated, and children deliberately or accidentally ingest more soil than adults (De Miguel et al. 2007). More recently, however, arsenic and cadmium have emerged as contaminants of concern, including issues of soil contamination with As and Cd in urban environments (De Miguel et al. 2007; Rodríguez-Eugenio et al. 2018).

Organic Contaminants There is a very wide range of organic contaminants to which humans may be exposed if they are present in soils. Several organic substances (PAH, PCB, PCDD, etc.) generate considerable concern due to their known potential as human carcinogens (Oliver and Gregory 2015; Rodríguez-Eugenio et al. 2018). Similarly to metals, exposure of humans to organic contaminants in soils is also believed to have several other adverse health effects; for example, exposure to organic pesticide compounds can cause hormone disruption, asthma, allergies, hypersensitivity, and even cancers. The wide range of persistent organic pollutants (POPs) have long residence times in soils; combined with their known carcinogenic and toxic properties, and widespread occurrence in urban soils worldwide, they pose significant risk to human health (Rodríguez-Eugenio et al. 2018).

Asbestos and Other Mineral Contaminants The fibrous minerals in the asbestos group are serious threats to human health, because of their ability to be ingested by inhalation, the very long lifetimes of asbestos minerals in soils, and the severity of their adverse effects. The severe effects of asbestos exposure include forms of cancer such as lung cancer and mesothelioma (and other cancers), and other potentially fatal respiratory ailments such as asbestosis (Frank and Joshi 2014). The only direct exposure route for asbestos is by inhalation, so contact with asbestos-contaminated soil and consumption of plants grown on such soil are not necessarily indirect pathways. Asbestos adhering to skin or clothing can, however, generate airborne asbestos fibers, as can soil exposed to wind erosion or soil disturbance (USEPA 2008).

Chronic (long-term) exposure to non-asbestos silicate minerals can also lead to human health effects. Inhalation of quartz particles less than about 4 μm in size can lead eventually to silicosis, a non-cancer lung disease similar to asbestosis (Derbyshire 2007; Pepper 2013).

Radionuclides The radionuclides that may be present in soil can be transferred to humans via inhalation (as radon gas, or atmospheric particles), or by ingestion, since radioactive elements may leach into groundwater; this has been documented in some urban environments (Lee 2011). The most likely human exposure route is through the seepage of radon gas from underlying soil material into confined living spaces such as buildings (U.S. EPA 2008). Human exposure to radionuclides can result in serious adverse health effects; radon and uranium can both cause cancers, and uranium can also cause kidney damage. Some of the toxic effects of uranium are chemical rather than derived from the radiation it emits (Bjørklund et al. 2017).

Establishing a causative relationship between urban soil contamination and human health is difficult, because indirect exposure and the common time lags between exposure and symptoms serve to decouple soil and humans in space and time.

Pathogens A range of potentially pathogenic organisms can be found in urban soils, especially if organic wastes containing fecal material such as biosolids, animal manures, or incompletely treated wastewater have been applied to soil (Alloway 2004; Amoah et al. 2005). There has been a resurgence in interest in using biosolids as a "zero-waste" recycling strategy; one of the strategies to minimize the incidence of pathogens is with more rigorous pre-treatment of waste materials (Alvarez-Campos and Evanylo 2019; Chakravorty 2019). This is of particular relevance for many poor people in developing countries, where residential structures often lack constructed floors (Pickering et al. 2012); the soil floors can be contaminated with bacteria and viruses. Pathogens can also be transmitted to humans who consume vegetables grown on contaminated urban soil (Amoah et al. 2005).

Health Benefits Community and individual well-being is considered to derive benefits from soil-related activities such as gardening and urban agriculture. These benefits are manifested as improvements in physical health, related to improvements in nutrition and increases in physical exercise. In addition, emotional, and/or psychological, and/or relational health improvements occurred for a range of reasons including the calming, meditative nature of gardening; increased connection to nature; increased mental activity; and connections with like-minded others (Wakefield et al. 2007; Kim et al. 2014).

10.1.4 Education and Urban Soils

Soil science is a highly integrative subject, requiring skills across the *STEM* spectrum including literacy and numeracy, and specialized skills and knowledge relevant to geography, geology, hydrology, biology, chemistry, and physics. The access to urban soils afforded by the concentration of population in urban centers therefore means that Sustainable Development Goal 4 focus on "…inclusive and equitable quality education…" and "…lifelong learning…" can be facilitated by learning experiences based around soils. In particular, urban soils can be used as a focal point for environmental, health, and related education (Wortman and Lovell 2013; Kim et al. 2014; Gregory et al. 2016).

In the absence of appropriate soil-focused education, the risks from urban soil contamination may be exacerbated (Fett et al. 1992; Dietz et al. 2004; Lioy 2010), or adoption of sustainable practices may be less (Dhakal and Chevalier 2017). Some of the types of information required by individuals and communities who use urban soil to grow food are summarized in Table 10.2.

Table 10.2 The main types of information, related to contamination risks in urban soil, needed by people conducting urban agriculture (from Kim et al. 2014, used under terms of CC-BY-4.0 license)

Category	Specific information
Site history	How to find information about past uses of a plot of land
	Which contaminants to test for, given specific past land uses
	Geographic areas of the city where there are likely to be high levels of contamination
Soil testing	Importance of obtaining a soil test prior to gardening
	Which contaminants to test for
	Why to test for certain contaminants and not others
	Where to get soil testing done
	How much soil testing costs
	How to correctly take a soil sample for a soil test
Remediation	Best practices for remediating contaminated urban soils
Minimizing exposure	How to reduce exposure risks when gardening
	Contamination risks associated with imported materials such as compost or mulch

10.1.5 Gender Equality and Empowerment and Urban Soils

Sustainable Development Goal 5 contains Target 5.5, to "Ensure women's full and effective participation and equal opportunities for leadership…," and Target 5.a to "Undertake reforms to give women equal rights to economic resources, as well as access to ownership and control over land…." In this context, it is important to recognize that women are commonly leaders of urban soil-related enterprises (e.g., Hovorka et al. 2009; Orsini et al. 2013; Wozniacka 2019). Conversely, poverty related to urban soil loss may have more impact on women and girls, who may bear a disproportionate burden of soil-related work while still expected to perform domestic duties, or who provide mainly the labor for urban agriculture while men retain financial control (Hovorka et al. 2009).

10.1.6 Healthy Water and Urban Soils

Sustainable Development Goal 6 is to "Ensure availability and sustainable management of water and sanitation for all." In an urban soil context, Target 6.3 which addresses water quality and pollution reduction is clearly relevant, as is Target 6.6 to protect water-related ecosystems.

Permeable urban soils can act as physical/chemical/biological filters for groundwater (Pepper 2013). The filtration ability of soils is partly physical; the transport of particulates and microorganisms (including pathogens) can be suppressed by soil pores which are too fine to allow passage of solid particles or microbial cells (Voisin et al. 2018). Chemical immobilization, such as ion exchange, adsorption, or precipitation of contaminants, can also occur (Abiye et al. 2009), along with biological processes to remove contaminants, such as denitrification (Bettez and Groffman 2012). *Managed Aquifer Recharge* (Fig. 10.1), from surface infiltration basins (not injection wells), deliberately relies on the filtration ability of urban soils and the underlying regolith to safely recharge groundwater (Misra 2014). The input water for Managed Aquifer Recharge can either be stormwater (Voisin et al. 2018) or treated wastewater (Abiye et al. 2009).

It is usually thought that the amount of impervious surface cover in urban areas decreases recharge to groundwater, but this is not always the case. In an urban catchment with shallow groundwater and very permeable soils, recharge increased following urban development, since the impervious surface cover served to substantially decrease evaporation losses of water (Barron et al. 2013). In other urban environments, excess runoff caused by impervious surface cover can be decreased by rainfall interception by the canopies of large street trees (Livesley et al. 2014). Other factors related to urbanization can alter hydrological processes; for example, the loss of wetlands due to urbanization decreases hydrological buffering and can lead to flooding and soil erosion (Rashid and Aneaus 2019).

Urban soils, if contaminated, can also represent a threat to supplies of safe drinking water. Transfers of contaminants such as nutrients, metals, and organic pollutants can occur between urban soil and potable surface- or ground-water sources by leaching (Carpenter et al. 1998; Zhang et al. 2001; Imperato et al. 2003; Rodríguez-Eugenio et al. 2018). Leaching of contaminants is more pronounced if they are present in forms which are minimally retained by soils. For example, nitrate is very weakly retained by under most soil conditions and is a commonly encountered groundwater contaminant with potentially serious health consequences as described

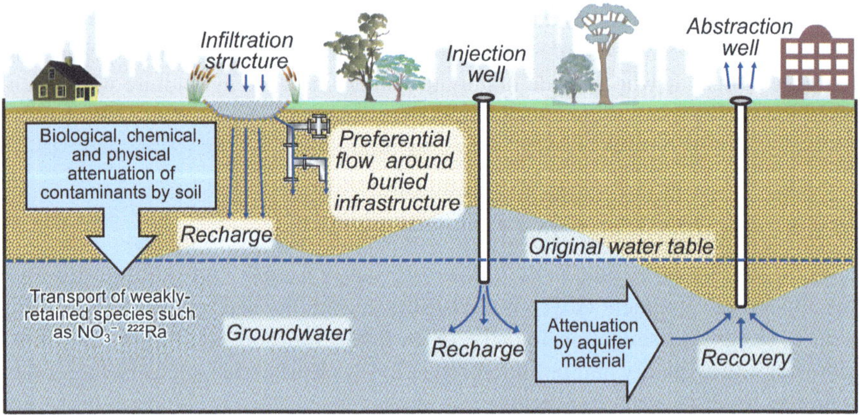

Fig. 10.1 Idealized schematic of Managed Aquifer Recharge. Structures such as infiltration basins or trenches (but not wells) utilize urban soil properties for passive treatment of wastewater and stormwater, causing concentration decreases (i.e., attenuation) for contaminants such as nutrients, metals, organics, and pathogens. (Based on Department of Water and Environmental Regulation n.d.)

above ("Physical health and well-being and urban soils" section). Similarly the radionuclide ^{222}Ra (radon) can leach into groundwater (Lee 2011). The physical properties of urban soils can also influence leaching of contaminants to groundwater. Preferential flow around the smooth surfaces of underground infrastructure such as pipework can increase leaching in urban soils, a phenomenon known as the "urban karst" effect (Bonneau et al. 2017), named after the preferential water flow observed in soils in dissected limestone, or "karst" landscapes (see Chap. 5).

Climate change is also predicted to increase flooding, which may severely affect some urbanized areas and have detrimental effects on water quality and access to safe water (Whitehead et al. 2015). Since urban flooding can be managed to some extent by modification of urban soils and landscapes, soils have a role to play in responses to climate change. We will discuss the future of urban soils, specifically in relation to climate change and other environmental threats, in the final chapter (Chap. 12).

10.1.7 Urban Soils and Employment

Sustainable Development Goal 8 is to promote sustained, inclusive, and sustainable economic growth, full and productive employment, and decent work for all. In particular, soil-related employment can help to achieve Target 8.5, to "...achieve full and productive employment and decent work for all women and men, including for young people and persons with disabilities."

Employment opportunities which relate to urban soils exist in field such as urban agriculture and forestry, extension, environmental consultancy, and urban soil remediation. As with other SGDs, one of the main barriers to employment related to use and management of urban soil is that soil continues to be lost to urbanization.

10.1.8 Climate Change and Urban Soils

Sustainable Development Goal 13 exhorts humanity to take urgent action to combat climate change and its impacts; the urgency reflects the status of anthropogenic climate change as possibly the most serious hazard to affect life on Earth. The adverse effects of climate change are more likely to be experienced in faster-growing cities, mainly present in the developing world (Verisk Maplecroft 2018). Urban areas are major contributors of greenhouse gases, with soils playing a part in emissions (Bellucci et al. 2012). Climate change is predicted to cause severe adverse health effects (WHO 2018), for example, increases in deaths from cardiovascular and respiratory disease, and longer transmission seasons for diseases transmitted via other organisms such as mosquitoes.

It is often assumed (Lal 2011) that soils, including those in urban environments, can provide a sink for carbon (with intentional management). The success or otherwise of soil management to address the primary cause of anthropogenic climate change, the emission of greenhouse gases into Earth's atmosphere, depends on the answers to these questions:

1. **Can we add carbon to [urban] soils that otherwise would have entered the atmosphere as CO_2 or CH_4?**

For example, diverting organic waste streams from landfill to compost (or even biochar) destined for urban soils; could a strategy as seemingly mundane as composting be a viable *carbon sequestration* technique? It is certainly true that urban environments are significant sources of organic waste materials which could be re-used beneficially (Lehmann 2011). Other ways to add carbon are to modify urban land uses, such as developing urban agriculture or urban forestry (FAO 2019). A special report of the Intergovernmental Panel on Climate Change relating to land-based processes (IPCC 2019) concluded that urban and peri-urban agriculture, along with other forms of "green infrastructure," can contribute to mitigation of climate change. Urban forests, another important form of urban green infrastructure, can store up to three times more carbon in the underlying soil than in the trees themselves (Lorenz and Lal 2012). In addition, lawns and other urban turf grass environments can store considerable quantities of carbon as soil organic matter (Brown et al. 2012).

It also has been suggested that inorganic carbon sequestration could represent a carbon capture mechanism in urban soils, since urban soils commonly contain finely particulate silicate minerals in the form of construction and demolition dusts which can consume atmospheric CO_2 during chemical weathering reactions (Jorat et al. 2015; Kolosz et al. 2019).

2. **Will the added carbon remain in urban soils for long enough to represent its removal from the short-term carbon cycle, and/or promote other mechanisms for removal of CO_2 or CH_4 from the atmosphere (such as increased carbon fixation by plants)?**

A common process used to stabilize organic waste material produced in urban environments is *composting*. Composting of urban organic wastes, followed by application to soils, has considerable advantages over disposal of organic wastes in landfill (Biala 2011). Composting is ideally an aerobic process for decomposition and stabilization of organic materials; this is in contrast to the predominantly anaerobic decomposition of putrescible wastes in landfills. Composting should emit only CO_2 and not result in significant emissions of methane and nitrous oxide to the atmosphere, as landfills do (Lou and Nair 2009). This is an advantage since the global warming potentials of both CH_4 and N_2O are substantially greater than for CO_2. Another stabilized carbon product applied to soils, *biochar*, is also believed to be able to sequester carbon (Singh et al. 2014). The use of biochar in some urban soils may provide additional benefits, for example, in maintaining soil fertility and an adequate soil water content range in the soil-limited "*green roof*" substrate studied by Chen et al. (2018). In the cases of both composts and biochar, rigorous quality control of the initial organic substrate is important, given the contamination of some urban organic waste sources with potentially toxic inorganic and organic contaminants (Rodríguez-Eugenio et al. 2018).

Urban soils can also represent suitable environments to grow trees which, apart from being able to sequester carbon, also offer several other benefits to urban residents (Fig. 10.2). Globally, it is possible that sufficient tree planting could sequester enough carbon, in combination with other strategies such as reduction in fossil fuel consumption, to capture more than two-thirds of all historical anthropogenic carbon emissions (Bastin et al. 2019). The other benefits of increased urban tree cover include shading, food production, wildlife habitat, filtration of particulates and pollutants, and even improved mental health, so urban and peri-urban forestry (FAO 2019) is an excellent use of urban soil resources. Of course, people living in urban environments also enjoy open spaces, and lawns and other turf grass environments such as sports facilities are also known to be able to sequester soil carbon, with the soil organic carbon persisting since turf grass environments are subject to minimal disturbance (Zirkle et al. 2012). In some urban environments, the carbon storage in soils exceeds that in vegetation (Fig. 10.3).

The sequestration of atmospheric carbon by abiotic mineral weathering has not been widely considered in the context of soils. This is despite the finding that silicate-based waste materials in soils containing calcium and magnesium, such as demolition and construction wastes, can remove considerably more carbon from the atmosphere than is possible by biotic processes alone (Washbourne et al. 2012). Enhancement of CO_2 concentration in the air-filled pore space of soils by microbial and root respiration suggest that shallow burial of construction and demolition wastes would promote more rapid chemical weathering of the silicate minerals by carbonic acid. Interestingly some researchers suggest that cement-based wastes, which also consume CO_2 as they chemically weather, are less important since the cement manufacturing process produces CO_2 emissions equivalent to consumption by weathering (Jorat et al. 2015). While this is true, the opportunity to improve the carbon footprint of cement manufacture should not be ignored.

The amount of atmospheric carbon realistically able to be sequestered by soils is controversial, and some scientists argue it be minimal, partly due to socioeconomic and political barriers (e.g., Amundson and Biardeau 2018). However, there are numerous additional benefits when urban enterprises, communities, and individuals are involved in soil carbon sequestration. For example, Christie and Waller (2019) outline how composting projects in residential apartment buildings generated a desire to create positive global change toward sustainability in participants, who also felt more connected to each other and nature. These environmental education outcomes can also provide strong momentum toward more widespread activism and adoption of sustainable practices by regulators and the wider community (Waller et al. 2018). In addition, the biodiversity and productivity of urban soils are also improved by deliberately increasing soil organic matter content (e.g., Basta et al. 2016; Huang et al. 2019).

Soils can also be involved in modifying urban microclimates, moderating the effect of urban heat islands (Alcoforado and Andrade 2008). Coutts et al. (2013) describe how increased soil water contents, in the context of *Water Sensitive Urban*

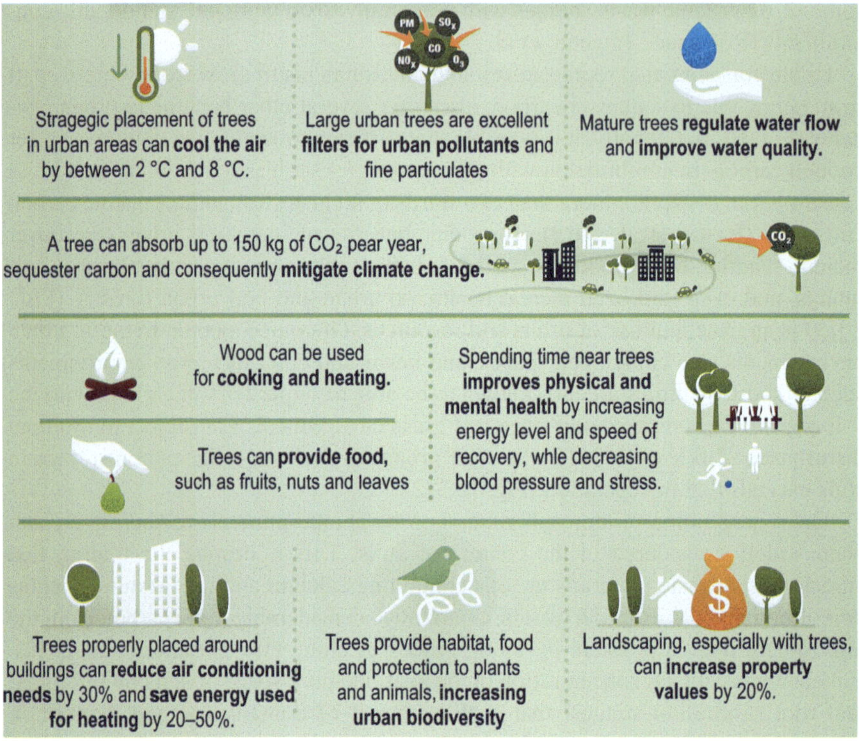

Fig. 10.2 Benefits of urban trees (modified detail of an infographic by FAO 2016)

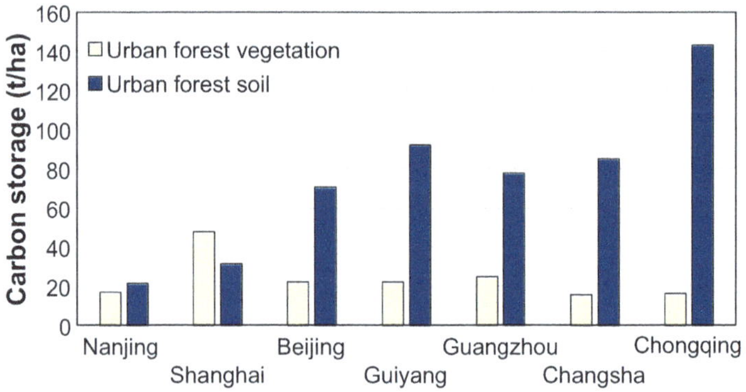

Fig. 10.3 Carbon storage in vegetation and soils in urban forests in Chinese cities (re-drawn from data tabulated in Wang et al. 2013)

Design, provide a cooling effect. Wetter soils, where the additional water may come from sustainable re-use of urban stormwater or appropriately treated wastewater, increase the capacity of the soil to absorb heat and allow for greater evaporative cooling (Coutts et al. 2013).

Urban soils overlying landfills are well known to emit greenhouse gases, mainly carbon dioxide and methane (Blume 1989). Greenhouse gas (GHG) emissions in urban soils are not limited to those on or near landfills, however; soils under urban turf can be net greenhouse gas emitters, especially of N_2O and CH_4 (Livesley et al. 2010; Townsend-Small and Czimczik 2010). In particular, urban soil respiration causes CO_2 emissions much greater than surrounding non-urbanized soils, and emissions depend on land use. In urban soils of Boston, USA, urban forest soils had the lowest carbon losses by soil respiration, with more CO_2 loss from lawns and the greatest losses from garden and landscaped soils (Decina et al. 2016). The losses of greenhouse gases from urban soils are also dependent on management; for example, where agricultural soils are very intensively managed, as in eastern China, urbanization may actually result in an increase in soil carbon storage (Xu et al. 2011).

10.1.9 Terrestrial Life and Urban Soils

The overall aim of Sustainable Development Goal 15 is to "Protect, restore and promote sustainable use of terrestrial ecosystems, sustainably manage forests, combat desertification, and halt and reverse land degradation and halt biodiversity loss." There are many ways that urban soils are related to the targets within SDG 15, such as (among others) Target 15.1 "…conservation, restoration and sustainable use of terrestrial and inland freshwater ecosystems…"; Target 15.5 "…reduce the degradation of natural habitats, halt the loss of biodiversity…"; and so on. SDG 15 is not specifically related to human health, but we will take the holistic view that human health is closely linked to ecosystem health, especially for the terrestrial and inland freshwater ecosystems specifically referred to in Target 15.1. If this is a valid viewpoint, then any function or management outcome related to urban soil that promotes healthier terrestrial ecosystems, therefore, promotes human health. There is certainly evidence that a connection to nature, of which a large proportion would be soil-dependent ecosystems, is beneficial for people's mental health and also physical health by providing places for exercise (Wakefield et al. 2007; Soga et al. 2017; Laidlaw et al. 2018).

The functioning of terrestrial ecosystems is vital to their resilience to external pressures such as climate change and soil pollution (Pavao-Zuckerman 2008; Nero and Anning 2018). Healthier urban ecosystems can promote better human health by diversifying dietary intake through urban agricultural ecosystems (Werner et al. 2019), or by allowing foraging for "wild" food in urban environments (McLain et al. 2014). The landscape diversity of cities is also important for human health;

Arnold and Gibbons (1996) assert that one indicator of urban quality of life and environmental health is the proportion of impervious surface cover. A low proportion of impervious surface cover means greater proportions of a city's land with unsealed soil, and therefore the existence of urban ecosystems such as gardens, grassland, and urban forests. Urban quality of life is therefore greater in areas of cities having less impervious surfaces.

Keesstra et al. (2018) explicitly address Target 15.3, which has the general aim of land restoration, in the context of soils. They discuss a number of ways in which soils (considered in this chapter to mean urban soils) can be part of Nature-Based Solutions to land degradation, including promoting infiltration of water and increasing soil organic matter which are expected to have multiple benefits including increases in biodiversity.

Biodiversity in urban environments depends to some extent on the soil, and commonly the extent to which soil is degraded or restored. Urban soils are often considered to support a less diverse plant community, which is related to the differences in ability of plant species to adapt to urban conditions (Vallet et al. 2010). In contrast, some anthropogenically modified urban soils may also provide specific habitats, not present in undisturbed sites, in which threatened plant species can survive (Albrecht and Haider 2013). An important issue for urban biodiversity is that urban environments can be hotbeds of invasive non-native species such as weeds, with urban soils containing large weed seed banks from their prior existence as degraded areas (Lake and Leishman 2004; Pavao-Zuckerman 2008). Even soil restoration in the form of urban agriculture can, in some instances, increase urban biodiversity (Orsini et al. 2013). Similarly, urban forests may also contain highly diverse plant communities, as shown in a study by Stewart et al. (2009) in the city of Christchurch, New Zealand.

Fewer studies have investigated the biodiversity of soil fauna in urban soils. Fountain and Hopkin (2004) found that for Collembola (springtails) in a range of contaminated and uncontaminated urban soils, the contaminated sites typically had a few dominant species with many rare species. Soils on contaminated sites sometimes contained more species than at uncontaminated sites. Pavao-Zuckerman and Coleman (2007) found that, while the functions performed by soil nematodes differed with varying degrees of urbanization, the taxonomic diversity of nematodes was similar for urbanized and non-urbanized soils. In contrast, Uno et al. (2010) found that the diversity of ant species was greater in forested areas than in urban environments; the lower urban ant diversity may have been related to the colonization of urban soils by an introduced ant species. We have discussed soil biodiversity and related issues in more detail in Chap. 8.

Soil continues to be lost to urbanization (compaction, sealing, contamination, etc.) causing loss of fertility, biodiversity, and soil itself. Poorly implemented, even urban agriculture may degrade soil (Taylor and Lovell 2015).

10.2 Other Human Health Issues Related to Urban Soils

10.2.1 Urban Soil Remediation

Remediation of urban soils, a topic we will address in detail in Chap. 11, offers multiple direct and indirect benefits to human health. The direct benefits are lower exposure to contaminants such as potentially toxic elements ("metals") and persistent organic pollutants (Thornton et al. 2008), and consequent lower risk of ingestion, especially for children (Ottesen et al. 2008). Urban soil remediation has benefits from several perspectives. Remediated soil should result in cleaner ground water and surface water (Van Wezel et al. 2008); there may also be improved infiltration and reduced runoff of precipitation (Olson et al. 2013). Cleaner soil, or soil with improved physical properties, facilitates urban agriculture (Wortman and Lovell 2013) – community gardening itself is a practice that can be used to remediate soil (Al-Delaimy and Webb 2017). Soil biological properties are also affected by remediation, which improves soil ecosystem functioning (Kumar and Hundal 2016). Further indirect benefits may be educational (Kim et al. 2014), for example, the opportunity to promote phytoremediation and other "nature-based" solutions (Song et al. 2019) which also lack the adverse effects of more conventional soil remediation, such as dust generation from excavation.

Urban forests. We have discussed the benefits of urban trees earlier in this chapter; remediated urban soils improve the growth of trees (Layman et al. 2016). Urban forests can be an outcome of soil remediation and rehabilitation, or may represent remnant vegetation. The soils supporting remnant forests sequester more carbon than soil under other urban land uses (Pouyat et al. 2002), which is also an indirect benefit for human health. Livesley et al. (2016) summarized multiple ecosystem services provided by urban forests: at the scale of individual trees, the street scale, and for whole cities. The benefits to human health included cooling, carbon sequestration, energy savings, increased biodiversity, lower water losses by runoff, and reductions in particulate air pollution.

10.2.2 Acid Sulfate Soils and Human Health

There are a number of human health issues associated with acid sulfate soils, with public concern about living in areas known to contain acid sulfate soils (Thomas et al. 2016). The low pH in acid sulfate soils can cause declines in quality, or contamination, of groundwater (Appleyard et al. 2004; Salmon et al. 2014; see also Chap. 6). Acid sulfate soils cannot support a dense plant community, and the resulting susceptibility to wind erosion and generation of atmospheric dust can have health effects (Ljung et al. 2009). The acidic water associated with acid sulfate soils may be toxic to aquatic life such as fish (Powell and Martens 2005), but mosquitoes are often tolerant of acidic conditions and may thrive in acid sulfate soil landscapes (Soukup and Portnoy 1986; Alsemgeest et al. 2005), and so the risk of mosquito-borne diseases can increase.

Contact with water acidified by acid soil processes can result in skin irritation (DER 2015). One study has shown some evidence for increased uptake of metals (arsenic, cadmium, lead, copper, and zinc) by individuals who consume groundwater from acid sulfate soil areas (Hinwood et al. 2008). The higher concentrations of bioavailable aluminum and other metals may have the potential to cause human health problems (Fältmarsch et al. 2008). Food production, such as the growth and survival of agricultural crops (Khuong et al. 2018) and fish or crustaceans grown in aquaculture (Widyatmanti and Sammut 2017), can also be adversely affected.

10.2.3 Environmental Justice Issues

Environmental Justice emerged in the 1980s as a grassroots social activist movement, in response to concerns about the tendency for waste facilities to be located in areas populated with ethnic minorities. Such issues are compounded in low socio-economic neighborhoods, since the residents' overriding concerns are often economic, and the regulation or closure of polluters may result in loss of employment opportunities (Checker 2002). Inequitable access to a healthy environment, however, consists of more than just exposure to water, air, or soil pollution, however. In many cases, people of low socio-economic status, or those outside the dominant ethnic group, have less access to environmental benefits such as good-quality urban infrastructure, "green" infrastructure, urban green space, or opportunity to grow food locally (Rowan and Fridgen 2003; Baker et al. 2019; Siegner et al. 2019). The United States Environmental Protection Agency (USEPA 2008) defines environmental justice broadly to allow application of the concept to all aspects of healthy environments, stating that environmental justice is:

> … the fair treatment and meaningful involvement of all people regardless of race, color, national origin, or income, with respect to the development, implementation, and enforcement of environmental laws, regulations, and policies.

A number of international studies have shown that environmental inequity occurs in urban areas. For example, Zhuo et al. (2012) and Aelion et al. (2013) found that concentrations of potentially toxic elements (Pb, As) in soils were spatially correlated with socioeconomic status and/or predominant ethnic background in urban areas in the USA. Similar relationships between soil pollution and socioeconomic measures have been identified in several other countries, including cities in Australia (Cooper et al. 2018), Canada (Lambert et al. 2006), and the UK (Morrison et al. 2014). The risk of environmental injustice was also recognized as a possible consequence of introducing more stringent environmental regulations for soils in China (Hou and Li 2017) and in an analysis of the aftermath of the Fukushima nuclear accident in Japan (Otsuki 2016). These and other studies reinforce the need for the principles of environmental justice to be incorporated into environmental legal frameworks.

The United Nations' Sustainable Development Goals address issues of social and environmental justice on a global scale, with clear implications for local instances of inequity. Despite the SDGs including aims to achieve both "...sustained, inclusive and sustainable economic growth" (SDG 8) and "Build resilient infrastructure, promote inclusive and sustainable industrialization..." (SDG 9), the SDGs also mandate conservation and sustainable use of water and marine and terrestrial resources (SDGs 6, 14, and 15) and promote "...urgent action to combat climate change and its impacts" (SDG 13). The Sustainable Development Goals also clearly aim to achieve just societies, including equitable access to education (SDG 4), gender equality (SDG 5), reduction of inequality within and among countries (SDG 10), and "...access to justice for all..." (SDG 16). The common ground occupied by environmental justice and the SDGs was analyzed by the Center for International Environmental Law (2002), who identified the most important common issues to be the right to life, including the right to a healthy environment; property rights of indigenous communities; and the rights of communities to make decisions related to their livelihoods and survival (Newton 2009).

10.2.4 Case Studies of Environmental (in)Justice in Urban Soils

Warren County Landfill, North Carolina, USA. In 1982, actions which precipitated the birth of the environmental justice movement as a significant social force (at least in the USA) began in Warren County, North Carolina, USA (Fig. 10.4). The triggering event was the proposal to develop former farmland as a landfill for ca. 180,000 m^3 of soil material contaminated with polychlorinated biphenyls (PCBs), at concentrations up to 500 mg/kg (Hirschhorn 1998; Burnwell 2007). It is significant for this book that this incident involved soil, both as the source and disposal site of contaminated material, even though the environment was predominantly rural. The local residents, mostly African-Americans with low incomes, conducted a series of protests which greatly raised the profile of environmental justice issues in the USA, with a Presidential Executive Order establishing a national office of Environmental Justice in 1994 (Checker 2002). Although the protests were not successful in preventing the landfill, they did prompt a commitment from the state Governor to assess remediation options once technology became available. The site was fully remediated in 2003 using base-catalyzed decomposition (Burnwell 2007; Lyons 2007).

New Orleans: lead-contaminated soils and children's health. The relationships between lead (Pb) concentrations in soils and population variables, in New Orleans, Louisiana, USA, were studied by Campanella and Mielke (2008). The relevance of this study is that lead pollution is widespread in soils, potentially being derived from

Fig. 10.4 Scenes from the construction of the Warren County landfill, NC, USA: (**a**) dumping of PCB-contaminated soil (from Burnwell 2007); (**b**) protesters attempting to block access to the landfill site (from Lyons 2007)

a number of sources, including lead-based paints, lead additives in vehicle fuel, and other industrial uses such as roofing or batteries. The most significant sources in soils for human health are leaded gasoline due to the presence of added Pb in very fine particles, although fine paint fragments can also be important (Mielke and Reagan 1998). Humans can be exposed to lead by ingestion of soil, via soil adhering to plant produce, or inhalation of soil-derived house dusts; children are at greater risk since some younger children ingest soil directly (Clark et al. 2008). Lead poisoning causes neurological effects in humans, and in children has adverse effects on their cognitive and learning abilities.

Campanella and Mielke (2008) found that, in areas of New Orleans having soil lead concentrations greater than 100–200 mg/kg, people of African-American ("black") ethnicity represented the greatest proportion of the population (Figure 10.5a). The median income for residents was negatively correlated with soil lead concentration. Significantly, there were significant proportions of children exposed to soil Pb concentrations greater than the EPA guideline of 400 mg/kg applicable at the time (Figure 10.5b). The risk may have been compounded by significant resuspension of soil during flooding caused by the catastrophic Hurricane Katrina in 2005.

Soil metal pollution and deprivation in Glasgow, Scotland. Glasgow, a major city in Scotland (UK), has been urbanized and industrialized for several centuries, and was studied by Morrison et al. (2014) since its history has resulted in widespread soil contamination. They measured selected soil metal and metalloid concentrations (As, Cr, Cu, Ni, Pb, Se, Zn) which were combined into an average index value for each map polygon (see Fig. 10.6a). A "deprivation index" was also calculated from population demographic data in each map polygon, based on health, education, employment, housing, income, access to services, and crime (Fig. 10.6b). Each map polygon contained approximately equal numbers of households having similar socioeconomic attributes.

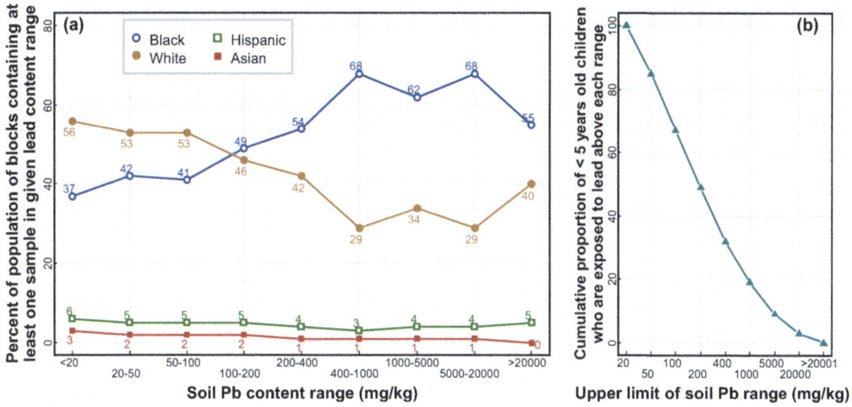

Fig. 10.5 (**a**) Racial composition of New Orleans population as a function of soil lead (Pb) content; (**b**) proportions of young children (< 5 years old) in New Orleans exposed to different concentrations of lead (redrawn from Campanella and Mielke 2008; used with permission from Springer)

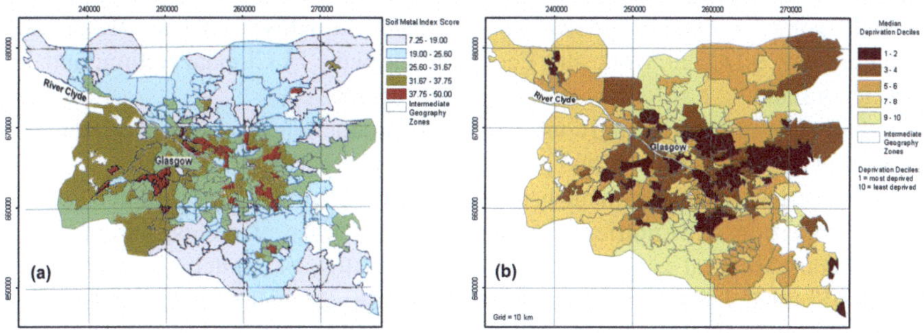

Fig. 10.6 Comparison of (**a**) soil metal contamination index (red/darker colors are most contaminated) and (**b**) socioeconomic deprivation index (darker colors are most deprived) in Glasgow, Scotland, UK (redrawn from Morrison et al. 2014 and used with permission from Springer)

Morrison et al.'s (2014) data showed a statistically significant correlation between soil metal index and deprivation index, which is illustrated in map format in Fig. 10.6. The correlation was explained in terms of the lower cost of rehabilitated former industrial land, which supported lower cost housing, but on which the soil also contained a legacy of trace element contamination from historical metal-processing industries. The authors recommended that assessments of communities in the context of deprivation and environmental justice should include information on soil chemical quality.

10.3 Further Reading

Brevik EC, Burgess LC (eds) Soils and Human Health. CRC Press (Taylor & Francis), Boca Raton, FL, USA

Li G, Sun G-X, Ren Y, Luo X-S, Zhu Y-G (2018) Urban soil and human health: a review. European Journal of Soil Science 69:196–215. https://doi.org/10.1111/ejss.12518

10.4 Summary

- With appropriate use and management, urban soils have crucial roles to play in complete achievement of the United Nations' Sustainable Development Goals. The links between the Sustainable Development Goals and urban soils represent a link between urban soils and human health in multiple contexts.
- The main Sustainable Development Goals requiring an understanding of urban soil processes are as follows: 1 No Poverty; 2 Zero Hunger; 3 Good Health and Well-Being; 4 Quality Education; 5 Gender Equality; 6 Clean Water and Sanitation; 8 Decent Work and Economic Growth; 13 Climate Action; and 15 Life on Land.
- Urban soils are increasingly used for food production by urban communities, with an opportunity to address poverty and food security.
- Many of the adverse human health effects of urban soils are related to soil contamination with nutrients, potentially toxic trace elements, organic contaminants, mineral contaminants such as asbestos, radionuclides, and pathogens. Soil has a role in maintaining water quality as well, where contaminants are also an issue.
- An urgent human health issue in a holistic sense is that of the effects of climate change. Urban soils have the capacity to directly (through sequestration of carbon) and indirectly (by supporting urban vegetation especially trees) affect processes relevant to climate change.
- Environmental (in)justice issues are highly relevant in cities; for example, many studies show that communities having lower socioeconomic status also live in environments where soil contamination is more severe.

10.5 Questions

10.5.1 Checking Your Understanding

1. What are the risks and benefits, in term of the Sustainable Development Goals, of growing food for human consumption in urban soils?

2. How can soils in cities affect water quality?
3. Recall the various adverse human health effects that might occur if urban soils are contaminated with potentially toxic trace elements, organic contaminants, asbestos, radionuclides, and pathogens.
4. What are the ways in which urban communities can be involved with urban soils?
5. List some benefits to urban communities of being involved with urban soils. Which of the Sustainable Development Goals could be addressed from community involvement with soils?

10.5.2 Thinking About the Topics more Deeply

6. Discuss the benefits and risks of managed aquifer recharge in the context of human and ecosystem health.
7. Identify the mechanisms by which urban soils can be either a source of, or a sink for, atmospheric carbon.
8. What do you think the factors are, which result in environmental injustice involving soil or land in urban environments?
9. What are the barriers to urban communities having access to the information that they need about urban soils? How might these barriers be overcome?

10.5.3 Thinking Creatively About Urban Soils

10. Design a soil-based activity for an urban community, and show how the proposed activity addresses three or more of the Sustainable Development Goals.

References

Abiye TA, Sulieman H, Ayalew M (2009) Use of treated wastewater for managed aquifer recharge in highly populated urban centers: a case study in Addis Ababa, Ethiopia. Environ Geol 58:55–59. https://doi.org/10.1007/s00254-008-1490-y

Aelion CM, Davis HT, McDermott S, Lawson AB (2009) Soil metal concentrations and toxicity: associations with distances to industrial facilities and implications for human health. Sci Total Environ 407:2216–2223

Aelion CM, Davis HT, Lawson AB, Cai B, McDermott S (2012) Associations of estimated residential soil arsenic and lead concentrations and community-level environmental measures with mother-child health conditions in South Carolina. Health Place 18:774–781. https://doi.org/10.1016/j.healthplace.2012.04.005

Aelion CM, Davis HT, Lawson AB, Cai B, McDermott S (2013) Associations between soil lead concentrations and populations by race/ethnicity and income-to-poverty ratio in urban and rural areas. Environ Geochem Health 35:1–12. https://doi.org/10.1007/s10653-012-9472-0

Albrecht H, Haider S (2013) Species diversity and life history traits in calcareous grasslands vary along an urbanization gradient. Biodivers Conserv 22:2243–2267. https://doi.org/10.1007/s10531-013-0437-0

Alcoforado MJ, Andrade H (2008) Global warming and the urban heat island. In: Marzluff J et al (eds) Urban ecology: an international perspective on the interaction between humans and nature. Springer, New York, pp 249–262. https://doi.org/10.1007/978-0-387-73412-5_14

Al-Delaimy WK, Webb M (2017) Community gardens as environmental health interventions: benefits versus potential risks. Curr Environ Health Rep 4:252–265. https://doi.org/10.1007/s40572-017-0133-4

Alloway BJ (2004) Contamination of soils in domestic gardens and allotments: a brief overview. Land Contamination Reclamation 12:179–187

Alsemgeest G, Dale P, Alsemgeest D (2005) Evaluating the risk of potential acid sulfate soils and habitat modification for mosquito control (runneling) in coastal salt marshes: comparing methods and managing the risk. Environ Manag 36:152–161. https://doi.org/10.1007/s00267-003-0112-4

Alvarez-Campos O, Evanylo GK (2019) Plant available nitrogen estimation tools for a biosolids-amended, clayey urban soil. Soil Sci Soc Am J 83:808–816. https://doi.org/10.2136/sssaj2018.11.0441

Amoah P, Drechsel P, Abaidoo RC (2005) Irrigated urban vegetable production in Ghana: sources of pathogen contamination and health risk elimination. Irrig Drain 54:S49–S61. https://doi.org/10.1002/ird.185

Amundson R, Biardeau L (2018) Opinion: soil carbon sequestration is an elusive climate mitigation tool. Proceed Natl Acad Sci 115:11652–11656. https://doi.org/10.1073/pnas.1815901115

Appleyard S, Wong S, Willis-Jones B, Angeloni J, Watkins R (2004) Groundwater acidification caused by urban development in Perth, Western Australia: source, distribution, and implications for management. Aust J Soil Res 42:579–585

Arnold CL, Gibbons CJ (1996) Impervious surface coverage: the emergence of a key environmental indicator. J Am Plan Assoc 62:243–258. https://doi.org/10.1080/01944369608975688

Baker A, Brenneman E, Chang H, McPhillips L, Matsler M (2019) Spatial analysis of landscape and sociodemographic factors associated with green stormwater infrastructure distribution in Baltimore, Maryland and Portland, Oregon. Sci Total Environ 664:461–473. https://doi.org/10.1016/j.scitotenv.2019.01.417

Barron OV, Barr AD, Donn MJ (2013) Effect of urbanisation on the water balance of a catchment with shallow groundwater. J Hydrol 485:162–176. https://doi.org/10.1016/j.jhydrol.2012.04.027

Basta NT, Busalacchi DM, Hundal LS, Kumar K, Dick RP, Lanno RP et al (2016) Restoring ecosystem function in degraded urban soil using biosolids, biosolids blend, and compost. J Environ Qual 45:74–83. https://doi.org/10.2134/jeq2015.01.0009

Bastin J-F, Finegold Y, Garcia C, Mollicone D, Rezende M, Routh D et al (2019) The global tree restoration potential. Science 365:76–79. https://doi.org/10.1126/science.aax0848

Bellucci F, Bogner JE, Sturchio NC (2012) Greenhouse gas emissions at the urban scale. Elements 8:445–450. https://doi.org/10.2113/gselements.8.6.445

Bettez ND, Groffman PM (2012) Denitrification potential in stormwater control structures and natural riparian zones in an urban landscape. Environ Sci Technol 46:10909–10917. https://doi.org/10.1021/es301409z

Biala J (2011) The benefits of using compost for mitigating climate change. NSW Office of Environment and Heritage Report OEH 2011/0385, Offi ce of environment and heritage, Department of Premier and Cabinet, Sydney, Australia

Bjørklund G, Albert Christophersen O, Chirumbolo S, Selinus O, Aaseth J (2017) Recent aspects of uranium toxicology in medical geology. Environ Res 156:526–533. https://doi.org/10.1016/j.envres.2017.04.010

Blume HP (1989) Classification of soils in urban agglomerations. Catena 16:269–275. https://doi.org/10.1016/0341-8162(89)90013-1

Bonneau J, Fletcher TD, Costelloe JF, Burns MJ (2017) Stormwater infiltration and the 'urban karst' – a review. J Hydrol 552:141–150. https://doi.org/10.1016/j.jhydrol.2017.06.043
Brown S, Miltner E, Cogger C (2012) Carbon sequestration potential in urban soils. In: Lal R, Augustin B (eds) Carbon sequestration in urban ecosystems. Springer, Dordrecht, pp 173–195. https://doi.org/10.1007/978-94-007-2366-5_7
Burnwell DB (2007) Warren County, NC past and present: "a story of community involvement and empowerment", US Environmental Protection Agency, Warren County, NC, USA. https://semspub.epa.gov/work/HQ/174263.pdf
Campanella R, Mielke HW (2008) Human geography of New Orleans' high-lead geochemical setting. Environ Geochem Health 30:531–540. https://doi.org/10.1007/s10653-008-9190-9
Carpenter SR, Caraco NF, Correll DL, Howarth RW, Sharpley AN, Smith VH (1998) Nonpoint pollution of surface waters with phosphorus and nitrogen. Ecol Appl 8:559–568. https://doi.org/10.1890/1051-0761(1998)008[0559:nposww]2.0.co;2
Center for International Environmental Law (2002) One species, One Planet: Environmental Justice and Sustainable Development CIEL, Washington, DC, USA
Chakravorty A (2019) Nitrogen from biosolids can help urban soils and plant growth, https://www.soils.org/discover-soils/story/nitrogen-from-biosolids-can-help-urban-soils-and-plant-growth (accessed 20190910)
Checker MA (2002) 1999 Peter K. New prize recipient "It's in the air": redefining the environment as a new metaphor for old social justice struggles. Hum Organ 61:94–105
Chen H, Ma J, Wang X, Xu P, Zheng S, Zhao Y (2018) Effects of biochar and sludge on carbon storage of urban green roofs. Forests 9. https://doi.org/10.3390/f9070413
Christie B, Waller V (2019) Community learnings through residential composting in apartment buildings. J Environ Educ 50:97–112. https://doi.org/10.1080/00958964.2018.1509289
Clark HF, Hausladen DM, Brabander DJ (2008) Urban gardens: lead exposure, recontamination mechanisms, and implications for remediation design. Environ Res 107:312–319. https://doi.org/10.1016/j.envres.2008.03.003
Cooper N, Green D, Sullivan M, Cohen D (2018) Environmental justice analyses may hide inequalities in indigenous people's exposure to lead in Mount Isa. Queensland Environ Res Lett 13. https://doi.org/10.1088/1748-9326/aad295
Coutts AM, Tapper NJ, Beringer J, Loughnan M, Demuzere M (2013) Watering our cities:the capacity for water sensitive Urban Design to support urban cooling and improve human thermal comfort in the Australian context. Progr Phys Geography: Earth Environ 37:2–28. https://doi.org/10.1177/0309133312461032
Croll BT, Hayes CR (1988) Nitrate and water supplies in the United Kingdom. Environ Pollut 50:163–187. https://doi.org/10.1016/0269-7491(88)90190-X
De Miguel E, Iribarren I, Chacón E, Ordoñez A, Charlesworth S (2007) Risk-based evaluation of the exposure of children to trace elements in playgrounds in Madrid (Spain). Chemosphere 66:505–513. https://doi.org/10.1016/j.chemosphere.2006.05.065
Decina SM, Hutyra LR, Gately CK, Getson JM, Reinmann AB, Short Gianotti AG, Templer PH (2016) Soil respiration contributes substantially to urban carbon fluxes in the greater Boston area. Environ Pollut 212:433–439. https://doi.org/10.1016/j.envpol.2016.01.012
Department of Water and Environmental Regulation (n.d.) Managed aquifer recharge. Government of Western Australia, Perth, Australia, https://www.water.wa.gov.au/urban-water/water-recycling-efficiencies/managed-aquifer-recharge (accessed 2020.06.03)
DER (2015) Treatment and management of soil and water in acid sulfate soil landscapes, acid sulfate soils guideline series, Department of Environment Regulation, State of Western Australia, Perth. https://www.der.wa.gov.au/images/documents/your-environment/acid-sulfate-soils/guidelines/Identification_and_investigation_of_acid_ss_and_acidic_landscapes.pdf
Derbyshire E (2007) Natural minerogenic dust and human health. Ambio 36:73–77. https://doi.org/10.1579/0044-7447(2007)36[73:NMDAHH]2.0.CO;2

Dhakal KP, Chevalier LR (2017) Managing urban stormwater for urban sustainability: barriers and policy solutions for green infrastructure application. J Environ Manag 203:171–181. https://doi.org/10.1016/j.jenvman.2017.07.065
Dietz ME, Clausen JC, Filchak KK (2004) Education and changes in residential nonpoint source pollution. Environ Manag 34:684–690. https://doi.org/10.1007/s00267-003-0238-4
Fältmarsch RM, Åström ME, Vuori KM (2008) Environmental risks of metals mobilised from acid sulphate soils in Finland: a literature review. Boreal Environ Res 13:444–456
FAO (2016) Benefits of urban trees (infographic), Food and Agriculture Organization of the United Nations. http://www.fao.org/resources/infographics/infographics-details/en/c/411348/ (accesed 20210618)
FAO (2019) Urban and Peri-Urban Forestry, http://www.fao.org/forestry/urbanforestry/en/ (accessed 20210618)
Fett MJ, Mira M, Smith J, Alperstein G, Causer J, Brokenshire T et al (1992) Community prevalence survey of children's blood lead levels and environmental lead contamination in inner Sydney. Med J Aust 157:441–445
Fountain MT, Hopkin SP (2004) Biodiversity of collembola in urban soils and the use of Folsomia candida to assess soil 'quality'. Ecotoxicology 13:555–572. https://doi.org/10.1023/B:ECTX.0000037192.70167.00
Frank AL, Joshi TK (2014) The global spread of asbestos. Ann Glob Health 80:257–262. https://doi.org/10.1016/j.aogh.2014.09.016
Gathuru K, Njenga M, Karanja N, Munyao P (2009) Gender perspectives in organic waste recycling for urban agriculture in Nairobi, Kenya. In: Hovorka A, de Zeeuw H, Njenga M (eds) Women feeding cities - mainstreaming gender in urban agriculture and food security. Practical Action Publishing, Rugby, pp 141–155
Gregory MM, Leslie TW, Drinkwater LE (2016) Agroecological and social characteristics of New York city community gardens: contributions to urban food security, ecosystem services, and environmental education. Urban Ecosyst 19:763–794. https://doi.org/10.1007/s11252-015-0505-1
Hinwood A, Horwitz P, Rogan R (2008) Human exposure to metals in groundwater affected by acid sulfate soil disturbance. Arch Environ Contam Toxicol 55:538–545. https://doi.org/10.1007/s00244-007-9076-3
Hirschhorn JS (1998) Comparative evaluation of demonstration testing of two PCB detoxification technologies. Remediation 1998:81–93. https://doi.org/10.1002/rem.3440090109
Hou D, Li F (2017) Complexities surrounding China's soil action plan. Land Degrad Dev 28:2315–2320. https://doi.org/10.1002/ldr.2741
Hovorka A, de Zeeuw H, Njenga M (eds) (2009) Women feeding cities - mainstreaming gender in urban agriculture and food security. Practical Action Publishing, Rugby, 377 pp
Huang XF, Li SQ, Li SY, Ye GY, Lu LJ, Zhang L et al (2019) The effects of biochar and dredged sediments on soil structure and fertility promote the growth, photosynthetic and rhizosphere microbial diversity of Phragmites communis (Cav.) Trin. ex Steud. Sci Total Environ 697. https://doi.org/10.1016/j.scitotenv.2019.134073
Imperato M, Adamo P, Naimo D, Arienzo M, Stanzione D, Violante P (2003) Spatial distribution of heavy metals in urban soils of Naples city (Italy). Environ Pollut 124:247–256
IPCC (2019) Climate change and land: an IPCC special report on climate change, desertification, land degradation, sustainable land management, food security, and greenhouse gas fluxes in terrestrial ecosystems, intergovernmental panel on climate change, Geneva, Switzerland. https://www.ipcc.ch/site/assets/uploads/2019/11/SRCCL-Full-Report-Compiled-191128.pdf
Jorat E, Goddard M, Kolosz B, Sohi S, Manning D (2015) Sustainable urban carbon capture: engineering soils for climate change (SUCCESS). In: Winter MG, Smith DM, Eldred PJL, Toll DG (eds) Proceedings of the XVI European conference on soil mechanics and geotechnical engineering: geotechnical engineering for infrastructure and development. ICE Publishing, London, pp 2559–2563. https://doi.org/10.1680/ecsmge.60678

Keesstra S, Mol G, de Leeuw J, Okx J, Molenaar C, de Cleen M, Visser S (2018) Soil-related sustainable development goals: Four concepts to make land degradation neutrality and restoration work. Land 7. https://doi.org/10.3390/land7040133

Khuong NQ, Kantachote D, Onthong J, Xuan LNT, Sukhoom A (2018) Enhancement of rice growth and yield in actual acid sulfate soils by potent acid-resistant Rhodopseudomonas palustris strains for producing safe rice. Plant Soil 429:483–501. https://doi.org/10.1007/s11104-018-3705-7

Kim BF, Poulsen MN, Margulies JD, Dix KL, Palmer AM, Nachman KE (2014) Urban community gardeners' knowledge and perceptions of soil contaminant risks. PLoS One 9. https://doi.org/10.1371/journal.pone.0087913

Kolosz B, Sohi S, Manning D (2019) CASPER: a modelling framework to link mineral carbonation with the turnover of organic matter in soil. Comput Geosci 124:58–71. https://doi.org/10.1016/j.cageo.2018.12.012

Kumar K, Hundal LS (2016) Soil in the city: sustainably improving urban soils. J Environ Qual 45:2–8. https://doi.org/10.2134/jeq2015.11.0589

Laidlaw MAS, Alankarage DH, Reichman SM, Taylor MP, Ball AS (2018) Assessment of soil metal concentrations in residential and community vegetable gardens in Melbourne, Australia. Chemosphere 199:303–311. https://doi.org/10.1016/j.chemosphere.2018.02.044

Lake JC, Leishman MR (2004) Invasion success of exotic plants in natural ecosystems: the role of disturbance, plant attributes and freedom from herbivores. Biol Conserv 117:215–226. https://doi.org/10.1016/s0006-3207(03)00294-5

Lal R (2011) Soil Carbon Sequestration. SOLAW background thematic report - TR04B. Food and Agriculture Organization of the United Nations, Rome and Earthscan, London

Lal R (2017) Urban agriculture in the 21st century. In: Lal R, Stewart BA (eds) Urban soils. Advances in soil science. CRC Press/Taylor & Francis Inc., Boca Raton, pp 1–13

Lambert TW, Guyn L, Lane SE (2006) Development of local knowledge of environmental contamination in Sydney, Nova Scotia: environmental health practice from an environmental justice perspective. Sci Total Environ 368:471–484. https://doi.org/10.1016/j.scitotenv.2006.03.012

Layman RM, Day SD, Mitchell DK, Chen Y, Harris JR, Daniels WL (2016) Below ground matters: urban soil rehabilitation increases tree canopy and speeds establishment. Urban For Urban Green 16:25–35. https://doi.org/10.1016/j.ufug.2016.01.004

Lee JY (2011) Environmental issues of groundwater in Korea: implications for sustainable use. Environ Conserv 38:64–74. https://doi.org/10.1017/S0376892911000087

Lehmann S (2011) Optimizing urban material flows and waste streams in urban development through principles of zero waste and sustainable consumption. Sustainability 3:155–183. https://doi.org/10.3390/su3010155

Li G, Sun G-X, Ren Y, Luo X-S, Zhu Y-G (2018) Urban soil and human health: a review. Eur J Soil Sci 69:196–215. https://doi.org/10.1111/ejss.12518

Lioy P (2010) Exposure science: a view of the past and milestones for the future. Environ Health Perspect 118:1081–1090. https://doi.org/10.1289/ehp.0901634

Livesley SJ, Dougherty BJ, Smith AJ, Navaud D, Wylie LJ, Arndt SK (2010) Soil-atmosphere exchange of carbon dioxide, methane and nitrous oxide in urban garden systems: impact of irrigation, fertiliser and mulch. Urban Ecosyst 13:273–293. https://doi.org/10.1007/s11252-009-0119-6

Livesley SJ, Baudinette B, Glover D (2014) Rainfall interception and stem flow by eucalypt street trees – the impacts of canopy density and bark type. Urban For Urban Green 13:192–197. https://doi.org/10.1016/j.ufug.2013.09.001

Livesley SJ, McPherson EG, Calfapietra C (2016) The urban forest and ecosystem services: impacts on urban water, heat, and pollution cycles at the tree, street, and city scale. J Environ Qual 45:119–124. https://doi.org/10.2134/jeq2015.11.0567

Ljung K, Maley F, Cook A, Weinstein P (2009) Acid sulfate soils and human health-a millennium ecosystem assessment. Environ Int 35:1234–1242

Lorenz K, Lal R (2012) Carbon storage in some urban forest soils of Columbus, Ohio, USA. In: Lal R, Augustin B (eds) Carbon sequestration in urban ecosystems. Springer, Dordrecht, pp 139–158. https://doi.org/10.1007/978-94-007-2366-5_7

Lou XF, Nair J (2009) The impact of landfilling and composting on greenhouse gas emissions – a review. Bioresour Technol 100:3792–3798. https://doi.org/10.1016/j.biortech.2008.12.006
Lyons T (2007) Base catalyzed decomposition (BCD) of PCB and dioxin contaminated condensate oil from the remediation of the Warren County landfill, NC, USEPA Office of Research and Development, National Risk Management Research Laboratory, Cincinatti, OH, USA. https://cfpub.epa.gov/si/si_public_file_download.cfm?p_download_id=461179&Lab=NRMRL (accessed 20210618)
McClintock N (2015) A critical physical geography of urban soil contamination. Geoforum 65:69–85. https://doi.org/10.1016/j.geoforum.2015.07.010
McLain RJ, Hurley PT, Emery MR, Poe MR (2014) Gathering "wild" food in the city: rethinking the role of foraging in urban ecosystem planning and management. Local Environ Int J Justice Sustainability 19:220–240. https://doi.org/10.1080/13549839.2013.841659
Mielke HW, Reagan PL (1998) Soil is an important pathway of human lead exposure. Environ Health Perspect 106:217–229
Mielke HW, Gonzales CR, Smith MK, Mielke PW (1999) The urban environment and children's health: soils as an integrator of lead, zinc, and cadmium in New Orleans, Louisiana, U.S.A. Environ Res 81:117–129. https://doi.org/10.1006/enrs.1999.3966
Misra AK (2014) Climate change and challenges of water and food security. Int J Sustain Built Environ 3:153–165. https://doi.org/10.1016/j.ijsbe.2014.04.006
Morgan R (2012) Soil, heavy metals, and human health. In: Brevik EC, Burgess LC (eds) Soils and human health. CRC Press (Taylor & Francis), Boca Raton, pp 59–82
Morrison S, Fordyce FM, Scott EM (2014) An initial assessment of spatial relationships between respiratory cases, soil metal content, air quality and deprivation indicators in Glasgow, Scotland, UK: relevance to the environmental justice agenda. Environ Geochem Health 36:319–332. https://doi.org/10.1007/s10653-013-9565-4
Nero BF, Anning AK (2018) Variations in soil characteristics among urban green spaces in Kumasi, Ghana. Environ Earth Sci 77:317. https://doi.org/10.1007/s12665-018-7441-3
Newton DE (2009) Environmental justice : a reference handbook. Contemporary World Issues Series. ABC-CLIO, LLC
Oladimeji Y, Adepoju SA, Abdulsalam Z (2015) Reviving pottery enterprise: an impetus to poverty alleviation and self-reliance among women folks in Ilorin, Kwara state, Nigeria. Int J Dev Sustain 4:145–160
Oliver MA, Gregory PJ (2015) Soil, food security and human health: a review. Eur J Soil Sci 66:257–276. https://doi.org/10.1111/ejss.12216
Olson NC, Gulliver JS, Nieber JL, Kayhanian M (2013) Remediation to improve infiltration into compact soils. J Environ Manag 117:85–95. https://doi.org/10.1016/j.jenvman.2012.10.057
Orsini F, Kahane R, Nono-Womdim R, Gianquinto G (2013) Urban agriculture in the developing world: a review. Agron Sustain Dev 33:695–720. https://doi.org/10.1007/s13593-013-0143-z
Otsuki K (2016) Procedural equity and corporeality: imagining a just recovery in Fukushima. J Rural Stud 47:300–310. https://doi.org/10.1016/j.jrurstud.2015.12.012
Ottesen RT, Alexander J, Langedal M, Haugland T, Høygaard E (2008) Soil pollution in day-care centers and playgrounds in Norway: national action plan for mapping and remediation. Environ Geochem Health 30:623–637. https://doi.org/10.1007/s10653-008-9181-x
Pavao-Zuckerman MA (2008) The nature of urban soils and their role in ecological restoration in cities. Restor Ecol 16:642–649. https://doi.org/10.1111/j.1526-100X.2008.00486.x
Pavao-Zuckerman MA, Coleman DC (2007) Urbanization alters the functional composition, but not taxonomic diversity, of the soil nematode community. Appl Soil Ecol 35:329–339. https://doi.org/10.1016/j.apsoil.2006.07.008
Pepper IL (2013) The soil health-human health nexus. Crit Rev Environ Sci Technol 43:2617–2652. https://doi.org/10.1080/10643389.2012.694330
Pickering AJ, Julian TR, Marks SJ, Mattioli MC, Boehm AB, Schwab KJ, Davis J (2012) Fecal contamination and diarrheal pathogens on surfaces and in soils among Tanzanian households

with and without improved sanitation. Environ Sci Technol 46:5736–5743. https://doi.org/10.1021/es300022c
Pouyat R, Groffman P, Yesilonis I, Hernandez L (2002) Soil carbon pools and fluxes in urban ecosystems. Environ Pollut 116:S107–S118. https://doi.org/10.1016/S0269-7491(01)00263-9
Powell B, Martens M (2005) A review of acid sulfate soil impacts, actions and policies that impact on water quality in great barrier reef catchments, including a case study on remediation at east trinity. Mar Pollut Bull 51:149–164
Rashid I, Aneaus S (2019) High-resolution earth observation data for assessing the impact of land system changes on wetland health in Kashmir Himalaya, India. Arab J Geosci 12:453. https://doi.org/10.1007/s12517-019-4649-9
Rodríguez-Eugenio N, McLaughlin M, Pennock D (2018) Soil pollution – a hidden reality, food and agriculture Organization of the United Nations Rome, Italy. http://www.fao.org/3/I9183EN/i9183en.pdf
Rowan GT, Fridgen C (2003) Brownfields and environmental justice: the threats and challenges of contamination. Environ Pract 5:58–61. https://doi.org/10.1017/S1466046603030163
Salmon SU, Rate AW, Rengel Z, Appleyard SJ, Prommer H, Hinz C (2014) Reactive transport controls on sandy acid sulfate soils and impacts on shallow groundwater quality. Water Resour Res 50:4924–4952. https://doi.org/10.1002/2013WR014404
Siegner A, Sowerwine J, Acey C (2018) Does urban agriculture improve food security? Examining the nexus of food access and distribution of urban produced foods in the United States: a systematic review. Sustainability 10:2988. https://doi.org/10.3390/su10092988
Siegner AB, Acey C, Sowerwine J (2019) Producing urban agroecology in the East Bay: from soil health to community empowerment. Agroecol Sustain Food Syst. https://doi.org/10.1080/21683565.2019.1690615
Singh B, Macdonald LM, Kookana RS, van Zwieten L, Butler G, Joseph S et al (2014) Opportunities and constraints for biochar technology in Australian agriculture: looking beyond carbon sequestration. Soil Res 52:739–750. https://doi.org/10.1071/SR14112
Soga M, Gaston KJ, Yamaura Y (2017) Gardening is beneficial for health: a meta-analysis. Prev Med Rep 5:92–99. https://doi.org/10.1016/j.pmedr.2016.11.007
Song Y, Kirkwood N, Maksimović Č, Zheng X, O'Connor D, Jin Y, Hou D (2019) Nature based solutions for contaminated land remediation and brownfield redevelopment in cities: a review. Sci Total Environ 663:568–579. https://doi.org/10.1016/j.scitotenv.2019.01.347
Soukup MA, Portnoy JW (1986) Impacts from mosquito control-induced Sulphur mobilization in a Cape Cod estuary. Environ Conserv 13:47–50. https://doi.org/10.1017/S0376892900035864
Stewart GH, Meurk CD, Ignatieva ME, Buckley HL, Magueur A, Case BS et al (2009) URban biotopes of Aotearoa New Zealand (URBANZ) II: floristics, biodiversity and conservation values of urban residential and public woodlands, Christchurch. Urban For Urban Greening 8:149–162. https://doi.org/10.1016/j.ufug.2009.06.004
Taylor JR, Lovell ST (2015) Urban home gardens in the global north: a mixed methods study of ethnic and migrant home gardens in Chicago, IL. Renewable Agric Food Syst 30:22–32. https://doi.org/10.1017/S1742170514000180
Thomas M, Richardson C, Durbridge R, Fitzpatrick R, Seaman R (2016) Mobilising citizen scientists to monitor rapidly changing acid sulfate soils. Trans R Soc S Aust 140:186–202. https://doi.org/10.1080/03721426.2016.1203141
Thornton I, Farago ME, Thums CR, Parrish RR, McGill RAR, Breward N et al (2008) Urban geochemistry: research strategies to assist risk assessment and remediation of brownfield sites in urban areas. Environ Geochem Health 30:565–576. https://doi.org/10.1007/s10653-008-9182-9
Townsend-Small A, Czimczik CI (2010) Carbon sequestration and greenhouse gas emissions in urban turf. Geophys Res Lett 37. https://doi.org/10.1029/2009GL041675
U.S. EPA (2008) EPA assessment of risks from radon in homes. EPA 402-R-03-003, Office of Radiation and Indoor Air, United States Environmental Protection Agency, Washington, DC, USA. https://www.epa.gov/sites/production/files/2015-05/documents/402-r-03-003.pdf
UNEP (n.d.) Why do the sustainable development goals matter? United Nations Environment Programme, Nairobi, Kenya (accessed 20200115)

United Nations (2015) Transforming our world: the 2030 agenda for sustainable development. Resolution 70/1 UN general assembly, New York, USA. https://www.un.org/ga/search/view_doc.asp?symbol=A/RES/70/1&Lang=E (accessed 20210618)

Uno S, Cotton J, Philpott SM (2010) Diversity, abundance, and species composition of ants in urban green spaces. Urban Ecosyst 13:425–441. https://doi.org/10.1007/s11252-010-0136-5

USEPA (2008) Framework for investigating Asbestos-contaminated superfund sites. OSWER DIRECTIVE #9200.0-68, United States Environmental Protection Agency, Washington, DC, USA. https://semspub.epa.gov/work/HQ/175329.pdf

USEPA (2021) Environmental Justice. United States Environmental Protection Agency, Washington, DC, USA, https://www.epa.gov/environmentaljustice (accessed 20210618)

Vallet J, Daniel H, Beaujouan V, Rozé F, Pavoine S (2010) Using biological traits to assess how urbanization filters plant species of small woodlands. Appl Veg Sci 13:412–424. https://doi.org/10.1111/j.1654-109X.2010.01087.x

Van Wezel AP, Franken ROG, Drissen E, Versluijs KCW, Van Den Berg R (2008) Societal cost-benefit analysis for soil remediation in the Netherlands. Integr Environ Assess Manag 4:61–74. https://doi.org/10.1897/IEAM_2007-034.1

Verisk Maplecroft (2018) 84% of world's fastest growing cities face 'extreme' climate change risks, https://www.maplecroft.com/insights/analysis/84-of-worlds-fastest-growing-cities-face-extreme-climate-change-risks/ (accessed 20210618)

Voisin J, Cournoyer B, Vienney A, Mermillod-Blondin F (2018) Aquifer recharge with stormwater runoff in urban areas: influence of vadose zone thickness on nutrient and bacterial transfers from the surface of infiltration basins to groundwater. Sci Total Environ 637-638:1497–1507. https://doi.org/10.1016/j.scitotenv.2018.05.094

Wakefield S, Yeudall F, Taron C, Reynolds J, Skinner A (2007) Growing urban health: community gardening in south-east Toronto. Health Promot Int 22:92–101. https://doi.org/10.1093/heapro/dam001

Waller V, Blackall L, Newton P (2018) Composting as everyday alchemy: producing compost from food scraps in twenty-first century urban environments. In: Crocker R, Chiveralls K (eds) Subverting consumerism: reuse in an accelerated world. Routledge, ProQuest Ebook Central, pp 186–203

Wang Z, Cui X, Yin S, Shen GR, Han YJ, Liu CJ (2013) Characteristics of carbon storage in Shanghai's urban forest. Chin Sci Bull 58:1130–1138. https://doi.org/10.1007/s11434-012-5443-1

Washbourne CL, Renforth P, Manning DAC (2012) Investigating carbonate formation in urban soils as a method for capture and storage of atmospheric carbon. Sci Total Environ 431:166–175. https://doi.org/10.1016/j.scitotenv.2012.05.037

Werner S, Akoto-Danso EK, Manka'abusi D, Steiner C, Haering V, Nyarko G et al (2019) Nutrient balances with wastewater irrigation and biochar application in urban agriculture of northern Ghana. Nutr Cycl Agroecosyst. https://doi.org/10.1007/s10705-019-09989-w

Whitehead PG, Barbour E, Futter MN, Sarkar S, Rodda H, Caesar J et al (2015) Impacts of climate change and socio-economic scenarios on flow and water quality of the Ganges, Brahmaputra and Meghna (GBM) river systems: low flow and flood statistics. Environ Sci Processes Impacts 17:1057–1069. https://doi.org/10.1039/c4em00619d

WHO (2018) Climate change and health. World Health Organization, Geneva, Switzerland, https://www.who.int/news-room/fact-sheets/detail/climate-change-and-health (accessed 20210618)

Widyatmanti W, Sammut J (2017) Hydro-geomorphic controls on the development and distribution of acid sulfate soils in Central Java, Indonesia. Geoderma 308:321–332. https://doi.org/10.1016/j.geoderma.2017.08.024

Wortman SE, Lovell ST (2013) Environmental challenges threatening the growth of urban agriculture in the United States. J Environ Qual 42:1283–1294. https://doi.org/10.2134/jeq2013.01.0031

Wozniacka G (2019) Soil generation is saving community gardens in Philadelphia, https://civileats.com/2019/07/23/soil-generation-is-saving-community-gardens-in-philadelphia/ (accessed 20210618)

Xu N, Zhang T, Wang X, Liu H (2011) Soil organic carbon storage changes in Yangtze Delta region, China. Environ Earth Sci 63:1021–1028. https://doi.org/10.1007/s12665-010-0778-x

Zhang GL, Burghardt W, Lu Y, Gong ZT (2001) Phosphorus-enriched soils of urban and suburban Nanjing and their effect on groundwater phosphorus. J Plant Nutr Soil Sci 164:295–301. https://doi.org/10.1002/1522-2624(200106)164:3<295::AID-JPLN295>3.0.CO;2-T

Zhuo X, Boone CG, Shock EL (2012) Soil lead distribution and environmental justice in the Phoenix Metropolitan Region. Environ Justice 5:206–213. https://doi.org/10.1089/env.2011.0041

Zirkle G, Lal R, Augustin B, Follett R (2012) Modeling carbon sequestration in the U.S. residential landscape. In: Lal R, Augustin B (eds) Carbon sequestration in urban ecosystems. Springer, Dordrecht, pp 265–276. https://doi.org/10.1007/978-94-007-2366-5_7

Chapter 11
Urban Soil Remediation

Andrew W. Rate

Abstract The high incidence of contamination and other forms of degradation of urban soils means that soil remediation is an important issue in cities. In this chapter, we first discuss the factors affecting choice of remediation method, including the common types of objectives for soil remediation. Numerous categories of soil remediation techniques are described and discussed, in the order of methods based on physical, chemical, and finally biological processes. Where data are available, each remediation technique is presented in terms of the underlying principles, applicability to different environments and contaminants, time frame, and effectiveness. We then cover soils themselves as treatment media for water in applications such as constructed wetlands and Water Sensitive Urban Design. Case studies for the remediation of asbestos, an organic pollutant, and metals in urban soils are presented, with reference to the techniques described in previous sections. The chapter closes with a description of some examples of regulatory frameworks for urban soil remediation, in the USA, Europe, and Australia.

Keywords Soil remediation · Urban soils · Remediation criteria · Containment · Soil amendments · Chemical treatment · Natural attenuation · Bioremediation · Phytoremediation · Regulatory frameworks

---

A. W. Rate (✉)
School of Agriculture and Environment, University of Western Australia, Crawley, WA, Australia
e-mail: andrew.rate@uwa.edu.au

A. W. Rate (ed.), *Urban Soils*, Progress in Soil Science,
https://doi.org/10.1007/978-3-030-87316-5_11

What you could learn from this chapter:

- The range and categories of soil remediation techniques available, and the criteria we need to consider to choose the most suitable method.
- The science behind physical, chemical, and biological remediation methods, the contaminants they are effective for, and any constraints imposed by soil properties. The methods include bioremediation, phytoremediation, and in situ physical and chemical remediation methods.
- Some ways in which urban soils can be utilised to treat urban water.
- Real examples of where urban soil remediation has occurred, and how.
- Some examples of regulatory frameworks for soil remediation from around the world.

11.1 Introduction

There is much more to soil remediation than simply cleaning up ‘brownfields’ or contaminated land. This chapter discusses issues relating to remediation projects, the factors affecting the choice of remediation technology that is applied, some of the remediation methods that are available, and whether the soil remediation methods being considered have been used successfully in actual urban soil remediation projects. This chapter will also consider how some regulatory frameworks for soil remediation are constructed.

Soil remediation, in its most general sense, is treatment or management of degraded soil in order to make it suitable for an intended land use (Hodson 2010). In the common case of contaminated soil, soil remediation involves (the partial or complete) removal, or containment and/or immobilization, of contaminants to acceptable levels (Meuser 2012). Caliman et al. (2011) and the Federal Remediation

Technologies Roundtable (2020) identify different ways in which remediation technologies are applied, and we will consider the following three categories:

a) *In situ*, where treatment occurs at the actual location of contamination without removal of soil or groundwater
b) *Ex situ*, where soil is removed, treated, and replaced or contained
c) *Off-site*, where soil is removed to another location for treatment or disposal

Remediation can also be categorised as being *active*, where external energy (such as excavation or pumping) or materials (such as chemical reactants) are imposed on soil to achieve remediation, or *passive*, in which natural processes alone such as solar energy or biological metabolism provide a soil remediation effect.

Contaminant removal is achieved, in general, by chemical separation, phase (physical) separation, chemical destruction, or biodegradation (Rulkens et al. 1998). Remediation may also, of course, be necessary for uncontaminated soil, for example, if it has lost its organic matter, is severely compacted, or if the soil surface is sealed.

11.2 Remediation Criteria

11.2.1 Factors Affecting the Selection of Remediation Method(s)

There are a range of factors which form the basis for choosing one soil remediation method over another (Table 11.1). The compilation in Table 11.1 is based on information presented by the Danish Environmental Protection Agency (2002), Meuser, (2012), Cappuyns (2013), and the Department of Environment Regulation (2014).

More recently the concept of life cycle assessment has been applied to soil and groundwater remediation, where all of the resources used, and environmental impacts generated, are accounted for in the total lifespan of remediated land, from planning to closure (Favara et al. 2011). For example, a life cycle assessment would account for the environmental benefits of reducing or eliminating contaminants, but also account for the environmental harm from vehicle emissions and energy usage during the active phases of a project (Lemming et al. 2009). In addition, other contaminants can be released during remediation of target substances, as conditions such as anoxia change in the soil or groundwater being treated (Mulligan and Yong 2004). The carbon footprint is of particular contemporary concern in soil remediation, with the most immediately straightforward or economical method not necessarily being the process that minimises carbon emissions to the atmosphere in the long-term (for an example, see Jing et al. 2018). The trend in many European states, for example, has been towards in situ biological remediation and away from more energy-intensive methods such as excavation and removal (Fig. 11.1). Implementation of life cycle assessment is related to the practices of 'Sustainable Remediation', for

Table 11.1 Factors affecting the choice of soil remediation method in urban environments

Environmental factors	Examples of factors or constraints
Soil and landscape properties	Soil: hydraulic conductivity; organic matter content Landscape: slope, proximity to water bodies
Geological and hydrological properties of site	Depth to groundwater; groundwater flow rate and direction; rock type and fracturing; depth to rock
The type of contaminant(s) present, or type of degradation	Hydrocarbons, metals, nutrients, persistent organic pollutants, compaction, loss of fertility
The concentration of contaminants present, especially in relation to guideline concentrations	Threshold concentrations for specific soil conditions (e.g. pH, CEC) and proposed land use
Logistical factors	
Previous performance of the options for remediation	If the preferred option is landfarming, but this has not been proven in the environment considered
The availability or feasibility of suitable remediation methods (including technical and logistical constraints)	If the best theoretical option is ISCO, but expertise is unavailable or the site properties do not allow it
The time frame over which remediation must be implemented	Phytoremediation may be unsuitable if cleanup is required within several months
Social factors	
The history or traditions of use of different remediation techniques by practitioners, or in the contaminated site's geographic or regulatory environment	A technique is not considered because it has not been used by a company, or in that city, before (whether or not the technique is effective)
Cultural or social values of the land (e.g. for indigenous peoples)	Sacred sites such as burial grounds precluding *ex situ* remediation
The impacts of the rehabilitation process itself on environments, amenity, health, social issues, etc.	Emissions of fine dust during excavation of contaminated soil; evacuation of residents
The current and intended land users	Current or future land users are committed environmentalists, and prefer a 'green' biological remediation approach
Risk factors	
The calculated risk of various options (including doing nothing) to both human health and different environmental compartments (air quality, surface water, groundwater, and so on)	A scenario where doing nothing may create immediate risk to a few people, whereas *in situ* remediation may generate long-term risk to a sensitive environment
Economic and regulatory factors	
The current and intended land use ('fit for purpose' principle)	The intended use of the land is a car park, which itself serves as remediation by containment
The cost of suitable methods	Encapsulation or vitrification may be the most effective options, but are uneconomic or wasteful
The remediation targets (e.g. the permissible concentration of contaminant(s) remaining in soil)	Some remediation methods (e.g. natural attenuation) may not be able to decrease contaminant concentrations sufficiently

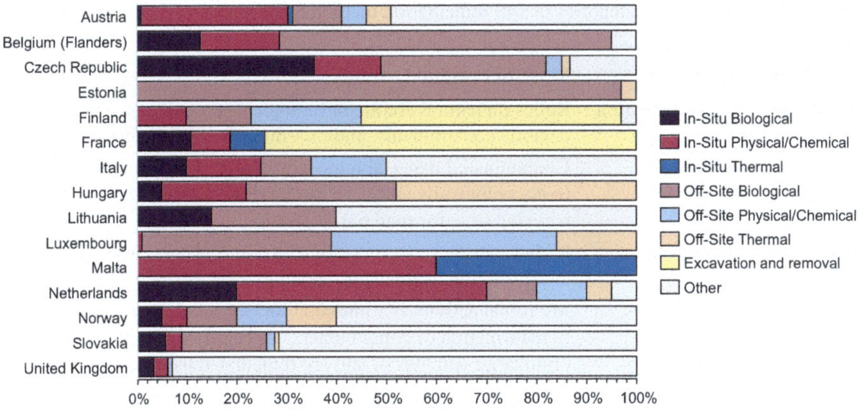

Fig. 11.1 Soil remediation techniques applied at contaminated sites in various European countries in 2006 or 2011. (European Environment Agency 2020)

the promotion of which various multiple-stakeholder organisations exist worldwide. Smith and Nadebaum (2016) provide an excellent summary of the framework for a sustainable remediation approach, focused on Australia and New Zealand where well-developed regulatory frameworks control remediation and management of contaminated sites.

11.2.2 Objectives for Urban Land/Soil Remediation/ Rehabilitation

Depending on the factors listed in Sect. 11.2.1, an urban soil rehabilitation project will have objectives or targets which indicate the success of, or progress towards, a soil remediation end-point (EPA WA 2006). These end-points may be numerical targets such as:

- Decreasing the concentration or bioavailability of a contaminant (or soil physical property like compaction) below a maximum threshold
- Increasing a soil fertility parameter such as porosity, organic matter, or phosphorus content above a minimum threshold
- A biodiversity parameter such as the number of plant species successfully established on the soil

There may also be qualitative objectives, such as creating an aesthetically pleasing landscape.

There are numerous ways in which urban soil can be, and has been, remediated (Table 11.2). The field of contaminated soil remediation in urban areas is continually evolving, and the relative importance of the drivers listed above changes with changes in community, practitioner, and regulatory attitudes towards environmental protection.

Table 11.2 Generalised categories of soil remediation methods

Remediation technology/ management strategy	Mode of operation
Monitored natural attenuation	Relies on natural abiotic or biological processes to volatilise or degrade contaminants, and/or reduce their bioavailability
Excavation and removal	Contaminated material removed to lower-risk environments (e.g. geologically stable secure landfill, disused mine shafts) or recycled (e.g. for road base)
Barrier and containment methods	Include surface covers, sealing, grouting, lateral barriers, encapsulation using clays, geotextiles, etc.
Bioremediation	Microbial activity is stimulated for degradation or transformation of contaminants, especially hydrocarbons – often large volumes of soil are remediated using biopiles or landfarming
Phytoremediation	Plants that accumulate toxic metals, or biodegrade organics in root zone
Vapour extraction	Removal of volatile hydrocarbons in soil, sometimes simultaneously with groundwater remediation by pump-and-treat
Soil washing	Treating soil with liquid or dissolved chemicals to remove contaminants in situ, or by excavation-washing-replacement
Solidification-stabilisation	Amendment of soil with cementing substances to decrease contaminant mobility and bioavailability, or (electro)thermal solidification/melting of soil
Electrokinetic	Application of a low-intensity current that creates a gradient for ions to move from either cathode to anode or vice versa
Ultrasonic	Ultrasonic energy applied to soils may enhance the performance of some other remediation techniques, or enhance degradation of some organic compounds
Thermal	Soil can be heated to high temperatures for melting/solidification, or heat can be applied to enhance volatilization or desorption of contaminants
Administrative	Thorough documentation, restricting human activities on or near a site, and/or placing constraints on future land use

Source: Meuser (2012), Naidu (2013), CRC CARE (2018b)

11.3 Physical Soil Remediation Methods

11.3.1 Natural Physical Attenuation

Physical processes such as dilution are recognised as important processes to consider in the context of remediation of contaminated soils (Mulligan and Yong 2004). The outdated adage that 'the solution to pollution is dilution' has, however, been refuted widely in recent times by highly reputable organisations (such as the UN Environment Programme (see Carr and Neary 2006); US Department of State 2011). In the context of urban soils, therefore, practices that simply allow contaminants to leach into surface water or groundwater, or to spread more widely throughout a larger volume of soil, are unacceptable. Monitored natural attenuation which relies on chemical or biological processes can, however, be effective, and these approaches are discussed in the following sections.

11.3.2 Soil Excavation and Removal

Excavation of contaminated soil, followed by transport to a disposal at another site, is generally considered to be one of the least sustainable soil remediation options (Smith and Nadebaum 2016). Nevertheless, excavation, sometimes called 'dig and dump', is probably the most common method for soil remediation worldwide, possibly because there were no feasible alternatives prior to approximately 1990 (Danish Environmental Protection Agency 2002; U.S. EPA 2012a; CRC CARE 2018e).

Although it is superficially a simple procedure, there are several issues that need to be considered if excavation of contaminated soil is performed. Contaminated dust or vapours may be released into the surrounding environment, and excavated soil is at risk of leaching or erosion by rainfall (U.S. EPA 2012a). These secondary processes, which are caused by excavation, represent new migration and exposure pathways for contamination. Buried infrastructure needs to remain undamaged, and deep excavations can themselves represent a safety hazard. In addition, in most regulatory frameworks consider contaminated soil to be a potentially hazardous waste, requiring purpose-built facilities for transport and disposal (Meuser 2012; CRC CARE 2018a).

Soil excavation and removal is a viable option where in situ remediation methods will be too expensive or too slow to implement, or where small volumes of contaminated soil are present (U.S. EPA 2012a). In some cases, excavated material can be re-used rather than disposed of; for example, moderately contaminated material, which is too contaminated for residential land use, may be appropriately used as fill material for industrial development (Meuser 2012).

11.3.3 Soil Containment: Covering, Surface Sealing, and Encapsulation

Covers prevent transport of contaminants from soil by wind erosion, or vertical and lateral leaching (Meuser 2012; CRC CARE 2018d). There may be varying complexity of covers depending on the severity and type of risk(s) present. For example, if dust generation is the only anticipated risk, simply allowing the contaminated soil to remain at depth, or promoting additional vegetation, may be sufficient. The more severe the contamination, the more technically sophisticated the cover above contaminated material needs to be. For the most severe contamination covers need to prevent water (and sometimes oxygen) entry, vertical transport (upward or downward), lateral migration (in any direction), and soil deformation by thermal stress or bioturbation. Figure 11.2 shows some examples of barriers pre- or post-construction.

Cover construction may involve (from bottom up): a synthetic impermeable membrane; compacted swelling clay (especially close to the groundwater table); a drainage layer; a permeable geotextile; clean fill; and finally clean topsoil with growing plants (see Fig. 11.3). The thickness of clean material depends on the following land use; e.g. while for parks or sports fields, 10 cm might be sufficient, 60–100 cm can be required for soil in which food plants are grown. Covers may require a positive (constructional) anthropogenic landform, depending on the closeness of the contaminated material to the current land surface.

Surface sealing is mainly used for highly contaminated soil; the impermeable cap needs to direct water off-site. Capping of contaminated soil seals the soil surface completely. If there is no risk of vapour generation, building construction following surface sealing (or the building itself forming the surface seal) is an appropriate land use (Meuser 2012; CRC CARE 2018d).

Encapsulation Sealing of soil in more than one dimension, including complete encapsulation of contaminated soil, is used if there are risks of vertical and/or lateral transport of contaminants. Encapsulation can range from lateral barriers (which may include a cap), to systems where contaminated soil is enclosed on all sides. Lateral barriers alone extending above the soil surface may be used to suppress soil erosion by water. The lower boundary of contaminated soil can be con-

Fig. 11.2 Examples of soil containment (**a**) lateral 'silt' barrier to prevent erosion (from epa.vic.gov.au); (**b**) installing a geotextile barrier at a residential development in Perth, Western Australia (from der.wa.gov.au; used with permission); (**c**) large containment cell at a contaminated railway site, Hunter Valley, Australia purecontracting.com.au

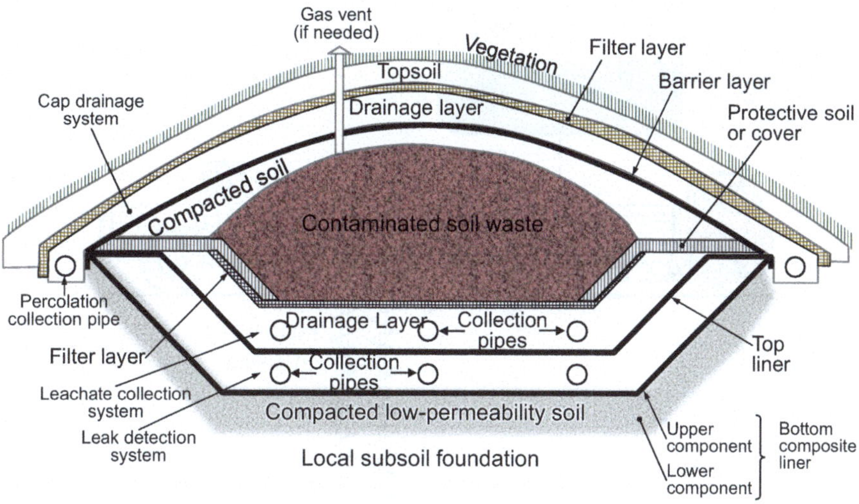

Fig. 11.3 Schematic cross-section through encapsulation for contaminated soil (or secure landfill). (Graphic by Andrew W. Rate)

strained by pre-existing impermeable layers below the contamination, by injection of a cementing agent, or by excavation and replacement after construction of a lower barrier (Mulligan et al. 2001a; Meuser 2012). A form of encapsulation is effectively achieved by 'surcharging', or adding excess fill material over contaminated material so that compression and settlement causes compaction and dewatering of contaminated soil (Patel 2019). Surcharging relies on the simultaneous removal of pore water using vertical drains (PVDs or 'wick drains'), which may be enhanced by applying a partial vacuum, as shown by Indraratna et al. (2011) in Brisbane, Australia. The lower porosity and hydraulic conductivity caused by surcharging creates a lens or aquitard of low-permeability soil, suppressing contaminant migration.

11.3.4 Soil Solidification

Cementation Contaminated urban soil can be solidified to immobilise contaminants by mixing with a cementing agent after complete or partial excavation, followed by replacement or capping. In situ cementation methods can also be used (e.g. grouting, which is also used to produce lateral and lower barriers for in situ containment of soil, or vertical injection as shown in Fig. 11.4) (Meuser 2012). Cementation can be used to increase soil strength and stability as well as decrease the mobility and risk of transfer of contaminants (Wang et al. 2014). A wide range of cementing materials can be used, including Portland cement, lime, ash from power generation, clays, thermoplastics, resins, or combinations of these. Lime and

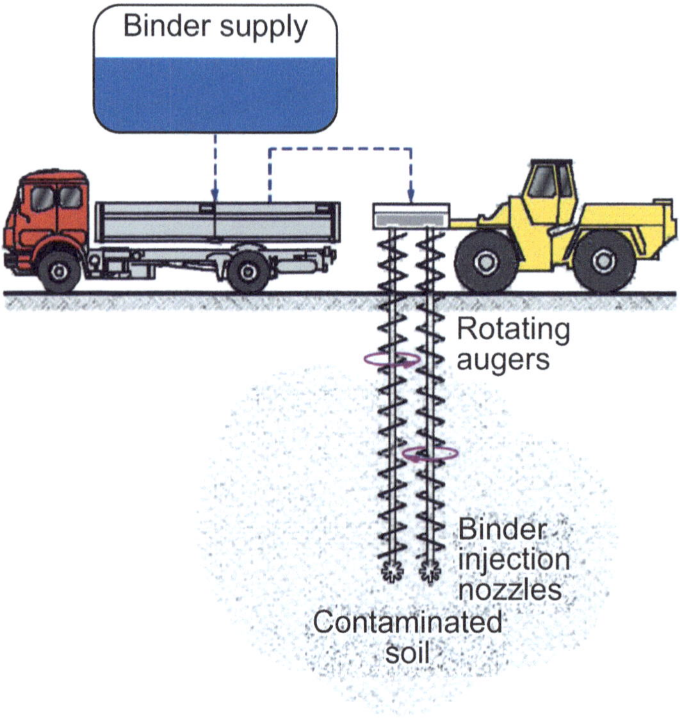

Fig. 11.4 In situ soil solidification by cementation. (Redrawn from Burlakovs and Vircavs 2012, CC-BY licence)

cement-based solidification is applicable to contamination with cationic metals or asbestos; polymer or resin-based solidification can be used for persistent organic pollutants (Meuser 2012; Wang et al. 2014).

Thermal Solidification Also termed *vitrification*, heating of soil to produce a glassy solid mass can be used in cases where the risks associated with transfer of soil contamination are high, for example, if high concentrations of radionuclides are present (Oma et al. 1982). Solidification by heating can be performed on excavated soil (e.g. the vitrification plant at the Hanford Nuclear site in the USA; see hanfordvitplant.com). In situ vitrification can be achieved by application of high electrical voltage and current via two or more electrodes inserted into soil, with addition of a flux material to increase the initial electrical conductivity. Another in situ soil vitrification method involves insertion of a plasma torch, capable of generating temperatures up to 7000 °C, into a series of vertical bores through contaminated soil. Volatile materials emitted by the intense heating are captured, and the vitrified soil is then capped and covered with clean fill (Fig. 11.5).

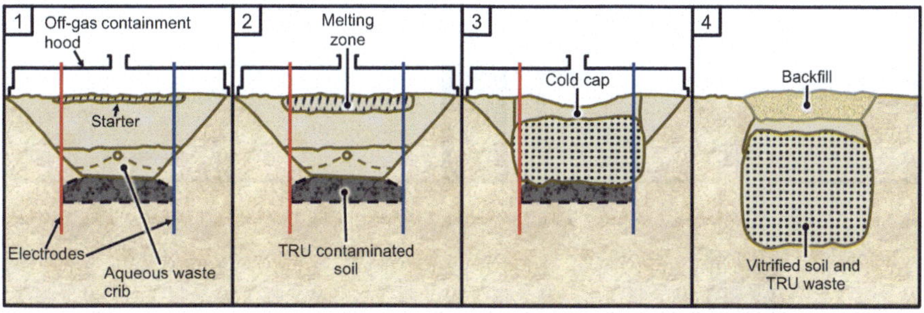

Fig. 11.5 Sequence of stages in an in situ vitrification procedure for contaminated soil. (Redrawn from Oma et al. 1982). TRU = transuranic elements (e.g. plutonium)

11.3.5 Phase Separation

Vapour Extraction and Sparging (also for groundwater) Soil vapour extraction is conducted where a volatile contaminant (e.g. petroleum hydrocarbons) is present in soil (related contamination is most commonly also present in the underlying groundwater). The soil vapour extraction technique is based on *in situ* extraction of the soil gas phase by mechanical vacuum, which is passed through a treatment plant which may have two streams, one to treat the contaminated vapour and the other to treat any condensed liquid (CRC CARE 2018g). The efficiency of soil vapour extraction can be increased by heating soil by various methods, as described below. In addition, pumping of air into contaminated soil (usually below the water table), or *air sparging*, is used as a supplement to soil vapour extraction to increase extraction efficiency. As with some other pumping or vacuum-based methods, air sparging is more effective in permeable soils, and preferential flow paths can decrease the removal of contaminants. Additional flow paths for contaminant removal can be created by hydraulic or pneumatic fracturing (U.S. ACE 2002).

Ongoing production of volatile or gas-phase products such as methane by landfills is a related but separate issue, encompassing the emission of landfill gases in general (mainly methane and carbon dioxide, but also many other volatile organic compounds). Landfill gas may be captured for re-use and energy generation, or released safely into the atmosphere by flaring or venting away from enclosed structures (Omar and Rohani 2015). A more thorough discussion of landfill gas treatment is beyond the scope of this textbook.

Pump-and-Treat (Groundwater Pumping) Groundwater and soil remediation can be combined using multi-phase extraction (MPE), which can target contaminants in separate liquid phases, in the aqueous phase (dissolved or emulsified/suspended in water), and in the vapour phase (U.S. ACE 2002). It is applicable to soils with mod-

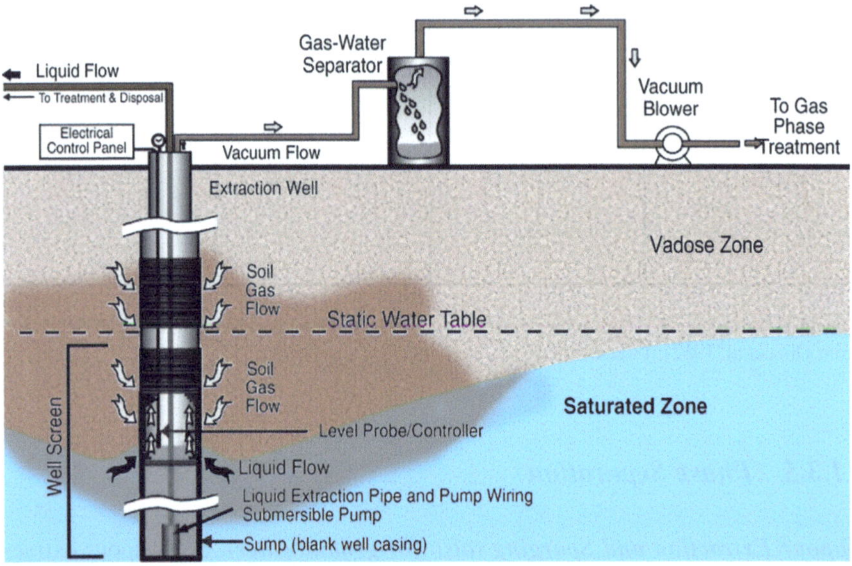

Fig. 11.6 Dual-phase pump-and-treat system, for simultaneous removal of volatile contaminants from soil and groundwater (modified from U.S. ACE 2002). The system combines soil vapour extraction in the unsaturated (vadose) zone and the cone of depression of groundwater caused by pumping, with groundwater pumping in the saturated zone below the water table

erately high permeability, so some clay and/or highly compacted soils are not suitable for treatment by MPE. A high vacuum is applied to an extraction well as in soil vapour extraction, with mechanical separation of gas and liquid phases at the surface followed by separate treatment streams (Fig. 11.6).

Thermal Desorption

Soil heating can be achieved by steam injection, direct application of heat using heating elements, radio-frequency radiation, or by electrical resistive heating (U.S. ACE 2002). Heating is effective in increasing the efficiency of soil vapour extraction, since heat increases the vapour pressure of volatile compounds, promotes desorption of organics from soil particle surfaces, and is one factor that can stimulate biological degradation (see also the section on Biological Soil Remediation Methods below). Steam injection creates an upward convective flux which also assists vapour extraction.

11.4 Chemical Soil Remediation Methods

11.4.1 Abiotic Natural Attenuation

Removal of contaminants from soil can occur naturally if there is transfer to another environmental compartment, such as groundwater or the atmosphere – neither being a preferred or desirable outcome, since the pollution is then simply shifted from one medium to another. Volatilisation of organic compounds in soils is well-known to result in soil-to-atmosphere transfers, particularly for volatile organic compounds (VOC) which have significant vapour pressures in the vadose zone (see Chap. 7; Mulligan and Yong 2004; Shepherd et al. 2006). Even less volatile organic contaminants such as PAHs can be transferred from soil to atmosphere, particularly in warmer climatic zones (Liu et al. 2011). Transfer into groundwater is also common, especially for dense organic liquids (DNAPLs) (Shepherd et al. 2006).

Volatilisation can also occur for inorganic contaminants such as arsenic, selenium, or mercury (Mulligan and Yong 2004). All three of these elements (As, Se, and Hg) can occur as volatile methylated species such as dimethyl arsine, dimethyl selenide, or dimethylmercury which form mainly under reducing conditions (Alloway 2012). In the extremely reducing conditions created in landfills, other elements such as Sn, Sb, Te, Pb, Bi, Mo, and W may form volatile compounds which are released into the atmosphere (Feldmann and Cullen 1997).

While fire is usually considered to provide a source of organic contaminants such as PAHs and PCBs to soils *via* incomplete combustion of organic materials, fire may also result in the emission of some inorganic contaminants. Nriagu (1989) calculated that forest fires were a source of trace metals to the atmosphere, and several inorganic forms of various trace elements (e.g. Hg, Pb, Se, Tl) are volatile at high temperatures (Alloway 2012).

Abiotic attenuation processes, in contrast with biological natural attenuation, do not always decrease the total amount of a contaminant in soil. *Transformations* of both organic and inorganic compounds can occur abiotically, but in the case of inorganic contaminants this predominantly does not remove the contaminant, because they exist mainly as elements. The bioavailability of elemental contaminants, though, may be changed by chemical transformations, as discussed below. An exception among inorganic contaminants is cyanide, for which toxicity depends on the existence of a specific molecule or ion (CN^-); cyanides can be degraded abiotically under both aerobic and anoxic conditions (Kjeldsen 1999).

The toxicity of organic compounds is always related to specific molecules rather than elements, and so abiotic degradation removes organic pollutants without transfer to other environmental compartments. Organic compounds, depending on the properties of the molecules, can be degraded abiotically by photolysis, hydrolysis, dehydrohalogenation, or reduction-oxidation reactions (Yong and Mulligan 2003; Duarte et al. 2017). Hydrocarbons containing only C and H are less vulnerable to chemical degradation than are organic molecules containing other atoms in addition to carbon and hydrogen (commonly halogens such as chlorine, oxygen, or nitrogen). Photolysis (or photodegradation) occurs only at the soil surface, and ultraviolet light is more effective than visible or longer-wavelength radiation.

Abiotic Decreases in Contaminant Bioavailability The attenuation of the effects of a range of pollutants by natural, abiotic (non-biological) processes is related to the common observation of a decrease in bioavailability with increasing time since contamination – the 'ageing effect'.

For organic pollutants, Alexander (2000) argued that resistance to biodegradation of compounds, such as PAHs and DDT, showed evidence of a decrease in bioavailability with age since pollutant addition. The ageing effect for organic pollutants has been explained by movement of their molecules into ultra-fine soil pores which excluded microorganisms, or to partitioning of organic compounds into soil organic matter (Semple et al. 2003, and see also Chap. 7). In both cases the process leading to the decreased bioavailability would need to be effectively irreversible, so that re-equilibration did not continue to expose pollutant molecules to accessibility by soil microorganisms.

Smolders et al. (2009) showed compellingly that ageing effects are important for metals and metalloids, with decreasing bioavailability with increasing time since contamination related to slow continuing chemical reactions and/or soil diagenetic processes (e.g. Backes et al. 1995). An example of this effect is shown in Fig. 11.7 where the toxicity of nickel in soil to springtails (Collembola) is lower in aged soil than in soil in which contamination was added recently (Liu et al. 2015).

11.4.2 Soil Amendments

Liming Metal contaminants existing as cations (e.g. Cd, Pb, Zn) generally become less bioavailable and less mobile as soil pH increases (Harrison and Alloway 1999), and this assumption is implicit in some regulatory guidelines for assessing contamination which allow greater contaminant concentrations at higher soil pH (e.g. National Environment Protection Council 2013). The risk to humans and other urban ecosystem components, of pollution with cationic metals, can therefore be decreased by amending soil with a substance that increases soil pH, such as lime (usually in the form of crushed calcitic limestone, mostly composed of $CaCO_3$). The chemical reactions involved in treatment with lime are based on addition of alkalinity and carbonate ions to soil, and include (see Chap. 6):

- More favourable adsorption of cationic metals on soil clays, oxide minerals, carbonate minerals, and organic matter at higher soil pH
- Conditions favouring formation of insoluble oxides, hydroxides, and carbonates of cationic metals (more likely at very high contaminant metal concentrations)

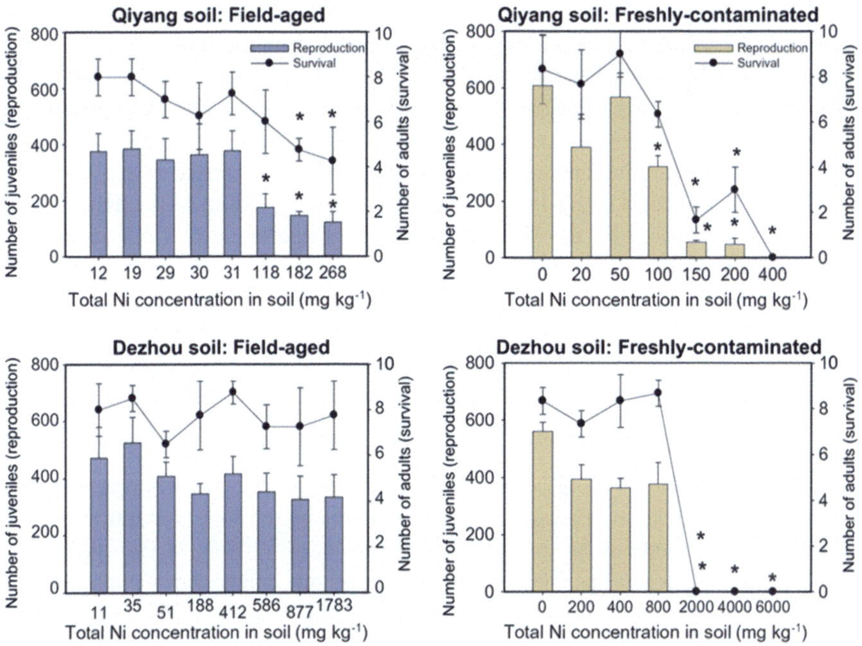

Fig. 11.7 Reproduction (juveniles) and survival (adults) of the springtail *Folsomia candida* in soils contaminated with nickel (Ni) which has been either field-aged (left) or freshly added (right). The lower numbers of juveniles and adults in the freshly contaminated soil, compared with equivalent concentrations in aged soil, indicate that ageing decreases Ni toxicity. (Redrawn from Liu et al. (2015); used with permission from Springer)

Raising soil pH is not a remedial option if the main soil contaminants are elements present as oxyanions, such as arsenic (As), hexavalent chromium (Cr), molybdenum (Mo), or selenium (Se). The bioavailability of these elements may increase as soil pH is raised to values achievable by liming. For organic contaminants, chemical and biological reactivity is affected by pH, but whether bioavailability and mobility increase or decrease is highly dependent on the particular compounds involved. The behaviour of non-ionisable contaminants such as hydrocarbons is relatively insensitive to changes in soil pH, but many other compounds such as pesticides, surfactants, etc., are weakly acidic or basic. Such compounds change their ionisation and therefore water-solubility, adsorption, and photo- or bio-degradation depending on their degree of ionisation (see Chap. 7 and McBride 1994).

Box 11.1 Liming and other treatments to prevent acidification in potential acid sulphate soils

When potential acid sulphate soils (PASS) are disturbed, for example, by installation of buried infrastructure or other construction activities, the preferred option is to prevent acidification of the material exposed to the atmosphere. The concepts in this Information Box are mainly based on the thorough guidelines prepared by the Western Australian Department of Environment Regulation (2015).

The most commonly used neutralizing agent, as discussed in Sect. 11.4.2, is limestone ($CaCO_3$, or $[Ca,Mg]CO_3$) ground to a fine particle size. The application rate necessary to avoid the production of acid must be based on an accurate measurement of the net total acidity of the PASS. This is given by:

Net acidity = potential acidity + existing acidity – acid neutralizing capacity

The potential and existing acidity needs to be measured by techniques that account for oxidation of reduced sulphur minerals in the soil, and options for measurement methods are presented by Ahern et al. (2004). In very poorly buffered soils, as little as 0.005% sulphur may result in net acidification. The acid neutralizing capacity accounts for any alkaline materials such as calcite already present in the soil. Sufficient samples must be taken at the site to assess the range of net acidity values, and a conservative measure (e.g. the maximum net acidity measurement) is used in calculations.

The lime requirement is calculated by:

Lime requirement = Net acidity × 1.02 × SF × 100/ENV

where:

Lime requirement is in kg lime / 10^3kg soil (the weight of actual liming material used)

Net acidity is in kg of H_2SO_4 / 10^3kg soil

The factor of **1.02** converts the mole ratios of $CaCO_3$ and H_2SO_4 to mass ratios

SF is a safety factor (recommended to be $\geq$ 1.5)

ENV is the effective neutralizing value of the liming material relative to pure $CaCO_3$, expressed as a percentage, which takes into account grain size and purity

Lime is applied to PASS material as a powder or slurry and thoroughly mixed mechanically, preferably on an impermeable treatment pad. Re-sampling and analysis of the limed material is performed to assess treatment effectiveness.

(continued)

Box 11.1 (continued)

Other options for treatment of potential acid sulphate soils

Other less common methods to manage acid generation by PASS include:

Submergence: placement of ASS in an excavated void below the water table, which excludes oxygen

Hydraulic mechanical separation: techniques such as sluicing and cycloning to separate fine sulfidic materials from sandy soil or sediment (usually applied in dredging procedures)

Stockpiling: short-term storage of excavated PASS material before oxidation can occur

Offsite treatment and disposal

Immobilising Chemicals Like lime, a wide range of other chemical amendments can be added to contaminated soil to decrease contaminant bioavailability. The chemical mechanisms allowing these amendments to remediate soil contamination include decreasing bioavailability/mobility by ion exchange, adsorption, or (co)precipitation (see Chap. 6). Mineral phases added as solids such as aluminosilicates (clays and zeolites), zero-valent iron, or ferric oxides are strong adsorbents for inorganic contaminants such as metals (both metal cations and metal(loid) oxyanions). Activated carbon and charcoal, and in some cases zeolites, are strong adsorbents for many organic contaminants. Apart from lime, a commonly used immobilising chemical is phosphate, which forms solid precipitates with some metal cations, particularly lead (Pb) which forms the very insoluble mineral pyromorphite ($Pb_5(PO_4)_3[OH,F,Cl]$). Addition of phosphate to Pb-contaminated soil has been shown to significantly reduce lead bioavailability in animals and humans (Scheckel et al. 2013).

Manufactured nanomaterials such as nanoscale zero-valent iron or carbon nanotubes have also been considered as amendments to remediate contaminated soils (Mueller and Nowack 2010; Unrine et al. 2010; Rodríguez-Eugenio et al. 2018). Nanomaterials have the potential advantage of very high surface area available for reactions, due to the very small size of the particles. The ecological and human health effects, however, of the manufactured nanomaterials themselves are not well-known, so there may be considerable unknown risk associated with using such materials for soil amendments.

Organic Amendments Manufactured organic materials such as activated carbon can be used as amendments for soil contaminated with organic compounds (O'Day and Vlassopoulos 2010). The use of organic waste materials, however, can be beneficial from the perspective of both waste minimization (since the organic waste is re-used) and soil remediation (since many organic materials can decrease the bioavailability and mobility of inorganic and organic contaminants) (Jones and Healey 2010) including chemical reduction of toxic chromium (VI) to non-toxic chromium (III).

In situ ***Chemical Oxidation*** Resistant or non-volatile organic contaminants in soils, which are not remediated using techniques like soil vapour extraction, can be oxidised using *in situ* chemical oxidation (ISCO) with various reagents. Solutions containing permanganate (MnO_4^-), ozone (O_3), catalysed hydrogen peroxide (e.g. Fenton's reagent = hydrogen peroxide, H_2O_2, plus ferrous iron, Fe^{2+}), and persulfate ($S_2O_8^{2-}$) are injected into soil via a well or bore (Santos and Rosa 2016). Hydrogen peroxide can oxidise organic contaminants without a catalyst if ferrous iron minerals are present in the soil environment. The efficiency of oxidation can be enhanced by applying ultrasonic energy or heat. In situ chemical oxidation is considered to be a rapid soil remediation technique (CRC CARE 2018f), with time frames for treatment from months to a few years (U.S. EPA 2012b; Regenesis 2019), depending on factors like soil properties such as hydraulic conductivity, textural homogeneity, groundwater flow or water table, and the rate of consumption of the chemical oxidant injected into soil.

Monitoring bores are usually installed to verify contaminant removal, but there is no need for removal of a contaminated solution. The products of the oxidation reactions in ISCO are generally not considered to be harmful; the organic contaminants are oxidised to yield carbon dioxide and water (and chloride in the case of chlorinated hydrocarbons). Most of the reagents also produce non-harmful products such as water, sulphate, and iron oxides; the exception is permanganate, since manganese is not common in all soils, but the MnO_2 which is ultimately produced is largely benign. An emerging issue with ISCO is the potential for oxidation of PFAS, if these contaminants are present. The oxidation of PFAS by conventional ISCO is ineffective, and the oxidation products may have greater solubility in water and thus greater mobility (NGWA 2018).

In situ chemical reduction (ISCR) is used for a few particular contaminants, usually hexavalent chromium (CrO_4^{2-}/$Cr_2O_7^{2-}$, or 'chromium 6') and trichloroethylene (C_2HCl_3, a DNAPL). The reducing agent most commonly used is zero-valent iron (nanoparticulate metallic iron, Fe^0, or ZVI) usually injected into soil as an aqueous slurry. Hexavalent chromium is reduced to Cr^{3+}, which forms mineral solid phases such as Cr_2O_3 of very low solubility. Trichloroethylene is reductively dechlorinated to eventually form ethane (C_2H_6) via a series of intermediate reaction products; the ethane is lost from soil as a gas. Like ISCO, in situ chemical reduction is a relatively rapid remediation technique, with significant decreases in contaminant concentrations within months. The ISCR technique can be implemented as a permeable reactive barrier (see Sect. 11.4.6), in which case remediation is slower, requiring time frames of several years (U.S. EPA 2012c).

11.4.3 Soil Washing

Excavation of soil, and extraction of contaminants with different cleaning fluids or solutions, comprises the technique of *soil washing*. Derived from processes developed for the mineral processing industry, soil washing is usually performed with mechanical agitation in a purpose-built washing plant (Meuser 2012). The solutions used are usually aqueous (water-based); solutions of acids and complexing agents are generally used to remove metals from soils. Non-polar organic compounds such as hydrocarbons, PAHs and PCBs (including NAPLs) have low water-solubility, so their dissolution or emulsion in aqueous solutions is increased using surfactants, co-solvents, cyclodextrins, and other agents. The substances added to enhance contaminant removal should ideally be of low toxicity and/or rapidly biodegradable, and have minimal interaction with the soil matrix (Mulligan et al. 2001b; Mousset et al. 2014). The effectiveness of soil washing theoretically can be enhanced with ultrasonication by generating locally high pressures and temperatures as cavitation occurs in the washing fluid (Kim and Wang 2003). Large volumes of soil can be treated relatively rapidly using soil washing technology; at a site in New Jersey, USA, contaminated with Cu, Cr, and Ni, approximately 180 tonnes/day were treated over a period of approximately 3 months (U.S. EPA 1995a).

11.4.4 Soil Flushing (In Situ Soil Washing)

Soil washing conducted *in situ* is also termed *soil flushing* to distinguish it from washing of excavated soil, but the principles are similar except for constraints imposed by the *in situ* soil properties. The flushing solution is injected into one or more input wells and captured in an extraction well (Fig. 11.8). The recovered contaminants are removed from the solution (e.g. by adsorption or precipitation), and the purified solution is recycled back into the injection well.

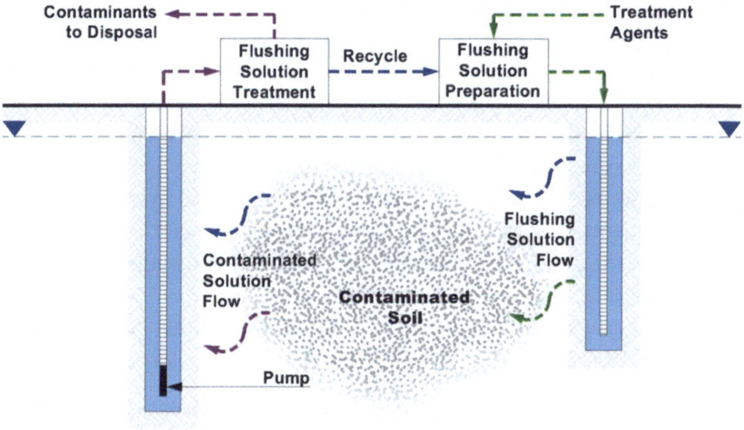

Fig. 11.8 Schematic of an *in situ* soil flushing procedure. (From Burlakovs and Vircavs 2012, CC-BY licence)

The effectiveness of soil flushing depends on conduction of water through the soil matrix, and therefore fine-textured soils (≥ 30–40% silt + clay) are likely to be unsuitable. Similarly, soils with high capacity to adsorb target contaminants (e.g. high organic matter, clay or iron oxide content) will significantly reduce soil flushing remediation effectiveness (Lombi and Hamon 2005; Meuser 2012). Soil flushing has been conducted at very large sites, such as landfills. Completion of remediation at large sites by soil flushing can be a very lengthy process; in part of a Superfund landfill site in New Jersey, USA, remediation was completed in approximately 15 years (U.S. EPA 2017).

11.4.5 Electrokinetic Migration

The migration of ions in an electric field has been used in the soil remediation technique of electroremediation, or electrokinetic migration (Meuser 2012). A direct current is applied to an array of electrodes in soil (e.g. between a single cathode and multiple anodes, with a potential between 20–500 V/m and current of a few mA/cm^2) (see also Naidu 2013). The soil water content needs to be high enough (≥ 50–70% of field capacity; see Chap. 4) to maintain an electrical circuit; however, electroremediation can be used in low-permeability (fine-textured and/or dense) soils, unlike some other *in situ* remediation methods (Kim et al. 2011). Cationic contaminants such as metal ions or ammonium (NH_4^+) will migrate towards the negatively charged cathode, and anions such as cyanide, chromate, or nitrate towards the anode. The electrical migration also causes mass flow and 'electro-osmosis', which allows non-ionic contaminants to be treated. In situ electroremediation also includes bores within hollow electrodes which are pumped out to remove and treat the resulting contaminated solutions. The relatively high electrical potential applied also causes electrolysis of water, with migration of the resulting hydrogen ions towards the cathode and hydroxide ions towards the anode. The changes in soil pore water pH help to maintain the ions in solution (e.g. there will be low pH at the cathode due to migration of H^+, which helps to suppress precipitation or adsorption of metal cations which also accumulate at the cathode). The pH changes are, however, undesirable side effects of electroremediation technology which have potentially adverse effects on the soil ecosystem, and require additional management (Meuser 2012). The effectiveness of electroremediation can be enhanced by adding solubilising chemicals to the solution around the electrodes, or with ultrasonication (Kim et al. 2011). The time required for completion of electroremediation of soil varies with amount and type of contamination and soil properties, but can range from 2 to 18 months (U.S. EPA 1997).

11.4.6 Permeable Reactive Barriers

Permeable reactive barriers combine elements of chemical and biological treatment of contaminants. They are mainly used for treatment of contaminants such as DNAPL and nitrate in groundwater, but are included here since they are installed in soil materials through which contaminated groundwater passes and is treated by the barrier. The barrier may be a simple trench filled with reactive material such as wood waste or compost; more complex designs include a well to inject a carbon source such as vegetable oil or molasses to stimulate microbial activity (Fig. 11.9), or use of *in situ* chemical remediation such as zero-valent iron (U.S. EPA 2012c and Sect. 11.4.2).

The contaminant removal processes in permeable reactive barriers usually depend, at least indirectly, on soil microbial processes. For example, removal of nitrate can be achieved by denitrification, by providing denitrifying organisms with a carbon source following depletion of oxygen by anaerobic decomposers. A carbon source can also be used to promote sulphate reduction in permeable reactive barriers used to decrease cationic metal ion bioavailability/mobility, by formation of insoluble metal sulphides (Patterson et al. 2005). In theory, permeable reactive barriers can provide ongoing treatment of low-level groundwater contamination, particularly if resupply of a carbon source is possible. The life span of permeable reactive barriers has been estimated conservatively at ≥ 10 years (Higgins and Olson 2009), and treatment effectiveness may continue for 20–30 years.

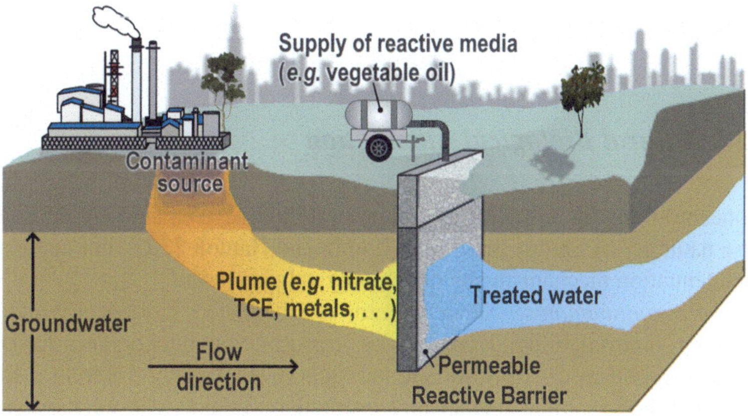

Fig. 11.9 Simplified permeable reactive barrier with an injection well design, constructed in soil down-gradient of contaminated groundwater. (Redrawn based on U.S. EPA 2012d)

11.5 Biological Soil Remediation Methods

Organisms present in soil – in particular, the various groups of microorganisms (Chap. 8) – but also plants, are able to interact with a wide range of potential contaminants. These biological interactions include processes such as transformations and degradation of organic compounds; changes in chemical speciation; and uptake and sequestration. The synthetic organic compounds which are potential contaminants are particularly susceptible to transformations mediated by heterotrophic bacteria and fungi in soils and sediments (Alexander 1981). For remediation to occur, the products of biological transformations must have low or zero toxicity to receptors in soil and receiving environments. The following categories of biological soil degradation, or *bioremediation*, methods are recognised (with examples expanded on in the following sections):

Natural bioremediation: where the indigenous soil microorganisms and/or plants are able to treat contamination *in situ* without human modification.

Enhanced bioremediation: in which soil microorganisms and/or plants are supplemented with essential substances, such as water, nutrients, or oxygen, that might be limiting their ability to treat contamination. Enhanced bioremediation can be performed *in situ,* or on-site using excavated soil. In some cases the bioavailability of organic contaminants can be enhanced by the addition of surfactants (Mulligan et al. 2001b).

Augmented bioremediation: where specific organisms, which are known to be able to treat the contaminant(s) of concern, are added to soil or groundwater. Augmented bioremediation most commonly refers to addition of microorganisms, but phytoremediation (Sect. 11.5.5) using a specific plant species would fall into this category as well.

11.5.1 Natural Biological Attenuation

The concentrations, or mobility and bioavailability, of contaminants in soil can decrease naturally by abiotic processes (Lombi and Hamon 2005), but many of the natural attenuation mechanisms involve biological processes.

Inorganic contaminants: Nutrients can be attenuated biologically by plant uptake, their bioavailability decreased by incorporation into organic forms via microbial metabolism or, for nitrogen and sulphur, or lost by different forms of volatilisation. For example, plants in constructed wetlands can be harvested to remove nitrogen, phosphorus, and other nutrients taken up from wetland soils or water (Malaviya and Singh 2012). The ionic, bioavailable forms of nitrogen and phosphorus, ammonium (NH_4^+), nitrate (NO_3^-), and phosphate (HPO_4^{2-}) can all be assimilated by microorganisms and recycled into organic forms which have lower bioavailability and potential environmental impact. In addition, plants can convert phosphate to phytates which enter soil in plant residues and have very low bioavailability (Lung and Lim 2006). Finally, nitrate present in soils subject to spatially heterogeneous or temporally fluctuating redox conditions can be lost to the atmosphere as gaseous species such as dinitrogen (N_2) or nitrous oxide (N_2O) (Groffman et al. 2002).

Inorganic contaminants: Metals and metalloids are clearly not biodegradable, being elemental contaminants, and are not always considered in biological natural attenuation (Mulligan and Yong 2004). A few elements, notably arsenic and mercury, can undergo biological methylation in reducing (anoxic) soil environments, following which they can be lost from soils by volatilization (Alexander 1981).

Organic contaminants: Organic pollutants in soils are mainly attenuated by natural biological processes, rather than by the abiotic mechanisms described in Sects. 11.3.1 and 11.4.1 (Megharaj et al. 2011). As described by Alexander (1981), the processes involved in remediation of organic pollutants by microorganisms include direct metabolism, where the heterotrophic microorganisms use organic compounds as a source of carbon and energy, and co-metabolism, where some organic compounds are degraded along with another organic substrate, but cannot themselves sustain microbial growth. In most soils, degradation of organic compounds occurs aerobically, with oxygen as the electron acceptor (oxidizing agent) for metabolism; aerobic degradation is faster than anoxic (anaerobic) processes (Yong and Mulligan 2019).

The existence and progress of natural biological attenuation of degradable organic contaminants (e.g. BTEX) in groundwater can be measured by monitoring chemical changes relating to oxidation-reduction reactions. The degradation of organic compounds to produce CO_2 is an oxidative process, so that degradation will initially deplete dissolved oxygen which is required for metabolism of carbon compounds by aerobic microorganisms. Once oxygen is depleted, microorganisms which can use nitrate as an electron acceptor for metabolism of carbon compounds (i.e. the denitrifiers) become the predominant degraders of contaminants. Similarly, manganese oxides (if present), sulphate, then ferric iron oxides are consumed (in that order) as electron acceptors, until finally bicarbonate is consumed and methanogenesis occurs (Yong and Mulligan 2019).

11.5.2 In Situ *Enhanced Bioremediation*

The conditions required to optimise contaminant treatment soils by microbial bioremediation reflect the conditions required generally for microbial growth and survival of microorganisms. Ideally, then, enhanced bioremediation aims to achieve: a suitable range of soil temperature, pH, and redox potential; adequate water, oxygen (or alternative electron acceptors), and nutrients; and the amounts and bioavailability of organic substrates including the contaminant(s) present (Megharaj et al. 2011). Enhanced bioremediation therefore relies on the stimulation of the existing indigenous soil microbial community by artificial addition of water, nutrients, and sometimes amendments such as composts or lime. Amendments other than water can be applied as solutions, or as solid fertiliser and/or lime materials which may be mixed into surface soil by cultivation, which also allows enhanced entry of oxygen (N.S.W. EPA 2014). Temperature can be controlled by the installation of bores with heat exchangers (e.g. solar-heated water) or soil covers such as temporary plastic tunnel houses (Moradi et al. 2018).

Since *in situ* bioremediation relies on the indigenous microorganisms in the soil, it can be a slow process with remediation typically being completed in 1–4 years. It has been shown to be effective for volatile organic compounds, including halogenated hydrocarbons, but enhanced bioremediation methods can also degrade persistent organic pollutants such as PAHs (FRTR (Federal Remediation Technologies Roundtable), 2020). For any application of enhanced bioremediation, a site-specific feasibility study (usually laboratory-based) is required to demonstrate that contaminant treatment is possible, and monitoring during and after treatment is essential (EPA 2010).

The microbial population changes during the course of enhanced bioremediation (Wang et al. 2016). The microbial community structure (i.e. the balance of microorganism species) changes in response to the selection pressure imposed by the contaminant (Kozdrój and Van Elsas 2001). Microorganism species also show different responses to the changes in trophic state and soil properties introduced by addition of amendments such as nutrients, lime, surfactants, and organic materials (Sarkar et al. 2016).

11.5.3 Bioventing and Biosparging

Bioventing and biosparging are further examples of *in situ* enhanced bioremediation, in which the main strategy is to supplement the indigenous microbial population with oxygen. This is to switch the microbial processes to become fully aerobic, which increases degradation rates.

Bioventing is implemented by using pumping to apply pressure or suction to the soil and/or vadose zone to allow the entry of atmospheric oxygen (Fig. 11.10). Since the objective is to treat contamination with potentially volatile organic compounds, suction may be preferred so that gas-phase contaminants can be captured (FRTR (Federal Remediation Technologies Roundtable), 2020). Pumping is most commonly mechanical, but passive pumping which relies on changes in atmospheric pressure has also been successful, and minimises energy consumption (Larson 2000). In all cases, wells are installed in the soil to allow air entry, and the objective of mechanical or passive pumping is to increase the oxygen concentration in soil pores space to $\geq$ 5%. The pumping needs to occur at rates which allow stimulation of indigenous microorganisms, but are not so high that volatile contaminants are lost by direct volatilization, but instead are degraded *in situ* (U.S. ACE 2002).

Bioventing requires soils or vadose zone materials with relatively high permeability to gases, and is not applicable for water-saturated materials including

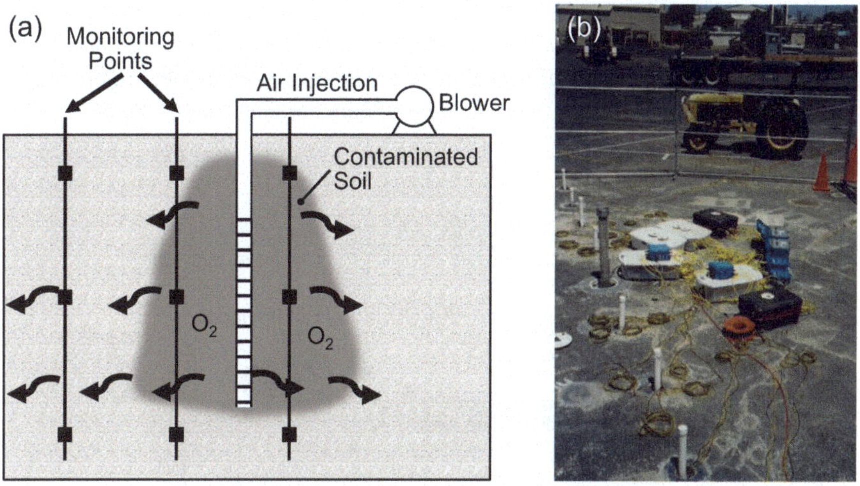

Fig. 11.10 Simplified bioventing setup for promoting oxidative degradation of soil contaminated with organic compounds (redrawn from U.S. ACE 2002); (b) ambient-pressure bioventing setup at an airport in California, USA. (Public domain image from Larson 2000)

groundwater. Soil permeability can decrease during treatment due to increased growth of microorganisms in pores, in which case pulsed pumping can be beneficial (CRC CARE 2018c). Since, unlike soil vapour extraction (Sect. 11.3.5) bioventing does not rely on volatilization of contaminants, it can be effective for both volatile and non-volatile organic contaminants such as PAHs (Larson 2000).

Biosparging mainly relies on accelerating volatilization for its effectiveness, and is therefore similar to the soil vapour extraction technique discussed in Sect. 11.3.5.

11.5.4 Bioremediation Using Biopiles, Landfarming, or Bioreactors

Biopiles and landfarming systems are both examples of techniques where soil is not treated *in situ*; since soil is excavated before treatment, they are categorised as *ex situ* methods (Sect. 11.1). They are similar to enhanced bioremediation in that the excavated soil is treated under conditions which optimise biological degradation of contaminants by controlling air entry, water content, and other factors.

A biopile is a mound or row of contaminated soil installed on an impermeable pad to minimise and capture leachate, and in mounded configuration may have perforated pipes installed to pump air into the pile or capture the evolved gases (FRTR (Federal Remediation Technologies Roundtable) 2020). Windrows are rows of contaminated material (again on an impermeable base) which are aerated periodically by turning using specialised mechanical equipment.

Landfarming is configured as a thin layer of contaminated soil (e.g. ≤30 cm) spread over a large area (N.S.W. EPA 2014). The soil is cultivated to introduce oxygen into the soil to promote degradation of contaminants by aerobic microorganisms (Figs. 11.11 and 11.12a). Volatilization of contaminants and odours cannot be

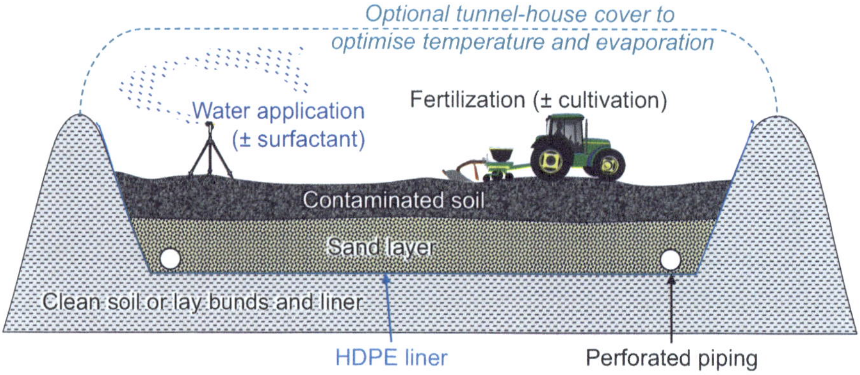

Fig. 11.11 Landfarming schematic. (Graphic by Andrew W. Rate)

Fig. 11.12 (**a**) Hydrocarbon-contaminated soil being placed in a land-farming bioremediation cell lined with an impermeable geomembrane, Casey Station, Antarctica (used with permission from Australian Antarctic Division 2012); (**b**) bioreactor being constructed to promote denitrification for nitrate removal in SE Queensland, Australia. (From Wegscheidl et al. 2018 used under CC-BY-4.0)

controlled easily in landfarming operations, so given the large land area required, this method may not be as suitable as a biopile for all urban scenarios.

Amendments as well as oxygen are added to landfarming systems and biopiles: they are watered to a specified fraction (e.g. 40–85%) of field capacity, and fertilisers or organic amendments are added to maintain fertility. The temperature within the pile usually increases due to the metabolism of microorganisms and the thermal insulation of the pile, but without reaching the temperatures achieved in conventional composting (CRC CARE 2018c). It is possible to employ landfarming or biopiles with less permeable soils, often with addition of organic materials to improve permeability to water and gases. Treatment times may be up to 2 years, depending on site-specific factors (N.S.W. EPA 2014).

Bioreactors combine aspects of biopiles and permeable reactive barriers (see Fig. 11.12b and Sect. 11.4.6) and are an anaerobic *in situ* treatment method for groundwater or surface water in drains (Wegscheidl et al. 2018). We discuss bioreactors here because they are excavated into soil and may be composed of soil materials – essentially they are a below-ground volume of materials such as mixtures of mulch, compost, sand, and/or ZVI with a microbial growth substrate such as vegetable oil (FRTR (Federal Remediation Technologies Roundtable) 2020). The substrates are selected to maintain anoxic conditions, so only applies to contaminants which can be treated by anoxic processes, such as halogenated hydrocarbons, PAHs, nitrate, or chromium (VI). Pumped groundwater is recirculated into the top of the bioreactor to maximise treatment of contaminants; surface water treatment relies on a flow-through design. Treatment times are long, but in many cases bioreactors are designed for long-term treatment of ongoing sources (plumes or drains) and can be effective for decades with periodic replenishment of growth substrate (FRTR (Federal Remediation Technologies Roundtable) 2020).

11.5.5 Phytoremediation

Phytoremediation refers to the use of plants to treat contaminated soil, either by uptake or by decreasing the bioavailability of contaminants. Most attention related to phytoremediation has been focused on remediation of trace element contaminants such as metals, although plants and their associated rhizosphere can also decrease the concentrations or bioavailability of organic pollutants. Phytoremediation is often an attractive option for remediation because of its relatively low cost (Tack and Meers 2010), and because the use of living plants gives an impression of being a *green technology*, ostensibly less harmful to the environment, compared with other remediation techniques (Cundy et al. 2016).

Phytoremediation Without Chemical Treatment The use of plant uptake to decrease total concentrations, that is, *phytoextraction*, of trace element contaminants relies on one of two scenarios (Dickinson et al. 2009). The first relates to the ability of certain plant species to be *hyperaccumulators* (i.e., to take up unusually high concentrations) of potentially toxic trace elements such as arsenic, nickel, cadmium, or zinc from soils. Alternatively, the plant species used are rapidly growing, with high biomass and moderate tissue concentrations of potentially toxic trace elements, which therefore have relatively high trace element uptake from soil. Successful remediation of contaminated soils using phytoextraction requires that the plants be harvested and removed from the site. The resulting plant biomass can be used for energy generation (Bert et al. 2013), depending on the ability to dispose of ash which is enriched in trace elements. For very metal-rich plant biomass, *phytomining* has been proposed, where the biomass is first combusted for energy production, and the ash used as feedstock for a smelter or other metallurgical extraction process (Anderson et al. 1999).

Plants may be used to reduce the risk of contaminant transfer by removing metals from bioavailable forms (Hamon and McLaughlin 1999) (a process called 'bioavailable contaminant stripping'), leaving only less mobile or available forms of metals in the soil. For example, Jacobs et al. (2017) showed that one harvest of the hyperaccumulator plant *Noccaea caerulescens* was able to decrease exchangeable Cd and Zn concentrations by about 20% in contaminated urban soils.

Alternatively, *phytostabilisation* aims to reduce contaminant bioavailability *via* combinations of plants, amendments, and rarely microorganisms (Bert et al. 2013). An important effect of phytostabilisation is to reduce export of contaminated soil by wind erosion (Dickinson et al. 2009). The amendments used to establish plant cover on contaminated land may also act to immobilise contaminants: for example, organic composts adsorb both metals and organic pollutants, or lime and phosphate compounds such as fertilisers can immobilise metals (see Sect. 11.4.2 above).

Enhanced Phytoextraction Very few species of plants have the combination of high uptake and large biomass yields that would make phytoextraction occur over reasonable time frames (e.g. a few years). One solution to this is to increase the bioavailability of metals using compounds (*ligands*) which can form complexes with metal ions and thereby increase the total dissolved metal concentration. The ligands used are often those which can form *chelates*, containing multiple binding sites and able to form very stable complexes; this is the 'chelate-assisted phytoextraction' technique (Huang et al. 1997). Chelating ligands disrupt the equilibria between metal ions in soil solid phases and metal ions in solution, by reducing free metal ion activities to very low values. This promotes dissolution of metal-bearing solid phases, or desorption of adsorbed metal ions, which form a stable dissolved complex with the ligand which can be taken up by plants (see Tack and Meers 2010). Chelating ligands such as EDTA have been shown to increase plant metal uptake in contaminated urban soils (an example is shown in Fig. 11.13).

Effectiveness of Phytoextraction Even though considerable attention has been given to phytoremediation in the research community, particularly phytoextraction, there are also realistic doubts about its practical effectiveness. For example, McGrath and Zhao (2003) calculated that only hyperaccumulator plants would significantly reduce concentrations of metals in soils in authentic soil remediation, since non-

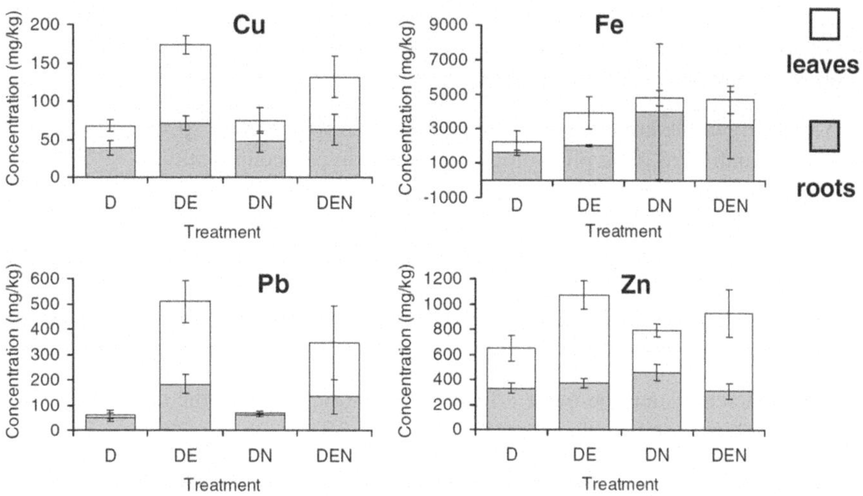

Fig. 11.13 Uptake of copper, iron, lead, and zinc from soils at former chemical manufacturing sites in the UK by common dandelion (*Taraxacum officinale*) plants with different soil treatments: D = plants only; DE = plants plus the chelating ligand EDTA; DN = plants plus NPK fertiliser; DEN = plants plus EDTA plus fertiliser. EDTA increased uptake of Cu, Pb, and Zn, but not Fe. (Diagram redrawn from Mossop et al. (2009); used with permission from Springer)

hyperaccumulator species would require tens or even hundreds of crop cycles to cause significant reductions, depending on yield. Many authors recognise shortcomings with the use of hyperaccumulator plants as well; Tack and Meers (2010) list limitations such as a restricted range of hyperaccumulated elements, slow growth, and low biomass production. Other issues include the commonly observed decreases in metal bioavailability with time since contamination, and potentially poor survival of plants in contaminated soils. Many reviews of phytoextraction conclude that this technique is not yet sufficiently developed to have wide application for remediation of contaminated soils — particularly with the contaminant concentrations found in many urban soils (see McGrath et al. 2002; Dickinson et al. 2009; Wenzel 2009; Mench et al. 2010).

The early stage of development of phytoextraction means that there are very few published accounts of authentic field applications in which successful remediation of metal contamination has been achieved using a phytoextraction technique. One report of successful field soil remediation by phytoextraction was made by Blaylock (2000), who found that, in 1 year, three crops of *Brassica juncea* (Indian mustard – not a hyperaccumulator) with EDTA addition were able to decrease soil Pb concentrations by a factor of < 0.5. Another evaluation of field phytoextraction showed the ability of tobacco and sunflower plants to remove bioavailable fractions of zinc in soil by 45–70% (Herzig et al. 2014).

A number of practical issues with phytoextraction remain unresolved (Blaylock 2000; McGrath et al. 2002; Dickinson et al. 2009; Marques et al. 2009; Wenzel 2009; Tack and Meers 2010), including:

- Variability in the amounts of metal uptake, even within plant species
- Understanding of plant physiology relevant to hyperaccumulation
- Heterogeneity of metal concentrations and root distributions in soils
- Food chain risk of increased metal concentrations in plants
- Toxicity of chelating ligands used to enhance phytoextraction (e.g. EDTA) to plants
- Enhanced leaching of metals, as aqueous complexes, if ligands are used to enhance phytoextraction

Phytoextraction may be more effective in treating water in the context of constructed wetlands, and we discuss this in Sect. 11.6.1.

'Rhizostabilisation' and 'Rhizovolatilisation' of Inorganic Contaminants These are both applications in the general category of rhizoremediation. Plants are able to stabilise soil against erosion by intercepting rainfall, and decreasing the kinetic energy of wind and water, reducing their velocity at the soil surface. Additional benefits occur as plants improve infiltration rates in surface soils, further decreasing runoff (Durán Zuazo and Rodríguez Pleguezuelo 2008). These benefits in soil stabilisation mean that contaminants are much less likely to be transferred from contaminated soils to receiving environments. In addition, the bioavailability of trace elements can be decreased in the rhizosphere, as some plants and their associated microbial community can cause precipitation of immobile solid phases in soil (Wenzel 2009).

The reactions which form volatile species of inorganic contaminants, such as dimethyl arsine, in soils (see Sect. 11.4.1) are facilitated by rhizosphere processes. According to Wenzel (2009), rhizosphere microorganisms can convert As, B, Sb, Se, Sn, Te, Pb, and Hg to volatile methylated or hydride forms. Given that the volatilised species must be lost to the atmosphere for soil remediation to occur, however, it is unlikely that rhizovolatilisation would be an acceptable technique for remediation of soils in urban environments due to the potential for airborne human health risks.

Plant-Based Remediation of Organic Contaminants The *rhizosphere* is the thin zone of soil which is directly influenced by the functioning of plant roots, including gradients in the concentrations of chemical compounds exuded by roots, and populations of root-specific microorganisms including rhizobial bacteria and mycorrhizal fungi (McNear Jr. 2013). There are numerous features of, and processes in, the rhizosphere which contribute to *rhizoremediation* (the attenuation of organic contaminants), including (Wenzel 2009; Mench et al. 2010):

- The ability of plants to take up organic pollutants from dissolved or vapour phases in soil (Bell and Failey 1991), including *bioconcentration* of organic pollutants above soil concentrations by some species (McKone and Maddalena 2007)

- Modification, by plants and/or associated microorganisms, of chemical conditions such as pH or E_h in rhizosphere soil which affect the rates and amount of abiotic degradation processes such as hydrolysis
- Degradation of organic pollutants *via* metabolic processes of plants and/or their associated microorganisms (Wenzel 2009)

11.6 Urban Soils as Media for Remediation of Water

The ability of soils to improve water quality by various mechanisms was discussed in Chapter 10 in relation to human health and the Sustainable Development Goals. Soil can remove a range of contaminants from water by physical filtration, chemical immobilization and transformation, and biological (mainly microbial) transformation or degradation. In some cases, the water-treatment capacity of soils is utilised in intentional remediation techniques, and we discuss some key examples of these below. In addition, water treatment technologies such as permeable reactive barriers (Sect. 11.4.6) and anaerobic bioreactors (Sect. 11.5.4) have already been considered in this chapter. Treatment of wastewater by soil-based media is also one strategy within Water Sensitive Urban Design (Wong 2006).

11.6.1 Constructed Wetlands

The soil in wetlands can perform the same physical, chemical, and biological functions as upland soils, and therefore the soil substrate in wetlands can decrease contaminant concentrations in water. Wetlands can be constructed to enhance or optimise the surface water flow pathways to enable this treatment to occur. For example, adding curvature or meanders to an otherwise straight stormwater drain (Fig. 11.14a) increases drain surface area and decreases water velocity, enhancing contaminant removal. Surface-flow wetlands mimic natural wetland environments, and subsurface-flow wetlands (Fig. 11.14b) are also constructed to maximise contact between water and the substrate.

Some treatment processes in wetlands are related to their hydrological characteristics, and we will not cover those in this book (see Kadlec and Wallace 2009 for detailed coverage of constructed wetland hydrology). As discussed in Chap. 2, however, wetland **soils** have distinctive properties that favour the immobilisation of some types of contaminants. For example, the chemically reducing conditions common in wetland sediments promote both the formation of non-bioavailable metal sulphides, and the microbiological reduction of nitrate with subsequent loss of gas-phase forms of nitrogen such as N_2O and N_2 to the atmosphere (Vepraskas et al. 2016). Uptake by wetland vegetation may be able to remove excess nutrients (Nagabhatla and Metcalfe 2017), and the wetland soil and/or plant root network

Fig. 11.14 (**a**) Constructed surface wetland in Welshpool, Western Australia, formerly part of a stormwater main drain, and re-formed from a deep compensating basin to reduce nutrient fluxes to the estuary downstream; (**b**) constructed subsurface wetland in Cannington, Western Australia, in the early stages of vegetation establishment. (Photographs by Andrew W. Rate)

may act as a filter for particulate contaminants including undesirable microorganisms such as pathogens (Vymazal 2011).

11.6.2 Infiltration Beds

Infiltration beds constructed from local or imported soil have been used for a long time to capture the excess water from septic tank storage of urban household sewage. Their effectiveness in treating water may be limited, however, unless they are installed with additional treatment features, such as the horizontal permeable reactive barrier systems described by Robertson and Cherry (1995). Similar systems have also been used to remove excess nitrogen from landfill leachate (Robertson and Anderson 1999).

11.6.3 WSUD (Swales, Rain Gardens, Wetlands, Green Roofs/Walls, Etc.)

Water-sensitive urban design (WSUD), also called 'sustainable urban drainage systems' (UK) or 'low impact development' (USA), aims to minimise the adverse hydrological effects of urbanization by capture or re-use of stormwater (Coutts et al. 2013). The structural elements of WSUD which include a soil component include rain gardens, sediment ponds, natural and constructed wetlands, swales, and green roofs/walls. All of these soil-containing systems are also bio-retention systems which depend to some extent on the phytoremediation or rhizoremediation ability of the urban soil-plant environment.

Green roofs primarily capture precipitation before it enters stormwater, but in some cases have been used to receive and treat domestic wastewater such as septage (Zapater-Pereyra et al. 2016). By necessity, green roofs have limited soil volume, and so recycling of water in this way not only treats wastewater but meets some of the water requirements of the green roof vegetation and microbiota. In many environments, water retention by manufactured soil in green roofs is a limiting factor for growth and survival of vegetation. Amendments such as organic waste materials (or their derivatives such as biochar) can be added to improve water storage and availability, which both improves the manufactured soil and provides a final treatment for the waste material (Gascó et al. 2018).

Rain gardens are examples of 'bioinfiltration systems' in which a topographic (micro)depression is re-filled with permeable material (e.g. a mixture of existing soil and coarse sand). They vary in size from < 10 to > 5000 m^2 and are planted with suitable vegetation, often waterlogging tolerant species since the rain garden may remain inundated for long periods. They have a dual function of controlling stormwater flow, and treatment of stormwater contaminants. For example, Komlos and Traver (2012) showed that a larger (0.53 ha) rain garden was successful at removing phosphate from urban stormwater for at least 9 years after establishment.

11.7 Case Studies in Brief: Effectiveness of Urban Soil Remediation

This section presents the choices of remediation method for various urban soil contamination scenarios, some of the issues involved, and their outcomes. The choice of remediation method will clearly depend on the preliminary and detailed site investigations, and the evolution of the conceptual site model for each contaminated urban soil environment. We do not present that 'back-story' here, so care must be taken not to generalise the case studies to other sites, since urban soil remediation is always highly site-specific.

11.7.1 Asbestos in Soil in an Urban Land Development, Canberra, Australia

This case study is taken from a detailed summary in Genever et al. (2017).

The site was an 8.5 ha parcel of land in the suburb of Campbell in central Canberra. Most of the contamination, originally thought to contain about 10,000 m^3 of asbestos-contaminated soil up to 2.4 m deep, was in the 4.2 hectares of the site to be redeveloped. The remainder of the site was largely reserved for open space. Remediation was mainly by excavation and disposal, which reflected the human health risks associated with asbestos and the sensitivity of the urban environment, and also the proximity of licensed secure landfill 22 km from the site. Some of the low-risk contaminated soil material was excavated and placed in an on-site 10,000 m^3 containment cell covered with a geotextile marker layer and ≥ 0.5 m of clean fill (mainly from excavation of the cell). In addition, a small volume of asbestos-contaminated soil close to infrastructure could not be remediated by excavation, and was capped and covered. During remediation, however, more asbestos-contaminated soil was discovered, and the final excavation volume was 40,000 m^3 of soil to a depth of 5 m. The remediation process took 70 weeks at a cost of $AUD 15 million in 2015. The site is now being developed for commercial use.

11.7.2 Polycyclic Aromatic Hydrocarbons (PAHs) in Timber Treatment Site Soil, Florida, USA

This case study is taken from a detailed summary in a report by the US Environmental Protection Agency (U.S. EPA 1995b).

This 21 ha site, in a peri-urban area of Live Oak, northern Florida, USA, operated as a timber treatment business prior to 1978, using creosote (a source of PAHs) as a preservative. Approximately 6,200 m^3 of stockpiled soil was contaminated with PAHs ($\sum_{car}$PAHs in the range 100–208 mg/kg; see Chap. 7). The target for remediation was $\sum_{car}$PAHs $\leq$ 100 mg/kg. The method selected for treatment was landfarming, on a 1.6 ha section of the site with a 30–90-cm-thick clay liner and clay berm. A bacterial inoculum containing PAH-degrading organisms was applied to soil in the first of three stages of the landfarming treatment, so bioaugmentation was also part of the remediation strategy. Each stage added 4–30 cm depth of contaminated soil, which was watered to 10% by volume and cultivated every 2 weeks. Additional stages were not added until the previous soil had reached the target of $\sum_{car}$PAHs $\leq$ 100 mg/kg. Remediation was complete in 18 months over 1988 and 1989, with $\sum_{car}$PAHs in the remediated soil between 23 and 92 mg/kg. Total costs were $US 565,400 in 1989. The site was removed from the National Priorities List in 1995; it is now designated for unrestricted use but remained undeveloped in 2020.

11.7.3 Trace Elements (As, Cd, Cu, Pb, Tl, Zn) from a Mine Tailings Spill, Guadiamar River valley, Spain

This case study is based on various sources cited in the text below. A large spill of sulfidic tailings (approximately 6 million m^3 of water/mud) from the Aznalcóllar mine, Spain, in 1988 contaminated about 2700 ha of riparian, peri-urban, and agricultural land. The tailings were acidic and acid-producing, decreasing the local soil pH by up to 2 pH units, and increasing trace element concentrations to up to 80 cm depth depending on soil texture (Cabrera et al. 1999). Remediation was in a conservation area, so the goals were to ensure that ecosystem services were maintained, and that cultural and recreational value was preserved. The remediation methods could not be harmful or invasive. Soil and vegetation were rehabilitated by removing tailings where possible, using soil amendments of sugar beet lime (a food processing by-product), iron-oxide-rich soil, and biosolids compost, and by planting native tree species (Frelih-Larsen et al. 2018; Madejón et al. 2018). The remediation led to significant ecosystem recovery in terms of increased vegetation cover and soil microbial functioning, although concentrations of some trace elements remained above regulatory thresholds. Continued monitoring of trace element uptake by vegetation, even 20 years after the tailings spill, was recommended. The total costs of remediation were approximately € 170 million (Madejón et al. 2018).

11.8 Regulatory Frameworks for Contaminated Sites and Soil Remediation

A comparison of international frameworks for management of contaminated sites (excluding African countries, Australia and New Zealand) has been presented by Gong (2010). In general, regulatory frameworks are consistent with or based on scientific procedures to assess and manage contaminated sites (Fig. 11.15). For example, many guideline documents include a Conceptual Site Model approach similar to that discussed in Chap. 9. Preparation of a Conceptual Site Model commonly occurs in parallel with preparation of some form of site investigation, for example, the initial, largely desktop-based, Preliminary Site Investigation approach presented by the Australian National Environment Protection Council (2011). If

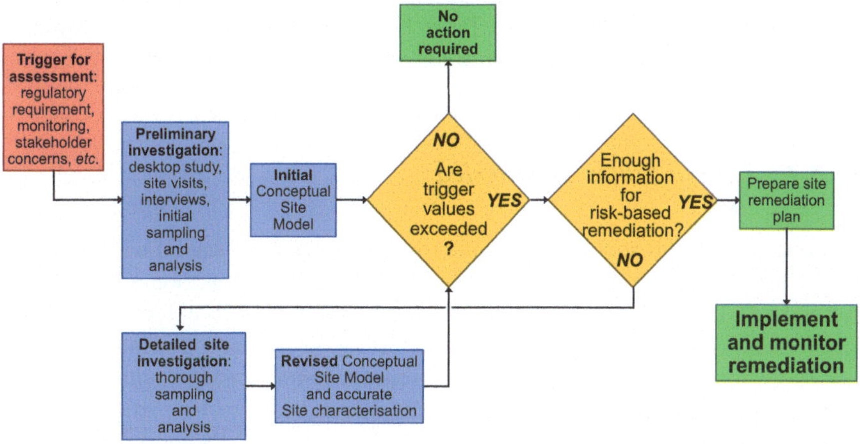

Fig. 11.15 Generalised and simplified flow diagram for assessment and remediation of contaminated sites

necessary sampling and analysis of contaminated soils occurs in a subsequent stage, the Detailed Site Investigation. The results of sampling and analysis may be used to update and refine the Conceptual Site Model, specifically to identify if plausible exposure pathways between the receptor and the source exist, which will drive the need for remediation if the risks are deemed unacceptable. The 'best' Conceptual Site Model will ultimately inform practitioners whether or not remediation is required, and the choice of soil remediation method. The choice of remediation method is seldom prescribed, since it will be dependent on multiple factors including the nature of the site and contamination, receiving environment(s) and socioeconomic factors. More recently the concept of *life cycle assessment* has been applied to soil and groundwater remediation, where all of the resources used, and environmental impacts generated, are accounted for in the total lifespan of remediated land, from planning to closure (Favara et al. 2011). For example, a life cycle assessment would account for the environmental benefits of reducing or eliminating contaminants, but also account for the environmental harm from vehicle emissions and energy usage during the active phases of a project (Lemming et al. 2009). In addition, other contaminants can be released during remediation of target substances, as conditions such as anoxia change in the soil or groundwater being treated (Mulligan and Yong 2004). The *carbon footprint* is of particular contemporary concern in soil remediation, with the most immediately straightforward or economical method not necessarily being the process that minimises carbon emissions to the atmosphere in the long-term (for an example see Jing et al. 2018). The trend in many European states, for example, has been towards *in situ* biological remediation and away from more energy-intensive methods such as excavation and removal (Fig. 11.1). Implementation of life cycle assessment is related to the practices of 'Sustainable Remediation', for the promotion of which various multiple-stakeholder organisations exist worldwide. Smith and Nadebaum (2016) provide an excellent summary of the framework for a sustainable remediation approach, focused on Australia and New Zealand where well-developed regulatory frameworks control remediation and management of contaminated sites, Table 11.1). In most cases, regulatory frameworks also require ongoing monitoring and reporting to be conducted.

11.8.1 United States of America

A formal process for soil remediation in the USA effectively started with the passing of the Comprehensive Environmental Response, Compensation and Liability Act in 1980 (the 'Superfund' Act) (Gong 2010). This federal-level (i.e. nationwide) legislation provided a framework to fund remediation efforts, and identify the parties liable for and contamination and consequent removal and/or remediation. A number of amendments in subsequent years allowed for greater participation by state-level authorities, while still providing federal funding – the most recent legislation being the Brownfields Utilization, Investment, and Local Development Act

(BUILD Act) in 2018 (U.S. EPA 2019). While still upholding the 'polluter-pays' principle that underpinned the Superfund Act, the BUILD Act allows for management of contaminated sites by state and local governments while still being eligible for federal funding.

The United States Environmental Protection Agency (US EPA) maintains overall responsibility for contaminated sites guidelines in the USA. The key guidelines are the Soil Screening Guidance (U.S. EPA 1996) and the various Preliminary Remediation Goals, which may be based on the relevant Soil Screening Levels. The Soil Screening Levels are a set of tiered values which can trigger a definite response or, at lower concentrations, site-specific actions. They are based on a structured risk analysis procedure for human exposure which includes (see Chap. 9):

- Potential exposure pathways, such as ingestion or inhalation
- Source size (i.e. volume of contaminated soil)
- Identification of site-specific factors
- The nature of the contaminant(s) (e.g. whether or not they are carcinogens)

11.8.2 Europe

The European 'Thematic Strategy on Soil Protection' was a key guiding document which was designed to give soils the same degree of protection as air and water (Commission of the European Communities 2006). The main objectives are, firstly, to prevent further soil degradation in order to maintain soil functions and, secondly, to restore degraded soils to allow them to support the existing or planned land use. The Thematic Strategy recognises urbanization as one of the multiple causes of soil degradation, against a background of increasing urbanization across the EU. The Thematic Strategy attempted to provide consistency in soil-related practices across the member states (European Commission 2006). The Directive (legislation) proposed on the basis of the Thematic Strategy, however, was opposed by key member states in the EU and was withdrawn in 2014, leaving the EU without a consistent soil protection strategy. The European Union (EU) consists of member states which do not necessarily delegate legislation for environmental protection to EU governance, and the specific legislation of member states is now still the most relevant.

Example: Germany In Germany, a Federal Act, the 'Federal Soil Conservation Act 1998 (Bundes-Bodenschutzgesetz)' and its associated 'Federal Soil Protection and Contaminated Sites Ordinance 1999 (Bundes-Bodenschutz- und Altlastenverordnung)' govern protection and remediation of soils and are based on essentially the same principles as the EU Thematic Strategy. That is, there are two main objectives: to protect the soil resource and to remediate soil at degraded sites (Umweltbundesamt 2009). Soil remediation is performed as a result of a phased assessment process similar to others in use worldwide. The phases of this investigation include:

- Validity check and historical investigation
- Exploratory investigation
- Detailed investigation
- Remedial investigation (which may include a remediation plan)

The investigations are informed by contaminant trigger values set by the Ordinance, which also defines acceptable sampling and analysis methods for soils and related materials. The trigger values have been established by a procedure similar to that used by the US Environmental Protection Agency in Sect. 11.8.1 above. Measures to prevent or remediate contamination may be initiated at any phase of the investigation process. The choice of remediation measure(s) is required to be based on the 'principle of proportionality', which balances the necessity, suitability, and reasonableness of available methods (Umweltbundesamt 2009).

11.8.3 Australia

In Australia the states and territory governments are responsible for regulating the assessment and management (including remediation) of contaminated sites. In all cases, state-level guidelines are consistent with the National Environment Protection (Assessment of Site Contamination) Measure (National Environment Protection Council 2013). Broadly, this Measure follows similar procedures for assessment of site contamination to the US and German systems described above, and while providing an assessment framework, does not prescribe any remediation methods. For example, the German validity check and historical/exploratory investigations, and initial stages of the USA's structured risk analysis procedure, are matched in Australia by the requirement to conduct a Preliminary Site Investigation and prepare a Conceptual Site Model (Chap. 9). As well as the legislative framework, the National Remediation Framework (CRC CARE 2018b and associated documentation) attempts to define a nationally consistent set principles by which contaminated site remediation occurs, and these are:

- The Precautionary Principle (erring on the side of caution if necessary; always minimizing risk)
- Prevention (of further contamination and risk)
- Risk Management (for all environmental compartments, including human exposure)
- An Options Hierarchy (i.e. an order of preference of remediation options)
- Sustainability (avoiding impacts on future generations)
- National and International Obligations (e.g. agreements, treaties, etc., including compatibility with the national legislation defined by the National Environment Protection Council (2013))

11.9 Further Reading

Antizar-Ladislao B (2010) Bioremediation: working with bacteria. Elements 6: 389–394. Doi: https://doi.org/10.2113/gselements.6.6.389.

Gong Y (2010) *International Experience in Policy and Regulatory Frameworks for Brownfield Site Management*, Sustainable Development – East Asia and Pacific Region: Discussion Papers, The World Bank, Washington, DC, USA. worldbank.org/.../578900ESW0P11910BOX353785B01PUBLIC1.pdf

Mulligan CN, Yong RN, Gibbs BF (2001) Remediation technologies for metal-contaminated soils and groundwater: an evaluation. Engineer Geol 60: 193–207. Doi: https://doi.org/10.1016/S0013-7952(00)00101-0.

Tack FMG, Meers E (2010) Assisted phytoextraction: helping plants to help us Elements 6: 383–388. Doi: https://doi.org/10.2113/gselements.6.6.383.

Yong RN, Mulligan CN (2019) Natural and Enhanced Attenuation of Contaminants in Soils, Second edition. CRC Press, Taylor & Francis Group, Boca Raton, FL, USA.

11.10 Summary

Multiple factors, environmental, logistical, social, risk-based, and economic/regulatory, affect the choice of remediation method at any one urban site containing contaminated soil. The overarching objective for urban soil remediation is always to render land fit for an intended use.

Many remediation techniques are available, based on some combination of physical, chemical, and biological processes. Remediation may be active or passive and performed *in situ* or with soil removed for treatment.

Physical soil remediation methods are based on either restricting the movement of contaminants to receptors, or based on spatial separation of contaminants by removal or various phase separation techniques.

Chemical soil remediation methods may be based on degradation of contaminants; transformation of contaminants to immobile or non-bioavailable forms; or chemical separation methods.

Biological soil remediation techniques rely on microorganisms and/or plants to degrade, immobilise, or transform contaminants.

Soils in urban environments may also provide important environments for remediation of water such as stormwater or wastewaters.

11.11 Review and Study Questions

11.11.1 Checking Your Understanding

1. Categorise each remediation technology in Table 11.2 as being either *in situ*, ex situ, or off-site (also try categorising each technology as active or passive).
2. List the various types of physical barrier remediation methods from least to most sophisticated.
3. What types of contaminants can be treated using (i) soil vapour extraction; (ii) abiotic natural attenuation; (iii) liming; and (iv) *in situ* chemical oxidation.
4. List the various materials that can be added to soil for chemical treatment of contaminants, with a brief explanation of how each amendment works.
5. Explain the difference between natural, enhanced, and augmented bioremediation.
6. Identify the various mechanisms by which plants can contribute to the remediation of contaminated soils.

11.11.2 Thinking About the Topics More Deeply

7. Which aspects of a soil remediation process are involved in a comprehensive life cycle assessment, but are not usually considered in a conceptual site model for remediation?
8. What are some scenarios for soil remediation which do not involve treatment of contaminants? How would we judge the need for such measurements (e.g. is risk still an issue)?
9. Identify the similarities and differences between the regulatory frameworks for soil remediation described in Sect. 11.8. If your home country is not included in this Section, think about how and or if your country's soil remediation framework resembles the examples given.

11.11.3 Thinking About Urban Soil Remediation with Your 'Left Brain'

10. How do the concepts of environmental justice apply to soil remediation in cities?

References

Ahern CR, McElnea AE, Sullivan LA (2004) Acid sulfate soils laboratory methods guidelines. Queensland Department of Natural Resources, Mines and Energy, Indooroopilly, Queensland, Australia.

Alexander M (1981) Biodegradation of chemicals of environmental concern. Science **211**:132–138. https://doi.org/10.1126/science.7444456

Alexander M (2000) Aging, bioavailability, and overestimation of risk from environmental pollutants. Environ Sci Technol **34**:4259–4265. https://doi.org/10.1021/es001069+

Alloway BJ (Ed.) (2012) Heavy metals in soils: trace metals and metalloids in soils and their bioavailability (3rd ed.). Environmental Pollution, Vol. 22. Springer, Dordrecht.

Anderson CWN, Brooks RR, Chiarucci A, Lacoste CJ, Leblanc M, Robinson BH et al (1999) Phytomining for nickel, thallium and gold. J Geochem Explor **67**:407–415. https://doi.org/10.1016/S0375-6742(99)00055-2

Australian Antarctic Division (2012) Remediation work: this week at Casey: 13 January 2012, https://www.antarctica.gov.au/news/stations/casey/2012/this-week-at-casey-13-january-2012/ (accessed 2020.06.03).

Backes CA, McLaren RG, Rate AW, Swift RS (1995) Kinetics of cadmium and cobalt desorption from iron and manganese oxides. Soil Sci Soc Am J **59**:778–785

Bell RM, Failey RA (1991) Plant uptake of organic pollutants. In: Jones, K.C. (Ed.), Organic Contaminants in the Environment: Environmental Pathways & Effects. Elsevier Applied Science, pp. 189–206. https://doi.org/10.1007/978-94-009-4329-2_6.

Bert V., Douay F, Faure O, Cadière F (2013) Les Phytotechnologies Appliquées aux Sites et Sols Pollués, Agence de l'Environnement et de la Maîtrise de l'Énergie (ADEME), Angers Cedex, France.

Blaylock MJ (2000) Field demonstrations of phytoremediation of lead-contaminated soils. In: Terry N, Bañuelos GS (eds) Phytoremediation of contaminated soil and water. CRC Press, Boca Raton, pp 1–12

Burlakovs J, Vircavs M (2012) Heavy metal remediation technologies in Latvia: possible applications and preliminary case study results. Ecol Chem Eng S **19**:533–547. https://doi.org/10.2478/v10216-011-0038-3

Cabrera F, Clemente L, D'iaz Barrientos E, López R, Murillo JM (1999) Heavy metal pollution of soils affected by the Guadiamar toxic flood. Sci Total Environ **242**:117–129. https://doi.org/10.1016/s0048-9697(99)00379-4

Caliman FA, Robu BM, Smaranda C, Pavel VL, Gavrilescu M (2011) Soil and groundwater cleanup: benefits and limits of emerging technologies. Clean Technol & Environ Policy **13**:241–268. https://doi.org/10.1007/s10098-010-0319-z

Cappuyns V (2013) Environmental impacts of soil remediation activities: quantitative and qualitative tools applied on three case studies. J Clean Prod **52**:145–154. https://doi.org/10.1016/j.jclepro.2013.03.023

Carr GM, Neary JP (2006) Water quality for ecosystem and human health, United Nations Environment Programme Global Environment Monitoring System (GEMS)/Water Programme, Burlington, Ontario, Canada.

Commission of the European Communities 2006. Thematic strategy for soil protection. COM(2006)231, Brussels, Belgium. https://eur-lex.europa.eu/legal-content/EN/TXT/PDF/?uri=CELEX:52006DC0231&from=EN (accessed 20210618).

Coutts AM, Tapper NJ, Beringer J, Loughnan M, Demuzere M (2013) Watering our cities: the capacity for Water Sensitive Urban Design to support urban cooling and improve human thermal comfort in the Australian context. Prog Phys Geog: Earth & Environ **37**:2–28. https://doi.org/10.1177/0309133312461032

CRC CARE (2018a) Guideline on Regulatory Considerations, National Remediation Framework. Version 0.1, CRC for Contamination Assessment and Remediation of

the Environment, Callaghan, NSW, Australia. https://www.crccare.com/files/dmfile/Guidelineonregulatoryconsiderations_Rev2.pdf
CRC CARE (2018b) Introduction to the National Remediation Framework, National Remediation Framework. Version 0.1, CRC for Contamination Assessment and Remediation of the Environment, Callaghan, NSW, Australia. https://www.crccare.com/files/dmfile/IntroductiontotheNRF_Rev3.pdf
CRC CARE (2018c) Technology Guide: Bioremediation, National Remediation Framework. Version 0.1, CRC for Contamination Assessment and Remediation of the Environment, Callaghan, NSW, Australia. https://www.crccare.com/files/dmfile/CTechguide_Bioremediation_Rev0.pdf.
CRC CARE (2018d) Technology Guide: Containment, National Remediation Framework. Version 0.1, CRC for Contamination Assessment and Remediation of the Environment, Callaghan, NSW, Australia. https://www.crccare.com/files/dmfile/ATechguide_Containment_Rev0.pdf.
CRC CARE (2018e) Technology Guide: Excavation, National Remediation Framework. Version 0.1, CRC for Contamination Assessment and Remediation of the Environment, Callaghan, NSW, Australia. https://www.crccare.com/files/dmfile/FTechguide-Excavation_Rev0.pdf.
CRC CARE (2018f) Technology Guide: In-situ chemical oxidation, National Remediation Framework. Version 0.1, CRC for Contamination Assessment and Remediation of the Environment, Callaghan, NSW, Australia. https://www.crccare.com/files/dmfile/ITechguide_ISCO_Rev0.pdf.
CRC CARE (2018g) Technology Guide: Soil vapour remediation, National Remediation Framework. Version 0.1, CRC for Contamination Assessment and Remediation of the Environment, Callaghan, NSW, Australia. https://www.crccare.com/files/dmfile/GTechguide_SVR_Rev0.pdf.
Cundy AB, Bardos RP, Puschenreiter M, Mench M, Bert V, Friesl-Hanl W et al (2016) Brownfields to green fields: Realising wider benefits from practical contaminant phytomanagement strategies. J Environ Manage **184**:67–77. https://doi.org/10.1016/j.jenvman.2016.03.028
Danish Environmental Protection Agency (2002) Guidelines on Remediation of Contaminated Sites. Miljø- og Fødevareministeriet, København. http://mst.dk/service/publikationer/publikationsarkiv/2002/sep/guidelines-on-remediation-of-contaminated-sites/ (accessed 20210618)
Department of Environment Regulation (2014) Assessment and management of contaminated sites. Contaminated Sites Guidelines, Government of Western Australia, Perth
Department of Environment Regulation (WA) (2015) Treatment and management of soils and water in acid sulfate soil landscapes. Acid Sulfate Soils Guideline Series, DER, Government of Western Australia, Perth
Dickinson NM, Baker AJM, Doronila A, Laidlaw S, Reeves RD (2009) Phytoremediation of inorganics: Realism and synergies. Int J Phytoremediation **11**:97–114. https://doi.org/10.1080/15226510802378368
Duarte AC, Cachada A, Rocha-Santos TAP (2017) Soil pollution. Academic Press, San Diego
Durán Zuazo VH, Rodríguez Pleguezuelo CR (2008) Soil-erosion and runoff prevention by plant covers. A review. Agron Sustain Dev **28**:65–86. https://doi.org/10.1051/agro:2007062
EPA, U.S. (2010) Green remediation best management practices: bioremediation, United States Environmental Protection Agency, Office of Solid Waste and Emergency Response. https://semspub.epa.gov/work/HQ/147895.pdf
EPA WA (2006) Guidance for the assessment of environmental factors – rehabilitation of terrestrial ecosystems. Environmental Protection Authority, Government of Western Australia, Perth, Western Australia
European Commission (2006) Soil protection – the story behind the strategy. European Communities, Luxembourg. https://ec.europa.eu/environment/archives/soil/pdf/soillight.pdf
European Environment Agency (2020) Indicator Assessment - Progress in management of contaminated sites, https://www.eea.europa.eu/data-and-maps/indicators/progress-in-management-of-contaminated-sites-3/assessment using data from https://esdac.jrc.ec.europa.eu/content/progress-management-contaminated-sites-europe-0 (accessed 20201026).

Favara PJ, Krieger TM, Boughton B, Fisher AS, Bhargava M (2011) Guidance for performing footprint analyses and life-cycle assessments for the remediation industry. Remediation (New York, NY) **21**:39–79. https://doi.org/10.1002/rem.20289

Feldmann J, Cullen WR (1997) Occurrence of volatile transition metal compounds in landfill gas: synthesis of molybdenum and tungsten carbonyls in the environment. Environ Sci Technol **31**:2125–2129. https://doi.org/10.1021/es960952y

Frelih-Larsen A, Ittner S, Marañón T, Vrinceanu N (2018) Policy brief: remediating historical soil contamination – effective measures and policy solutions. RECARE Project, Wageningen. https://www.ecologic.eu/sites/files/publication/2018/2730_recare_contamination_web.pdf

FRTR (Federal Remediation Technologies Roundtable) (2020) Technology screening matrix. Government of the United States of America, https://frtr.gov/matrix/ (accessed 2020-09-28).

Gascó G, Paz-Ferreiro J, Álvarez ML, Saa A, Méndez A (2018) Biochars and hydrochars prepared by pyrolysis and hydrothermal carbonisation of pig manure. Waste Manag **79**:395–403. https://doi.org/10.1016/j.wasman.2018.08.015

Genever M, Allan M, Bos S, (Asbestos Safety and Eradication Agency) (2017) Case studies of asbestos contaminated land: final report, Commonwealth of Australia, Rawtec Pty. Ltd., Reincarnate and Prensa Pty. Ltd., Sydney, Australia. https://www.asbestossafety.gov.au/sites/default/files/documents/2018-02/ASEA_Reports_Asbestos_land_contamination_case_studies_final_Nov17-1.pdf

Gong Y (2010) International experience in policy and regulatory frameworks for brownfield site management. Sustainable Development — East Asia and Pacific Region: Discussion Papers, The World Bank, Washington, DC. http://documents1.worldbank.org/curated/ar/105931468218106615/pdf/578900ESW0P11910BOX353785B01PUBLIC1.pdf

Groffman PM, Boulware NJ, Zipperer WC, Pouyat RV, Band LE, Colosimo MF (2002) Soil nitrogen cycle processes in urban riparian zones. Environ Sci Technol **36**:4547–4552. https://doi.org/10.1021/es020649z

Hamon RE, McLaughlin MJ (1999) Use of the hyperaccumulator Thlaspi caerulescens for bioavailable contaminant stripping. Fifth International Conference on the Biogeochemistry of Trace Elements (ICOBTE), Vienna, pp 908–909

Harrison RM, Alloway BJ (1999) Land contamination and reclamation. In: Harrison RM (ed) Understanding our environment: an introduction to environmental chemistry and pollution. Royal Society of Chemistry, Cambridge, UK, pp 199–236. https://doi.org/10.1039/9781847552235-00199

Herzig R, Nehnevajova E, Pfistner C, Schwitzguebel J-P, Ricci A, Keller C (2014) Feasibility of labile Zn phytoextraction using enhanced tobacco and sunflower: results of five- and one-year field-scale experiments in Switzerland. Int J Phytoremediation **16**:735–754. https://doi.org/10.1080/15226514.2013.856846

Higgins MR, Olson TM (2009) Life-cycle case study comparison of permeable reactive barrier versus pump-and-treat remediation. Environ Sci Technol **43**:9432–9438. https://doi.org/10.1021/es9015537

Hodson ME (2010) The need for sustainable soil remediation. Elements **6**:363–368. https://doi.org/10.2113/gselements.6.6.363

Huang JW, Chen J, Berti WR, Cunningham SD (1997) Phytoremediation of lead-contaminated soils: role of synthetic chelates in lead phytoextraction. Environ Sci Tech **31**:800–805

Indraratna B, Rujikiatkamjorn C, Ameratunga J, Boyle P (2011) Performance and prediction of vacuum combined surcharge consolidation at Port of Brisbane. J Geotech Geoenviron Eng **137**:1009–1018. https://doi.org/10.1061/(ASCE)GT.1943-5606.0000519

Jacobs A, Drouet T, Sterckeman T, Noret N (2017) Phytoremediation of urban soils contaminated with trace metals using Noccaea caerulescens: comparing non-metallicolous populations to the metallicolous 'Ganges' in field trials. Environ Sci Pollut Res **24**:8176–8188. https://doi.org/10.1007/s11356-017-8504-9

Jing R, Fusi S, Kjellerup BV (2018) Remediation of polychlorinated biphenyls (PCBs) in contaminated soils and sediment: State of knowledge and perspectives. Front Environ Sci **6**. https://doi.org/10.3389/fenvs.2018.00079

Jones D, Healey J (2010) Organic amendments for remediation: putting waste to good use. Elements **6**:369–374. https://doi.org/10.2113/gselements.6.6.369

Kadlec RH, Wallace SD (2009) Treatment wetlands, 2nd edn. CRC Press, Boca Raton, 1016 pp

Kim YU, Wang MC (2003) Effect of ultrasound on oil removal from soils. Ultrasonics **41**:539–542. https://doi.org/10.1016/S0041-624X(03)00168-9

Kim B-K, Baek K, Ko S-H, Yang J-W (2011) Research and field experiences on electrokinetic remediation in South Korea. Sep Purif Technol **79**:116–123. https://doi.org/10.1016/j.seppur.2011.03.002

Kjeldsen P (1999) Behaviour of cyanides in soil and groundwater: A review. Water Air Soil Pollut **115**:279–307. https://doi.org/10.1023/A:1005145324157

Komlos J, Traver RG (2012) Long-term orthophosphate removal in a field-scale storm-water bioinfiltration rain garden. J Environ Eng **138**:991–998. https://doi.org/10.1061/(ASCE)EE.1943-7870.0000566

Kozdrój J, Van Elsas JD (2001) Structural diversity of microorganisms in chemically perturbed soil assessed by molecular and cytochemical approaches. J Microbiol Methods **43**:197–212. https://doi.org/10.1016/S0167-7012(00)00197-4

Larson S (2000) Natural pressure-driven passive bioventing. ER-199715, U.S. Department of Defence, Naval Facilities Engineering Service Center, Port Hueneme, California, USA. https://www.serdp-estcp.org/content/download/3690/59034/file/CU-9715-FR-01.pdf.

Lemming G, Hauschild MZ, Bjerg PL (2009) Life cycle assessment of soil and groundwater remediation technologies: literature review. Int J Life Cycle Assess **15**:115. https://doi.org/10.1007/s11367-009-0129-x

Liu G, Yu L, Li J, Liu X, Zhang G (2011) PAHs in soils and estimated air-soil exchange in the Pearl River Delta, South China. Environ Monit Assess **173**:861–870. https://doi.org/10.1007/s10661-010-1429-0

Liu YR, Li J, He JZ, Ma YB, Zheng YM (2015) Different influences of field aging on nickel toxicity to Folsomia candida in two types of soil. Environ Sci Pollut Res **22**:8235–8241. https://doi.org/10.1007/s11356-014-3968-3

Lombi E, Hamon RE (2005) Remediation of polluted soils. In: Hillel D (ed) Encyclopedia of soils in the environment. Elsevier, Oxford, pp 379–385. https://doi.org/10.1016/B0-12-348530-4/00087-4

Lung SC, Lim BL (2006) Assimilation of phytate-phosphorus by the extracellular phytase activity of tobacco (Nicotiana tabacum) is affected by the availability of soluble phytate. Plant and Soil **279**:187–199. https://doi.org/10.1007/s11104-005-1009-1

Madejón P, Domínguez MT, Madejón E, Cabrera F, Marañón T, Murillo JM (2018) Soil-plant relationships and contamination by trace elements: A review of twenty years of experimentation and monitoring after the Aznalcóllar (SW Spain) mine accident. Sci Total Environ **625**:50–63. https://doi.org/10.1016/j.scitotenv.2017.12.277

Malaviya P, Singh A (2012) Constructed wetlands for management of urban stormwater runoff. Crit Rev Environ Sci Technol **42**:2153–2214. https://doi.org/10.1080/10643389.2011.574107

Marques APGC, Rangel AOSS, Castro PML (2009) Remediation of heavy metal contaminated soils: phytoremediation as a potentially promising clean-up technology. Crit Rev Environ Sci Technol **39**:622–654. https://doi.org/10.1080/10643380701798272

McBride MB (1994) Environmental chemistry of soils. Oxford University Press, New York

McGrath SP, Zhao F-J (2003) Phytoextraction of metals and metalloids from contaminated soils. Curr Opin Biotechnol **14**:277–282. https://doi.org/10.1016/S0958-1669(03)00060-0

McGrath SP, Zhao J, Lombi E (2002) Phytoremediation of metals, metalloids, and radionuclides. In: Advances in agronomy, Vol. 75. Academic Press, Cambridge, MA, pp 1–56. https://doi.org/10.1016/S0065-2113(02)75002-5

McKone TE, Maddalena RL (2007) Plant uptake of organic pollutants from soil: bioconcentration estimates based on models and experiments. Environ Toxicol Chem **26**:2494–2504. https://doi.org/10.1897/06-269.1

McNear DH Jr (2013) The rhizosphere – roots, soil and everything in between. Nat Educ Knowledg **4**:1

Megharaj M, Ramakrishnan B, Venkateswarlu K, Sethunathan N, Naidu R (2011) Bioremediation approaches for organic pollutants: a critical perspective. Environ Int **37**:1362–1375. https://doi.org/10.1016/j.envint.2011.06.003

Mench M, Lepp N, Bert V, Schwitzguébel JP, Gawronski SW, Schröder P, Vangronsveld J (2010) Successes and limitations of phytotechnologies at field scale: Outcomes, assessment and outlook from COST Action 859. J Soil Sediment **10**:1039–1070. https://doi.org/10.1007/s11368-010-0190-x

Meuser H (2012) Soil remediation and rehabilitation: treatment of contaminated and disturbed land, 23. Springer, Dordrecht. https://doi.org/10.1007/978-94-007-5751-6_2

Moradi AM, Smits KO, Sharp J (2018) Coupled thermally-enhanced bioremediation and renewable energy storage system: conceptual framework and modeling investigation. Water **10**:1288

Mossop KF, Davidson CM, Ure AM, Shand CA, Hillier SJ (2009) Effect of EDTA on the fractionation and uptake by Taraxacum officinale of potentially toxic elements in soil from former chemical manufacturing sites. Plant and Soil **320**:117–129. https://doi.org/10.1007/s11104-008-9875-y

Mousset E, Oturan MA, Van Hullebusch ED, Guibaud G, Esposito G (2014) Soil washing/flushing treatments of organic pollutants enhanced by cyclodextrins and integrated treatments: state of the art. Crit Rev Environ Sci Technol **44**:705–795. https://doi.org/10.1080/10643389.2012.741307

Mueller NC, Nowack B (2010) Nanoparticles for remediation: Solving big problems with little particles. Elements **6**:395–400. https://doi.org/10.2113/gselements.6.6.395

Mulligan CN, Yong RN (2004) Natural attenuation of contaminated soils. Environ Int **30**:587–601. https://doi.org/10.1016/j.envint.2003.11.001

Mulligan CN, Yong RN, Gibbs BF (2001a) Remediation technologies for metal-contaminated soils and groundwater: an evaluation. Eng Geol **60**:193–207. https://doi.org/10.1016/S0013-7952(00)00101-0

Mulligan CN, Yong RN, Gibbs BF (2001b) Surfactant-enhanced remediation of contaminated soil: a review. Eng Geol **60**:371–380. https://doi.org/10.1016/S0013-7952(00)00117-4

N.S.W. EPA (2014) Best Practice Note: Landfarming, New South Wales Environmental Protection Authority, Sydney, Australia. https://www.epa.nsw.gov.au/-/media/epa/corporate-site/resources/clm/140323landfarmbpn.pdf.

Nagabhatla N, Metcalfe CD (2017) Multifunctional wetlands: pollution abatement and other ecological services from natural and constructed wetlands. Environmental contamination remediation and management. Springer, Cham, Switzerland, https://doi.org/10.1007/978-3-319-67416-2.

Naidu R (2013) Recent advances in contaminated site remediation. Water Air Soil Pollut **224**:1705. https://doi.org/10.1007/s11270-013-1705-z

National Environment Protection Council (2011) Schedule B (2): Guideline on Site Characterisation, National Environment Protection (Assessment of Site Contamination) Measure (Amended). Commonwealth of Australia, Canberra.

National Environment Protection Council (2013) Schedule B (1): Guideline on the Investigation Levels for Soil and Groundwater, National Environment Protection (Assessment of Site Contamination) Measure (Amended). Commonwealth of Australia, Canberra.

NGWA (2018) Groundwater and PFAS: state of knowledge and practice. National Ground Water Association, Westerville

Nriagu JO (1989) A global assessment of natural sources of atmospheric trace metals. Nature **338**:47–49. https://doi.org/10.1038/338047a0

O'Day PA, Vlassopoulos D (2010) Mineral-based amendments for remediation. Elements **6**:375–381. https://doi.org/10.2113/gselements.6.6.375

Oma KH, Farnsworth RK, Rusin JM (1982) In situ vitrification: application analysis for stabilization of transuranic waste, Pacific Northwest Laboratory for U.S. Department of Energy, Richland, WA, USA. https://www.osti.gov/servlets/purl/6725441 (accessed 20210618).
Omar H, Rohani S (2015) Treatment of landfill waste, leachate and landfill gas: a review. Front Chem Sci & Eng **9**:15–32. https://doi.org/10.1007/s11705-015-1501-y
Patel A (2019) Geotechnical investigations and improvement of ground conditions. Woodhead Publishing Series in Civil and Structural Engineering. Elsevier Science & Technology, San Diego
Patterson BM, Ma Y, Grassi ME, Robertson BS, Davis GB, Lipman M et al (2005) Assessment of different carbon sources and delivery techniques to promote an in situ reactive zone for bioprecipitation of metals in groundwater. In: Boshoff GA, Bone BD (eds) Permeable reactive barriers. IAHS Publication. IAHS Press, Wallingford, pp 97–104
Regenesis (2019) 9 reasons to consider ISCO to treat your site, REGENESIS®, San Clemente, CA, USA.
Robertson WD, Anderson MR (1999) Nitrogen removal from landfill leachate using an infiltration bed coupled with a denitrification barrier. Ground Water Monitor & Remediation **19**:73–80. https://doi.org/10.1111/j.1745-6592.1999.tb00242.x
Robertson WD, Cherry JA (1995) In situ denitrification of septic-system nitrate using reactive porous media barriers: field trials. Groundwater **33**:99–111. https://doi.org/10.1111/j.1745-6584.1995.tb00266.x
Rodríguez-Eugenio N, McLaughlin M, Pennock D (2018) Soil pollution – a hidden reality. Food and Agriculture Organization of the United Nations, Rome. http://www.fao.org/3/I9183EN/i9183en.pdf
Rulkens WH, Tichy R, Grotenhuis JTC (1998) Remediation of polluted soil and sediment: perspectives and failures. Water Sci Technol **37**:27–35
Santos A, Rosa JM (2016) In Situ Chemical Oxidation (ISCO). In: de Albergaria JTVS, Nouws HPA (eds) Soil remediation: applications and new technologies. CRC Press, Boca Raton, pp 75–94
Sarkar J, Kazy SK, Gupta A, Dutta A, Mohapatra B, Roy A et al (2016) Biostimulation of indigenous microbial community for bioremediation of petroleum refinery sludge. Front Microbiol **7**. https://doi.org/10.3389/fmicb.2016.01407
Scheckel KG, Diamond GL, Burgess MF, Klotzbach JM, Maddaloni M, Miller BW et al (2013) Amending soils with phosphate as means to mitigate soil lead hazard: a critical review of the state of the science. J Toxicol & Environ Health, Part B **16**:337–380. https://doi.org/10.1080/10937404.2013.825216
Semple KT, Morriss AWJ, Paton GI (2003) Bioavailability of hydrophobic organic contaminants in soils: fundamental concepts and techniques for analysis. Eur J Soil Sci **54**:809–818
Shepherd KA, Ellis PA, Rivett MO (2006) Integrated understanding of urban land, groundwater, baseflow and surface-water quality – the City of Birmingham, UK. Sci Total Environ **360**:180–195. https://doi.org/10.1016/j.scitotenv.2005.08.052
Smith G, Nadebaum P (2016) The evolution of sustainable remediation in Australia and New Zealand: a storyline. J Environ Manage **184**:27–35. https://doi.org/10.1016/j.jenvman.2016.05.010
Smolders E, Oorts K, van Sprang P, Schoeters I, Janssen CR, McGrath SP, McLaughlin MJ (2009) Toxicity of trace metals in soil as affected by soil type and aging after contamination: using calibrated bioavailability models to set ecological soil standards. Environ Toxicol Chem **28**:1633–1642. https://doi.org/10.1897/08-592.1
Tack FMG, Meers E (2010) Assisted phytoextraction: helping plants to help us. Elements **6**:383–388. https://doi.org/10.2113/gselements.6.6.383
U.S. ACE (2002) Soil vapor extraction and bioventing, Engineer Manual. EM_1110-1-4001, U.S. Army Corps of Engineers, Washington, DC, USA. https://www.publications.usace.army.mil/Portals/76/Publications/EngineerManuals/EM_1110-1-4001.pdf

U.S. Department of State (2011) Global water issues: a compendium of articles. Bureau of International Information Programs, Washington DC

U.S. EPA (1995a) Cost and performance report: soil washing at the King of Prussia Technical Corporation Superfund Site Winslow Township, New Jersey. U.S. Environmental Protection Agency, Office of Solid Waste and Emergency Response, Washington, DC, USA. https://clu-in.org/products/costperf/SOILWASH/Kop.htm (accessed 20191003).

U.S. EPA (1995b) Land treatment at the brown wood preserving superfund site, Live Oak, Florida, U.S. Environmental Protection Agency Office of Solid Waste and Emergency Response, Technology Innovation Office, Cincinnati, OH, USA. https://clu-in.org/download/remed/brown.pdf.

U.S. EPA (1996) Soil screening guidance: user's guide (second edition) – attachment a: conceptual site model summary, Superfund, United States Environmental Protection Agency.

U.S. EPA (1997) Resource guide for electrokinetics laboratory and field processes applicable to radioactive and hazardous mixed wastes in soil and groundwater from 1992 to 1997. EPA 402-R-97-006, U.S. Environmental Protection Agency, Office of Radiation and Indoor Air Radiation Protection Division, Washington, DC, USA. https://www.epa.gov/radiation/resource-guide-electrokinetics-laboratory-and-field-processes-applicable-radioactive-and (accessed 20210618).

U.S. EPA (2012a) A Citizen's Guide to Excavation of Contaminated Soil. EPA 542-F-12-007, United States Environmental Protection Agency, Office of Solid Waste and Emergency Response, Washington, DC, USA.

U.S. EPA (2012b) A Citizen's Guide to In-situ Chemical Oxidation. EPA 542-F-12-011, United States Environmental Protection Agency, Office of Solid Waste and Emergency Response, Washington, DC, USA. https://clu-in.org/download/Citizens/a_citizens_guide_to_in_situ_chemical_oxidation.pdf.

U.S. EPA (2012c) A Citizen's Guide to In-situ Chemical Reduction. EPA 542-F-12-012, United States Environmental Protection Agency, Office of Solid Waste and Emergency Response, Washington, DC, USA. https://clu-in.org/download/Citizens/a_citizens_guide_to_in_situ_chemical_reduction.pdf.

U.S. EPA (2012d) A Citizen's Guide to Permeable Reactive Barriers. EPA 542-F-12-015, United States Environmental Protection Agency, Office of Solid Waste and Emergency Response, Washington, DC, USA.

U.S. EPA (2017) Fourth Five-Year Review Report for the Lipari Landfill Superfund Site Gloucester County, New Jersey. Document 528294, U.S. Environmental Protection Agency, Region 2, New York, NY, USA. https://semspub.epa.gov/work/02/528294.pdf.

U.S. EPA (2019) Brownfields Laws and Regulations: 2018 BUILD Act, https://www.epa.gov/brownfields/brownfields-laws-and-regulations (accessed 20201016).

Umweltbundesamt (2009) Current state and future prospects of remedial soil protection – background paper, Umweltbundesamt (German Environment Agency), Dessau-Roßlau, Germany. https://www.umweltbundesamt.de/sites/default/files/medien/publikation/long/4041.pdf.

Unrine JM, Tsyusko OV, Hunyadi SE, Judy JD, Bertsch PM (2010) Effects of particle size on chemical speciation and bioavailability of copper to earthworms (Eisenia fetida) exposed to copper nanoparticles. J Environ Qual **39**:1942–1953

Vepraskas MJ, Polizzotto M, Faulkner SP (2016) Redox chemistry of hydric soils. In: Vepraskas MJ, Craft CB (eds) Wetland soils: genesis, hydrology, landscapes, and classification, 2nd edn. CRC Press LLC, Boca Raton, pp 105–131

Vymazal J (2011) Constructed wetlands for wastewater treatment: five decades of experience. Environ Sci Tech **45**:61–69

Wang F, Wang H, Al-Tabbaa A (2014) Leachability and heavy metal speciation of 17-year old stabilised/solidified contaminated site soils. J Hazard Mater **278**:144–151. https://doi.org/10.1016/j.jhazmat.2014.05.102

Wang L, Li F, Zhan Y, Zhu L (2016) Shifts in microbial community structure during in situ surfactant-enhanced bioremediation of polycyclic aromatic hydrocarbon-contaminated soil. Environ Sci Pollut Res **23**:14451–14461. https://doi.org/10.1007/s11356-016-6630-4

Wegscheidl C, Layden I, Bryce I (2018) Bioreactors: key aspects for effective design, operation and monitoring., Department of Agriculture and Fisheries, State of Queensland, Toowoomba, QLD, Australia. https://publications.qld.gov.au/dataset/c6f486aa-30a1-4fe9-b5ea-c8894394f989/resource/c83ff8e2-024c-4e0a-8974-10c5fd5994f4/download/bioreactor-interim-guidance-final-accessible-2018.pdf

Wenzel WW (2009) Rhizosphere processes and management in plant-assisted bioremediation (phytoremediation) of soils. Plant and Soil **321**:385–408. https://doi.org/10.1007/s11104-008-9686-1

Wong THF (2006) Water sensitive urban design – the journey thus far. Aus J Water Resource **10**:213–222. https://doi.org/10.1080/13241583.2006.11465296

Yong RN, Mulligan CN (2003) Natural attenuation of contaminants in soils. CRC Press LLC, Baton Rouge

Yong RN, Mulligan CN (2019) Natural and enhanced attenuation of contaminants in soils, 2nd edn. CRC Press, Taylor & Francis Group, Boca Raton

Zapater-Pereyra M, Lavrnić S, van Dien F, van Bruggen JJA, Lens PNL (2016) Constructed wetroofs: a novel approach for the treatment and reuse of domestic wastewater. Ecol Eng **94**:545–554. https://doi.org/10.1016/j.ecoleng.2016.05.052

Chapter 12
The Future of Urban Soils

Andrew W. Rate

Abstract Urban soils are a global resource which presents many challenges and opportunities for human populations in cities. This chapter addresses several of these opportunities and threats, and highlights where scientific knowledge is uncertain or incomplete with implications for the direction of further research. Climate change and global warming is discussed in terms of soil resilience, carbon cycling and storage, pollutant fluxes, and changes in water inputs. We review studies finding that biodiversity change in the context of urban soils may be negative or positive and suggest that some urban environments can promote conservation of flora and fauna. Urban agriculture is a significant opportunity for beneficial use of urban soils, and the benefits and constraints of growing food and trees in cities are examined. Water Sensitive Urban Design is unevenly implemented globally and offers multiple benefits for sustainable development which are not fully realised. Urban soil contamination is presented as an ongoing issue, with discussions of both legacy contamination and emerging contaminants. We highlight the potential for urban soil remediation to be performed more sustainably by widespread adoption of life cycle assessment and emphasise the need to promote environmental justice in the context of urban soils worldwide. Finally, we draw attention to the opportunities to include indigenous, traditional, and local soil knowledge in parallel with scientific and technical understanding of urban soils.

Keywords Climate change · Urban soils · Biodiversity · Urban agriculture · Water Sensitive Urban Design · Soil contamination · Life cycle assessment · Environmental justice · Soil knowledge · Ethnopedology

A. W. Rate (✉)
School of Agriculture and Environment, University of Western Australia, Crawley, WA, Australia
e-mail: andrew.rate@uwa.edu.au

A. W. Rate (ed.), *Urban Soils*, Progress in Soil Science,
https://doi.org/10.1007/978-3-030-87316-5_12

What you can learn from this chapter:

- How climate change, arguably the most pressing issue for humanity, could affect urban soils and the processes and ecosystem services occurring in them.
- How soil-based land use may change in cities as a result of increasing urbanisation and environmental change.
- What the ongoing and emerging concerns for urban soils are now, and are likely to be in the future.
- Some of the issues relating to urban agriculture, gardening, and forestry.
- Why urban soil remediation is likely to require a thorough assessment of the complete life cycle of remediation activities.
- Why issues related to urban soils will remain an important component of environmental justice.
- That there are different ways of knowing about urban soils, not restricted to the purely scientific.

12.1 Introduction to the Future of Urban Soils

There is slow, but increasing recognition of the importance of soil in urban environments. Soil allows the existence of worthwhile public open spaces; supports the practice of urban agriculture, horticulture, and private and public gardens; modifies urban hydrology; and is crucial in urban carbon cycling. In many cases, official recognition of urban soils has not considered soil in the context of ecological functions or sustainability (Teixeira da Silva et al. 2018; Calzolari et al. 2020). The gradually emerging awareness of soil's essential functions, however, is demonstrated by the increased consideration of soil resources in official urban planning documents. A good analysis of the awareness of soils in urban planning was conducted by Blanchart et al. (2019), who found that reference to a "soil resource" in urban planning documents for 15 cities in France increased significantly in the period from 2000 to 2015. Of course, soils have been considered for a long time in formal urban planning schemes, mainly from the perspective of their suitability for infrastructure development (Morris 1966). We also know from preceding chapters that soil contamination limits urban development in many jurisdictions.

The need to conserve urban soils for their ability to provide essential ecosystem services is an even more pressing issue given the continuing and projected global trend for human populations to increase in cities relative to non-urban areas (United Nations 2018). As cities grow in geographical extent, soil and associated green spaces are replaced in many instances by impermeable surfaces, and extraction of water increases (Alcoforado and Andrade 2008). Larger areas of soil become dumping grounds for urban wastes (Asabere et al. 2018), and fertile land used for food production is commonly lost (Schneider et al. 2012; Du et al. 2014). There is increasing evidence that urban green space, underpinned by functioning urban soil, has many beneficial effects on human health and well-being (Li et al. 2018), so a clear argument exists for ethical stewardship of urban soil. There are many ways of preserving ecosystem functioning, in its most holistic sense, in urban soils; Fig. 12.1 and the following sections address some of these.

In many cases the awareness of soils by urban authorities is biased towards the engineering properties of soils, and their potential risks such as those from contaminated sites and acid sulphate soils. The future of urban soil lies in our ability to move beyond seeing soil as an inert substrate or a threat, and building a general awareness of the ecosystem services provided by soils and the opportunities that soils create for more healthy and harmonious urban communities. In Chap. 1 we discussed the idea that soil knowledge was not widespread, nor was it widely used, in urban communities. It seems, then, that the story of living, functioning soils in cities will need to be told by soil enthusiasts and educators. This narrative will come in many forms; formalised in academic literature and textbooks like this one; passed down from the original, indigenous inhabitants of the lands our cities are built on; and in the years of practical experience of home gardeners and (peri-) urban farmers.

Fig. 12.1 A selection of entities and activities sustaining the future of soils in urban environments

12.2 Climate Change Effects

Anthropogenic climate change is arguably the most pressing environmental challenge currently facing humanity (Steffen et al. 2015). The potential for global climate change to affect humans is likely to be exacerbated in cities, which present environmental challenges of their own in the form of increasing urbanisation and urban population growth (Grimm et al. 2008).

Urbanisation is known to have caused historical declines in soil fertility and organic matter content (based on a study of Mayan cities by Douglas et al. 2018). Although these declines were not caused by climate change, they represent processes which decrease the resilience of urbanised ecosystems to climate change effects. The effects of urbanisation on soils persist and continue to occur into contemporary times, and climate change is likely to cause soil fertility and organic carbon decline as warmer temperatures promote microbial decomposition of soil organic matter (Lal 2017). There is evidence from several studies, though, that soil organic carbon can increase with urbanisation. For example, Asabere et al. (2018) measured greater soil carbon contents in Kumasi, Ghana, in long-term urbanised environments than in recently urbanised or even rural soils, related to waste disposal practices (Fig. 12.2). While Asabere et al. (2018) focused on urban horticulture, similar effects were found by Pouyat et al. (2002) for forest soils, with urban oak forests having greater organic carbon contents than their suburban and rural counterparts. The greater soil carbon storage in urban forests was attributed to lower leaf litter quality in urban forests, leading to less decomposition by soil fauna and microorganisms. In addition, we have already mentioned (in Chap. 10) how urban soils can accumulate large amounts of inorganic carbon, in the form of carbonate

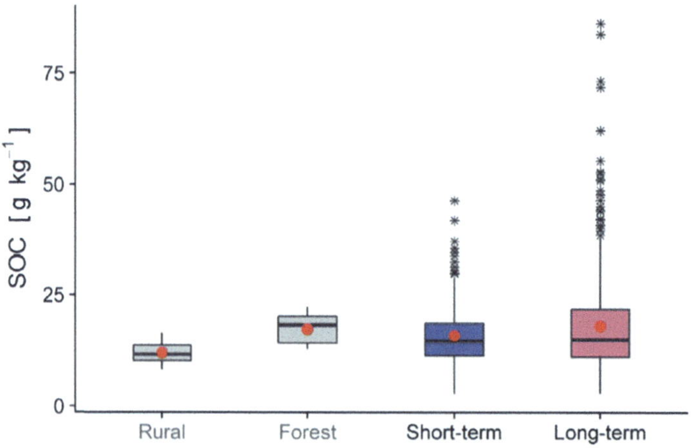

Fig. 12.2 Comparison of soil organic carbon content in the ≤ 2 mm fraction of soils from rural, forest, short-term urban, and long-term urban locations in Kumasi, Ghana (from Asabere et al. 2018, used under terms of CC-BY license). Solid red circles on box plots are arithmetic means (standard errors of mean are smaller than symbols)

minerals from construction and demolition wastes (Rawlins et al. 2015; Kolosz et al. 2019). In some city environments, however, urban forest soils may be more sensitive to the effects of climate change, and may lose carbon and nitrogen more rapidly than forests in rural areas (Hosseini Bai et al. 2015). Overall, urban soils collectively can contain large amounts of carbon, which should be accounted for and which are important in terms of the direction of their carbon fluxes to or from the atmosphere (Pouyat et al. 2006; Dorendorf et al. 2015; Calzolari et al. 2020). In particular, urban parklands can accumulate soil carbon (Wang et al. 2013).

The effects of a warming climate may be exacerbated in urban environments, due to the urban heat island effect which we discussed in Chap. 5 (Coutts et al. 2013). Warmer temperatures are also predicted to promote increased transfer of lower-volatility pollutants such as many POPs and Hg from soil to atmosphere. This increased volatilisation may in turn promote increased pollutant deposition into soils at higher latitudes. Higher temperatures may also, however, result in increased degradation rates of organic pollutants in soils (Nadal et al. 2015). Climate change is also expected to affect the behaviour of inorganic contaminants such as metals, such as increased fluxes of metal-bearing dust, or increases in metal bioavailability in drying soils (Paltseva and Neaman 2020).

Climate change is not restricted to increased temperatures, and changes in precipitation patterns are expected to alter hydrology and soil water contents. Most future climate scenarios are characterised by more frequent storm events which may be of greater intensity. Intense storm events have been known to increase the risk of pollutant transfer, for example, by flooding contaminated sites or increasing soil erosion (Maco et al. 2018). Combined with the observed and expected rises in sea level, flooding in coastal cities may also salinise soils, or result in longer seasonal or even permanent inundation of soils in low-lying areas.

The likely effects of climate change in some regions are decreased precipitation, and the effects of a drying climate on the water balance in soils are also important. Drying of soils and sediments and lowering of groundwater levels will obviously decrease plant productivity. Soil drying also allows greater aeration of soils, and may lead to increased acidification of potential acid sulphate soils (Devito and Hill 1999). The situation of many cities in coastal zones underlain by recent marine or estuarine sediments, the frequent disturbance of urban soils, and the extraction of groundwater may make acid sulphate soil formation an even more likely outcome in urban environments (for an example see Appleyard et al. 2004). Drying of soils and sediments may also have favourable outcomes, in that greater rates of aerobic decomposition of organic pollutants are possible (Noyes et al. 2009).

It is clear that more needs to be known about climate change and its effects before we can predict its effects on urban soil environments with any certainty. Nevertheless, it is very likely that climate change presents risks to urban environments, and that proper management of urban soils can have various roles in limiting those risks.

12.3 Urban Soils and Biodiversity

The loss of biodiversity on Earth is sufficiently critical that it exceeds the "safe operating space for humanity" defined by the planetary boundary concept (Steffen et al. 2015). Urban environments, including soils, are commonly thought to have less biodiversity than comparable natural environments (Foley et al. 2005; Albrecht and Haider 2013). These losses in biodiversity are related to habitat loss from urban land use change, altered hydrology, food consumption with its concomitant land requirements for production, and waste generation (McDonald et al. 2019). The conclusion of lower biodiversity of soil organisms in urban or contaminated environments is supported by some studies (e.g. Kozdrój and Van Elsas 2001; Uno et al. 2010), but not by others (e.g. Pavao-Zuckerman and Coleman 2007); the large number of potential controls on soil biodiversity means that it is hard to generalise results. Some studies have found that while the total numbers or biomass of organisms was lower in urban soils, the taxonomic diversity was not significantly different from non-urban soils (Pavao-Zuckerman and Coleman 2007; Santorufo et al. 2012). It may also be true, however, that urban environments present opportunities for conservation of biodiversity (Knapp et al. 2009). In some urban areas, biodiversity can increase in situations such as residential gardens or urban agriculture which include non-native plant species (Low 2003; Orsini et al. 2013), although whether or not this extends to soil organisms is uncertain.

Vegetated, unsealed urban soil is clearly more common in sports grounds, parks, gardens and reserves (Calzolari et al. 2020), and even wastelands (Bonthoux et al. 2014), and so these land use categories represent sanctuaries or *refugia* for soil biodiversity. For example, as mentioned in Chap. 8, Ramirez et al. (2014) found a rich diversity of soil microorganisms and invertebrates in the soil of Central Park, New York City. The distribution of soil organisms in Central Park was also significantly related to soil properties such as soil pH. In Paris, France, the diversity of soil environments in urban public gardens was found to contribute to the diversity of above-ground plants and animals (Shwartz et al. 2013). It should be noted, however, that soil preparation prior to creation of parklands has significant effects on soil biodiversity; for example, imported topsoil can result in greater diversity (Vergnes et al. 2017).

Urban soil biodiversity can even be linked to human health outcomes, in terms of its effect on the diversity of human microbiomes and the consequent immune system functioning (Li et al. 2018). In contrast, undesirable biodiversity exists in urban soils in the form of increased populations of potentially pathogenic organisms in areas where inadequate sanitation exists or where wastewaters are used for irrigation (Pickering et al. 2012; see Chap. 8).

While plant and animal biodiversity in urban environments has received considerable attention, with much being known about how these organisms are affected by urbanisation, much less is known about soil organisms. Nevertheless, soils are unquestionably an essential part of any urban ecosystem, and their biodiversity affects their functioning. Since humans are also intimately linked to the urban ecosystem, there will be an ongoing need to learn about and understand both the effects of urbanisation on soil organism diversity, and the consequences of soil biodiversity and its dynamics for other organisms and compartments in urban ecosystems.

12.4 Urban Agriculture and Gardening

12.4.1 Urban Gardening for Food Production and Wellbeing

Urban agriculture – used in a general sense here to mean the growing of plants in cities to produce food crops – has significant potential to contribute to food production, particularly for urban inhabitants (Edmondson et al. 2020). The land area available in cities worldwide is sufficient to meet plant-based food requirements for the global urban population (Martellozzo et al. 2014), but there may be limitations in terms of the availability of a suitable water supply (Mawois et al. 2012), and balancing sustainable energy and water usage (Eriksen-Hamel and Danso 2010; O'Sullivan et al. 2019). Urban agriculture has the potential to increase urban biodiversity in the form of agro-biodiversity (Orsini et al. 2013; Taylor and Lovell 2015).

Apart from water availability, the main constraints on the expansion of urban agriculture may be the possibility of urban soils already being contaminated with potentially harmful chemicals or pathogens, and a need for soil information in urban communities. Since urban soils are commonly contaminated or degraded in some way (Kessler 2013; Wortman and Lovell 2013), the concern about contamination is a real one, for example, with metals such as Pb (Brown et al. 2016; Jacobs et al. 2017) or organic pollutants such as pesticides (Margenat et al. 2018). Importing soil materials and amendments can address this problem if the imported materials have low or negligible concentrations of contaminants (Jones and Healey 2010; Chakravorty 2019), but this is not always the case (Gómez-Sagasti et al. 2018). Amendment of soil used for urban agriculture with solid wastes can also improve soil properties, particularly organic wastes (Chap. 11, and Anikwe and Nwobodo 2002), but solid waste re-use can also result in risks of contamination (Clarke and Smith 2011). Similarly, although beneficial re-use of wastewater to irrigate crops is a strategy to improve sustainability, Amoah et al. (2005) found that, in urban agriculture in Ghana, wastewater re-use resulted in contamination of vegetables with bacteria and parasites. Wastewater irrigation to support urban agriculture can also cause leaching of nutrients to groundwater (Werner et al. 2019). Another option for increasing the sustainability of water supply to soils used for urban agriculture is to capture precipitation from the roofs of buildings. For a large sample of urban gardens in Roma, Italy, Lupia et al. (2017) showed that rainwater harvesting from rooftops could supply between 19 and 44% of urban food garden water requirements, depending on garden type and water use efficiency.

Fig. 12.3 Examples of urban green spaces and green infrastructure: (**a**) **urban agriculture**, New York, USA; (**b**) **suburban garden**, Blackheath, UK; **urban forest**, Berlin, Germany; (**d**) **roadside stormwater swale**, Greenfield, WI, USA; (**e**) **suburban rain garden**, Chicago, USA; (**f**) **meandered and revegetated stormwater drain**, Perth, Australia. (Photo credits (all flickr images are CC-BY-2.0): (**a**) Preston Keres, USDA, public domain; (**b**) monoclepix on flickr; (**c**) Tomasz Baranowski on flickr; (**d**) Aaron Volkening on flickr; (**e**) Linda on flickr; (**f**) Andrew W. Rate on flickr)

In Chap. 10 we discussed the benefits of urban agriculture and gardening for human health and the wellbeing of individuals and urban communities. Although the most critical need globally is food production, other common factors also appear to inspire urban gardening worldwide, and some examples of beneficial urban green spaces are presented in Fig. 12.3. For example, Home and Vieli (2020) who studied a selection of cities in Switzerland and Chile found that, in order of importance, the common factors which motivate urban residents to tend gardens were ecological restoration, social connection, and food production. In other words, as well as growing food, urban citizens are inspired by desires to (re)create natural environments, and connect with one another around a soil-plant-nature-based activity.

Soil-less urban agriculture is possible (e.g. variations on hydroponic systems) but is unlikely to offer the same community cohesion and mental health benefits as soil-based gardening. The soil-less gardening systems which are proposed, or even highly water-efficient partially soil-based systems such as vertical gardens, seem to be more suited to commercial urban food production (Bradley and Marulanda 2001). As such they fulfil a need to provide food, but are not likely to foster ecological restoration or community-building.

12.4.2 Urban Forestry

In addition to urban food production, one of the most beneficial uses of vegetated urban soils is for different forms of urban forestry, ranging from isolated street trees to larger nature reserves. The "urban forest" refers to the collective tree cover in a city environment. Trees in urban environments fulfil multiple functions, many of which are addressed in Chap. 10. The constraints common to many urban soils may, however, restrict tree growth and survival, and particular attention needs to be paid to making urban soils suitable for trees (Jim 1998). Urban forests favour conservation of biodiversity by providing habitat, food sources, and travel corridors for urban wildlife, from insects to birds and mammals. Humans benefit from the urban forest as well, since trees provide cooling by evapotranspiration and shading. The interception and infiltration of stormwater is also improved by trees in urban environments, reducing runoff and decreasing the likelihood of flooding. The future of urban forests seems to be hopeful in more economically prosperous nations, with widespread and increasing adoption of urban forest strategies by city-wide and local governments. There is less information about urban forestry or urban greening in developing nations, and the economic constraints in less prosperous countries often lead to an emphasis of development over protection of ecosystems (Jim 2013). Despite this, urban trees of many so-called developing nations are important assets.

One of the ways in which the urban forest provides services is by direct provision of food in the form of fruit and nuts. Street and garden trees form an important food source, especially for poorer residents in developing countries and also in the developed world (Kaoma and Shackleton 2014). Some studies have shown that as well as producing large yields, trees may take up less contaminants into their fruit than do vegetables from soil in urban areas (Colinas et al. 2019).

Since soils in cities are most commonly left unsealed in areas such as parks, gardens, sports grounds, and so on, a wide range of tree and other plant species are cultivated, including those native to an area but also introduced species. There is some concern about the disadvantages of non-native tree species, for example, their invasiveness or the changes in soil properties that they can cause (Barker 2008; Useni Sikuzani et al. 2019). In contrast, tree species which are not native to an area can also represent significant food sources, and have more rapid growth and consequently earlier achievement of urban cooling and rain interception effects. It should be noted that the needs of urban forestry and urban agriculture are not always compatible, since trees compete with food crops for solar radiation and water (Johnson et al. 2015). Urban trees themselves, however, can provide food for urban residents, through deliberate harvesting or informal foraging (McLain et al. 2014; Colinas et al. 2019).

12.4.3 Manufactured Soils

As cities grow and physical space becomes increasingly limited, green infrastructure features such as green roofs, vertical gardens, rain gardens, and constructed plant beds are likely to become more common features of urban environments. In many cases these green infrastructure features will contain manufactured soils – materials which may behave like, or be derived from, natural soils but which have different composition or layering from in situ soils. Such soils are classified as

Isolatic Technosols (IUSS Working Group WRB 2014), meaning that they are emplaced by humans, contain anthropogenic materials, and are located in some sort of container (i.e. the infrastructure).

Urban soil science will need to expand to more fully understand how such materials behave and change, in the situations in which they are placed. Some research has already addressed green infrastructure materials (e.g. Komlos and Traver 2012; Bouzouidja et al. 2018), and both commercial suppliers of manufactured soil materials and designers of green infrastructure also maintain a body of knowledge on this topic. Many issues will need clarifying; for example, whether or not organic soil amendments, which are commonly used in green infrastructure, will need screening for contamination (Gómez-Sagasti et al. 2018; Rodríguez-Eugenio et al. 2018). Green infrastructure projects will probably need to be subject to the same level of scrutiny (including life cycle assessment) as currently is required for brownfield redevelopment.

12.5 Water-Sensitive Urban Design

Water-Sensitive Urban Design (WSUD) aims to allow, as far as possible, all the components of natural water cycles to occur in urban environments (Wong 2006). One of the most important components of WSUD is that of maintaining a more natural balance between runoff and infiltration, given that impermeable surfaces are an unavoidable feature of urban systems (Jacobson 2011). This hydrological balance, and other components of urban water cycles such as evapotranspiration and groundwater recharge, will be dependent to a large extent on the exposed area of urban soils and their properties and management.

The principles of Water-Sensitive Urban Design are being incorporated into planning guidelines by government entities, particularly in Perth, Western Australia, where Water-Sensitive Urban Design originated in the 1990s (Whelans et al. 1994). Perth is a seasonally dry city (*Köppen-Geiger* Csa) having a mean annual excess of potential evapotranspiration over precipitation of 716 mm, and the city is reliant on groundwater for a large proportion of its water supply. In addition, Perth's location on a sandy coastal plain is reflected in its mainly highly permeable soils, with only about 12% of urban stormwater generating excess runoff (Cargeeg et al. 1987). Since Perth has such permeable soils, it is an urban environment that facilitates WSUD, but the low retention capacity of its soils mean that contaminant transport is likely. The drivers of WSUD in Perth include the need to decrease export of nutrients to sensitive environments such as rivers and estuaries, and the benefits of increasing infiltration to recharge groundwater (which is a valuable resource for metropolitan water supply). Favouring infiltration of stormwater into soil rather than exporting stormwater as runoff also increases evapotranspiration, cooling urban environments (Coutts et al. 2013) – a desirable outcome for seasonally hot cities such as Perth and many others worldwide.

In cities with less permeable soils, WSUD strategies are potentially even more essential, since the excess runoff generated by impermeable surfaces can cause soil and stream-bank erosion, affecting receptors such as waterways (Paul and Meyer 2001). Large proportions of impermeable surfaces also lead to increased risks of flooding, with subsequent public health and contaminant mobilisation issues (Jacobson 2011).

There is considerable awareness of the beneficial effects of WSUD (or "Sustainable Drainage Systems" (SuDS), or Low Impact Development (LID)) in many cities worldwide (Zhou 2014). This awareness has not yet led to full implementation of WSUD and related strategies, even though it can be more cost-effective than conventional stormwater infrastructure (Eckart et al. 2017). In the developing world, even more barriers to the adoption of green infrastructure such as WSUD exist (Jiusto and Kenney 2016), even though adoption of WSUD offers substantial improvements to many aspects of urban populations and environments (Mguni et al. 2016).

In terms of urban soils, green stormwater management strategies such as WSUD are about retaining (or re-introducing) soil processes into urban water cycles. Through benefits such as flood control, groundwater recharge, increased green space, and increased evapotranspiration, such strategies offer options to improve public health and liveability in cities worldwide. As discussed in Chap. 10, WSUD and related green urban water management systems address several of the Sustainable Development Goals, such as Goal 3 to ensure health, Goal 6 relating explicitly to sustainable water management, and Goal 9 which includes building of resilient infrastructure (United Nations 2015). Implementation of green stormwater management, which depends to a large extent on urban soils, will not necessarily fix global issues such as climate change and biodiversity loss, but it presents numerous opportunities to improve the quality of life for the world's increasing urban population.

12.6 Soil Contamination

Contamination of urban soils is, of course an ongoing problem by itself, independent of the imperatives of environmental justice. Many pollutants, as we have seen in previous chapters, persist in soils for very long times or even indefinitely, and soil remediation is costly and may not be performed at all. As a result, historical soil contamination continues to be a concern in urban environments, especially as land uses change to accommodate burgeoning urban populations. Humanity is also very accomplished at unearthing or releasing existing hazardous materials, creating new contaminants, or simply recognising that substances we previously thought were harmless are almost certainly not. The topic of "emerging contaminants" has effectively become a research field in its own right.

12.6.1 Ongoing and Legacy Contamination and Brownfields

Legacy contamination is contamination that continues to be an issue after long time periods have elapsed. Of course, some legacy contamination has been used to generate archaeological information, as discussed in Chap. 2. The phenomenon of longevity means that legacy contaminants are persistent, such as trace elements or persistent organic pollutants (POPs), and the term particularly applies to substances where regulation has substantially restricted their use, but where they still present a potential risk. The legacy contaminant of most ongoing concern is probably lead (Pb), but legacy issues also exist for persistent organic compounds such as polycyclic aromatic hydrocarbons (PAHs), and asbestos.

Brownfields are sites, commonly in urban areas, where an industrial source of contamination such as a smelter or factory previously operated but is currently derelict or has been demolished, leaving vacant but contaminated land. As described by Albanese and Cicchella (2012), increases in urban populations also increase the demand for land for residential purposes. Effective remediation then needs to be conducted to avoid putting residents at risk of exposure to brownfield pollutants. Even before development, brownfields have been used for activities such as gardening, in which cases risks still existed. For example, concentrations of As and Pb in allotment gardens at a former industrial site in Newcastle, UK, exceeded environmental guidelines (Pless-Mulloli et al. 2004).

As we discussed above, in the context of environmental justice, lead pollution in soils still persists. This is despite complete bans on the use of leaded paint beginning in 1909 in France and in most other countries by the early twenty-first century, and similar bans on the leaded fuel additives such as tetraethyl lead by the late 1990s in the USA (with later complete bans in Europe and other countries). Soils in many cities worldwide have been found to have legacy lead contamination, and several examples are shown in Table 12.1. (Note that Table 12.1 has notable omissions from developing countries, such as on the African continent, partly since use of leaded fuels is still widespread (Maas et al. 2010; Sellami et al. 2020)).

Although we have focused on lead, legacy contamination with other pollutants is also, unfortunately, relatively common (Nadal et al. 2015). For example, Qu et al. (2019) found substantial contamination of urban soils of Napoli, Italy, with organochlorine pesticides, polychlorinated biphenyls (PCBs), and polycyclic aromatic hydrocarbons (PAHs). From Table 12.1 it is also apparent that, while the expected sources of legacy contaminants may be common (traffic and pigments for Pb), there are other sources which need to be considered, some of which (like shooting ranges or archaeological artefacts) may be unexpected. The same caveat would apply to contamination of urban soil with any persistent substance, and that is one good reason that a Preliminary Site Investigation (Chap. 11), which would include a site history, is essential for management of any projects involving use of disturbance of urban soil or sediment.

Table 12.1 Examples of contamination of urban soils with lead with sources identified as "legacy" or "historical"

Urban area	Country	Assumed legacy Pb source	References
Melbourne	Australia	Lead-based paints in older buildings; traffic sources of leaded fuel emissions	Laidlaw et al. (2018)
Sydney	Australia	Lead-based paints in older buildings; traffic sources of leaded fuel emissions	Rouillon et al. (2017)
Copiapó	Chile	Mining (e.g. tailings) and smelting; industries	Carkovic et al. (2016)
Beijing	China	Pigments; traffic sources of leaded fuel emissions	Xia et al. (2011)
Nanjing	China	Industries, pigments (cultural layer); industry, traffic (contemporary surface)	Zhang et al. (2005)
Lefkosia	Cyprus	Older developed areas in the city	Zissimos et al. (2018)
Athens	Greece	Former shooting range	Urrutia-Goyes et al. (2017)
Napoli, Roma	Italy	Road traffic	Cicchella et al. (2015)
Mexico City	Mexico	Traffic sources of leaded fuel emissions	Morton-Bermea et al. (2011)
Dunedin	New Zealand	Lead-based paints in older buildings; traffic sources of leaded fuel emissions	Turnbull et al. (2019)
London	UK	Lead-based paints in older buildings; traffic sources of leaded fuel emissions	Kelly et al. (1996)
Newcastle	UK	Ash from power-from-waste generation	Pless-Mulloli et al. (2004)
Multiple urban areas in California	USA	Lead-based paints in older buildings; highway sources of leaded fuel emissions	Mielke et al. (2010)
New York	USA	Lead-based paints in older buildings; automobile emissions	Mitchell et al. (2014)

12.6.2 Emerging Contaminants

A wide range of synthetic chemicals are manufactured in contemporary human societies to meet the demand for health and medical products, personal care, packaging, industrial uses, and so on. As a result, many of these compounds are entering receiving environments such as soil, water, and air. In many cases involving recently developed compounds or materials, though, there is currently insufficient knowledge of their human health or ecological effects. Advances in technology have introduced many new materials into widespread use, and a proportion of these are potential contaminants. In parallel, advances in chemical analysis techniques have made it possible to detect and measure a wide range of new and existing compounds at trace concentrations. The term *emerging contaminants* (or "contaminants of emerging concern") was used as early as the 1980s (and somewhat later for soils) to describe substances that were not usually considered when assessing contamination of environmental compartments. Of course, many of the early emerging contaminants are now well-known and, in some jurisdictions, regulated as pollutants (e.g. PBDE flame retardants, or neonicotinoid pesticides). The types of materials considered include plasticisers, pharmaceuticals including endocrine-disrupting chemicals, chemicals used in cosmetics, nanoparticles, preservatives, plastics and microplastics, flame retardants, antibiotic-resistant bacteria, by-products of water treatment, and numerous others (Sauvé and Desrosiers 2014).

Given the large concentrations of human population in urban centres, it is not surprising that many emerging contaminants have been found in urban soils. In some cases, an anthroposequence of contamination has been observed, with concentrations of PBDEs in soils decreasing from intense urban to rural areas (Mahmood et al. 2015). In other examples, there is no clear effect of land use on the distribution of emerging contaminants (Karpuzcu et al. 2014). Similarly, microplastics in urban soils have not, so far, shown a consistent effect of urban land use. Rafique et al. (2020) found similar concentrations of microplastic particles across a range of urban and peri-urban land uses in Lahore, Pakistan. In contrast, Choi et al. (2020) found that land use did have an effect on microplastic concentrations in urban and adjacent soils (Fig. 12.4). Similarly, Lutz et al. (2021) showed an effect of land use on the concentrations of microplastics in urban stormwater drains. The findings of

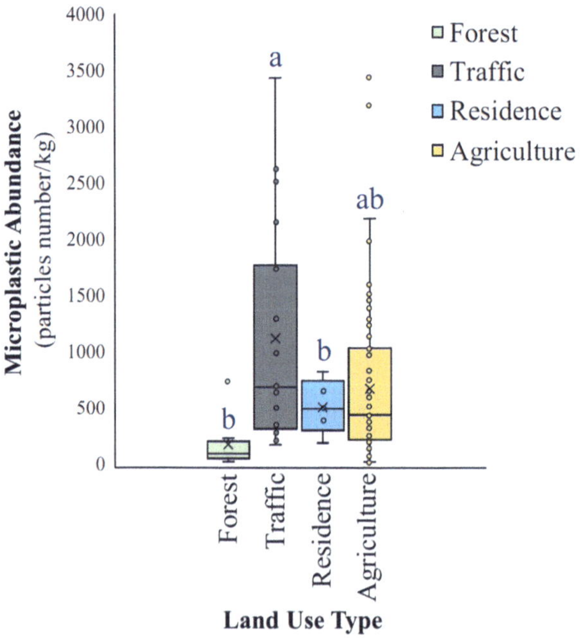

Fig. 12.4 Microplastic concentrations in soils for different land use types in Yeoju City, Republic of Korea. Different letters above each box mean a significant difference ($p \leq 0.05$) between mean values in each land use category (redrawn from Choi et al. (2020); used with permission from Springer)

Choi et al. (2020) and Lutz et al. (2021) suggest that more attention should be paid to examining the interaction of urban soils and land use as controls on terrestrial inputs of microplastics into oceans.

Considerable worldwide attention has been given recently to the issue of soil and groundwater contamination with poly- and perfluorinated alkyl sulfonates (PFAS). The concern with PFAS relates to their uncontained usage in large amounts in flame-retardant firefighting foams (especially at aviation facilities), a use for which they are very effective. The PFAS represent a large group of related compounds which resist environmental degradation, and therefore persist for long periods of time in soil and groundwater (EPA 2017). PFAS can also bioaccumulate and transfer up the food chain, and so behave like many other persistent organic pollutants (POPs) which were discussed in Chap. 7 (Conder et al. 2010). An additional concern with PFAS and related compounds is their high water-solubility – and for some PFAS compounds, relatively high volatility – both properties allow PFAS compounds to be mobile between environmental compartments (EPA 2017). PFAS-type compounds are known to affect the health of animals, but human health effects are poorly understood (U.S. EPA 2019).

The phenomena of legacy and ongoing contamination, and of continuing emergence of new contaminants, mean that urban soil science will evolve to meet these challenges. In order to manage urban soil contamination correctly, the risks need to be known. To understand these risks, scientific and regulatory communities will need to collect data on the occurrence, distribution, and controls on the environmental behaviour of the contaminants. Importantly, we will also need to be able to measure the toxicological responses and identify and quantify the exposure pathways, for all receptors. Given that we have an incomplete knowledge of soil biodiversity, we may also need to identify a more complete range of receptors.

12.7 Life Cycle Assessment of Soil Remediation

There is an opportunity for longer-term planning around urban soil remediation if the cost-benefit analysis extends over longer time frames, and includes a life cycle assessment. The relevance of considering the longer-term impacts of remediation options is related to fossil fuel and energy consumption by many commonly used remediation methods such as excavation and disposal. Energy usage obviously has implications for carbon budgets, atmospheric warming, and climate change, but there are other factors which are relevant as well. Urban soil remediation decisions are based on multiple, often opposing, factors such as environmental protection and potential for income from redevelopment. As with any decision having multiple standpoints and stakeholders, there will also be differing and potentially opposing values placed on the various forms of amenity involved, whether these be ecologically, commercially, or socially motivated.

The increasing awareness of the importance of soil remediation which is sustainable is illustrated by the US Sustainable Remediation Forum, which produced a guidance document for remediation practitioners to conduct life cycle analyses and *footprint analyses* for remediation projects (Favara et al. 2011). The guidance is structured into nine steps, of which one of the most relevant here is the establishment of system boundaries. What this means is to identify which components of, or processes in, the life cycle of a project are relevant to include. The life cycle components and processes considered include energy (including transport), materials, processing, and waste treatment factors. Spatial boundaries are also considered, under on-site, local, regional, and global categories. The time frame over which impacts are considered also needs to be defined, as do any restrictions on the choice of remediation technology (e.g. restrictions resulting from availability or regulatory constraints) (Favara et al. 2011). The main innovations of completing a life cycle assessment for soil remediation are that (i) the environmental impact of the remediation itself is considered, and (ii) the environmental effects considered are expanded in scope to account for impacts from a more comprehensive spatial and temporal influence of the remediation project.

Perhaps independently of the trends towards life cycle and footprint assessments, there are indications that more sustainable, less energy-intensive remediation methods have recently been chosen in favour of less advanced methods. For example, the compilation of European data summarised in Fig. 11.1 show that, in several European states, in situ remediation is used for contaminated soils (European Environment Agency 2020). It is a reasonable assumption that a substantial proportion of these sites would be in urban environments. Similarly, an analysis of groundwater remediation choices for the "Superfund" sites in the United States of America (Simon 2020) shows several interesting trends which are apparent in Table 12.2. Energy-intensive remediation methods such as pump-and-treat have declined in use (from approximately 98% of sites in 1982 to 19% in 2017), whereas in situ groundwater remediation techniques became more commonly used (from 0% of sites in 1982 to 53% in 2017). Similarly for soils, in situ remediation such as chemical treatment and amendments increased in use over time, whereas soil vapour extraction, soil flushing, and somewhat surprisingly ex situ bioremediation decreased in usage between 1988–2002 and 2003–2017 (European Environment Agency 2020).

If life-cycle assessment of urban soil remediation or urban development projects becomes the norm, a potential benefit may be to introduce and normalise processes for non-market valuation of urban soil resources. So far, this has not been attempted widely, perhaps due to the difficulties in assigning monetary values to ecosystem services in heterogeneous urban soil environments (Saad et al. 2011; Greenhalgh et al. 2017).

Table 12.2 Soil and groundwater remediation trends at Superfund contaminated sites in the United States of America (modified from Simon 2020 and used with permission from John Wiley and Sons)

Treatment category	Treatment technologies compared between 1988–2002 and 2003–2017 [a]	
	Increased frequency of use	Decreased frequency of use
Ex situ source treatment	Physical separation	Soil flushing
	Recycling	Incineration
	Solidification/stabilisation	Aeration
	–	Thermal desorption
In situ source treatment	Soil amendments	Flushing
	Chemical treatment	Bioremediation
	–	Soil vapour extraction
	–	Solidification/stabilisation
In situ groundwater treatment	Chemical treatment	Vapour extraction
	Thermal treatment	Air sparging
	Bioremediation	–
	Permeable reactive barriers	–

[a]The 4 technologies showing the greatest increases, and the 4 technologies showing the greatest decreases, but limited to technologies used to remediate ≥ 10 U.S Superfund sites between 1998 and 2017

12.8 Urban Soil and Environmental Justice

The issue of environmental justice issues is becoming increasingly important, even developing into a key concern during the United States of America's contentious Federal election campaign in 2020 (Redd et al. 2020). Instances of inequity in environmental quality or access are still emerging, however. For example, lead poisoning in children in the city of Baltimore, USA, was recognised in the 1940s (Schucker et al. 1965). Segregation of Baltimore neighbourhoods by race and income, however, still means that some socioeconomic groups – residents who are poor, and/or have non-European ethnic backgrounds – still suffer the most from lead pollution (Zaleski 2020). In the specific case of lead toxicity, soil pollution is just one component of the problem, with issues such as legacy infrastructure and household dust contaminated with historical lead-based paint also needing to be addressed.

Forms of environmental injustice other than soil contamination are also relevant, and the issues are not restricted to the global north. In the Limpopo Province, South Africa, increasing urbanisation is a factor contributing to soil erosion, which inequitably affects the rural poor (Musakwa et al. 2020). On a more global scale, indigenous people have historically been subjected to colonisation or land appropriation from other ethnic groups, and in many instances this has led to environmental injustice with urbanisation as one of the drivers. Pollution of soil and other media is known for many indigenous peoples, with poverty decreasing indigenous people's capacity to address environmental injustice (Fernández-Llamazares et al. 2020).

Urban biodiversity, which depends to some extent on the area of exposed urban soil and its properties, can be linked to socioeconomic measures such as a deprivation index (Stewart et al. 2009). Urban residents' access to urban green space (which represents vegetated urban soil) is also unevenly distributed among different socioeconomic zones of cities such as Berlin, Germany (as described by Kabisch and Haase 2014). Inequity in access to urban green space could potentially be addressed by provision of green stormwater infrastructure, which would logically be needed in all socioeconomic zones of a city (Wendel et al. 2011).

Environmental justice as a social movement, as discussed in Chap. 10, had its origins in soil contamination issues. Degraded or contaminated soils and their uneven socioeconomic distributions in cities can foster an awareness of environmental justice. Part of the mandate for soil enthusiasts and educators, then, could be to highlight environmental justice inequities at the same time as empowering citizens to address soil degradation and pollution issues.

12.9 Indigenous, Traditional, and Local Soil Knowledge

Practitioners of any scientific discipline can find it difficult to acknowledge that non-scientific knowledge can stand on an equal footing with the understanding gained from the "scientific method". At the same time, people untrained in science commonly believe scientific principles to require an unreachable level of intellect and erudition, or may even become suspicious of science and scientists themselves. The intellectual detachment required to achieve the scientific method's ideal of objectivity may also have contributed, ironically, to self-reflection by some scientists and a questioning of the primacy of scientific knowledge over knowledge obtained in other ways. Fortunately, a recent trend in academia is to value both scientific, technical understanding and local, indigenous, or traditional forms of knowing. The argument over which form of knowledge, or which way of obtaining it, is superior then becomes irrelevant, in a worldview that considers all forms of knowledge to have some validity.

Soil knowledge is an excellent example of a discipline in which information and understanding can have multiple valid origins. The various types of soil knowledge have been considered in academia for a few decades. In one of the earlier studies, Winklerprins (1999) concluded that sustainable land management could be planned more effectively if "local soil knowledge" was considered. She defined local soil knowledge as "...knowledge of soil ... possessed by people living in a particular environment for some period of time". The advantages of considering such knowledge reflect the close relationship that local inhabitants have with particular areas of land and the soils underlying them. In contrast, scientific or technical knowledge often has an overview of issues, and detailed process-based understanding, both of which may be generalisable to specific situations. It makes sense, then, to combine both local and scientific soil knowledge, and integrate the generalisable mechanistic understanding with the intimate knowledge of specific soil environments gained from a local perspective.

Local soil knowledge is also called "ethnopedology" (or "indigenous soil knowledge", "traditional soil knowledge", or "folk soil knowledge" (Winklerprins 1999)). Barrera-Bassols and Zinck (2003) studied how academia has reacted to diverse forms of soil knowledge, and many of the academic responses seem to try to fit such knowledge into a scientific mould, for example, by looking for soil or landscape classification schemes within a body of local soil knowledge. More recent work acknowledged the detail inherent in local soil knowledge (e.g. a soil or landform classification), and that the knowledge was both practically oriented and did not require technical inputs in the form of laboratory analyses. A technical soil classification scheme is based on soil-forming processes and rigorous identification of certain features, but use of classifications based on local knowledge are more pragmatic and easily implementable, having their origins in the lived experience of land use and soil management (Barrera-Bassols 2015). Indigenous people have, by definition, the longest history of inhabiting a particular land area and interacting with its soils. In particular, many indigenous peoples have a belief system that explicitly considers soils and their origins and role in the cosmos (Pauli et al. 2016). A cosmology which includes soils would presumably also favour soil conservation, but we do not yet know of any evidence for the existence, in urban environments, of a credo which explicitly includes soils.

Much of the published work on local soil knowledge relates, understandably, to rural agriculture. Some studies on the importance of local soil knowledge in urban agriculture and gardening are emerging. In Chap. 10 we discussed the multiple health, social, and ecological benefits of urban agriculture and gardening. These are often community-building activities: as Teuber et al. (2019) point out, gardens are "… social ecological systems …" in which "… humans interact with the ecological environment through soil and plant cultivation". Local soil knowledge allows urban gardeners to understand soil-plant relationships, conserve their soils, and implement novel soil management practices. A beneficial collaboration between stakeholders in urban soils would place local knowledge on an equal footing with scientific/technical knowledge and knowledge of urban policy (Teuber et al. 2019).

As an increasingly urbanised species, humanity still has an opportunity to develop a more holistic and compassionate attitude towards all of our urban neighbours: human, animate and inert, sentient and reflexive. Those of us who breathe, breathe the same air; we depend, every part of our ecosystems on the same water, the same land, the same soil. Urban soils stand at a multivariate intermingling of the traditional environmental compartments; and of human creativity, endeavours, follies, anxieties, and longings. The soils in cities, then, are one hub around which we can centre our collective efforts to preserve Earth's fragile yet exquisite ecology, and build a kind and just society.

> *The soil is the great connector of lives, the source and destination of all. It is the healer and restorer and resurrector, by which disease passes into health, age into youth, death into life. Without proper care for it we can have no community, because without proper care for it we can have no life. — Wendell Berry (1996), The Unsettling of America: Culture and Agriculture*

12.10 Further Reading

Coutts, A.M., Tapper, N.J., Beringer, J., Loughnan, M., Demuzere, M., 2013. Watering our cities: The capacity for Water Sensitive Urban Design to support urban cooling and improve human thermal comfort in the Australian context. *Progress in Physical Geography: Earth and Environment*, **37**: 2–28, https://doi.org/10.1177/0309133312461032.

Lal, R., Stewart, B.A. (Eds.), 2017. *Urban Soils*. Advances in Soil Science. CRC Press/Taylor & Francis Inc., Boca Raton, FL, USA, 416 pp.

Maco, B., Bardos, P., Coulon, F., Erickson-Mulanax, E., Hansen, L.J., Harclerode, M., . . . Wick, W.D., 2018. Resilient remediation: Addressing extreme weather and climate change, creating community value. *Remediation (New York, N.Y.)*, **29**: 7–18, https://doi.org/10.1002/rem.21585.

O'Sullivan, C.A., Bonnett, G.D., McIntyre, C.L., Hochman, Z., Wasson, A.P., 2019. Strategies to improve the productivity, product diversity and profitability of urban agriculture. *Agricultural Systems*, **174**: 133–144, https://doi.org/10.1016/j.agsy.2019.05.007.

Rawlins, B.G., Harris, J., Price, S., Bartlett, M., 2015. A review of climate change impacts on urban soil functions with examples and policy insights from England, UK. *Soil Use and Management*, **31**: 46–61, https://doi.org/10.1111/sum.12079.

12.11 Summary

- The future of urban soils will involve substantial change due to increasing urbanisation, larger urban populations, changes in climate, and changes in societal attitudes towards issues like food production and soil remediation.
- Climate change will change urban soils. There are likely to be some significant risks for the soils in cities, such as contaminant mobilisation and soil erosion, associated with climate change. There may be some positive effects as well, such as increased degradation of organic pollutants. The response of urban soils to climate change is very complex, and not enough is yet understood.
- There is not yet enough knowledge about urbanisation effects on soil biodiversity.
- Use of soils in cities for urban agriculture and urban forestry is likely to increase to meet humanity's requirements for food, sustainable water management, and climate moderation.
- Soil contamination will persist in urban environments, in many cases with new or as-yet undiscovered substances. Remediation of this contamination is likely to require thorough assessment of the complete life cycle of remediation and development projects, requiring a better knowledge of how to value urban soils.

- As humanity continues to address the Sustainable Development Goals, access to soil-based ecosystem services will remain an important component of strategies to achieve urban environmental justice.
- Knowledge about urban soils will need to reflect a functional partnership between all stakeholders including scientists and technologists, regulators and policy-makers, and indigenous and local communities.

12.12 Review and Study Questions

12.12.1 Checking Your Understanding

1. List biological, chemical, and physical changes that might be expected in urban soils as a result of climate change.
2. What are some direct and indirect effects of urban soil management on biodiversity?
3. List the possible constraints on the utilisation of urban soils for growing plants to feed humans.
4. What is meant by Water-Sensitive Urban Design? What are its advantages and disadvantages?
5. What is "legacy contamination" of urban soil? For which contaminants is it relevant, and why?
6. Why do "emerging contaminants" become apparent? List as many examples as you can of emerging contaminants which have (i) become mainstream pollutants, and (ii) are emerging now.

12.12.2 Thinking About the Topics More Deeply

7. Consider a typical urban soil remediation project. If you were asked to prepare a Life Cycle Assessment for the project, what inputs of materials and energy, and broader environmental impacts, would you need to consider?
8. What are some ways in which we could use soil knowledge to work towards achieving environmental justice in urban environments?
9. What points would you include in an argument to support ethical stewardship of urban soils?

12.12.3 Thinking About Urban Soil Remediation with Your "Left Brain"

10. In which instances might local or traditional knowledge about soils enhance a purely scientific approach in the context of urban environments?

References

Albanese S, Cicchella D (2012) Legacy problems in urban geochemistry. Elements 8:423–428. https://doi.org/10.2113/gselements.8.6.423

Albrecht H, Haider S (2013) Species diversity and life history traits in calcareous grasslands vary along an urbanization gradient. Biodivers Conserv 22:2243–2267. https://doi.org/10.1007/s10531-013-0437-0

Alcoforado MJ, Andrade H (2008) Global warming and the urban heat island. In: Marzluff J et al (eds) Urban ecology: an international perspective on the interaction between humans and nature. Springer, New York, pp 249–262. https://doi.org/10.1007/978-0-387-73412-5_14

Amoah P, Drechsel P, Abaidoo RC (2005) Irrigated urban vegetable production in Ghana: sources of pathogen contamination and health risk elimination. Irrig Drain 54:S49–S61. https://doi.org/10.1002/ird.185

Anikwe MAN, Nwobodo KCA (2002) Long term effect of municipal waste disposal on soil properties and productivity of sites used for urban agriculture in Abakaliki, Nigeria. Bioresour Technol 83:241–250. https://doi.org/10.1016/S0960-8524(01)00154-7

Appleyard S, Wong S, Willis-Jones B, Angeloni J, Watkins R (2004) Groundwater acidification caused by urban development in Perth, Western Australia: source, distribution, and implications for management. Aust J Soil Res 42:579–585

Asabere SB, Zeppenfeld T, Nketia KA, Sauer D (2018) Urbanization leads to increases in pH, carbonate, and soil organic matter stocks of arable soils of Kumasi, Ghana (West Africa). Front Environ Sci 6. https://doi.org/10.3389/fenvs.2018.00119

Barker K (2008) Flexible boundaries in biosecurity: Accomodating gorse in Aotearoa New Zealand. Environ Plan A 40:1598–1614. https://doi.org/10.1068/a4062

Barrera-Bassols N (2015) Linking ethnopedology and geopedology: A synergistic approach to soil mapping. Case study in an indigenous community of Central Mexico. In: Zinck JA, Metternicht G, Bocco G, Del Valle HF (eds) Geopedology: an integration of geomorphology and pedology for soil and landscape studies, pp 167–181. https://doi.org/10.1007/978-3-319-19159-1_9

Barrera-Bassols N, Zinck JA (2003) Ethnopedology: a worldwide view on the soil knowledge of local people. Geoderma 111:171–195. https://doi.org/10.1016/S0016-7061(02)00263-X

Berry W (1996) The unsettling of America: culture and agriculture. Sierra Club Books, San Francisco, 246 pp

Blanchart A, Consalès JN, Séré G, Schwartz C (2019) Consideration of soil in urban planning documents—a French case study. J Soils Sediments 19:3235–3244. https://doi.org/10.1007/s11368-018-2028-x

Bonthoux S, Brun M, Di Pietro F, Greulich S, Bouché-Pillon S (2014) How can wastelands promote biodiversity in cities? A review. Landsc Urban Plan 132:79–88. https://doi.org/10.1016/j.landurbplan.2014.08.010

Bouzouidja R, Rousseau G, Galzin V, Claverie R, Lacroix D, Séré G (2018) Green roof ageing or Isolatic Technosol's pedogenesis? J Soils Sediments 18:418–425. https://doi.org/10.1007/s11368-016-1513-3

Bradley P, Marulanda C (2001) Simplified hydroponics to reduce global hunger, Acta Horticulturae, pp. 289-295, https://doi.org/10.17660/actahortic.2001.554.31

Brown SL, Chaney RL, Hettiarachchi GM (2016) Lead in urban soils: a real or perceived concern for urban agriculture? J Environ Qual 45:26–36. https://doi.org/10.2134/jeq2015.07.0376

Calzolari C, Tarocco P, Lombardo N, Marchi N, Ungaro F (2020) Assessing soil ecosystem services in urban and peri-urban areas: from urban soils survey to providing support tool for urban planning. Land Use Policy 99:105037. https://doi.org/10.1016/j.landusepol.2020.105037

Cargeeg GC, Boughton GN, Townley L, Smith GR, Appleyard S, Smith RA (1987) The Perth urban water balance study, volume 1 - findings. Water Authority of Western Australia, Leederville

Carkovic AB, Calcagni MS, Vega AS, Coquery M, Moya PM, Bonilla CA, Pastén PA (2016) Active and legacy mining in an arid urban environment: challenges and perspectives for Copiapó, Northern Chile. Environ Geochem Health 38:1001–1014. https://doi.org/10.1007/s10653-016-9793-5

Chakravorty A (2019) Nitrogen from biosolids can help urban soils and plant growth, https://www.soils.org/discover-soils/story/nitrogen-from-biosolids-can-help-urban-soils-and-plant-growth (accessed 20190910)

Choi YR, Kim YN, Yoon JH, Dickinson N, Kim KH (2020) Plastic contamination of forest, urban, and agricultural soils: a case study of Yeoju City in the Republic of Korea. J Soils Sediments. https://doi.org/10.1007/s11368-020-02759-0

Cicchella D, Giaccio L, Dinelli E, Albanese S, Lima A, Zuzolo D et al (2015) GEMAS: spatial distribution of chemical elements in agricultural and grazing land soil of Italy. J Geochem Explor 154:129–142. https://doi.org/10.1016/j.gexplo.2014.11.009

Clarke BO, Smith SR (2011) Review of 'emerging' organic contaminants in biosolids and assessment of international research priorities for the agricultural use of biosolids. Environ Int 37:226–247. https://doi.org/10.1016/j.envint.2010.06.004

Colinas J, Bush P, Manaugh K (2019) The socio-environmental impacts of public urban fruit trees: a Montreal case-study. Urban For Urban Green 45:126132. https://doi.org/10.1016/j.ufug.2018.05.002

Conder, J.M., Wenning, R.J., Travers, M., Blom., M., 2010. Overview of the environmental fate of Perfluorinated compounds. In: Visser, E.-L. (Ed.), Report of the NICOLE technical meeting: emerging contaminants and solutions for large quantities of oil contaminated soil network for industrially contaminated land in Europe (NICOLE), Brussels, Belgium

Coutts AM, Tapper NJ, Beringer J, Loughnan M, Demuzere M (2013) Watering our cities: the capacity for water sensitive Urban Design to support urban cooling and improve human thermal comfort in the Australian context. Progr Phys Geogr Earth Environ 37:2–28. https://doi.org/10.1177/0309133312461032

Devito KJ, Hill AR (1999) Sulphate mobilization and pore water chemistry in relation to groundwater hydrology and summer drought in two conifer swamps on the Canadian shield. Water Air Soil Pollut 113:97–114. https://doi.org/10.1023/A:1005081505086

Dorendorf J, Eschenbach A, Schmidt K, Jensen K (2015) Both tree and soil carbon need to be quantified for carbon assessments of cities. Urban For Urban Greening 14:447–455. https://doi.org/10.1016/j.ufug.2015.04.005

Douglas PMJ, Pagani M, Eglinton TI, Brenner M, Curtis JH, Breckenridge A, Johnston K (2018) A long-term decrease in the persistence of soil carbon caused by ancient Maya land use. Nat Geosci 11:645–649. https://doi.org/10.1038/s41561-018-0192-7

Du S, Shi P, Van Rompaey A (2014) The relationship between urban sprawl and farmland displacement in the Pearl River Delta, China. Land 3:34–51. https://doi.org/10.3390/land3010034

Eckart K, McPhee Z, Bolisetti T (2017) Performance and implementation of low impact development – a review. Sci Total Environ 607-608:413–432. https://doi.org/10.1016/j.scitotenv.2017.06.254

Edmondson JL, Cunningham H, Densley Tingley DO, Dobson MC, Grafius DR, Leake JR et al (2020) The hidden potential of urban horticulture. Nat Food 1:155–159. https://doi.org/10.1038/s43016-020-0045-6

EPA, U.S. (2017) Technical fact sheet – Perfluorooctane sulfonate (PFOS) and Perfluorooctanoic acid (PFOA). EPA 505-F-17-001, United States Environmental Protection Agency, Office of

Land and Emergency Management, Washington, DC, USA. https://www.epa.gov/sites/production/files/2017-12/documents/ffrrofactsheet_contaminants_pfos_pfoa_11-20-17_508_0.pdf
Eriksen-Hamel N, Danso G (2010) Agronomic considerations for urban agriculture in southern cities. Int J Agric Sustain 8:86–93. https://doi.org/10.3763/ijas.2009.0452
European Environment Agency (2020) Indicator Assessment - Progress in management of contaminated sites, https://www.eea.europa.eu/data-and-maps/indicators/progress-in-management-of-contaminated-sites-3/assessment using data from https://esdac.jrc.ec.europa.eu/content/progress-management-contaminated-sites-europe-0 (accessed 20201026)
Favara PJ, Krieger TM, Boughton B, Fisher AS, Bhargava M (2011) Guidance for performing footprint analyses and life-cycle assessments for the remediation industry. Remediation (New York, N.Y.) 21:39–79. https://doi.org/10.1002/rem.20289
Fernández-Llamazares Á, Garteizgogeascoa M, Basu N, Brondizio ES, Cabeza M, Martínez-Alier J et al (2020) A state-of-the-art review of indigenous peoples and environmental pollution. Integr Environ Assess Manag 16:324–341. https://doi.org/10.1002/ieam.4239
Foley JA, DeFries R, Asner GP, Barford C, Bonan G, Carpenter SR et al (2005) Global consequences of land use. Science 309:570–574. https://doi.org/10.1126/science.1111772
Gómez-Sagasti MT, Hernández A, Artetxe U, Garbisu C, Becerril JM (2018) How valuable are organic amendments as tools for the phytomanagement of degraded soils? The knowns, known unknowns, and unknowns. Front Sustain Food Systems 2. https://doi.org/10.3389/fsufs.2018.00068
Greenhalgh S, Samarasinghe O, Curran-Cournane F, Wright W, Brown P (2017) Using ecosystem services to underpin cost–benefit analysis: is it a way to protect finite soil resources? Ecosyst Serv 27:1–14. https://doi.org/10.1016/j.ecoser.2017.07.005
Grimm NB, Faeth SH, Golubiewski NE, Redman CL, Wu J, Bai X, Briggs JM (2008) Global change and the ecology of cities. Science 319:756–760
Home R, Vieli L (2020) Psychosocial outcomes as motivations for urban gardening: a cross-cultural comparison of Swiss and Chilean gardeners. Urban For Urban Green 52:126703. https://doi.org/10.1016/j.ufug.2020.126703
Hosseini Bai S, Xu Z, Blumfield TJ, Reverchon F (2015) Human footprints in urban forests: implication of nitrogen deposition for nitrogen and carbon storage. J Soils Sediments 15:1927–1936. https://doi.org/10.1007/s11368-015-1205-4
IUSS Working Group WRB (2014) World Reference Base for soil resources 2014, world soil resources reports food and agriculture Organization of the United Nations, Rome
Jacobs A, Drouet T, Sterckeman T, Noret N (2017) Phytoremediation of urban soils contaminated with trace metals using Noccaea caerulescens: comparing non-metallicolous populations to the metallicolous 'Ganges' in field trials. Environ Sci Pollut Res 24:8176–8188. https://doi.org/10.1007/s11356-017-8504-9
Jacobson CR (2011) Identification and quantification of the hydrological impacts of imperviousness in urban catchments: a review. J Environ Manag 92:1438–1448. https://doi.org/10.1016/j.jenvman.2011.01.018
Jim CY (1998) Urban soil characteristics and limitations for landscape planting in Hong Kong. Landsc Urban Plan 40:235–249. https://doi.org/10.1016/S0169-2046(97)00117-5
Jim CY (2013) Sustainable urban greening strategies for compact cities in developing and developed economies. Urban Ecosyst 16:741–761. https://doi.org/10.1007/s11252-012-0268-x
Jiusto S, Kenney M (2016) Hard rain gonna fall: strategies for sustainable urban drainage in informal settlements. Urban Water J 13:253–269. https://doi.org/10.1080/1573062X.2014.991329
Johnson MS, Lathuillière MJ, Tooke TR, Coops NC (2015) Attenuation of urban agricultural production potential and crop water footprint due to shading from buildings and trees. Environ Res Lett 10. https://doi.org/10.1088/1748-9326/10/6/064007
Jones D, Healey J (2010) Organic amendments for remediation: putting waste to good use. Elements 6:369–374. https://doi.org/10.2113/gselements.6.6.369
Kabisch N, Haase D (2014) Green justice or just green? Provision of urban green spaces in Berlin, Germany. Landsc Urban Plan 122:129–139. https://doi.org/10.1016/j.landurbplan.2013.11.016

Kaoma H, Shackleton CM (2014) Collection of urban tree products by households in poorer residential areas of three South African towns. Urban For Urban Green 13:244–252. https://doi.org/10.1016/j.ufug.2014.02.002

Karpuzcu ME, Fairbairn D, Arnold WA, Barber BL, Kaufenberg E, Koskinen WC et al (2014) Identifying sources of emerging organic contaminants in a mixed use watershed using principal components analysis. Environ Sci Processes Impacts 16:2390–2399. https://doi.org/10.1039/c4em00324a

Kelly J, Thornton I, Simpson PR (1996) Urban geochemistry: a study of the influence of anthropogenic activity on the heavy metal content of soils in traditionally industrial and non-industrial areas of Britain. Appl Geochem 11:363–370. https://doi.org/10.1016/0883-2927(95)00084-4

Kessler R (2013) Urban gardening: managing the risks of contaminated soil.(news focus). Environ Health Persp 121:A326. https://doi.org/10.1289/ehp.121-A326

Knapp S, Kühn I, Bakker JP, Kleyer M, Klotz S, Ozinga WA et al (2009) How species traits and affinity to urban land use control large-scale species frequency. Divers Distrib 15:533–546. https://doi.org/10.1111/j.1472-4642.2009.00561.x

Kolosz B, Sohi S, Manning D (2019) CASPER: a modelling framework to link mineral carbonation with the turnover of organic matter in soil. Comput Geosci 124:58–71. https://doi.org/10.1016/j.cageo.2018.12.012

Komlos J, Traver RG (2012) Long-term orthophosphate removal in a field-scale storm-water bioinfiltration rain garden. J Environ Eng 138:991–998. https://doi.org/10.1061/(ASCE)EE.1943-7870.0000566

Kozdrój J, Van Elsas JD (2001) Structural diversity of microorganisms in chemically perturbed soil assessed by molecular and cytochemical approaches. J Microbiol Methods 43:197–212. https://doi.org/10.1016/S0167-7012(00)00197-4

Laidlaw MAS, Alankarage DH, Reichman SM, Taylor MP, Ball AS (2018) Assessment of soil metal concentrations in residential and community vegetable gardens in Melbourne, Australia. Chemosphere 199:303–311. https://doi.org/10.1016/j.chemosphere.2018.02.044

Lal R (2017) Urban agriculture in the 21st century. In: Lal R, Stewart BA (eds) Urban Soils. Advances in soil science. CRC Press/Taylor & Francis Inc., Boca Raton, pp 1–13

Li G, Sun G-X, Ren Y, Luo X-S, Zhu Y-G (2018) Urban soil and human health: a review. Eur J Soil Sci 69:196–215. https://doi.org/10.1111/ejss.12518

Low T (2003) The new nature. The new nature : winners and losers in wild Australia. Penguin, Camberwell, Vic

Lupia F, Baiocchi V, Lelo K, Pulighe G (2017) Exploring rooftop rainwater harvesting potential for food production in urban areas. Agriculture (Switzerland) 7. https://doi.org/10.3390/agriculture7060046

Lutz N, Fogarty J, Rate A (2021) Accumulation and potential for transport of microplastics in stormwater drains into marine environments, Perth region, Western Australia. Mar Pollut Bull 168:112362. https://doi.org/10.1016/j.marpolbul.2021.112362

Maas S, Scheifler R, Benslama M, Crini N, Lucot E, Brahmia Z et al (2010) Spatial distribution of heavy metal concentrations in urban, suburban and agricultural soils in a Mediterranean city of Algeria. Environ Pollut 158:2294–2301. https://doi.org/10.1016/j.envpol.2010.02.001

Maco B, Bardos P, Coulon F, Erickson-Mulanax E, Hansen LJ, Harclerode M et al (2018) Resilient remediation: Addressing extreme weather and climate change, creating community value. Remediation (New York, N.Y.) 29:7–18. https://doi.org/10.1002/rem.21585

Mahmood A, Malik RN, Syed JH, Li J, Zhang G (2015) Dietary exposure and screening-level risk assessment of polybrominated diphenyl ethers (PBDEs) and dechloran plus (DP) in wheat, rice, soil and air along two tributaries of the River Chenab, Pakistan. Chemosphere 118:57–64. https://doi.org/10.1016/j.chemosphere.2014.05.071

Margenat A, Matamoros V, Díez S, Cañameras N, Comas J, Bayona JM (2018) Occurrence and bioaccumulation of chemical contaminants in lettuce grown in peri-urban horticulture. Sci Total Environ 637-638:1166–1174. https://doi.org/10.1016/j.scitotenv.2018.05.035

Martellozzo F, Landry J-S, Plouffe D, Seufert V, Rowhani P, Ramankutty N (2014) Urban agriculture: a global analysis of the space constraint to meet urban vegetable demand. Environ Res Lett 9. https://doi.org/10.1088/1748-9326/9/6/064025

Mawois M, Le Bail M, Navarrete M, Aubry C (2012) Modelling spatial extension of vegetable land use in urban farms. Agron Sustain Dev 32:911–924. https://doi.org/10.1007/s13593-012-0093-x

McDonald RI, Mansur AV, Ascensão F, Colbert MI, Crossman K, Elmqvist T et al (2019) Research gaps in knowledge of the impact of urban growth on biodiversity. Nat Sustain. https://doi.org/10.1038/s41893-019-0436-6

McLain RJ, Hurley PT, Emery MR, Poe MR (2014) Gathering "wild" food in the city: rethinking the role of foraging in urban ecosystem planning and management. Local Environ Int J Justice Sustain 19:220–240. https://doi.org/10.1080/13549839.2013.841659

Mguni P, Herslund L, Jensen MB (2016) Sustainable urban drainage systems: examining the potential for green infrastructure-based stormwater management for sub-Saharan cities. Nat Hazards 82:241–257. https://doi.org/10.1007/s11069-016-2309-x

Mielke HW, Laidlaw MAS, Gonzales C (2010) Lead (Pb) legacy from vehicle traffic in eight California urbanized areas: continuing influence of lead dust on children's health. Sci Total Environ 408:3965–3975. https://doi.org/10.1016/j.scitotenv.2010.05.017

Mitchell RG, Spliethoff HM, Ribaudo LN, Lopp DM, Shayler HA, Marquez-Bravo LG et al (2014) Lead (Pb) and other metals in new York City community garden soils: factors influencing contaminant distributions. Environ Pollut 187:162–169

Morris JG (1966) The use of soils information in urban planning and implementation. In: Bartelli LJ, Klingebiel AA, Baird JV, Heddleson MR (eds) Soil surveys and land use planning. Soil Science Society of America and American Society of Agronomy, Madison, pp 37–41

Morton-Bermea O, Rodríguez-Salazar MT, Hernández-Alvarez E, García-Arreola ME, Lozano-Santacruz R (2011) Lead isotopes as tracers of anthropogenic pollution in urban topsoils of Mexico City. Chem Erde 71:189–195. https://doi.org/10.1016/j.chemer.2011.03.003

Musakwa W, Wang S, Wei F, Malapane OL, Thomas MM, Mavengahama S et al (2020) Survey of community livelihoods and landscape change along the Nzhelele and Levuvhu river catchments in Limpopo province. South Africa Land 9. https://doi.org/10.3390/land9030091

Nadal M, Marquès M, Mari M, Domingo JL (2015) Climate change and environmental concentrations of POPs: a review. Environ Res 143:177–185. https://doi.org/10.1016/j.envres.2015.10.012

Noyes PD, McElwee MK, Miller HD, Clark BW, Van Tiem LA, Walcott KC et al (2009) The toxicology of climate change: environmental contaminants in a warming world. Environ Int 35:971–986. https://doi.org/10.1016/j.envint.2009.02.006

Orsini F, Kahane R, Nono-Womdim R, Gianquinto G (2013) Urban agriculture in the developing world: a review. Agron Sustain Dev 33:695–720. https://doi.org/10.1007/s13593-013-0143-z

O'Sullivan CA, Bonnett GD, McIntyre CL, Hochman Z, Wasson AP (2019) Strategies to improve the productivity, product diversity and profitability of urban agriculture. Agric Syst 174:133–144. https://doi.org/10.1016/j.agsy.2019.05.007

Paltseva AA, Neaman A (2020) An emerging frontier: Metal(loid) soil pollution threat under global climate change. Environ Toxicol Chem 39:1653–1654. https://doi.org/10.1002/etc.4790

Paul MJ, Meyer JL (2001) Streams in the urban landscape. Annu Rev Ecol Syst 32:333–365

Pauli N, Abbott LK, Negrete-Yankelevich S, Andrés P (2016) Farmers' knowledge and use of soil fauna in agriculture: a worldwide review. Ecol Soc 21:19. https://doi.org/10.5751/ES-08597-210319

Pavao-Zuckerman MA, Coleman DC (2007) Urbanization alters the functional composition, but not taxonomic diversity, of the soil nematode community. Appl Soil Ecol 35:329–339. https://doi.org/10.1016/j.apsoil.2006.07.008

Pickering AJ, Julian TR, Marks SJ, Mattioli MC, Boehm AB, Schwab KJ, Davis J (2012) Fecal contamination and diarrheal pathogens on surfaces and in soils among Tanzanian households with and without improved sanitation. Environ Sci Technol 46:5736–5743. https://doi.org/10.1021/es300022c

Pless-Mulloli T, Air V, Vizard C, Singleton I, Rimmer D, Hartley P (2004) The legacy of historic land-use in allotment gardens in industrial urban settings: Walker road allotment in Newcastle upon Tyne, UK. Land Contamination Reclamation 12:239–251. https://doi.org/10.2462/09670513.638

Pouyat R, Groffman P, Yesilonis I, Hernandez L (2002) Soil carbon pools and fluxes in urban ecosystems. Environ Pollut 116:S107–S118. https://doi.org/10.1016/S0269-7491(01)00263-9

Pouyat RV, Yesilonis ID, Nowak DJ (2006) Carbon storage by urban soils in the United States. J Environ Qual 35:1566–1575. https://doi.org/10.2134/jeq2005.0215

Qu C, Albanese S, Lima A, Hope D, Pond P, Fortelli A et al (2019) The occurrence of OCPs, PCBs, and PAHs in the soil, air, and bulk deposition of the Naples metropolitan area, southern Italy: implications for sources and environmental processes. Environ Int 124:89–97. https://doi.org/10.1016/j.envint.2018.12.031

Rafique A, Irfan M, Mumtaz M, Qadir A (2020) Spatial distribution of microplastics in soil with context to human activities: a case study from the urban center. Environ Monit Assess 192. https://doi.org/10.1007/s10661-020-08641-3

Ramirez KS, Leff JW, Barberán A, Bates ST, Betley J, Crowther TW et al (2014) Biogeographic patterns in below-ground diversity in new York City's Central Park are similar to those observed globally. Proc R Soc B Biol Sci, 281. https://doi.org/10.1098/rspb.2014.1988

Rawlins BG, Harris J, Price S, Bartlett M (2015) A review of climate change impacts on urban soil functions with examples and policy insights from England, UK. Soil Use Manag 31:46–61. https://doi.org/10.1111/sum.12079

Redd JM, Jacobs H, Halliday SS (2020) The evolving landscape of environmental justice in 2020 and beyond. Beveridge & Diamond, New York, USA https://www.bdlaw.com/publications/the-evolving-landscape-of-environmental-justice-in-2020-and-beyond/ (accessed 20201123)

Rodríguez-Eugenio N, McLaughlin M, Pennock D (2018) Soil pollution - a hidden reality, food and agriculture Organization of the United Nations Rome, Italy. http://www.fao.org/3/I9183EN/i9183en.pdf

Rouillon M, Harvey PJ, Kristensen LJ, George SG, Taylor MP (2017) VegeSafe: a community science program measuring soil-metal contamination, evaluating risk and providing advice for safe gardening. Environ Pollut 222:557–566. https://doi.org/10.1016/j.envpol.2016.11.024

Saad R, Margni M, Koellner T, Wittstock B, Deschênes L (2011) Assessment of land use impacts on soil ecological functions: development of spatially differentiated characterization factors within a Canadian context. Int J Life Cycle Assess 16:198–211. https://doi.org/10.1007/s11367-011-0258-x

Santorufo L, Van Gestel CAM, Rocco A, Maisto G (2012) Soil invertebrates as bioindicators of urban soil quality. Environ Pollut 161:57–63. https://doi.org/10.1016/j.envpol.2011.09.042

Sauvé S, Desrosiers M (2014) A review of what is an emerging contaminant. Chem Cent J 8:15–15. https://doi.org/10.1186/1752-153X-8-15

Schneider A, Logan KE, Kucharik CJ (2012) Impacts of urbanization on ecosystem goods and services in the U.S. corn belt. Ecosystems 15:519–541. https://doi.org/10.1007/s10021-012-9519-1

Schucker GW, Vail EH, Kelley EB, Kaplan E (1965) Prevention of lead paint poisoning among Baltimore children. Hard-Sell Program Pub Health Rep 80:969–974. https://doi.org/10.2307/4592586

Sellami S, Zeghouan O, Lassaad M, Moussaoui Y, Kebabi B (2020) Determination of lead concentrations in the soils of Setif City, Eastern Algeria. Arab J Geosci 13:929. https://doi.org/10.1007/s12517-020-05977-5

Shwartz A, Muratet A, Simon L, Julliard R (2013) Local and management variables outweigh landscape effects in enhancing the diversity of different taxa in a big metropolis. Biol Conserv 157:285–292. https://doi.org/10.1016/j.biocon.2012.09.009

Simon JA (2020) Editor's perspective—U.S. EPA's 2020 superfund remedy report shows distinctive trends in remediation technologies. Remediation (New York, N.Y.) 30:3–5. https://doi.org/10.1002/rem.21665

Steffen W, Richardson K, Rockström J, Cornell SE, Fetzer I, Bennett EM et al (2015) Planetary boundaries: guiding human development on a changing planet. Science 347. https://doi.org/10.1126/science.1259855

Stewart GH, Meurk CD, Ignatieva ME, Buckley HL, Magueur A, Case BS et al (2009) URban biotopes of Aotearoa New Zealand (URBANZ) II: floristics, biodiversity and conservation values of urban residential and public woodlands, Christchurch. Urban For Urban Greening 8:149–162. https://doi.org/10.1016/j.ufug.2009.06.004

Taylor JR, Lovell ST (2015) Urban home gardens in the global north: a mixed methods study of ethnic and migrant home gardens in Chicago, IL. Renewable Agric Food Sys 30:22–32. https://doi.org/10.1017/S1742170514000180

Teixeira da Silva R, Fleskens L, van Delden H, van der Ploeg M (2018) Incorporating soil ecosystem services into urban planning: status, challenges and opportunities. Landsc Ecol 33:1087–1102. https://doi.org/10.1007/s10980-018-0652-x

Teuber S, Schmidt K, Kühn P, Scholten T (2019) Engaging with urban green spaces – a comparison of urban and rural allotment gardens in Southwestern Germany. Urban For Urban Greening 43. https://doi.org/10.1016/j.ufug.2019.126381

Turnbull R, Rogers K, Martin A, Rattenbury M, Morgan R (2019) Human impacts recorded in chemical and isotopic fingerprints of soils from Dunedin City, New Zealand. Sci Total Environ 673:455–469. https://doi.org/10.1016/j.scitotenv.2019.04.063

U.S. EPA (2019) Chemicals and toxics topics. United States Environmental Protection Agency, Arlington, VA, USA, https://www.epa.gov/environmental-topics/chemicals-and-toxics-topics (accessed 20190906)

United Nations (2015) Transforming our world: the 2030 agenda for sustainable development. Resolution 70/1 UN general assembly, New York, USA. https://www.un.org/ga/search/view_doc.asp?symbol=A/RES/70/1&Lang=E

United Nations (2018) World urbanization prospects: the 2018 revision. Custom data acquired via https://population.un.org/

Uno S, Cotton J, Philpott SM (2010) Diversity, abundance, and species composition of ants in urban green spaces. Urban Ecosyst 13:425–441. https://doi.org/10.1007/s11252-010-0136-5

Urrutia-Goyes R, Argyraki A, Ornelas-Soto N (2017) Assessing lead, nickel, and zinc pollution in topsoil from a historic shooting range rehabilitated into a public urban park. Int J Environ Res Public Health 14. https://doi.org/10.3390/ijerph14070698

Useni Sikuzani Y, Malaisse F, Cabala Kaleba S, Kalumba Mwanke A, Yamba AM, Nkuku Khonde C et al (2019) Tree diversity and structure on green space of urban and peri-urban zones: the case of Lubumbashi City in the Democratic Republic of Congo. Urban For Urban Green 41:67–74. https://doi.org/10.1016/j.ufug.2019.03.008

Vergnes A, Blouin M, Muratet A, Lerch TZ, Mendez-Millan M, Rouelle-Castrec M, Dubs F (2017) Initial conditions during Technosol implementation shape earthworms and ants diversity. Landsc Urban Plan 159:32–41. https://doi.org/10.1016/j.landurbplan.2016.10.002

Wang Z, Cui X, Yin S, Shen GR, Han YJ, Liu CJ (2013) Characteristics of carbon storage in Shanghai's urban forest. Chin Sci Bull 58:1130–1138. https://doi.org/10.1007/s11434-012-5443-1

Wendel HEW, Downs JA, Mihelcic JR (2011) Assessing equitable access to urban green space: the role of engineered water infrastructure. Environ Sci Technol 45:6728–6734. https://doi.org/10.1021/es103949f

Werner S, Akoto-Danso EK, Manka'abusi D, Steiner C, Haering V, Nyarko G et al (2019) Nutrient balances with wastewater irrigation and biochar application in urban agriculture of northern Ghana. Nutr Cycl Agroecosyst. https://doi.org/10.1007/s10705-019-09989-w

Whelans C, Halpern-Glick-Maunsell Pty. Ltd., Thompson-Palmer (1994) Planning and management guidelines for water sensitive urban (residential) design, report for the Department of Planning and Urban Development, Perth, Western Australia

Winklerprins AMGA (1999) Insights and applications local soil knowledge: a tool for sustainable land management. Soc Nat Resour 12:151–161. https://doi.org/10.1080/089419299279812

Wong THF (2006) Water sensitive urban design - the journey thus far. Aus J Water Resour 10:213–222. https://doi.org/10.1080/13241583.2006.11465296
Wortman SE, Lovell ST (2013) Environmental challenges threatening the growth of urban agriculture in the United States. J Environ Qual 42:1283–1294. https://doi.org/10.2134/jeq2013.01.0031
Xia X, Chen X, Liu R, Liu H (2011) Heavy metals in urban soils with various types of land use in Beijing, China. J Hazard Mater 186:2043–2050. https://doi.org/10.1016/j.jhazmat.2010.12.104
Zaleski A (2020) The unequal burden of urban Lead. Bloomberg CityLab https://www.bloomberg.com/news/articles/2020-01-02/undoing-the-legacy-of-lead-poisoning-in-america (accessed 20201123)
Zhang G-L, Yang F-G, Zhao Y-G, Zhao W-J, Yang J-L, Gong Z-T (2005) Historical change of heavy metals in urban soils of Nanjing, China during the past 20 centuries. Environ Int 31:913–919. https://doi.org/10.1016/j.envint.2005.05.035
Zhou Q (2014) A review of sustainable urban drainage systems considering the climate change and urbanization impacts. Water (Switzerland) 6:976–992. https://doi.org/10.3390/w6040976
Zissimos AM, Cohen DR, Christoforou IC (2018) Land use influences on soil geochemistry in Lefkosia (Nicosia) Cyprus. J Geochem Explor 187:6–20. https://doi.org/10.1016/j.gexplo.2017.03.005

Glossary

A

A priori A Latin phrase meaning 'from the former', often used to mean 'independent of later or additional information'

Acidification (soil acidification) The addition of acidic substances (containing either free or un-dissociated H^+), resulting in decreased soil pH and less acid buffering

Adsorption The accumulation of ions or molecules at the surface of a colloid by any mechanism (see *chemisorption*, and *exchangeable cations*)

Aeolian In soils, material transported and/or deposited by wind

Aerosol Solid or liquid particles suspended in air

Aggregates (peds, soil aggregates) Cohesive fragments of soil composed of many individual grains held together by a range of forces and separated by pores and planes of weakness

Aliphatic Organic compounds or fragments with only linear, branched, or non-*aromatic* ring structures

Alkane(s) Organic compounds or fragments of only C and H, where C atoms are connected in linear or branched geometry with single bonds

Alluvial Relating to rivers (e.g. an alluvial sediment has been transported by a river and deposited on a river floodplain)

Amino acids Simple organic compounds containing both $–NH_2$ and $–COOH$ functional groups, 22 of which combine to form proteins

Amphoteric (Of a chemical substance) able to accept or release hydrogen ions, i.e., act as either an acid or a base

Amplicons A fragment of DNA or RNA which is the source or product of natural or artificial amplification, e.g. by PCR

Annotate (in DNA analysis) To identify the locations and functions of genes, which contain biologically significant information, on a DNA sequence or genome

A. W. Rate (ed.), *Urban Soils*, Progress in Soil Science,
https://doi.org/10.1007/978-3-030-87316-5

Anoxic In the absence of oxygen, anoxic conditions occur in saturated soils in the presence of any decomposable organic compounds

Anthropocene The most recent geological period, defined by profound human impact on Earth systems

Anthropogenic Generated by human activity

Anthroposequence A *soil sequence* in a landscape with varying properties due mainly to a gradient in human influence (e.g., urbanisation)

Anthrosol, Anthroposol A soil having human influence(s) dominating its origins or formation processes

Aromatic Containing 5- or 6-carbon ring structures with delocalised (conjugated) electron bonding (relating to organic chemical compounds)

Artefact An object (or fragment of one) created by humans

Atmosphere The Earth sub-system comprised of gases above or near the Earth's surface

Attenuation Decreases in the concentration and/or bioavailability of contaminant(s) by natural or (un)intentional anthropogenic processes

Atterberg limits Soil water content thresholds between soil acting as a brittle solid, elastic solid, plastic solid, or viscous liquid

B

Background concentration(s) The concentration of a potential contaminant substance in an uncontaminated soil, against which a valid comparison can be made with a potentially contaminated soil

Basin, geological A major area of sedimentary rocks which is a current or historical sink for terrestrial sediment (may be terrestrial or submarine)

Binned (in DNA analysis) In relation to sub-sequences of DNA, grouped by fragment length and/or composition

Bioaccessible In relation to substances in soils, capable of being released into a bioavailable form

Bioaccumulate (bioaccumulation, biomagnification) (Of pollutants) To increase in amount/concentration in an organism, especially organisms higher in food chains; caused by faster intake than catabolism or excretion

Bioavailable (bioavailability) Immediately able to be taken up or to interact with organisms

Biochar A carbon-rich, solid material produced by thermal decomposition of organic material in the absence, or limited supply, of oxygen

Bioconcentration Accumulation of a substance (e.g. pollutant) in living biological tissue at concentrations greater than in the growth medium (e.g. soil)

Biodegradation Conversion of organic pollutant compounds to non-toxic products by microorganisms, which use the *pollutant* as a substrate

Biodiversity The range or number of different types (species, genotypes) of organisms in an environment or environmental compartment

Biogeochemical cycle The various transformations and fluxes by which an element moves within and between environmental compartments

Bioindicator An organism which has a measureable response to contamination or other change in an environment

Bioinformatics The acquisition, storage, analysis, and dissemination of biological data, specifically genetic (i.e. DNA and RNA) and protein (i.e. amino acid) sequence information

Biomagnification See *Bioaccumulate*

Bioremediation Degradation, transformation to non-toxic products, or uptake of contaminants by soil microorganisms or plants

Bioswale A shallow excavation in permeable material, which is vegetated and is designed for stormwater infiltration

Bioturbation Mixing or churning of soil by organisms

Black carbon Pyrogenic carbon-rich material comprising partially combusted solid materials such as soot or finely-divided charcoal

Brownfields Contaminated sites prior to any remediation (often vacant, disused industrial land)

C

Carbon footprint The equivalent amount of CO_2 released into the atmosphere as a result of a human activity (see *Life cycle assessment*)

Carbon sequestration Removal of atmospheric carbon (CO_2 or CH_4) into another ecosystem compartment (e.g. soil, trees)

Carboxylates Organic molecules containing the fragment ('functional group') –COOH which is a weak acid and dissociates to $-COO^- + H^+$

Chemical composition [Of a soil] The concentrations of chemical elements and compounds in a soil

Chemisorption The accumulation of ions or molecules at the surface of a colloid by formation of electron-sharing chemical bonds

Chronosequence A sequence of soils with similar soil-forming factors (parent material, climate, etc.), except having started forming at different times

Clay mineral Minerals in the *phyllosilicate* group

Co-metabolism Conversion of organic pollutant compounds to different compounds by microorganisms, while metabolizing a *non-pollutant substrate*

Colloid (colloidal) In soils, solid particles in the clay-sized range or smaller ($\leq 2 \times 10^6$ m)

Colony-forming units (CFUs) The number of microorganisms in a sample which form a colony on culture media, used for counting populations in an environment

Competition In ecology, an interaction between organisms in relation to scarce resource(s) which results in harm (e.g. population decrease) to one or both organisms

Composting Biological degradation and stabilization of organic waste materials for re-use, by incubation in conditions favouring aerobic mesophilic bacteria

Conceptual site model A formal description of a contaminated site which considers source and receptors of the contaminant(s), and the mechanisms for contaminant transfer

Contamination The presence of any substance in an environmental compartment (such as soil) in amounts of concentrations that could harm ecosystem or human health

Contigs Contiguous sequences of DNA constructed from a set of overlapping DNA sub-sequences. The outcome of *binning*

Coordination complex A molecule or molecular fragment formed by sharing of a donor 'ligand' atom's non-bonded electron pair with another atom (commonly a metal ion)

Critical zone The zone at the Earth's surface from the lowest accessible groundwater to the upper vegetation canopy responsible for sustaining terrestrial life

Cultural layer An anthropogenic soil horizon which contains artefacts derived from human occupation and disposal of materials

D

Denitrification The biochemical reduction of nitrogen species under oxygen-deficient conditions, ultimately leading to loss of dinitrogen gas

Depth profile The relationship between the amount or concentration of a soil component (including contaminants) and soil depth

Diffuse source(s) Sources of contamination which are not limited to a discrete location (e.g. road traffic, lawn fertilisers)

Diversity The range or number of different types of items in a category, including their abundance or *evenness* (see *biodiversity*)

Domains The highest rank of taxonomic classification of organisms: the Archaea, Bacteria, and Eukarya (formerly called superkingdoms)

Dredge spoil Sediment removed from submerged freshwater or marine environments to aid water flow or boat transport

E

e-waste Electronic and electrical waste (also called techwaste) – discarded appliances such as mobile phones, computers, televisions, etc.

Ecosystem engineers Organisms that create a measurable change in their habitat or environment

Ecosystem services The benefits for humans and other organisms derived from the natural environment and from properly functioning ecosystems

Effect size A standardised measure of (1) the difference between means, or (2) the relationship between variables

Environmental compartment A subset or sub-system of any environment being studied

Environmental DNA (eDNA) DNA collected for analysis from an environmental compartment such as soil or water, rather than directly from an organism

Environmental guideline A threshold concentration of a contaminant that triggers a regulatory action

Environmental justice Fair treatment of all people with respect to access to a healthy environment, and the development, implementation, and enforcement of environmental legislation

Enzymes Biochemicals (often proteins) which act as catalysts to regulate the rate of biochemical reactions

Eutrophic Having much greater than adequate concentrations of nutrients (in water or soil)

Evenness The similarity of population sizes or abundances of different organisms in an environment (see *richness*)

Exchangeable cations Cations held on negatively charged soil colloids by electrostatic forces

Exploratory data analysis The use of statistical and graphical methods to discover [often unexpected] patterns, trends, and relationships in a dataset

Extracellular polymeric substances Polymeric compounds (such as polysaccharides or proteins), produced and exuded by various microorganisms

F

Fine earth The soil material excluding grains greater than 2 mm in diameter (e.g. gravel)

Flagella Tiny hair-like organelles which can oscillate to provide motion for microorganisms

Footprint analysis An accounting of the ongoing resource requirements and environmental impacts of a human activity (e.g. soil remediation), often expressed as a land area

Fractionation (Of chemical substances) Changes in the relative amounts of related substances such as isotopes, the rare-earth elements, or PAHs due to environmental processes

Free ion activity model A theory of biological uptake where the amount of uptake of an element is related mainly to the activity (concentration) of its free ions in solution

Functional group(s) Specific arrangements of atoms in the structure of organic molecules or mineral crystal structures

G

Geochemical Relating to the chemical composition of materials that make up the Earth, including chemical reactions and cycles

Geochemical signature In soils: a chemical composition of soil (combination of concentrations of chemical elements) that is characteristic of a certain type of environment

Geogenic Derived from the soil parent material (referring to concentrations of a potential contaminant); not added to soil by human activity

Geomembrane (geotextile) A low-permeability manufactured polymer or textile sheet used to limit the transfer of liquids or gases in soil engineering. Sometimes refers to a porous textile used to suppress weeds or reduce erosion

Geomorphology The landforms or 'shape' of the Earth's surface, and the study of landforms and the phenomena which shape them

Green infrastructure Vegetation, soils, in constructions deliberately used to restore natural processes for water management, climatic cooling, and human health in urban environments

Green roof A soil substrate with planted vegetation installed on the roof of a building

Green technology Technology designed, or considered, to be less harmful to the environment than 'conventional' technology

H

Half-life The time taken for half of a reactant in a (bio)chemical reaction to be consumed; often applied to the time taken to halve the amount of an organic pollutant remaining in soil

Heat island (urban heat island) The warmer microclimate in cities due to absorption of solar radiation and emission of heat by constructed surfaces, and greater greenhouse gas concentrations in cities

Heterogeneity Variation in properties, such as composition, of an environmental compartment

Horizon (soil horizon) A recognizable layer in a soil *profile* caused by soil-forming processes

Hotspot (hot spot) An area in soil that is localised, or small, relative to the scale of observation in which the concentration of contaminant(s) is high (e.g. exceeds an environmental guideline)

Hydrosphere The sub-system of the Earth comprised of free water in any phase

Hyperaccumulator A plant species which is able to take up high concentrations of potentially toxic elements (e.g. As, Cd, Pb, Zn)

I

Immobilization The uptake of inorganic forms of elements (N, P, S) by soil microbiota and reincorporation into organic molecules

Isotope An atom with a specific atomic mass (the same element can have different mass atoms due to different neutron numbers, e.g. carbon-12, carbon-13, carbon-14)

K

Köppen-Geiger A widely used global climate classification system (sometimes just called Köppen)

L

Life cycle assessment An assessment accounting for environmental impacts associated with all stages in the life of a project, process, or product

Ligand An ion, molecule, or functional group that can form a coordination complex with chemical bonds to a central metal ion

Ligand exchange Adsorption of anions ('ligands') which displace another ligand such as OH^- already coordinated to a metal ion at the surface of an oxide or clay mineral structure

Lithosphere The sub-system of the Earth comprised of rocks and related materials

Lysimeter A device such as a specialised soil core or subsoil installation for measuring leaching of water or solutes in soil

M

Macronutrient The elements N, P, K, S, Ca, and Mg which are required by organisms in concentrations $\gtrsim 0.1\%$ by dry weight

Macropore Relatively large continuous soil pores, caused by root penetration, animal burrows, cracking, buried infrastructure, etc.

Magnetic susceptibility The extent to which a material will become magnetised in an applied magnetic field

Managed Aquifer Recharge (MAR) Deliberate addition of water, including stormwater or wastewater, to groundwater via infiltration structures or injection wells

Metabolites Products of biochemical (e.g. degradation) processes derived from primary substrates such as pollutants (e.g. DDE from DDT)
Metagenomics The study of a collection of genetic material (genomes) from a mixed community of organisms and/or from environmental samples
Metatranscriptomics The study of gene expression from a mixed community of organisms and/or from environmental samples by RNA sequencing
Microbial respiration Utilization of carbon compounds by microorganisms to produce energy and CO_2
Micronutrient The elements Fe, Mn, Zn, Cu, B, Mo, Cl, Ni Co, Cr, I, and Se which are required by organisms in concentrations $\lesssim 0.1\%$ by dry weight
Mineralization Release of elements contained in organic molecules (e.g. N, P, S) as inorganic forms during soil organic matter decomposition

N

Nitrification The oxidation of ammonium under oxygen-sufficient conditions to produce nitrate (through a nitrite intermediate) by bacteria and archaea
Nucleic acid A biopolymer such as DNA or RNA, composed off linked chains of nucleotides, which stores genetic information

O

Ordination analysis A numerical/statistical method which calculates new variables which contain information from multiple actual variables, such as factor, principal component, dissimilarity, or correspondence analyses
Organic amendment An organic (carbon-based) material added to soils to improve chemical and/or physical properties, and/or add nutrients
Oversaturation Concentrations of the ions in a solution, making up a potential solid, which exceed that solid's solubility product

P

PAH Polycyclic aromatic hydrocarbon (a class of persistent organic pollutants)
Parent material The starting material for a soil, such as rock, sediment, landfill, etc.
PCB Poly-chlorinated biphenyl (a class of persistent organic pollutants)
Pedogenesis The process of formation of soils which differentiates it from its parent material(s)

Pedology The study of soils in their [natural] environment: soil formation, soil characteristics, and soil distribution

Peds See *aggregates*

Perched water table A layer of saturated soil caused by a low-permeability layer (*e.g*. Fine-textured or compact soil, or a barrier) underneath

Peri-urban Within an urban-rural transition zone; the 'urban fringe'

Persistent organic pollutant(s) (POP(s)) Toxic organic chemicals which do not degrade readily in the environment

Photosynthesis The process by which plants synthesise more complex carbon compounds from CO_2 captured from the atmosphere, and soil water, using energy from sunlight

Phyllosilicate Aluminosilicate minerals characterised by a chemical structure comprising sheets of interlinked Al-O(-OH) octahedra and Si-O tetrahedra

Phylum (plural Phyla) A high-level taxonomic category that ranks above class and below Domain

Phytoextraction A form of phytoremediation where plant uptake and biomass removal decreases contaminant concentrations in soil

Phytolith A mineral particle formed within a living plant, tens of μm in size, which can persist in soils for millennia and preserve (historical) information on plant communities

Phytomining A proposed form of phytoremediation where the extraction of metals from plant residues is economically viable

Phytoremediation The use of green plants to treat soil contamination

Phytostabilisation A form of phytoremediation where growing plants decreases contaminant mobility or bioavailability in soil

Point source(s) Sources of contamination which are traceable to a discrete location (e.g. Smelters, chemical spills)

Pollution Introduction of undesirable substance(s) into and environment or environmental compartment

Porosity The space between grains and aggregates of solid material in a soil, usually expressed as a fractional (*e.g*. Percentage) volume

Predation In ecology, an interaction between organisms in which one organism kills and consumes another

Preferential flow Flow of soil water and solutes in zones of soil with greater hydraulic conductivity such as *macropores*

Profile (soil profile) The vertical cross-section of a soil differentiated into separate *horizons*

Protein A biopolymer composed off linked chains of amino acids which has a specific function within organisms

Priming effect An increase in *mineralization* of C and nutrients in pre-existing soil organic matter caused by addition of fertiliser or an *organic amendment*

Pyrogenic Generated by fire (see Black carbon)

R

Receiving environment An environmental compartment (e.g. Surface water) which receives material (*e.g.* Contaminants) from another (e.g. Soil)

Redox potential Reduction-oxidation potential: the ability of soil to act as a chemically reducing or oxidizing environment (more accurately, the ability of a soil to lose or accept electrons in chemical reactions)

Regolith The unconsolidated material overlying unweathered rock (may be developed in place, or deposited as a sediment)

Rhizosphere Soil adjacent to plant roots which is influenced by root exudation and specific root-related microorganisms

Ribosomes Complex intracellular structure which translates genetic code into chains of amino acids to form proteins

Richness See *species richness*

Root exudates A range of organic compounds released by plant roots into soil e.g. organic acids, *biopolymers*, root-cell material

S

Secondary mineral A mineral formed in the soil environment by chemical weathering or other mechanism

Sequencing Determination of the sequence of nucleotides (i.e. the order of base pairs) within nucleic acids such as DNA and RNA

Smectite (smectite clay) A group of related 2:1 phyllosilicate clay minerals having 2 SiO_n and 1 $Al[OOH]_n$ layers in their structure, high cation exchange capacity, and which swell when wet

Soil classification A system for categorizing soils based on their properties and [assumed] formation processes

Soil engineers *Ecosystem engineers*, in the specific context of soils, which change soil properties by modifying their habitats

Soil fluxes Changes in the location or composition of material in soils, such as additions, losses, transformations, and translocations

Soil pH The acidic or alkaline property of soil, defined as $-\log_{10}(H^+)$, where (H^+) is hydrogen ion activity in the soil pore water (*soil solution*)

Soil physical conditions Soil properties related to states of matter, arrangement of solid particles and pores, internal and external forces, energy balance, etc. (e.g. density, water retention)

Soil remediation Improvement of soil health to make it suitable for a desired purpose (may include contaminant removal, revegetation, etc.)

Soil sequence A sequence of geographically separated soils (e.g., along a transect), ideally which differ in only one state factor such as relief or time

Soil solution The water in soil pores containing dissolved ions and molecules

Solubility product The equilibrium constant for a solid substance dissolving in an aqueous solution: the product of the component ion concentrations each raised to the power of their stoichiometric coefficient

Spatial analysis A mathematical analysis of the relationships between measurements made in different locations

Speciation (chemical speciation) The existence of various chemical forms of an element or compound

Species richness The number of different species within a given environment or environmental compartment (see *evenness*)

State factor A descriptor of the state of an Earth subsystem (such as a soil or biome) which is expected to affect the properties of and processes in that subsystem. These descriptors include climate, organisms present, geological materials and age, etc.

Statistical analysis Examination, summarization, manipulation, and interpretation of numerical data to discover any underlying differences, patterns, relationships, and trends

STEM Science, technology, engineering, and mathematics

Stratified sampling Sampling within predefined areas, based on soil type, land use, census district, etc.

Structure (soil structure) Organisation of soil grains into aggregates, and the size and geometry of the aggregates and their associated pore spaces

Surface complexation Adsorption of ions at a colloid surface where the bonding has characteristics of a coordination complex

T

Taxa Units of any rank (domain, phylum, class, order, family, genus, species) designating an organism or a group of organisms

Technosol An anthropogenic soil having ≥20% by volume of artefacts in upper 1 m, or containing a hard layer of industrial origin

Terrestrial Relating to the **land** surface(s) of Earth

Texture (soil texture) A category based on the relative proportions of grains of different sizes in a soil

Texture triangle A triangular plot where the 3 axes show the proportions of sand-, silt-, and clay-sized grains in soils. The plot area is subdivided into polygons defining texture categories

Toposequence A *soil sequence* in a landscape with varying properties due mainly to a gradient in relief (e.g., position on a hillslope)

Transcripts Messenger RNA sequences produced by conversion of DNA to messenger RNA

Trophic level A level in a 'food chain'; the lowest level includes the photosynthetic organisms; energy is transferred to the next level when these organisms consume plants, etc.

U

Univariate Referring to a single variable in a dataset (usually in the context of statistical analysis)
Upland soil A soil which is never submerged during normal seasonal cycles
Urban heat island See *heat island*
Urbanisation The increase in human population in cities relative to non-urban areas
Urban karst The combination of impervious surfaces and below-ground infrastructure and cavities that result in preferential water flow
Urbosphere The sub-system of the Earth showing a dominant influence of urban development

V

Vadose zone The vertical extent of soil and sediment which is above groundwater and has unsaturated conditions
Volume relationships The proportions of the soil volume occupied by solids, liquids, and gases (and derived parameters such as density)

W

Water Sensitive Urban Design (WSUD) Integrated water management for urban environments to improve sustainable water use, environmental protection, and urban aesthetics
Weathering Chemical or physical changes to rocks or other solid materials due to exposure to air and water, heating/cooling etc.
Wetland soil A soil which is submerged for at least part of the year during normal seasonal cycles

Index

A. W. Rate (ed.), *Urban Soils*, Progress in Soil Science,
https://doi.org/10.1007/978-3-030-87316-5

The manufacturer's authorised representative in the EU is Springer Nature Customer Service Centre GmbH, Europaplatz 3, 69115 Heidelberg, Germany. If you have any concerns regarding our products, please contact ProductSafety@springernature.com

Printed and bound by CPI Group (UK) Ltd, Croydon, CR0 4YY
30/09/2026
02225687-0010